Properties and Applications of Nanoparticles and Nanomaterials: 2nd Edition

Properties and Applications of Nanoparticles and Nanomaterials: 2nd Edition

Guest Editor

Xiaogang Li

MDPI

Basel • Beijing • Wuhan • Barcelona • Belgrade • Novi Sad • Cluj • Manchester

Guest Editor
Xiaogang Li
Institute of Nuclear and New
Energy Technology
Tsinghua University
Beijing
China

Editorial Office
MDPI AG
Grosspeteranlage 5
4052 Basel, Switzerland

This is a reprint of the Special Issue, published open access by the journal *International Journal of Molecular Sciences* (ISSN 1422-0067), freely accessible at: https://www.mdpi.com/journal/ijms/special_issues/Y34FV4QW2H.

For citation purposes, cite each article independently as indicated on the article page online and as indicated below:

Lastname, A.A.; Lastname, B.B. Article Title. *Journal Name* **Year**, *Volume Number*, Page Range.

ISBN 978-3-7258-5845-3 (Hbk)
ISBN 978-3-7258-5846-0 (PDF)
https://doi.org/10.3390/books978-3-7258-5846-0

Contents

International Journal of
Molecular Sciences

MDPI

Editorial

Special Issue "Properties and Applications of Nanoparticles and Nanomaterials: 2nd Edition"

Xiaogang Li

Institute of Nuclear and New Energy Technology, Key Laboratory of Advanced Reactor Engineering and Safety of Ministry of Education, Tsinghua University, Beijing 100084, China; xiaogangli@mail.tsinghua.edu.cn

Nanomaterials have rapidly developed, and attention surrounding their use has increased in recent years. The emergence of various nanomaterials, i.e., nanoparticles, nanograined alloys, and gradient nanostructures, is expected to make it possible for materials with super or very special properties to be applied in unusual practical contexts. There is a wide range of applications for nanomaterials in biochemistry and molecular medicine, including fuel cells or metal-ion batteries, flexible electronics, and various components related to energy. The physical and chemical properties of nanostructures are determined by their chemical composition, structure and formation processes, which is critical for their reliability and use in practical applications.

The purpose of this Special Issue is to provide a research forum to report on the structure, properties, processing, and applications of nanoparticles and nanomaterials to explore more possibilities to address intractable challenges. Topics of interest include but are not limited to, the areas mentioned above. Other relevant studies, such as the design of novel nanostructures or the modification of nanoparticles, have also been included. The contributors to this Special Issue, comprising outstanding scholars and industrial researchers, have put forward cutting-edge and in-depth scientific views to help our colleagues deepen their understanding of the problems facing nanoparticles and nanomaterials and address these intractable challenges.

Cea-Callejo et al. contributed a paper entitled "Point-of-Care Diagnostic Test for Rapid Detection of Infectious Laryngotracheitis Virus by Loop-Mediated Isothermal Amplification and Nanoprobes", which details the development of a rapid, point-of-care test using LAMP and DNA–gold nanoparticle probes to detect the Infectious Laryngotracheitis Virus (ILTV) in chickens without DNA extraction, providing a visible colorimetric result. This cost-effective method achieves 100% sensitivity/specificity, detects 200 viral copies in 45 min, and enables early on-farm outbreak control to enhance poultry biosecurity.

Mohebbi et al. contributed a paper entitled "Tunable Optical Properties of Cu/VSe_2 from the Visible to Terahertz Spectral Range: A First-Principles Study", which reveals that while the monolayer's metallic characteristic persists, Cu placement (top/corner) shifts electronic states near the Fermi level and significantly alters optical properties—enhancing Re (ω) in the visible range and introducing strong THz absorption peaks (0.2 THz for Top, 2.9 THz for Corner) in Im (ω). This was determined using DFT/DFTB calculations on Cu/VSe_2 interfaces. The Cu layer also increased the refractive index (5.4 for Top and 4.6 for Corner vs. 2.9 for pure VSe_2) and boosted the electrical current in Au/Cu/VSe_2/Au and Ag/Cu/VSe_2/Ag nanodiodes compared to Cu-free devices.

Etefa et al. contributed a paper entitled "Applications of Green Carbon Dots in Personalized Diagnostics for Precision Medicine", which introduces the exceptional potential of Green Carbon Dots (GCDs) in precision medicine, enabling precise non-invasive diagnostics, real-time monitoring, and targeted therapy delivery via integration with genomic, proteomic, and bioinformatics platforms. The unique optical properties and biocompatibility of GCDs allow for tailored treatment strategies and enhanced patient-specific applications, significantly improving diagnostic accuracy and therapeutic outcomes. The ongoing integration of GCDs with advanced technologies promises the further revolutionization of personalized healthcare, paving the way for more effective individualized medical care.

Basu et al. contributed a paper entitled "Advancements in Lithography Techniques and Emerging Molecular Strategies for Nanostructure Fabrication", which reports that lithography enables semiconductor scaling through advanced techniques like high-resolution EUV, maskless EBL/IBL, cost-effective NIL, and synchrotron-based XRL—each addressing specific needs in resolution, throughput, or prototyping for ICs, photonics, and MEMS/NEMS. Despite persistent challenges (defect control and material limits), innovations in alignment marks, mask fabrication, and molecular strategies ensure that lithography remains indispensable for pushing nanoscale manufacturing boundaries.

Martínez-Cisterna et al. contributed a paper entitled "Chitosan-Coated Silver Nanocomposites: Biosynthesis, Mechanical Properties, and Ag+ Release in Liquid and Biofilm Forms", in which chitosan-coated silver nanoparticles (AgChNPs) were biosynthesized using the Galega officinalis extract and optimized (1% chitosan, pH 4), achieving a size of 207.88 nm, +42.30 mV zeta potential, and spherical morphology (5.54–61.46 nm), with FTIR confirming functionalization. AgChNP-integrated biofilms demonstrated tunable mechanical strength (3.48 MPa tensile and 24.99 mm elongation), controlled silver ionization (62.57 mg/L in liquid; 184.07 mg/kg in films), and thermal stability, highlighting their potential for controlled-release pest management in sustainable agriculture.

Broset et al. contributed a paper entitled "A Complete Approach for circRNA Therapeutics from Purification to Lyophilized Delivery Using Novel Ionizable Lipids". Optimized purification and delivery methods were developed for circular RNA (circRNA), including high-yield Oligo(dT) affinity chromatography and novel lipid nanoparticles (LNPs) formulated with ionizable lipids (CP-LC-0867/CP-LC-0729), which significantly enhanced protein expression with sustained activity over 14 days compared to conventional SM-102 lipids. The successful lyophilization of LNP-circRNA complexes using CP-LC-0729 maintained therapeutic efficacy without cold-chain storage, demonstrating scalable strategies to overcome critical stability, purification, and delivery barriers for practical circRNA therapies.

Taha et al. contributed a paper entitled "Silver Nanoparticles as a Novel Tissue Preservative: A Comparative Study with 10% Neutral Buffered Formalin". The authors claim that silver nanoparticles (AgNPs) outperform formalin in preserving DNA, RNA, and protein integrity across multiple murine tissues over 72 h but are less effective at maintaining tissue morphology. This supports AgNPs as a viable molecular-preserving fixative, pending improvements in structural preservation for broader histopathological use.

Kim et al. contributed a paper entitled "Cardiovascular Toxicity of Metal-Based Nanoparticles". The authors note that metal-based nanoparticles (MNPs) pose cardiovascular risks due to systemic translocation after inhalation, ingestion, or dermal exposure, accumulating in tissues and disrupting endothelial function via oxidative stress, apoptosis, ferroptosis, and barrier impairment. Their nanoscale properties and ion release synergistically drive toxicity—increasing endothelial permeability, compromising the blood–brain barrier, and enhancing procoagulant activity—contributing to vascular/cardiac dysfunction and cardiovascular diseases. This review synthesizes recent in vitro and in vivo evidence on MNP biodistribution and cytotoxic mechanisms, providing a comprehensive assessment of their pathological impact on the cardiovascular system.

Loskutova et al. contributed a paper entitled "Quantum Dots-Based Luminescent Sensors: Review from Analytical Perspective". This review critically analyzes luminescent quantum dot (QD)-based sensors, synthesizing data from 124 studies to reveal that chemiluminescent QDs achieve exceptional sensitivity (average LOD: 0.109 pM), outperforming fluorescent (38 nM) and phosphorescent (26 nM) variants. While AI integration (e.g., ATTBeadNet and SVM) enhances analytical performance, key challenges persist in recovery rates, biocompatibility, and stability, guiding the future development of robust QD sensors for diverse analyte detection.

Tomic et al. contributed a paper entitled "Targeting Osteosarcoma: The Dual Action of Halogenated Boroxine and Cerium Oxide Nanoparticles", which has been approved for publication by us and will be available online soon. Addressing the limitations of decades-old osteosarcoma therapies, the authors evaluate the novel agents of halogenated boroxine (HB) and dextran-coated cerium oxide nanoparticles (SD2), both individually and in combination in an in vitro model, assessing cytotoxicity, viability, oxidative stress, cell death, and the effects on 3D spheroids. The results of this work demonstrate dose-, time-, and cell type-dependent antitumor effects for both agents, primarily via cytotoxicity and specific ROS induction, with combination treatment yielding modulated responses (potentially synergistic/selective) and significantly reduced toxicity towards non-tumor cells in 3D models, suggesting a promising selective strategy warranting further mechanistic investigation at the transcriptomic/proteomic levels.

This Special Issue highlights novel approaches and views and is committed to the effective application of nanoparticles and nanomaterials. We sincerely thank the authors who submitted manuscripts to our Special Issue as well as our readers. We hope that this Special Issue will serve as a platform to promote dialog among scientists in the field of nanoparticles and nanomaterials. In addition, in view of our continued intake of high-quality articles on this topic, we have begun a new Special Issue, "Properties and Applications of Nanoparticles and Nanomaterials: 3rd Edition"; the submission address is https://www.mdpi.com/journal/ijms/special_issues/HWH2W8ZA94 (accessed on 5 July 2025). Topics of interest include but are not limited to the structure, properties, processing, and applications of nanoparticles and nanomaterials, the design of novel nanostructures, and the modification of nanoparticles. The submission of research articles and reviews in this area of study is welcome.

Funding: The author appreciates the financial support provided by the National Natural Science Foundation of China [No. 52405385].

Acknowledgments: This work was supported by the LingChuang Research Project of the China National Nuclear Corporation.

Conflicts of Interest: The author declares no conflict of interest.

List of Contributions:

1. Cea-Callejo, P.; Trenado, C.; El Mansouri, E.; Gomez-Lucia, E.; Doménech, A.; Biarnés, M.; Cuenca, J.M.; Sánchez-Llatas, C.J.; Madrid, R.; Benítez, L. Point-of-Care Diagnostic Test for Rapid Detection of Infectious Laryngotracheitis Virus by Loop-Mediated Isothermal Amplification and Nanoprobes. *Int. J. Mol. Sci.* **2025**, *26*, 1971. https://doi.org/10.3390/ijms26051971.
2. Mohebbi, E.; Pavoni, E.; Stipa, P.; Pierantoni, L.; Laudadio, E.; Mencarelli, D. Tunable Optical Properties of Cu/VSe_2 from the Visible to Terahertz Spectral Range: A First-Principles Study. *Int. J. Mol. Sci.* **2025**, *26*, 2527. https://doi.org/10.3390/ijms26062527.
3. Etefa, H.F.; Dejene, F.B. Applications of Green Carbon Dots in Personalized Diagnostics for Precision Medicine. *Int. J. Mol. Sci.* **2025**, *26*, 2846. https://doi.org/10.3390/ijms26072846.

4. Basu, P.; Verma, J.; Abhinav, V.; Ratnesh, R.K.; Singla, Y.K.; Kumar, V. Advancements in Lithography Techniques and Emerging Molecular Strategies for Nanostructure Fabrication. *Int. J. Mol. Sci.* **2025**, *26*, 3027. https://doi.org/10.3390/ijms26073027.
5. Martínez-Cisterna, D.; Chen, L.; Bardehle, L.; Hermosilla, E.; Tortella, G.; Chacón-Fuentes, M.; Rubilar, O. Chitosan-Coated Silver Nanocomposites: Biosynthesis, Mechanical Properties, and Ag+ Release in Liquid and Biofilm Forms. *Int. J. Mol. Sci.* **2025**, *26*, 4130. https://doi.org/10.3390/ijms26094130.
6. Broset, E.; Larraga, A.; Lampaya, V.; Navarro, V.; López-Gavín, A.; de Miguel, D.; Peña, Á.; Martínez-Oliván, J.; Casabona, D. A Complete Approach for circRNA Therapeutics from Purification to Lyophilized Delivery Using Novel Ionizable Lipids. *Int. J. Mol. Sci.* **2025**, *26*, 5138. https://doi.org/10.3390/ijms26115138.
7. Taha, S.; Ismaeel, A.; Aljishi, M.; Selvam, S.; Esther, A.; Greish, K. Silver Nanoparticles as a Novel Tissue Preservative: A Comparative Study with 10% Neutral Buffered Formalin. *Int. J. Mol. Sci.* **2025**, *26*, 5335. https://doi.org/10.3390/ijms26115335.
8. Kim, E.-H.; Park, S.; Bae, O.-N. Cardiovascular Toxicity of Metal-Based Nanoparticles. *Int. J. Mol. Sci.* **2025**, *26*, 5816. https://doi.org/10.3390/ijms26125816.
9. Loskutova, A.; Seitkali, A.; Aliyev, D.; Bukasov, R. Quantum Dot-Based Luminescent Sensors: Review from Analytical Perspective. *Int. J. Mol. Sci.* **2025**, *26*, 6674. https://doi.org/10.3390/ijms26146674.
10. Tomic, N.; Esmkhani, S.; Bayramova, J.; Dinc, A.; Morva, A.; Medic, B.-S.; Ramic, J.; Lojo-Kadric, N.; Gazouli, M.; Galic, B.; et al. Targeting Osteosarcoma: The Dual Action of Halogenated Boroxine and Cerium Oxide Nanoparticles. *Int. J. Mol. Sci.* **2025**.

International Journal of *Molecular Sciences*

MDPI

Article

Targeting Osteosarcoma: The Dual Action of Halogenated Boroxine and Cerium Oxide Nanoparticles

Nikolina Tomic [1], Sahra Esmkhani [2,3,4], Jamila Bayramova [2,3,4,5], Ahmet Dinc [2,3,4], Ahsen Morva [2,6], Belmina Saric Medic [1], Jasmin Ramic [1], Naida Lojo-Kadric [1], Maria Gazouli [7,8], Borivoj Galic [9], Lejla Pojskic [1,*] and Hilal Yazici [2,*]

1 Institute for Genetic Engineering and Biotechnology, University of Sarajevo, Zmaja od Bosne 8, 71000 Sarajevo, Bosnia and Herzegovina; belmina.saric@ingeb.unsa.ba (B.S.M.); jasmin.ramic@ingeb.unsa.ba (J.R.); naida.lojo@ingeb.unsa.ba (N.L.-K.)

2 TÜBİTAK Marmara Research Center, Climate Change and Life Sciences, Biotechnology Research Group, 41470 Kocaeli, Türkiye; sahra.esmkhaniyouvalari@ogr.iu.edu.tr (S.E.); jamila.bayramova@ogr.iu.edu.tr (J.B.); ahmet_dinc47@hotmail.com (A.D.); ahsen.morva@tubitak.gov.tr (A.M.)

3 Division of Cancer Genetics, Department of Basic Oncology, Oncology Institute, Istanbul University, 34093 Istanbul, Türkiye

4 Institute of Graduate Studies in Health Sciences, Istanbul University, 34093 Istanbul, Türkiye

5 Department of Molecular Biology and Medical Genetics, Scientific Research Center, Azerbaijan Medical University, AZ1022 Baku, Azerbaijan

6 Transplant Immunology Research Center of Excellence TIREX, School of Medicine, Koc University, 34010 Istanbul, Türkiye

7 Laboratory of Biology, Department of Basic Medical Sciences, Medical School, National and Kapodistrian University of Athens, 11725 Athens, Greece; mgazouli@med.uoa.gr

8 Medical Physics Unit, Department of Radiology, Medical School, National and Kapodistrian University of Athens, Attikon University Hospital, 11725 Athens, Greece

9 Faculty of Science, University of Sarajevo, Zmaja od Bosne 33-35, 71000 Sarajevo, Bosnia and Herzegovina

* Correspondence: lejla.pojskic@ingeb.unsa.ba (L.P.); hilal.yazici@tubitak.gov.tr (H.Y.)

Abstract: Current standard treatments for osteosarcoma have not been changed for decades and have limited and variable success. The advancement of precision medicine technologies, along with the drug-repurposing and fast drug-screening methodologies available, has opened new avenues for the development of more effective therapeutic strategies. In this study, we evaluated the effectiveness of halogenated boroxine (HB) and dextran-coated cerium oxide nanoparticles—DexCeNPs (SD2)—in an in vitro osteosarcoma model. Both agents were tested individually and in combination. The research encompassed assessments of treatment-related cytotoxicity and cell viability, oxidative stress, and apoptotic and necrotic responses, as well as the effects on 3D spheroid models. The results demonstrated that the effects of HB and SD2 were strongly influenced by the dose, exposure time, and cell type. Both exhibited distinguished antitumor activity through cytotoxicity and specific reactive oxygen species (ROS) induction. The combined treatment produced modulated responses that were dependent on the treatment ratio and cell line, suggesting potential synergistic or selective interactions. Notably, the outcomes of the analysis conducted in 3D models revealed reduced toxicity toward non-tumor cells. These findings suggest the improved efficacy of HB and SD2 used in combination as a selective and novel antitumor strategy and underscore the need for further mechanistic studies at the transcriptomic and proteomic levels to elucidate the underlying pathways and clarify the mechanisms of action.

Keywords: osteosarcoma; halogenated boroxine; cerium oxide nanoparticles

1. Introduction

Osteosarcoma is the predominant malignant bone tumor affecting mainly children and adolescents and is characterized by a strong propensity for metastasis and recurrence [1–3]. Current standard therapies have remained largely unchanged for nearly fifty years, typically combining surgery with multi-agent chemotherapy, most often high-dose doxorubicin, methotrexate, and cisplatin [4]. Although these regimens have improved outcomes, their effectiveness is limited by severe side effects and the development of drug resistance. Combination therapies have demonstrated enhanced efficacy and remain a promising strategy to overcome these limitations [5]. However, the survival rates remain suboptimal due to tumor aggressiveness and chemotherapy resistance [1]. Therefore, there is an urgent need for novel therapeutic approaches that can effectively target tumor heterogeneity and resistance while minimizing undesired side effects.

Boron and boronic acid compounds have long been used in the development of therapies for various diseases, including cancer [6–8]. Boron neutron capture therapy (BNCT) has been an emerging strategy in osteosarcoma therapy lately [9–11]. Notably, boronic acid used in BNCT significantly reduced osteolysis and the tumor burden in osteosarcoma-bearing rat models, with studies also demonstrating its selective accumulation in tumor tissue [11].

Halogenated boroxine $K_2(B_3O_3F_4OH)$ (HB), a product of the cyclic anhydride of boric acid, is patented as a substance with potential antitumor and antiproliferative effects on skin lesions [12–14]. In a series of in vitro and in vivo studies on the antitumor potential of HB, its tumor-suppressive effects on various types of tumors have been observed across a range of HB concentrations with minimal effects on non-malignant cells [15–22]. Results of the research on its effect suggest that it acts as a pro-apoptotic agent in leukemia cells [19,20] and modulates metabolic phenotype and autophagy in human bladder carcinoma cells [21]. Still, its mechanism of action has not been fully elucidated, and its effects on bone tumors have not yet been investigated.

Significant improvements in developing selective tumor-targeting drugs were achieved with the advances in nanotechnology and knowledge of cancer biology [2]. Several nanomedicine formulations have even been clinically approved in recent decades [23]. Their growing use is driven by multifunctional capabilities in drug delivery, diagnosis, and imaging, along with advantages such as an extended circulation time, improved tissue accumulation, and reduced toxicity [23]. Among nanomaterials, cerium oxide nanoparticles (CeNPs) have attracted significant attention due to their unique redox properties, high biocompatibility, and potential biomedical applications. CeNPs exhibit self-regenerative redox activity, enabling dual antioxidant and pro-oxidant effects that regulate ROS levels in biological systems [24]. By mimicking ROS-related enzymes, CeNPs protect normal cells from oxidative stress while inducing cancer cell death, enhancing tumor sensitization for various therapies. Additionally, CeNPs are used as nanocarriers to improve drug delivery and in combination with other drugs to achieve synergistic anticancer effects [24].

In the context of osteosarcoma, CeNPs have shown promising anticancer activity, with selective cytotoxicity enhanced under acidic conditions that mimic the tumor microenvironment [25]. Additionally, there has been a study indicating that higher dextran coating concentrations can improve the therapeutic efficacy of CeNPs while minimizing the toxicity to healthy bone cells [26]. These properties position dextran-coated CeNPs (Dex-CeNPs) as a potential candidate for targeted osteosarcoma treatment.

The strategy of combining or co-delivering multiple therapeutic agents is increasingly employed in cancer research to maximize efficacy through synergistic, additive, or potentiating effects [27,28]. While HB has shown selective antitumor activity in various cancer models, its effectiveness against osteosarcoma remains unknown. On the other hand, as noted above, Dex-CeNPs have also exhibited selective antitumor effects, particularly in the osteosarcoma model. Therefore, the present study aimed to evaluate the cytotoxic and oxidative effects of HB and recently described Dex-CeNPs [29] on osteosarcoma cells. Additionally, the combined effects of HB and Dex-CeNPs were assessed on osteosarcoma and osteoblast cell lines to explore their potential as a novel combinatorial therapeutic approach. To enhance the physiological relevance of these findings, in addition to standard 2D cultures, 3D tumor spheroid models were employed, as they better represent the tumor microenvironment. Thus, evaluating HB, Dex-CeNPs, and their combination in spheroid cultures provided deeper insight into their therapeutic potential.

2. Results

The results are presented in the same order as the experiment was conducted. First, the cytotoxicity and ROS profile of two testing treatments were established for all used cell lines. After that an apoptosis profile for single and combined treatment replications in osteosarcoma and non-tumor cell lines was carried out. The relative cell growth in the treated tumor and non-tumor cell lines after the single and combined HB and SD2 treatments was assessed in 3D spheroid models.

2.1. Physicochemical Properties

A detailed physicochemical characterization of SD2 was conducted in the study published before [29]. As it is described in that paper, UV–Vis spectrophotometry showed a peak for SD2 at approximately 287 nm, the hydrodynamic diameter was 100 nm (25% ratio), the zeta potential was slightly negative, and the XPS analysis revealed a higher concentration of Ce^{3+}, with a Ce^{3+}/Ce^{4+} ratio of 1.45. All these characteristics were confirmed in our study, as well for each batch of newly synthesized SD2.

2.2. The Cytotoxicity and ROS Profile Measurements

2.2.1. The Cytotoxicity and ROS Profiles of HB on the Osteosarcoma Cell Lines

HB demonstrated dose-dependent and time-dependent cytotoxic effects on the MG-63 osteosarcoma cells. Higher concentrations (150 μg/mL, 200 μg/mL, and 400 μg/mL) significantly decreased the cell viability at both 24 and 72 h post-treatment (Figures 1a and 2a). In terms of the exposure duration, the cells treated with the lowest concentration (50 μg/mL) exhibited a slight increase in the viability at 72 h, suggesting a potential recovery effect. Conversely, the cells exposed to higher concentrations showed a continued decline in viability over time. Similarly, in the Saos-2 cells, higher HB concentrations (150 μg/mL, 200 μg/mL, and 400 μg/mL) induced a statistically significant dose-dependent reduction in the cell viability at both time points, mirroring the trend observed in the MG-63 cells. Interestingly, lower concentrations (50 μg/mL and 75 μg/mL) exhibited inhibitory effects only at 72 h of incubation, with measurable declining viability (Figures 1b and 2b). The cytotoxic response was more pronounced in the non-tumor hFOB 1.19 cells compared with the osteosarcoma lines. This prompted the inclusion of a lower concentration (25 μg/mL), which was the only dose that did not significantly reduce the viability at either time point. However, the response in the hFOB 1.19 cells did not display a strictly dose-dependent pattern, e.g., 200 μg/mL had a milder effect than 75 μg/mL and 100 μg/mL. The impact of increasing the HB concentrations appeared more linear than exponential, as depicted in the bar charts and dose response curve (Figures 1c and 2c).

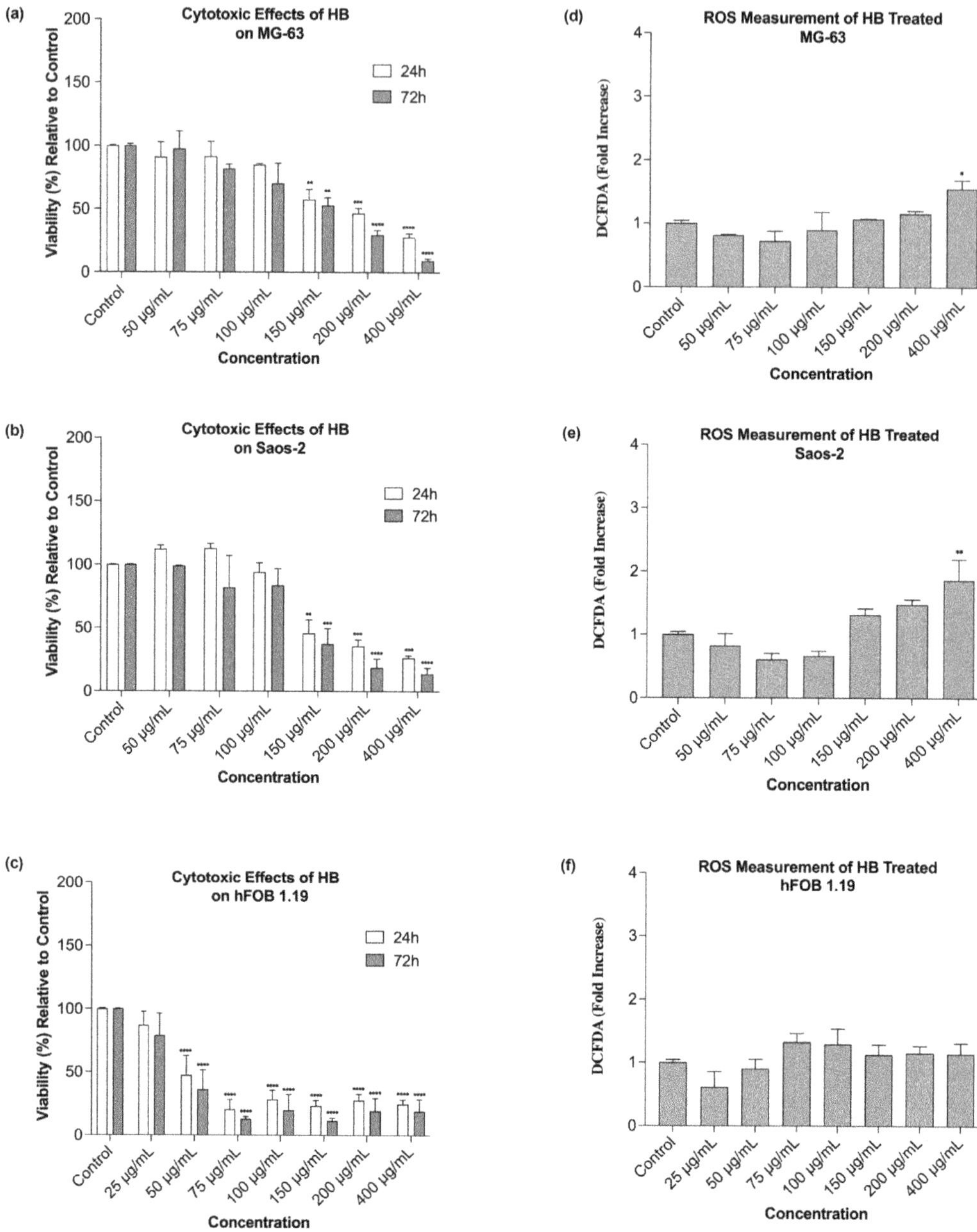

Figure 1. Cytotoxic effects of a series of HB concentrations on the (**a**) MG-63, (**b**) Saos-2, and (**c**) hFOB 1.19 cell lines after 24 h and 72 h of treatment. Dose- and time-dependent variabilities were observed across the tested cell lines. The reactive oxygen species (ROS) generation induced by the same HB concentrations in the (**d**) MG-63, (**e**) Saos-2, and (**f**) hFOB 1.19 cell lines after 24 h of treatment was found to be variable across the cell lines. Statistical significance is indicated as * $p < 0.05$, ** $p < 0.01$, *** $p < 0.001$, and **** $p < 0.0001$.

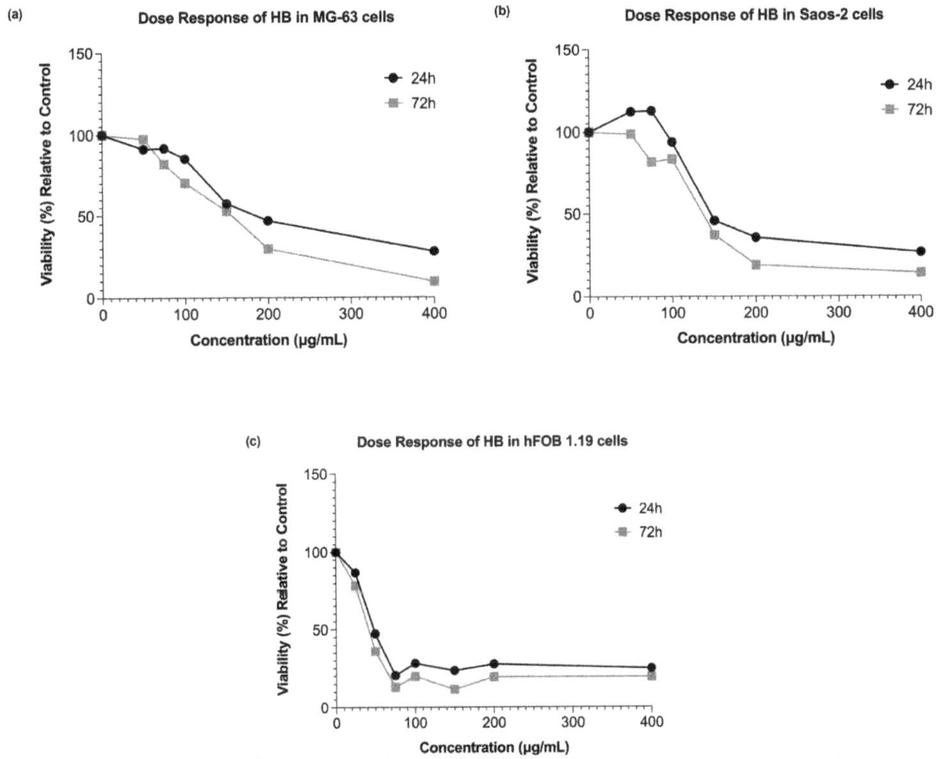

Figure 2. Effect of series of HB concentrations on cell viability in (**a**) MG-63, (**b**) Saos-2, and (**c**) hFOB 1.19 cell lines.

Depending on the cytotoxicity profile observed at 24 h post-treatment and reported above, HB's half-maximal inhibitory concentrations (IC_{50}) were calculated. The HB IC_{50} values for the osteosarcoma cell lines were similar to each other (203 µg/mL for the MG-63 and 192 µg/mL for the Saos-2), while the value for the non-tumor cell line hFOB 1.19 was notably lower (57 µg/mL).

The effects of HB treatment on the ROS production varied across the concentrations. In both the osteosarcoma cell lines, MG-63 and Saos-2, higher concentrations of HB led to increased ROS levels (Figure 1c,d), with the elevation at 400 µg/mL reaching statistical significance (p-values of 0.0409 and 0.098 for the MG-63 and Saos-2, respectively). Interestingly, lower concentrations (50 µg/mL, 75 µg/mL, and 100 µg/mL) resulted in reduced ROS levels compared with the untreated controls. In the hFOB 1.19 cell line, all concentrations—except 25 µg/mL and 50 µg/mL—elicited an increase in the oxidative stress relative to the control cells. However, these changes were not statistically significant (Figure 1f).

2.2.2. The Cytotoxicity and ROS Profiles of SD2 on the Osteosarcoma Cell Lines

SD2 exhibited cytotoxic effects in all three cell lines. In the MG-63 cells, all tested concentrations except 100 µg/mL resulted in a significant decrease in the cell viability at both time points compared with the control (Figure 3a). Interestingly, the treatment with 100 µg/mL led to an increase in the viability, although this effect was less pronounced after 72 h than at 24 h (Figure 4a). The cells treated with 250 µg/mL showed a slight recovery in viability after 72 h, a trend not observed with other concentrations, where extended exposure generally led to a slight decline in the cell viability. The cytotoxic response of the Saos-2 cells to SD2 differed from that observed in the MG-63 cells. At 24 h, a dose-dependent variability in the viability was noted, where higher concentrations produced comparable reductions in the cell viability (Figures 3b and 4b). These higher concentrations significantly decreased the viability at both time points. However, the longer exposure of 72 h appeared to result in a recovery in the viability, suggesting a possible cellular adaptation over time. The cytotoxic effects of SD2 on the hFOB 1.19 cells appeared to be dose-dependent at the 24 h time point, where higher concentrations (250, 500, and 1000 μg/mL) significantly reduced the cell viability compared with the control (Figure 3c). Interestingly, the cells treated with all concentrations showed a potential recovery effect, as the viability increased after 72 h compared with 24 h, although it remained significantly decreased relative to the control (Figures 3c and 4c).

In summary, SD2 induced significant dose- and time-dependent cytotoxicities in the MG-63 and Saos-2 cells starting from 250 μg/mL. The hFOB 1.19 cells were more sensitive to SD2, where they showed a marked decrease in the viability from 250 μg/mL, especially at 72 h ($p < 0.0001$). The SD2 IC_{50} values were calculated based on its cytotoxic profile generated 24 h post-treatment, which are described in the results above. Those values were 132 μg/mL for the MG-63 cell line, 222 μg/mL for the Saos-2, and 142 μg/mL for the hFOB 1.19.

ROS analysis revealed that SD2 significantly elevated ROS production in MG-63 cells at 500 and 1000 μg/mL ($p < 0.05$) (Figure 3d). Saos-2 cells showed a pronounced increase in the ROS levels, even at the lowest concentration (100 μg/mL), with the most significant changes observed at 250 μg/mL ($p < 0.0001$), indicating strong oxidative stress (Figure 3e). In contrast, the hFOB 1.19 cells demonstrated a notable decrease in the ROS generation across all SD2 concentrations, with statistically significant reductions compared with the control ($p < 0.001$ to $p < 0.0001$) (Figure 3f). These findings suggest that SD2 exerted cytotoxic effects on both the cancerous and non-cancerous osteoblast-like cells, potentially via ROS induction in the tumor cells but through different mechanisms in the normal cells. SD2 significantly increased the ROS production in the MG-63 and Saos-2 cells, where the Saos-2 showed strong oxidative responses, even at low concentrations. In contrast, the ROS levels were significantly reduced in the hFOB 1.19 cells at all SD2 doses, suggesting that SD2 induced cytotoxicity through ROS-dependent mechanisms in the cancer cells and ROS-independent pathways in the normal cells.

2.2.3. Time- and Dose-Dependent Efficacies of HB and SD2 Co-Treatment Through Cytotoxicity Analysis and ROS Profile

Co-treatment with HB and SD2 reduced the MG-63 cell viability across all tested combinations and time points (24 h and 72 h) (Figures 5a and 6a). However, the combinations that contained 80% of the HB IC_{50} with 20% of the SD2 IC_{50}, as well as 60% HB with 40% SD2, showed weaker effects compared with HB alone. The combination of 80% HB + 20% SD2 was also the only condition that did not result in statistically significant reduction in the viability. In contrast, the 20% HB + 80% SD2 combination significantly reduced the cell viability relative to the control ($p < 0.01$), although its effect was less pronounced than that of SD2 alone, suggesting a potentiated effect of SD2. Notably, this combination was more effective than the 80% HB + 20% SD2 treatment. The cytotoxic effects were more pronounced at 72 h, where almost all combinations significantly inhibited the cell viability at this time point. In the Saos-2 cells, the cytotoxic effects at 24 h were generally comparable across the treatment groups (Figures 5b and 6b). The 80% HB + 20% SD2, 50% HB + 50% SD2, and 40% HB + 60% SD2 combinations exhibited the weakest cytotoxicities. Conversely, their reciprocal combinations—60% HB + 40% SD2 and 20% HB + 80% SD2—resulted in significant reductions in the cell viability. By 72 h, all treatment combinations led to significant cytotoxic effects in the Saos-2 cells. In the hFOB 1.19 cells, all treatment combinations except 80% HB + 20% SD2 significantly reduced the cell viability at 24 h (Figure 6c). The combinations with a higher proportion of SD2 than HB produced stronger cytotoxic responses (Figure 5c). At 72 h, a partial recovery of cell viability was observed, with the most notable recovery in the 80% HB + 20% SD2 group. Similarly, the cells treated with HB at its exact IC_{50} concentration also showed signs of recovery over time. These results suggest a time-dependent response in the hFOB 1.19 cells, with reduced long-term cytotoxicity under certain treatment conditions.

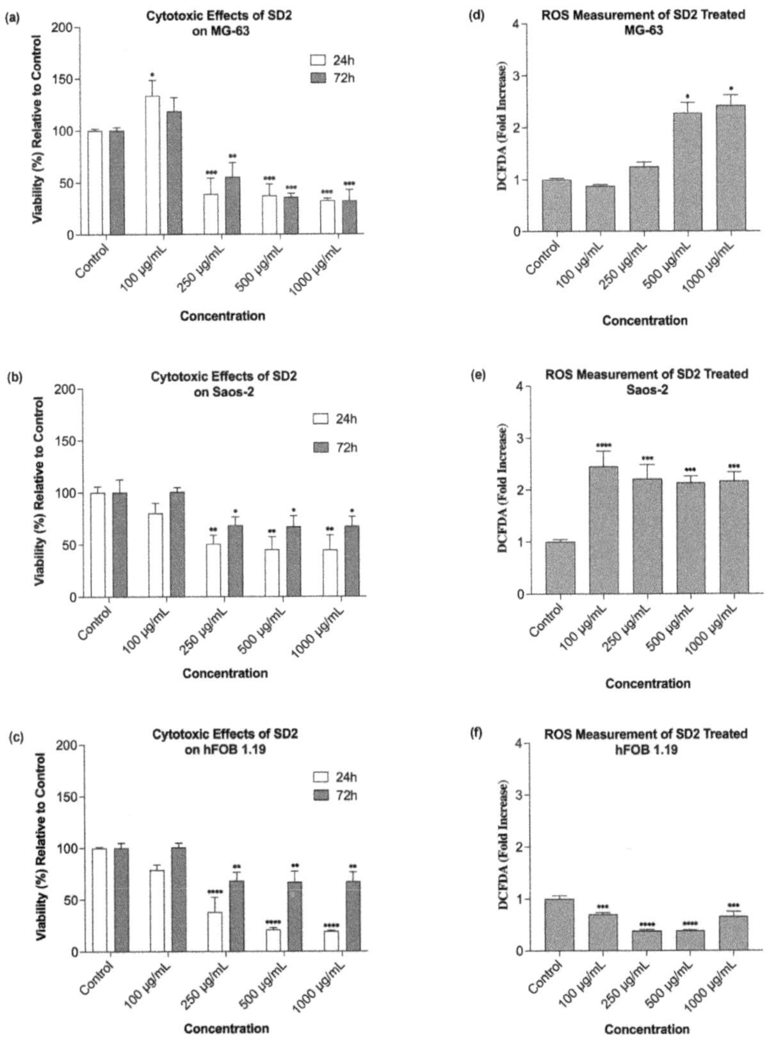

Figure 3. Cytotoxic effects of a series of SD2 concentrations on the (**a**) MG-63, (**b**) Saos-2, and (**c**) hFOB 1.19 cell lines after 24 h and 72 h of treatment. Dose- and time-dependent variabilities were observed across the tested cell lines. The reactive oxygen species (ROS) generation induced by the same SD2 concentrations in the (**d**) MG-63, (**e**) Saos-2, and (**f**) hFOB 1.19 cell lines after 24 h of treatment was found to be variable across the cell lines. Statistical significance is indicated as * $p < 0.05$, ** $p < 0.01$, *** $p < 0.001$, and **** $p < 0.0001$.

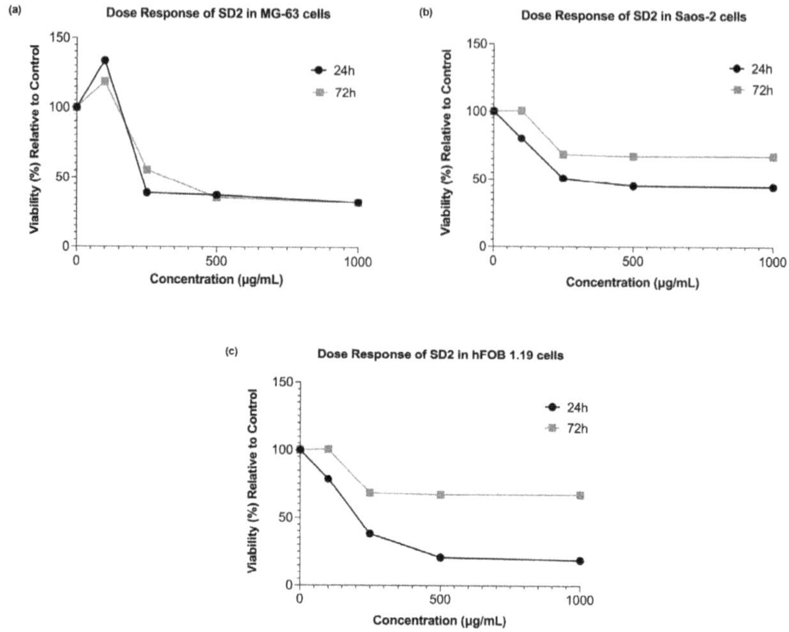

Figure 4. Effect of series of SD2 concentrations on cell viability in (**a**) MG-63, (**b**) Saos-2, and (**c**) hFOB 1.19 cell lines.

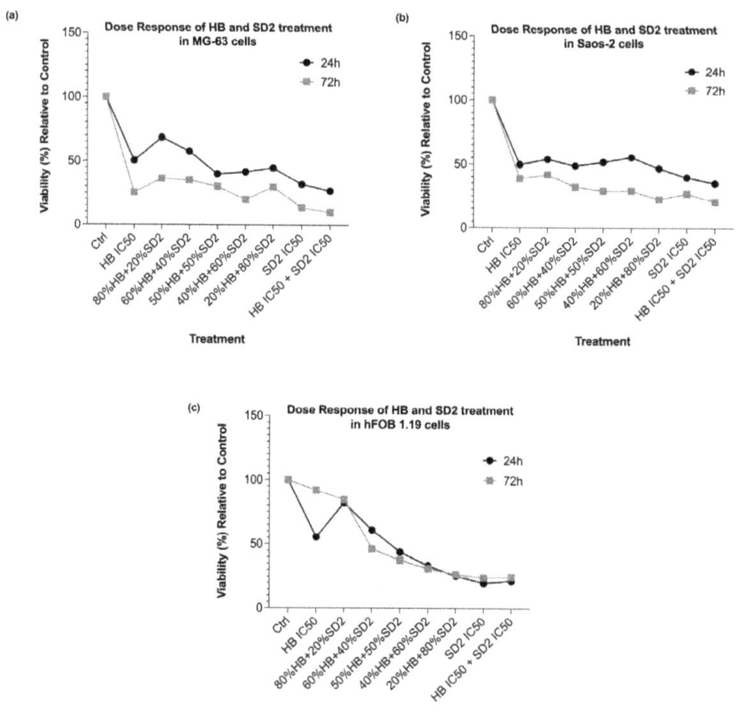

Figure 5. Effects of HB and SD2 treatment on cell viability in (**a**) MG-63, (**b**) Saos-2 and (**c**) hFOB 1.19 cell lines.

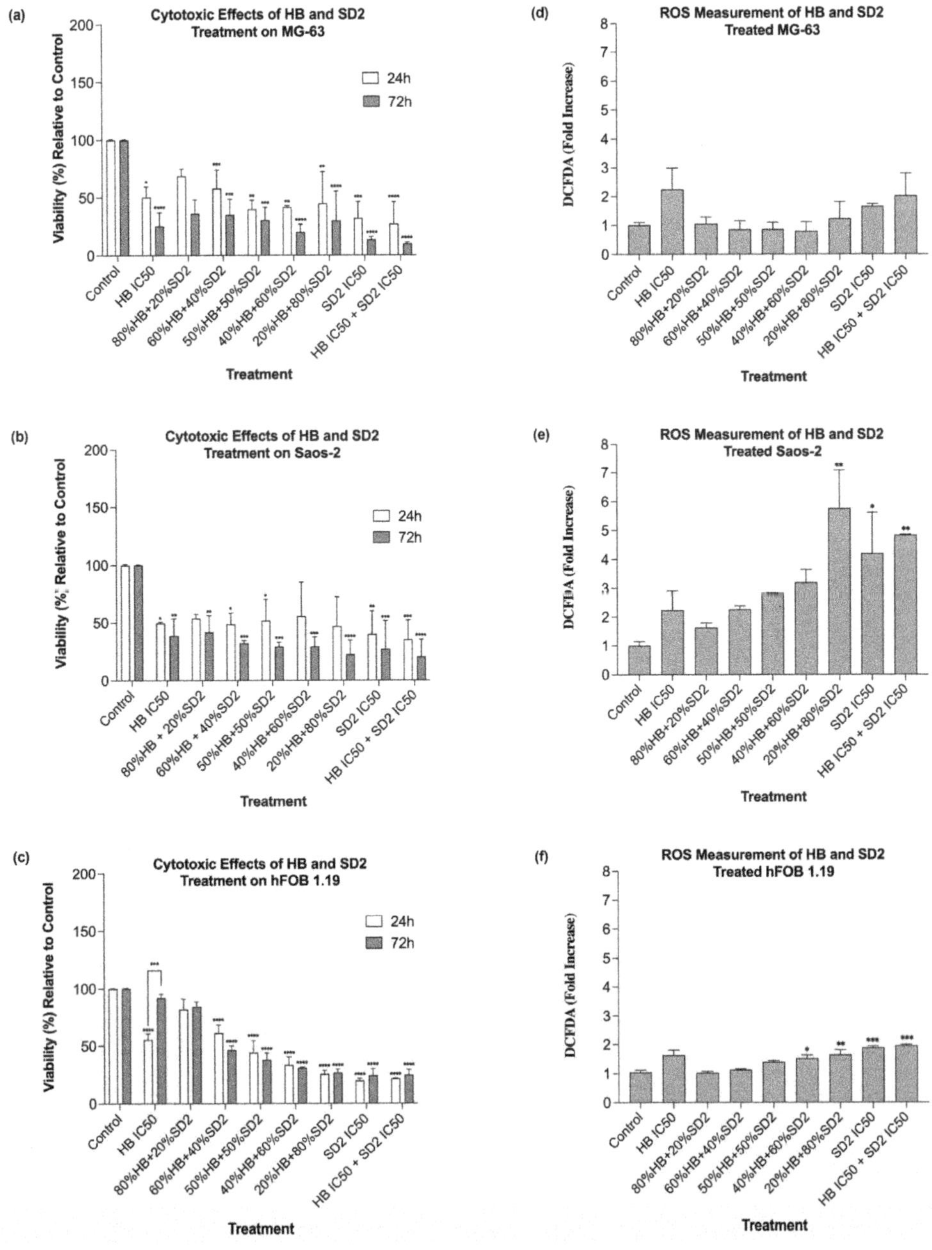

Figure 6. Effects of halogenated boroxine (HB) and Dex-CeNPs (SD2) co-treatment on the cell viability in the MG-63 (**a**), Saos-2 (**b**), and hFOB 1.19 (**c**) cell lines at 24 h and 72 h. Different HB and SD2 combinations showed variable effects. The reactive oxygen species (ROS) generation profile following 24 h treatment was cell-line-specific and variable across treatment combinations in the MG-63 (**d**), Saos-2 (**e**), and hFOB 1.19 (**f**) cell lines. Statistical significance is indicated as * $p < 0.05$, ** $p < 0.01$, *** $p < 0.001$, and **** $p < 0.0001$.

The effects of the HB and SD2 co-treatment on reactive oxygen species (ROS) production varied between the three cell lines. In the MG-63 cells, the combined treatments did not elicit significant oxidative stress (Figure 6d). In several cases, the ROS levels were lower than those observed in the untreated controls, suggesting a potential protective or antioxidative effect of certain HB–SD2 combinations. In contrast, the Saos-2 cells displayed markedly elevated ROS production following the co-treatment, particularly in combinations with higher proportions of SD2 (Figure 6e). The treatments with SD2 alone ($p < 0.05$), 20% HB + 80% SD2 ($p < 0.01$), and combined IC_{50} concentrations ($p < 0.01$) resulted in statistically significant increases in the ROS levels compared with the control, indicating a more pronounced oxidative response in this cell line. In the hFOB 1.19 cell line, the ROS levels remained relatively low across most treatment groups (Figure 6f). Nevertheless, some combinations with a higher SD2 content led to a statistically significant increase in the ROS production relative to the untreated cells.

2.3. *Apoptotic and Necrotic Responses of HB and SD2 Co-Treatment Through Flow Cytometry Analysis*

The effects of HB and SD2 on cell death were comprehensively analyzed in terms of the apoptotic and necrotic responses using flow cytometry following the treatments at varying doses and two time points.

Apoptotic profiles of the cell lines after 24 h and 72 h of treatment with HB, SD2 (at their respective IC_{50} concentrations), and their combination, alongside the profiles of untreated cells, are shown in Figures 7–9.

HB did not exhibit any apoptotic effect on the MG-63 cells after 24 h, while the SD2 treatment exerted a significant effect characterized by a decrease in healthy cells and an increase in the early and late apoptotic cells (Figure 7). In the Saos-2 cells, the treatment with HB at its IC_{50} concentration also did not cause a significant apoptotic effect after 24 h (Figure 8), and SD2 at its IC_{50} concentration had a significantly more potent effect, which sharply reduced the healthy cell percentage to 62.50%, which was accompanied by an increase in the early apoptosis to 12.80%, and a dramatic surge in the late apoptotic cells to 39.98%. Furthermore, a substantial proportion of cells (17.77%) were found to be necrotic. The treatment of the hFOB 1.19 cells with HB at its IC_{50} concentration reduced the healthy cell percentage to 74.35% after 24 h (Figure 9). This was accompanied by a weak increase in early apoptotic cells (4.97% $\pm$ 0.05%), late apoptotic cells (13.43%), and necrotic cells (9.75%). SD2 at its IC_{50} concentration had a more substantial impact on the hFOB 1.19 cells, which decreased the healthy cell proportion to 46.48% and was accompanied by increases in the early apoptotic (9.62%), late apoptotic (27.98%), and necrotic (15.93%) levels.

The HB treatment of MG-63 after 72 h resulted in minimal apoptotic activity (Figure 7). SD2 slightly reduced the proportion of healthy cells, with corresponding increases in the early (5.24%) and late apoptosis (6.68%). The similarity between the 24 and 72 h profiles suggests that prolonged SD2 exposure did not enhance the apoptotic induction in the MG-63 cells, consistent with the cytotoxicity data. In Saos-2 cells, the HB treatment for 72 h maintained a high proportion of healthy cells (Figure 8), although a slight increase in late apoptotic cells (8.23%) was noted compared with the results acquired at 24 h. In contrast, SD2 reduced the proportion of viable cells to 45.64%, where the late apoptotic cells reached 30.38%—a value nearly identical to that at 24 h. The treatment with HB moderately reduced the proportion of healthy hFOB 1.19 cells to 72.20% after 72 h (Figure 9), whereas SD2 led to a more pronounced decline to 46.42%, which was accompanied by increases in both the apoptotic and necrotic populations. Notably, the apoptotic distribution remained largely consistent between the 24 and 72 h time points.

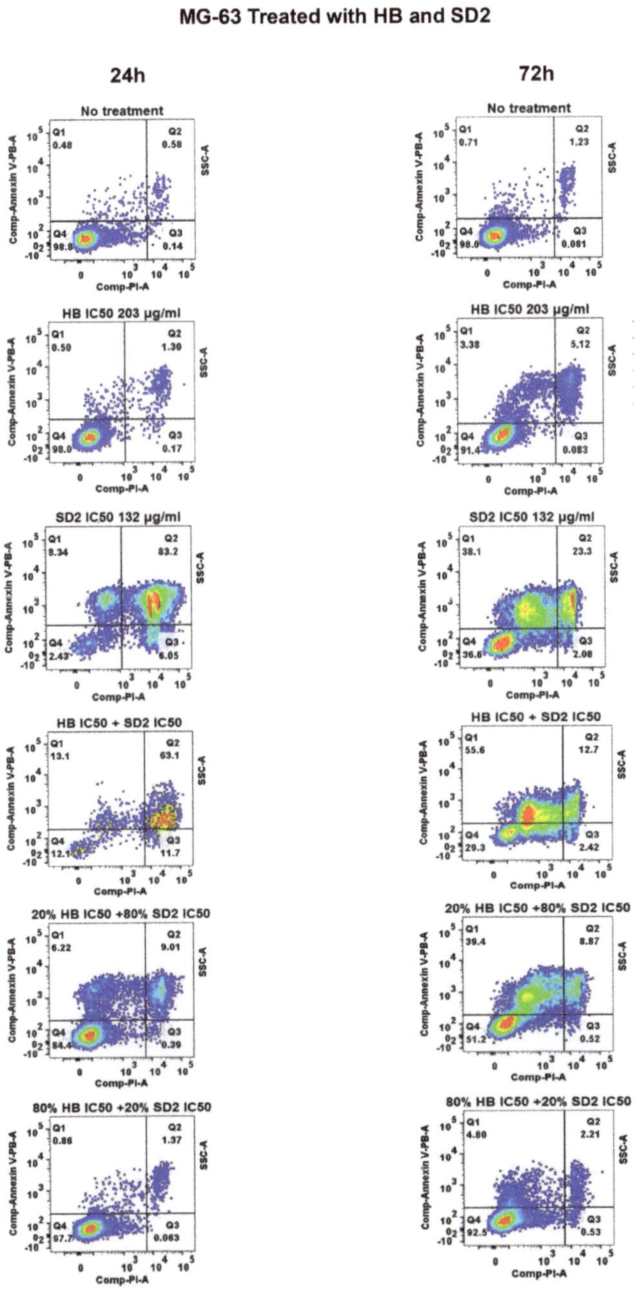

Figure 7. Effects of HB, SD2, or a combination of HB + SD2 on the apoptosis profiles of MG-63 cell lines. The no treatment panel corresponds to the untreated cells used as the negative control. The treatment panels correspond to 24 h and 72 h treatments with HB, SD2 or combination. Cell apoptosis was measured using Annexin V-PB/PI double staining. The results are expressed as the percentage of cells that corresponded to Q1: early apoptotic cells (Annexin V+/PI−), Q2: late apoptotic cells (Annexin V+/PI+), Q3: necrotic cells (Annexin V− PI+), and Q4: healthy cells (Annexin V−/PI−).

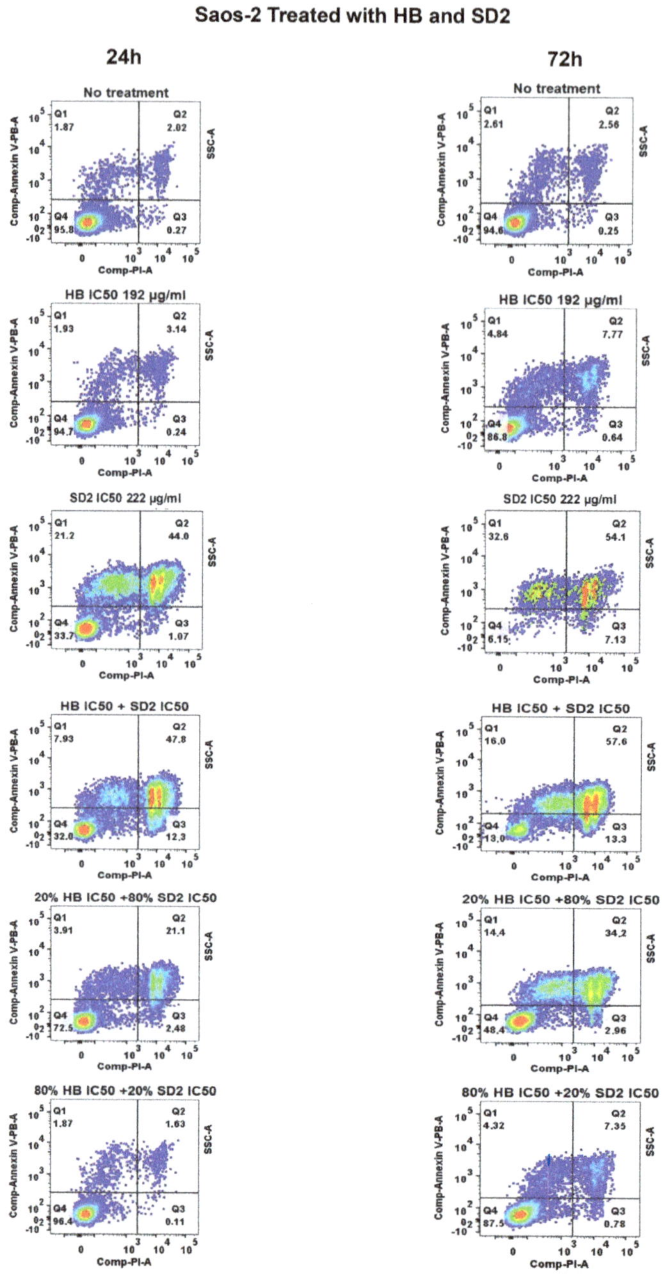

Figure 8. Effects of HB, SD2, or a combination of HB + SD2 on the apoptosis profiles of Saos-2 cell line. The no treatment panel corresponds to the untreated cells used as the negative control. The treatment panels correspond to the 24 h 72 h treatments with HB, SD2 or combination. Cell apoptosis was measured using Annexin V-PB/PI double staining. The results are expressed as the percentage of cells corresponding to Q1: early apoptotic cells (Annexin V+/PI−), Q2: late apoptotic cells (Annexin V+/PI+), Q3: necrotic cells (Annexin V− PI+), and Q4: healthy. Cells (Annexin V−/PI−).

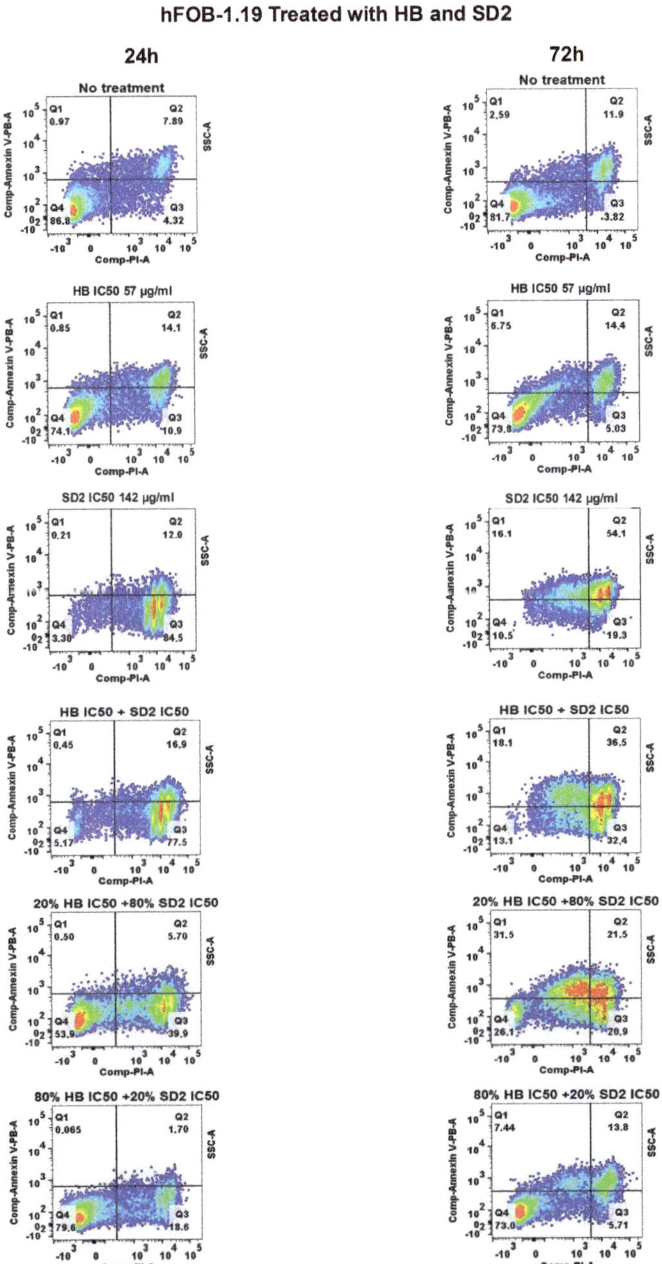

Figure 9. Effects of HB, SD2, or a combination of HB + SD2 on the apoptosis profiles of hFOB 1.19 cell line. The no treatment panel corresponds to the untreated cells used as the negative control. The treatment panels correspond to the 24 h and 72 h treatment with HB, SD2 or combination. Cell apoptosis was measured using Annexin V-PB/PI double staining. The results are expressed as the percentage of cells corresponding to Q1: early apoptotic cells (Annexin V+/PI−), Q2: late apoptotic cells (Annexin V+/PI+), Q3: necrotic cells (Annexin V− PI+), and Q4: healthy. Cells (Annexin V−/PI−).

The apoptotic profiles of the MG-63 cell line after 24 and 72 h of treatment with various combinations of HB and SD2 showed distinct cellular responses (Figure 7, Supplementary Figure S2). The combination of 20% HB + 80% SD2 exerted the strongest effect, where it reduced the proportion of healthy cells to 79.08% and slightly increased the percentages of early and late apoptotic cells to 4% and 6%, respectively, after 24 h of treatment. Conversely, the treatment with 80% HB + 20% SD2 largely preserved the cell viability, suggesting that a higher proportion of HB may have mitigated the cytotoxic effects of SD2 in the MG-63 cells. After 72 h of treatment, the combination of HB IC_{50} and SD2 IC_{50} continued to demonstrate a strong synergistic effect in the MG-63 cells, which reduced the proportion of healthy cells to 22.63%. This was accompanied by a substantial increase in the early apoptotic cells (59.73%), while late apoptosis and necrosis remained relatively low at 7.01% and 1.12%, respectively. This shift toward a predominant early apoptotic profile suggests a more defined apoptotic commitment over time. The evaluation of combination ratios revealed dose-dependent modulation. The combination 20% HB + 80% SD2 showed moderate cytotoxicity, where it reduced the percentage of healthy cells to 47.85%, and 29.95% underwent early apoptosis and 6.46% underwent late apoptosis. Necrosis remained minimal at 0.66%. In contrast, the 80% HB + 20% SD2 combination preserved the cell viability at 92.43%, indicating a marked reduction in cytotoxicity compared with the equimolar or SD2-dominant combinations. Collectively, these findings reinforce the time- and ratio-dependent natures of the HB–SD2 co-treatment and suggest that a higher proportion of SD2 drives apoptotic activity, particularly in the early stages, while HB may attenuate this effect when present in excess.

The impact of the combined treatment of HB and SD2 on the apoptotic profiles of the Saos-2 cell line after 24 and 72 h is shown in Figure 8 (and Supplementary Figure S3). The different combinations of HB with SD2 presented the significant apoptotic effect in a SD2-dose-dependent manner on the Saos-2 cell lines after 24 h. The cotreatment of HB and SD2 at IC_{50} concentrations demonstrated a profound impact, which led to a further reduction in the Saos-2 cell healthy profile to 32.95%. The percentage of apoptotic cell subpopulations were measured as 13.96%, 48.45%, and 6.87% in the early, late, and necrotic phases, respectively. The 20% HB + 80% SD2 combinatory treatment decreased the proportion of healthy cells to 79.23% while increasing the early apoptotic cells percentage to 13.76%, late apoptotic cells to 33.44%, and necrotic cells to 4.78%. Once again, with the dominant proportion of HB, the 80% HB + 20% SD2 combination largely preserved cell viability. Similar to the MG-63 cells, a higher proportion of HB in the combination appeared to reduce the cytotoxic effects observed with the SD2-dominant treatments in the Saos-2 cells. At 72 h, the co-treatment of HB and SD2 continued to exert a significant apoptotic effect on the Saos-2 cells in a SD2-dose-dependent manner. The combination of HB IC_{50} and SD2 IC_{50} led to a dramatic decrease in the healthy cell population to 3.14%, accompanied by a pronounced increase in early apoptotic cells (41.95%) and late apoptotic cells (51.20%). Necrosis was also observed at 8.45%. The 20% HB + 80% SD2 combination maintained a moderate apoptotic response, which reduced the healthy cells to 52.35%, while increasing the early and late apoptotic populations to 8.09% and 24.23%, respectively, along with 22.28% necrosis. Consistent with previous observations, the 80% HB + 20% SD2 combination largely preserved the cell viability at 88.43%, with minimal apoptotic and necrotic activity. These results reinforce the trend observed at 24 h, where increasing the proportion of SD2 in the co-treatment enhanced the cytotoxic and apoptotic responses, while a dominant presence of HB appeared to attenuate these effects in the Saos-2 cells.

The apoptotic profiles of the hFOB 1.19 cell line after 24 and 72 h of treatment with various combinations of HB and SD2 are shown in Figure 9 and Supplementary Figure S4. The combination of HB IC_{50} and SD2 IC_{50} after 24 h of treatment demonstrated the most pronounced effect on the hFOB 1.19 cells, which resulted in a dramatic reduction in healthy cells to 18.09%, while the percentage of necrotic cells surged to 46.80% without a significant change in the apoptotic subpopulations, suggesting that this combination might have induced a significant level of cell death through necrosis in the hFOB 1.19 cells. Further analysis of the combination ratios showed that 20% HB + 80% SD2 led to 40.83% healthy cells, which was associated with elevated early apoptosis (8.43%), late apoptosis (21.95%), and necrosis (25.83%) proportions. In contrast, the 80% HB + 20% SD2 combination allowed 80% of the cells to remain healthy, with a negligible effect on the apoptotic components. Similar to the other cell lines, a higher proportion of HB in the combination generally mitigated the more severe cytotoxic effects observed with higher SD2 concentrations. At 72 h, the combination of HB IC_{50} and SD2 IC_{50} continued to exert a pronounced effect on the hFOB 1.19 cells, where it further reduced the proportion of healthy cells to approximately 17.5% and drove a notable increase in necrotic cells to around 30.0%, while the apoptotic subpopulations remained relatively unchanged. After the 20% HB + 80% SD2 treatment, the healthy cell percentages dropped to about 25.0%, accompanied by elevations in early apoptosis (approximately 15.0%) and late apoptosis (around 26.0%), with a moderate necrotic fraction observed (~28.0%). Conversely, the 80% HB + 20% SD2 combination largely preserved the cell viability, where 78% of the cells remained healthy, with only minimal changes in the apoptotic and necrotic components. Overall, as observed at 24 h, a higher proportion of HB in the combination mitigated the severe cytotoxic effects associated with SD2-dominant treatments, indicating that HB may have exerted a protective influence on the hFOB 1.19 cells, even after extended exposure.

The flow cytometry analysis revealed that the SD2 and HB agents elicited distinct biological responses in the healthy and cancerous cell lines. In the normal osteoblast cell line hFOB 1.19, the SD2 treatment led to a dose-dependent increase in apoptosis and necrosis, with marked cellular stress and toxicity observed at higher concentrations. In contrast, HB exhibited a milder cytotoxic effect on the hFOB 1.19 cells, with a higher proportion of viable cells retained. In the osteosarcoma cell lines MG-63 and Saos-2, SD2 caused a significant reduction in the viable cell populations and induced a strong pro-apoptotic effect, as evidenced by increased early and late apoptosis. While HB also triggered apoptosis in these cancer cells, its effect was less pronounced compared with SD2. Under combined treatment conditions, co-administration of SD2 and HB led to further increases in the apoptotic cell populations relative to single-agent treatments. This synergistic effect was particularly evident in the late apoptosis rates and suggested an enhanced therapeutic efficacy. In conclusion, the combination of SD2 and HB potentiated apoptosis in the cancer cells while eliciting a more controlled response in the non-tumor cells, which supported the selective antitumor potential of SD2.

2.4. Evaluation of the HB and SD2 Treatment Effects on the 3D Cell Cultures

All three cell lines were successfully cultivated in 3D cultures, with observable changes over the time of the exposure (Figures 10–12). These changes were more pronounced in the cancer cell lines, especially in the Saos-2 spheroids, where certain treatments—namely, HB alone at its IC_{50} concentration and the combination of 80% HB + 20% SD2—caused disruptions that made measurement impossible after 72 h (Figures 11 and 13b).

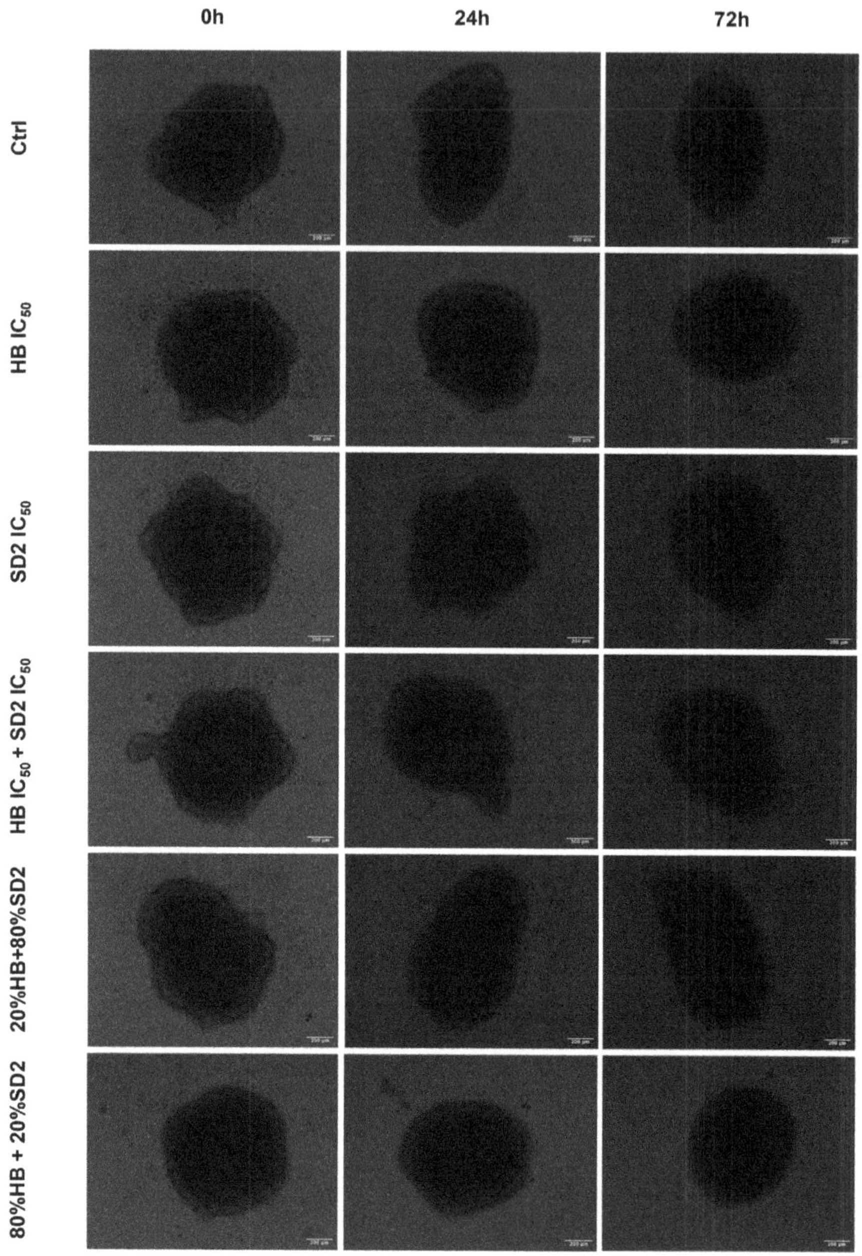

Figure 10. Representative transmitted light microscopy images of the MG-63 spheroids incubated with combined treatments of halogenated boroxine (HB) and Dex-CeNPs (SD2) for 24 h or 72 h, as indicated. The scale bar indicates a size of 200 μm.

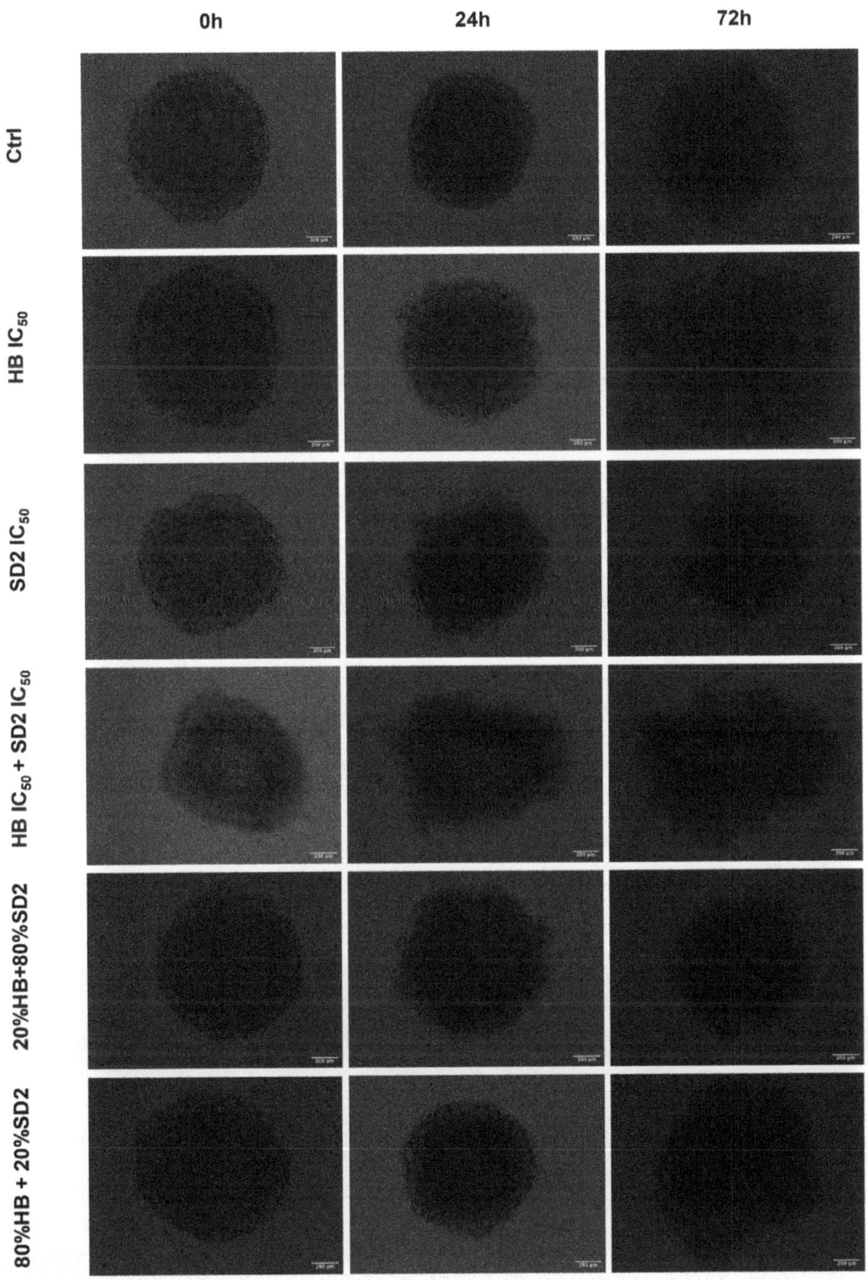

Figure 11. Representative transmitted light microscopy images of the Saos-2 spheroids incubated with combined treatments of halogenated boroxine (HB) and Dex-CeNPs (SD2) for 24 h or 72 h, as indicated. The scale bar indicates a size of 200 μm.

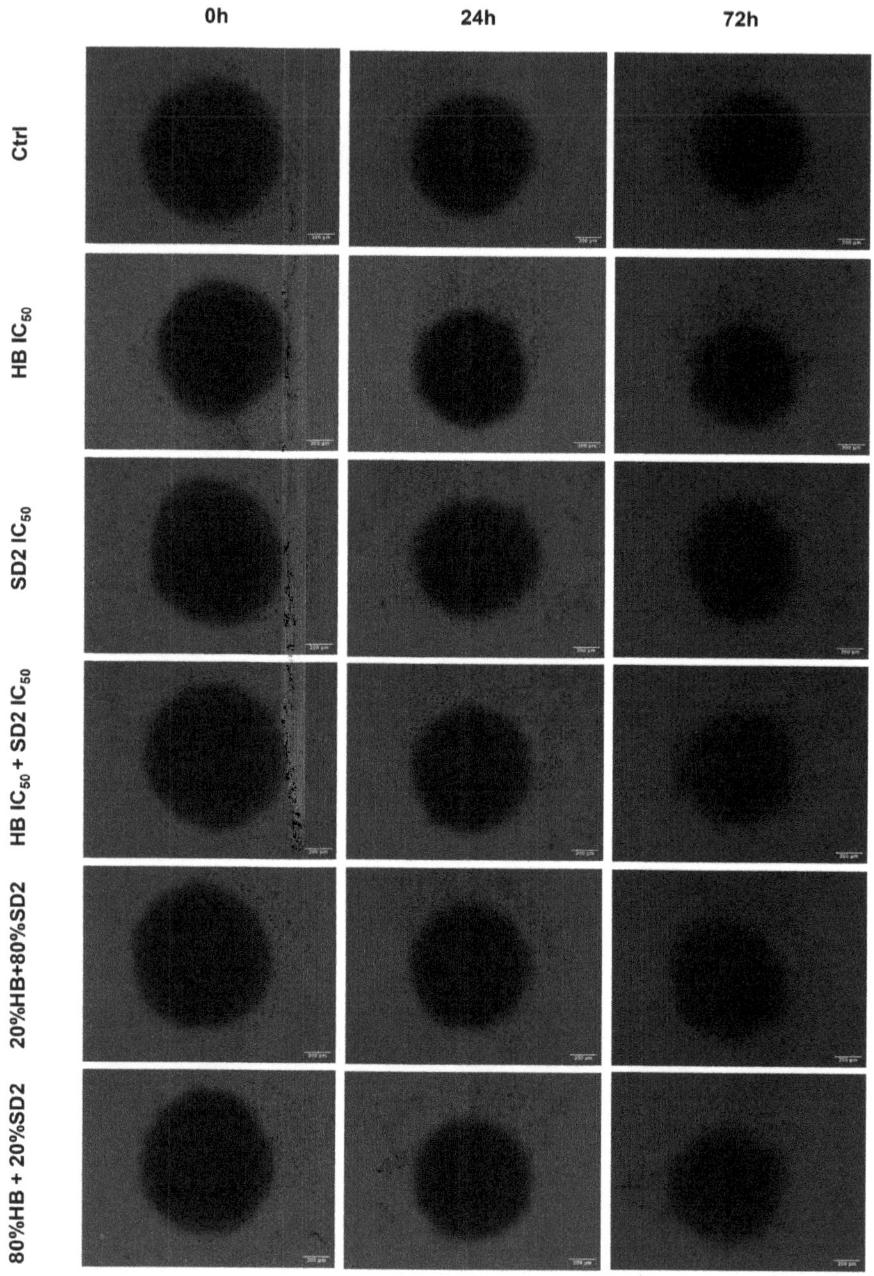

Figure 12. Representative transmitted light microscopy images of the hFOB 1.19 spheroids incubated with combined treatments of halogenated boroxine (HB) and Dex-CeNPs (SD2) for 24 h or 72 h, as indicated. The scale bar indicates a size of 200 μm.

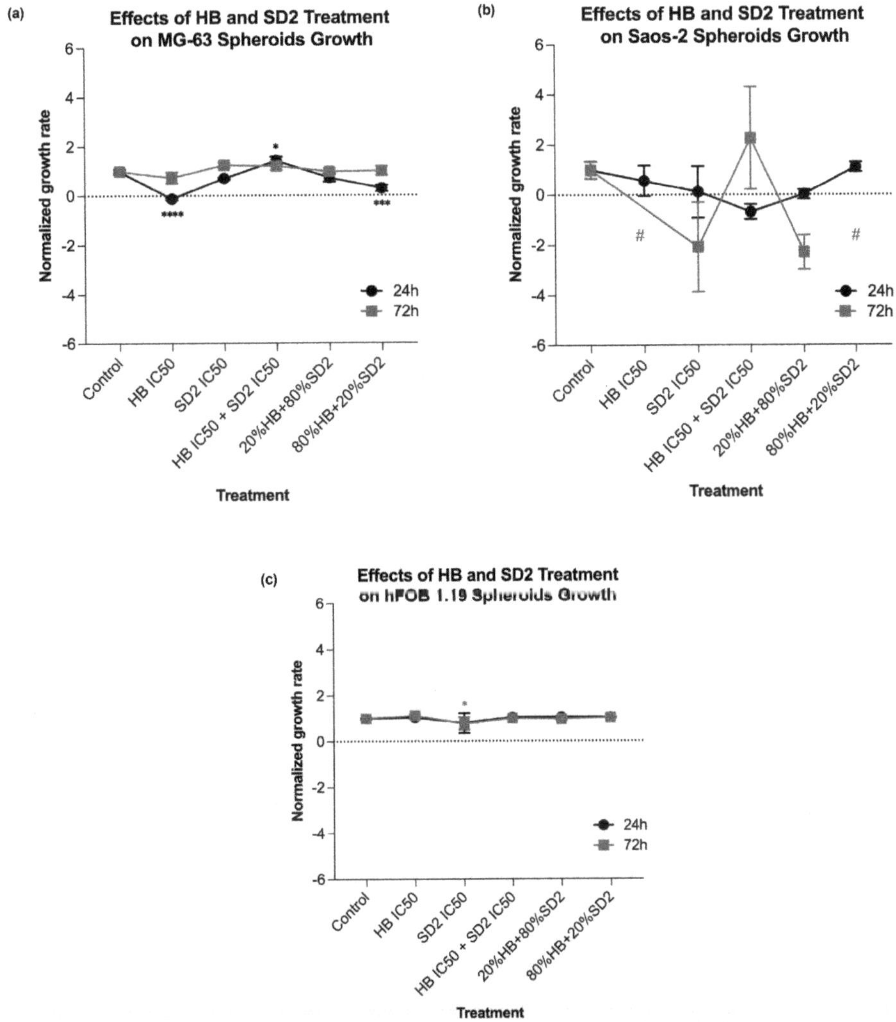

Figure 13. Effects of halogenated boroxine (HB) and Dex-CeNPs (SD2) co-treatments on the growth rates of MG-63 (**a**), Saos-2 (**b**), and hFOB 1.19 (**c**) spheroids. The Saos-2 spheroids treated with HB IC50 and 80% HB + 20% SD2 were disrupted at 72 h and could not be measured (marked with #). Statistical significance is indicated as * $p < 0.05$, *** $p < 0.001$, and **** $p < 0.0001$.

When evaluating the effects of the tested treatments on the spheroids, the growth rates of both osteosarcoma spheroid types were affected. In the MG-63 spheroids (Figure 13a), the treatment with HB alone and the combination of 80% HB + 20% SD2 significantly reduced its growth rate at 24 h; however, the growth appeared to recover after 72 h. Interestingly, the treatment with the combination of IC_{50} values of both compounds resulted in a slight but statistically significant increase in the growth rate compared with the control at 24 h (p-value = 0.0369), while a decrease was observed at 72 h. In the Saos-2 spheroids, no treatment showed a statistically significant effect compared with the control (Figure 13b). Nevertheless, based on the normalized growth rate profiles and morphological observations, the Saos-2 spheroids appeared to be the most affected. The treatment with the combination of HB IC_{50} and SD2 IC_{50} decreased the normalized growth rate after 24 h, but an increase was observed at 72 h. On the other hand, SD2 alone and the combination of 20% HB + 80% SD2 led to a time-dependent decrease in the normalized growth rate of the Saos-2 spheroids. Markedly, the hFOB 1.19 spheroids were the least affected. There were no substantial differences between the treated and control spheroids, nor were any time-dependent changes observed (Figure 13c).

3. Discussion

Using various experimental approaches, the results of this study integrate data on the effects of HB, SD2, and their co-treatments in tumorigenic (MG-63 and Saos-2) and non-tumor (hFOB 1.19) cell lines. Observations from individual and combined treatments are contextualized in the following section across multiple levels of analysis—starting with the cytotoxicity and ROS profiles, followed by the apoptotic and necrotic responses, and concluding with the 3D spheroid growth dynamics.

Our research showed that HB exerted dose- and time-dependent cytotoxic effects on both the MG-63 and Saos-2 osteosarcoma cell lines, which is in line with previous studies that established its pro-apoptotic and cytotoxic activities in hematological and solid malignancies [19–21]. In particular, the observed reduction in cell viability at higher HB concentrations corroborates findings in UT-7 leukemia cells, where HB triggered apoptosis via caspase activation and the deregulation of pro-survival genes [19,20]. Similarly, Elez-Burnjakovic et al. reported HB-induced cytotoxicity and the modulation of autophagy in melanoma and bladder carcinoma cells, implicating oxidative stress and metabolic reprogramming as possible underlying mechanisms [21,22].

Interestingly, the lowest concentration of HB (50 μg/mL) resulted in a slight increase in viability at 72 h in the MG-63 cells, suggesting a possible low-dose-induced adaptive response or mild cellular stimulation. This dual effect has been described in other cancer models exposed to boron-based agents, where low doses may activate compensatory survival pathways [30]. In contrast, the hFOB 1.19 cells exhibited a higher sensitivity to HB, with cytotoxicity apparent even at moderate concentrations. However, the response in this non-tumor line lacked a strict dose-dependent trend, indicating specific responses to HB, which may involve differences in the oxidative resilience, baseline metabolic activity, or other background mechanisms, which should be examined at the molecular level.

The cytotoxicity assessment of SD2 also revealed concentration- and cell-line-dependent responses across the osteosarcoma (MG-63 and Saos-2) and non-tumor osteoblastic (hFOB 1.19) cell models. Notably, the MG-63 cells demonstrated a biphasic, hormetic response, wherein a low concentration of SD2 (100 μg/mL) elicited a modest increase in viability at 24 h, suggesting the potential transient activation of survival pathways, followed by a pronounced reduction in viability at concentrations $\geq$ 250 μg/mL. This phenomenon is consistent with the hormetic behavior previously described for cerium oxide nanoparticles, wherein low-dose exposure activates cytoprotective antioxidant pathways, while higher concentrations shift the redox balance via Ce^{3+}/Ce^{4+} cycling, generating reactive oxygen species (ROS) that lead to mitochondrial dysfunction and apoptosis [31,32].

In contrast, the Saos-2 cells exhibited a relatively chemoresistant phenotype, where they showed less severe viability reduction and partial recovery over 72 h. This attenuated response likely reflects cell-line-specific differences in basal oxidative status, nanoparticle uptake, and redox buffering capacity [33]. Interestingly, the hFOB 1.19 line displayed the steepest early viability reduction (20–35% at 500–1000 μg/mL, 24 h), yet recovered to 60–70% by 72 h. Although the ROS levels were already suppressed below the control at all doses, the transient viability loss may stem from non-oxidative stresses, such as lysosomal overload or ionic imbalance. The rapid rebound concurs with SD2's catalase/SOD-mimetic activity that continually scavenges ROS in normal osteoblasts, allowing metabolic adaptation once the acute nanoparticle burden is resolved [34–36]. Mechanistically, the redox-switching property of cerium oxide is central to SD2's biological effects. At lower concentrations, cerium oxide nanoparticles scavenge free radicals and reduce oxidative stress, whereas higher doses promote oxidative damage through ROS accumulation, triggering apoptosis via mitochondrial and caspase-mediated pathways [35,36]. The dextran coating of SD2 significantly influences these interactions, possibly by enhancing the colloidal stability, facilitating endocytosis, and modulating the nanoparticle's intracellular fate through its specific branching configuration, which could alter the protein corona and cellular trafficking dynamics [37]. This may explain the steeper cytotoxic thresholds observed with SD2 compared with other surface coatings, such as polyethylene glycol (PEG) or polyvinylpyrrolidone (PVP), which have been shown to mitigate cerium-oxide-nanoparticle-induced toxicity [38].

Moreover, the observed intercellular heterogeneity in response to SD2 underscores the importance of considering tumor-specific oxidative metabolism and nanoparticle processing mechanisms. The heightened sensitivity of the MG-63 cells may reflect an increased nanoparticle uptake and a less competent antioxidant defense system, while the resilience of the hFOB 1.19 cells suggests efficient metabolic adaptation and the possible exploitation of cerium oxide nanoparticles' catalytic scavenging over time [39].

Overall, SD2 elicited a concentration-dependent cytotoxic response that was most pronounced in the MG-63 osteosarcoma cells, displayed moderate effects in the Saos-2, and caused only transient impairment in the non-tumor hFOB 1.19 cells. These findings, which are consistent with the redox-switching behavior of dextran-coated cerium oxide nanoparticles reported in the literature, underscore SD2's potential as a selective nanotherapeutic for osteosarcoma, while highlighting the necessity for dose optimization and mechanistic validation.

The reactive oxygen species (ROS) generation profiles revealed that high HB concentrations significantly increased the oxidative stress in osteosarcoma cells. This supports the hypothesis that HB's cytotoxicity is mediated, at least in part, by oxidative mechanisms, as previously demonstrated in UT-7 and GR-M cells, where HB altered the expression of redox- and apoptosis-related genes [19–21]. Conversely, low HB concentrations reduced ROS production, suggesting a threshold effect or ROS scavenging at sub-cytotoxic levels. In the hFOB 1.19 cells, the ROS induction was modest and not statistically significant across most conditions, which possibly reflected stronger antioxidant defenses in non-malignant osteoblasts.

SD2 caused markedly divergent reactive oxygen species (ROS) responses that tracked both the dose and cellular context. In the MG-63 osteosarcoma cells, the intracellular ROS levels remained near the baseline at 100–250 μg/mL but surged to nearly three-fold above the control levels at 500–1000 μg/mL, which mirrored the steep viability decline observed at these concentrations. This response aligns with previous findings where cerium oxide nanoparticles induced oxidative-stress-mediated cytotoxicity through mitochondrial ROS production and apoptosis [35,40].

The Saos-2 cells exhibited an even more pronounced pro-oxidant profile, with significant ROS elevation already detectable at 100 μg/mL and a plateau observed around 250 μg/mL, suggesting earlier saturation of nanoparticle-induced oxidative stress [39]. These differences likely stemmed from cell-line-specific variations in the basal redox state, nanoparticle internalization kinetics, and antioxidant capacity [33,34].

Strikingly, the hFOB 1.19 displayed a consistent reduction in ROS levels across all SD2 concentrations, with the DCFDA signals falling to nearly 30–40% of the control at 250–1000 μg/mL. This observation supports the proposed role of cerium oxide nanoparticles as potent ROS scavengers in normal cells, likely via catalase- and superoxide dismutase

(SOD)-mimetic mechanisms that become prominent under physiological pH and normoxic conditions [32,36]. Collectively, these findings underscore SD2's unique redox-modulatory properties, which are pro-oxidative in malignant osteosarcoma cells and antioxidative in non-transformed osteoblastic cells, thus supporting its potential utility as a context-sensitive nanotherapeutic agent in bone oncology [24,29].

The co-treatment with HB and SD2 produced complex, concentration-ratio-dependent effects on the cell viability and oxidative stress. In the MG-63 and Saos-2 cells, most HB–SD2 combinations induced cytotoxicity, though some combinations (notably 80% HB + 20% SD2 and 60% HB + 40% SD2) were less effective than HB alone. These findings suggest potential antagonistic interactions at certain ratios, possibly due to the ROS-scavenging properties of cerium oxide nanoparticles (CeNPs), which can neutralize the oxidative effects of HB [29]. In support of this, combined treatments in the MG-63 cells did not increase the ROS production, and in some cases, the ROS levels were lower than in untreated controls. These observations aligned with studies that have shown the dual redox nature of CeNPs, which can act as pro-oxidants in tumor microenvironments while protecting non-malignant cells from oxidative damage [24,29].

The Saos-2 cells responded to the co-treatment with elevated ROS levels, particularly for combinations rich in SD2. The increase in oxidative stress suggests that the balance between SD2's redox activity and the intrinsic oxidative state of the Saos-2 cells may differ from that in the MG-63 cells, potentially reflecting metabolic or mitochondrial differences between the lines. Furthermore, the pronounced ROS response in the Saos-2 cells following the 20% HB + 80% SD2 treatment supports the notion that CeNPs can shift from antioxidative to oxidative behavior depending on the cellular context and nanoparticle dose [29].

The behavior of the hFOB 1.19 cells under co-treatment conditions was notably different. While most combinations reduced the viability at 24 h, partial recovery was observed at 72 h. This suggests that the hFOB 1.19 cells may have engaged repair or adaptation mechanisms in response to early oxidative or cytotoxic stress, particularly when HB was present at lower ratios. Notably, the HB treatment at its IC_{50} concentrations alone also showed signs of reduced cytotoxicity over time in the hFOB 1.19 cells, indicating potential cellular resilience or activation of detoxification pathways that mitigate long-term damage. This recovery effect was not observed in the HB screening experiments; however, in those experiments, as mentioned before, the cells did not show a linear response to HB's cytotoxic effect, implying that healthy osteoblasts show concentration-specific and potentially transient cellular responses, possibly representing a narrow threshold at which normal cells can engage protective mechanisms.

Our findings in the 3D cell cultures revealed differential spheroid responses to HB, SD2, and their combinations across the tested cell lines. In the MG-63 spheroids, both HB alone and the 80% HB + 20% SD2 combination significantly reduced the growth at 24 h, followed by signs of recovery by 72 h. Notably, the IC_{50} combination paradoxically increased the growth at 24 h, which may indicate sub-lethal stress-induced proliferation or altered spheroid compactness, warranting further mechanistic investigation. Although the Saos-2 spheroid growth did not show statistically significant differences, the morphological indicators and normalized growth profiles suggested greater sensitivity to treatment, particularly with SD2 alone and the 20% HB + 80% SD2 combination. These treatments elicited a time-dependent decrease in the spheroid growth. The hFOB 1.19 spheroids demonstrated a remarkable resistance to all the treatments, with no significant changes in growth or morphology. This highlights and reconfirmed their relative robustness and corroborated the selective cytotoxicity of HB and SD2 toward malignant cells. Although the effects of HB in 3D culture systems have not yet been explored, its selective cytotoxicity has been extensively documented in 2D models [18–22], and our findings in spheroids appear to corroborate this selectivity. Overall, our data emphasize the importance of evaluating spheroid dynamics over multiple time points and integrating both quantitative and qualitative endpoints to comprehensively assess the therapeutic efficacy in 3D models.

4. Materials and Methods

4.1. Synthesis and Preparation of Halogenated Boroxine $K_2(B_3O_3F_4OH)$ (HB) and Dextran-Coated Cerium Oxide Nanoparticles (Dex-CeNPs)

HB is a white powder that is soluble in water (7.1% at 20 °C), ethanol, and DMSO. It was synthesized according to a previously described protocol with the slight modification [41]. The stock solutions were prepared by dissolving 20 mg of $K_2(B_3O_3F_4OH)$ in 1 mL of 1× PBS, and the final concentrations of the solutions to be tested were made in the cell culture medium.

The synthesis of dextran-coated cerium oxide nanoparticles (Dex-CeNPs), here labeled as SD2, was carried out according to the protocol published by Tarakci et al., 2025 [29]. In that study, the protocol was well optimized, and a detailed physicochemical characterization was conducted. Based on this, in the present study, cerium oxide nanoparticles coated with dextran from *Leuconostoc* spp. (Sigma, Cat# 31388, mol wt: ~6 kDa) (SD2) were used [29]. For each newly synthesized SD2 sample, basic characterization was once again performed, the concentration was determined, and solutions of a defined concentration were prepared in the cell culture medium for testing (Supplementary Figure S1).

4.2. Cell Lines and Cell Culture Maintenance

Two osteosarcoma cell lines—MG-63 (CRL-1427) and Saos-2 (HTB-85)—and one osteoblast cell line—hFOB 1.19 (CRL-3602)—were obtained from the American Type Culture Collection (ATCC, Manassas, VA, USA). The MG-63 cells were cultured in Eagle's Minimum Essential Medium (Sigma-Aldrich, Saint Louis, MO, USA), the Saos-2 cells were cultured in McCoy's 5a Medium Modified (Sigma-Aldrich), and the hFOB 1.19 cells were maintained in Dulbecco's Modified Eagle's Medium/Nutrient Mixture F-12 Ham (Sigma-Aldrich). All cell media were supplemented with 10% fetal bovine serum (FBS, Belize City, Belize) and 1% penicillin–streptomycin, and were incubated in a humidified atmosphere of 5% CO_2 in air and at a temperature of 37 °C. All were also tested on mycoplasma periodically using MycoAlert test (Lonza, Walkersville, MD, USA). The cell line identity was confirmed using PowerPlex® Fusion System (Promega, Madison, WI, USA) and STR profiles available in the ATCC database.

4.3. HB Treatments

The experiments were carried out in 96-well microtiter plates, with 5000 cells seeded per well. After 24 h, the cell cultures were treated with HB at final concentrations of 50, 75, 100, 150, 200, and 400 μg/mL. Due to the different sensitivity, the hFOB 1.19 cells were additionally treated with HB at a final concentration of 25 μL/mL. Each treatment was carried out in triplicate and experiments were repeated at least twice. Untreated cells were used as controls. The cell viabilities of the MG-63, Saos-2, and hFOB 1.19 cells cultured in the presence of HB were calculated as a percentage of the control cells, and the half-maximal inhibitory concentration (IC_{50}) values were obtained from the dose–response curves plotted for the results obtained after 24 h.

4.4. SD2 Treatments

To evaluate the effects of SD2, the MG-63, Saos-2, and hFOB 1.19 cells were also seeded at a density of 5000 cells per well in 96-well plates. Following a 24 h incubation period, the cells were exposed to SD2 at concentrations of 100, 250, 500, and 1000 μg/mL. All treatments were performed in triplicate and independently repeated at least three times to ensure reproducibility. Untreated cells served as negative controls. The cell viability was assessed relative to the control group, as in the HB screening experiments, and the dose–response curves were generated to determine the IC_{50} values for each cell line 24 h post-treatment.

4.5. HB and SD2 Co-Treatments

The IC_{50} values obtained from the single-drug cell viability/cytotoxicity assays were used to design subsequent drug combination experiments for each particular cell type. The following combined/synergistic treatment ratios were defined: 0.8-, 0.6-, 0.5-, 0.4-, and 0.2-fold of the HB IC_{50} with 0.2-, 0.4-, 0.5-, 0.6-, and 0.8-fold of the SD2 IC_{50}, respectively. In addition, a combination of both compounds' IC_{50} concentrations was tested. All treatments were performed in triplicate, with at least three independent experimental replicates.

4.6. Cytotoxicity Assay

The cytotoxic activity was determined using the WST-1 assay and it was performed at two time points: at day 1 (24 h after treatment) and day 3 (72 h after treatment). Upon the incubation of 24 h and 72 h after the treatments of cells, the medium containing the treatments was removed from each well, and 100 μL of a WST-1 reagent (1:10 ratio with cell culture medium) was added to each well and incubated for 2 h. At the end of the incubation, a color change from pink to light brown was observed, and the absorbances of each well were measured at 450 nm and 650 nm using a BioTek Cytation 5 Cell imaging multimode reader. The viability/cytotoxicity percentages were counted, and graphs were plotted by normalizing the data against the untreated control groups.

4.7. Reactive Oxygen Species (ROS) Detection

To assess the reactive oxygen species (ROS), a 2′, 7′- dichlorofluorescein diacetate (DCFDA) assay was used. The ROS levels were measured simultaneously with each cytotoxicity 24 h post-treatment, with the same treatments and controls (untreated cells) applied.

After the end of the incubation period of 24 h, the medium with treatments was removed, and each well was washed twice with 1× PBS. The remaining cells were then incubated with 10 μM DCFDA in the dark for 45 min at 37 °C in a 5% CO_2. Following the incubation, the fluorescence intensity was measured in the 485 nm/535 nm range using a BioTek Cytation 5 Cell imaging multi-mode reader. The ROS-related fluorescence intensity fold changes were calculated and normalized based on the cell viability data.

4.8. Flow Cytometry Analysis

The flow cytometry analysis was performed to evaluate the cell viability using the APC Annexin V Apoptosis Detection Kit with PI (BioLegend) specifically designed for the identification of healthy, apoptotic, and necrotic cells. Based on the staining patterns, the cells were classified as follows: healthy—Annexin-V^-/PI^-, early apoptotic—Annexin-V^+/PI^-, late apoptotic—Annexin-V^+/PI^+, and necrotic—Annexin-V^-/PI^+.

To perform the flow cytometry, 300,000 cells per well were seeded in a 6-well plate and incubated for 24 h. Following the incubation, each cell line was treated with HB at its IC_{50} concentration and SD2 at its IC_{50} concentration. Additionally, selected combination treatments were applied: 0.2- and 0.8-fold of the HB IC_{50} concentration with 0.8- and 0.2-fold of the nanoceria IC_{50} concentration, respectively. Also, a combination of both compounds at their IC_{50} concentrations was tested. The untreated cells served as a control.

After 24 or 72 h of treatment, the medium was removed, and the cells were washed with 1× PBS. Following the manufacturer's instructions, the cells were additionally washed, resuspended in Annexin Buffer and stained with Annexin V and PI at a 1:2 ratio, with the volume adjusted based on the cell count. The samples were then incubated for 15 min in the dark at room temperature, followed by the addition of another 100 μL of Annexin Buffer before measurement on a BD FACS Canto II instrument and analysis using FlowJo v10 software.

4.9. Spheroid Growth Analysis

In addition to the established standard 2D cultivation of the cell lines MG-63, Saos-2, and hFOB 1.19, 3D cultures of each cell line were established to conduct preliminary tests of treatment effects in three-dimensional growth models in the form of spheroids by the liquid layer or floating cells method [42]. Prior to the cell cultivation, a hot, sterile 1.5% agarose gel was poured in a 96-well plate, followed by cell seeding with a density of 5000 cells/well. Spheroids were usually formed within 24 h after the seeding when bright-field images of all spheroid replicas were acquired with a 10× objective of an EVOS M5000 imaging system. Afterwards, the same treatment regime was performed as previously performed in the 2D cell culture for flow cytometry analysis. Spheroids were imaged again at 24 h and 72 h after the treatment.

The spheroid area, as a morphological descriptor, was analyzed using ImageJ v0.5.7 image-processing software to determine the growth rate. The measurement of the spheroid area from transmitted-light images is described as a classical method to evaluate the activity of antiproliferative compounds in 3D models [43]. The growth rates at 24 h and 72 h were calculated for each spheroid relative to the spheroid area before the addition of treatments. The average growth rate at the HB and SD2 IC_{50} concentrations and combined treatments was normalized to the average growth rate of the untreated spheroids (controls) for each executed experiment.

4.10. Statistical Analysis

Statistical analyses were conducted using GraphPad Prism software (version 10.4.1). The differences between the treatment groups were estimated using one-way and two-way analyses of variance (ANOVA), followed by Tukey's post hoc test. A p-value of less than 0.05 was considered statistically significant, with significance levels denoted as * $p < 0.05$, ** $p < 0.001$, *** $p < 0.001$, and **** $p < 0.0001$.

5. Conclusions

In our work we aimed to investigate the single and combined treatment effects of HB and SD2 on selected osteosarcoma cell lines and corresponding healthy cell controls in 2D and 3D (spheroid) microenvironments. The presented results emphasize the importance of dose, exposure time, and cellular context in determining the effects of HB and SD2. While HB demonstrated clear antitumor activity through cytotoxicity and ROS induction, its combination with redox-active nanoparticles, such as SD2, could either enhance or attenuate these effects depending on the ratio and cell type. We found that the combined treatment of 20% HB + 80% SD2 produced replicable and reliable antitumor effects in both osteosarcoma cell lines. This can be explained by specific biochemical interactions and microenvironment generation of HB and SD2 that were previously described in independent studies. The tumor selectivity in this study was more evident in the 3D models, which better replicated the in vivo tissue structure and treatment diffusion limitations. However, one of the key limitations of this study was the exclusive use of established cell lines for both the 2D and 3D (spheroid) models. While these models provide valuable initial insights into the cytotoxic and redox effects of HB and SD2, they do not fully replicate the complexity of the tumor microenvironment or the heterogeneity observed in patient tumors. Cell lines often exhibit genetic and phenotypic drift from primary tumors and may not accurately represent inter-patient variability. Therefore, future studies should consider incorporating patient-derived organoids or spheroids, which more closely mimic the architecture, cellular diversity, and treatment responses of tumors in vivo. These advanced models would provide a more physiologically relevant platform for evaluating the therapeutic potential and selectivity of HB–SD2 combination treatments. The observed complexity of antitumor activities of HB and SD2 as single treatments and combined underscores the need for further mechanistic studies, particularly at the transcriptomic and proteomic levels, to delineate the exact pathways modulated by HB–SD2 co-treatments that drive the explicit and desired biological effect.

Supplementary Materials: The following supporting information can be downloaded from https://www.mdpi.com/article/10.3390/ijms26209837/s1.

Author Contributions: Conceptualization, H.Y., L.P. and N.T.; methodology, N.T., S.E., A.M., M.G. and B.G.; formal analysis, N.T., A.D., A.M. and J.B.; investigation, N.T., S.E. and J.B.; resources, N.T., S.E., J.R. and N.L.-K.; writing—original draft preparation, N.T., H.Y. and L.P.; writing—review and editing, N.T., H.Y., L.P., S.E. and B.S.M.; visualization, N.T., A.M. and J.B.; supervision, H.Y. and L.P.; project administration, H.Y. and L.P.; funding acquisition, L.P. and H.Y. All authors have read and agreed to the published version of the manuscript.

Funding: This study was supported by the International Centre for Genetic Engineering and Biotechnology (ICGEB) (CRP/TUR 18-03), TUBA-GEBIP Distinguished Young Scientist Award (Hilal Yazici, 2019), Federal Ministry of Education and Science (05-35-2446-1/23), and ICGEB SMART PhD Fellowship (S/BIH23-01).

Institutional Review Board Statement: Not applicable.

Informed Consent Statement: Not applicable.

Data Availability Statement: The original contributions presented in this study are included in the article/Supplementary Material. Further inquiries can be directed to the corresponding authors.

Conflicts of Interest: The authors declare no conflicts of interest.

References

1. Kansara, M.; Teng, M.W.; Smyth, M.J.; Thomas, D.M. Translational biology of osteosarcoma. *Nat. Rev. Cancer* **2014**, *14*, 722–735. [CrossRef]
2. Wang, S.Y.; Hu, H.Z.; Qing, X.C.; Zhang, Z.C.; Shao, Z.W. Recent advances of drug delivery nanocarriers in osteosarcoma treatment. *J. Cancer* **2020**, *11*, 69–82. [CrossRef]
3. Luo, Y.; Sun, M.; Tan, L.; Li, T.; Min, L. Nano-Based Drug Delivery Systems: Potential Developments in the therapy of Metastatic Osteosarcoma—A Narrative Review. *Pharmaceutics* **2023**, *15*, 2717. [CrossRef]
4. Sellers, G.S.; Poirier, M.A.; Mayberry, T.G.; Cowan, B.C.; Wakefield, M.R.; Fang, Y. From conventional to cutting edge: An exploration of osteosarcoma treatments. *Med. Oncol.* **2025**, *42*, 81. [CrossRef]
5. Broadhead, M.L.; Clark, J.C.M.; Myers, D.E.; Dass, C.R.; Choong, P.F.M. The Molecular Pathogenesis of Osteosarcoma: A review. *Sarcoma* **2011**, *2011*, 959248. [CrossRef] [PubMed]
6. Yang, W.; Gao, X.; Wang, B. Boronic acid compounds as potential pharmaceutical agents. *Med. Res. Rev.* **2003**, *23*, 346–368. [CrossRef]
7. Ban, H.S.; Nakamura, H. Boron-Based Drug Design. *Chem. Rec.* **2015**, *15*, 616–635. [CrossRef] [PubMed]
8. Dayal, N.; Chaudhuri, R.; Yeboah, K.S.; Brauer, N.R.; Sintim, H.O. Boronic Acid-Containing 3H- pyrazolo [4,3-f]quinoline Compounds as Dual CLK/ROCK Inhibitors with Anticancer Properties. *Pharmaceuticals* **2024**, *17*, 1660. [CrossRef] [PubMed]
9. Zhu, Y.; Hosmane, N.S. Nanostructured boron compounds for cancer therapy. *Pure Appl. Chem.* **2017**, *90*, 653–663. [CrossRef]
10. Delgrosso, E.; Scocozza, F.; Cansolino, L.; Riva, F.; Conti, M.; Loi, G.; Auricchio, F.; Postuma, I.; Bortolussi, S.; Cobianchi, L.; et al. 3D bioprinted osteosarcoma model for experimental boron neutron capture therapy (BNCT) applications: Preliminary assessment. *J. Biomed. Mater. Res. B Appl. Biomater.* **2023**, *111*, 1571–1580. [CrossRef]
11. Hsu, C.F.; Liu, H.M.; Peir, J.J.; Liao, J.W.; Chen, K.S.; Chen, Y.W.; Chuang, Y.J.; Chou, F.I. Therapeutic Efficacy and Radiobiological Effects of Boric-Acid-Mediated BNCT in an Osteosarcoma-Bearing SD Rat Model. *Life* **2023**, *13*, 514. [CrossRef]

12. Hall, D.C. *Boronic Acids*; John Wiley & Sons: New York, NY, USA, 2005.
13. Galić, B. Removal of Skin Changes. European Patent 1996514, 31 July 2013.
14. Galić, B. Boroxine Composition for Removal of Skin Changes. U.S. Patent 8,278,289, 2 October 2012.
15. Haverić, S.; Haverić, A.; Bajrović, K.; Galić, B.; Maksimović, M. Effects of dipotassium trioxohydroxytetrafluorotriborate (K2[B3O3F4OH]) on genetic material and inhibition of cell division in human cell cultures. *Drug. Chem. Toxicol.* **2011**, *34*, 250–254. [CrossRef]
16. Ivanković, S.; Stojković, R.; Galić, Z.; Galić, B.; Ostojić, J.; Marasović, M.; Milos, M. In vitro and in vivo antitumor activity of the halogenated boroxine dipotassium trioxohydroxytetrafluorotriborate (K2[B3O3F4OH]). *J. Enzyme Inhib. Med. Chem.* **2015**, *18*, 354–359. [CrossRef]
17. Ivanković, S.; Stojković, R.; Maksimović, M.; Galić, B.; Miloš, M. Impact of calcium ion on cytotoxic effect of the boroxine derivative, K2[B3O3F4OH]. *J. Enzyme Inhib. Med. Chem.* **2016**, *31*, 70–74. [CrossRef]
18. Pojskić, L.; Haverić, S.; Lojo-Kadrić, N.; Hadžić, M.; Haverić, A.; Galić, Z.; Galić, B.; Vullo, D.; Supuran, C.T.; Milos, M. Effects of dipotassium-trioxohydroxytetrafluorotriborate, K2[B3O3F4OH], on cell viability and gene expression of common human cancer drug targets in a melanoma cell line. *J. Enzyme Inhib. Med. Chem.* **2016**, *31*, 999–1004. [CrossRef] [PubMed]
19. Hadzic, M.; Pojskic, L.; Lojo-Kadric, N.; Haveric, A.; Ramic, J.; Galic, B.; Haveric, S. Novel boron-containing compound, halogenated boroxine, induces selective cytotoxicity through apoptosis triggering in UT-7 leukemia. *J. Biochem. Mol. Toxicol.* **2022**, *36*, e23005. [CrossRef] [PubMed]
20. Hadzic, M.; Sun, Y.; Tomic, N.; Tsirvouli, E.; Kuiper, M.; Pojskic, L. Halogenated boroxine increases propensity to apoptosis in leukemia (UT-7) but not non-tumor cells in vitro. *FEBS Open Bio* **2023**, *13*, 143–153. [CrossRef] [PubMed]
21. Elez-Burnjaković, N.; Pojskić, L.; Haverić, A.; Lojo-Kadrić, N.; Omanović, M.H.; Ramić, J.; Smajlović, A.; Maksimović, M.; Haverić, S. New in vitro findings about halogenated boroxine cytotoxicity and deregulation of cell death-related genes in GR-M melanoma cells. *Arh. Hig. Rada. Toksikol.* **2023**, *74*, 16–21. [CrossRef]
22. Elez-Burnjaković, N.; Pojskić, L.; Haverić, A.; Lojo-Kadrić, N.; Hadžić Omanović, M.; Smajlović, A.; Kalaydjiev, S.; Maksimović, M.; Joksimović, B.; Haverić, S. Halogenated Boroxine K2[B3O3F4OH] Modulates Metabolic Phenotype and Autophagy in Human Bladder Carcinoma 5637 Cell Line. *Molecules* **2024**, *29*, 2919. [CrossRef]
23. Jing, Z.; Du, Q.; Zhang, X.; Zhang, Y. Nanomedicines and nanomaterials for cancer therapy: Progress, challenge and Perspectives. *Chem. Eng. J.* **2022**, *446*, 137147. [CrossRef]
24. Tang, J.L.Y.; Moonshi, S.S.; Ta, H.T. Nanoceria: An innovative strategy for cancer treatment. *Cell. Mol. Life Sci.* **2023**, *80*, 46. [CrossRef]
25. Alpaslan, E.; Yazici, H.; Golshan, N.H.; Ziemer, K.S.; Webster, T.J. Ph-dependent activity of dextran-coated cerium oxide nanoparticles on prohibiting osteosarcoma cell proliferation. *ACS Biomater. Sci. Eng.* **2015**, *1*, 1096–1103. [CrossRef]
26. Yazici, H.; Alpaslan, E.; Webster, T.J. The role of dextran coatings on the cytotoxicity properties of Ceria nanoparticles toward bone cancer cells. *JOM* **2015**, *67*, 804–810. [CrossRef]
27. Jogdeo, C.M.; Panja, S.; Kanvinde, S.; Kapoor, E.; Siddhanta, K.; Oupický, D. Advances in Lipid-Based codelivery systems for cancer and inflammatory diseases. *Adv. Healthc. Mater.* **2022**, *12*, e2202400. [CrossRef]
28. Li, X.; Peng, X.; Zoulikha, M.; Boafo, G.F.; Magar, K.T.; Ju, Y.; He, W. Multifunctional nanoparticle-mediated combining therapy for human diseases. *Signal Transduct. Target. Ther.* **2024**, *9*, 1. [CrossRef]
29. Tarakci, E.; Esmkhani, S.; Bayramova, J.; Bilgin, F.M.; Kidik, K.; Adiguzel, S.; Tufan, Y.; Morva Yilmaz, A.; Yilmaz, H.; Duygulu, O.; et al. New insights of cerium oxide nanoparticles in head and neck cancer treatment. *Sci. Rep.* **2025**, *15*, 7665. [CrossRef] [PubMed]
30. Kulkarni, S.; Bhandary, D.; Singh, Y.; Monga, V.; Thareja, S. Boron in cancer therapeutics: An overview. *Pharmacol. Ther.* **2023**, *251*, 108548. [CrossRef] [PubMed]
31. Korsvik, C.; Patil, S.; Seal, S.; Self, W.T. Superoxide dismutase mimetic properties exhibited by vacancy engineered ceria nanoparticles. *Chem. Commun.* **2007**, *10*, 1056–1058. [CrossRef] [PubMed]
32. Das, S.; Dowding, J.M.; Klump, K.E.; McGinnis, J.F.; Self, W.; Seal, S. Cerium oxide nanoparticles: Applications and prospects in nanomedicine. *Nanomedicine* **2013**, *8*, 1483–1508. [CrossRef]
33. Hirst, S.M.; Karakoti, A.; Singh, S.; Self, W.; Tyler, R.; Seal, S.; Reilly, C.M. Bio-distribution and in vivo antioxidant effects of cerium oxide nanoparticles in mice. *Environ. Toxicol.* **2013**, *28*, 107–118. [CrossRef]
34. Xu, C.; Qu, X. Cerium oxide nanoparticle: A remarkably versatile rare earth nanomaterial for biological applications. *NPG Asia Mater.* **2014**, *6*, e90. [CrossRef]
35. Dowding, J.M.; Das, S.; Kumar, A.; Dosani, T.; McCormack, R.; Gupta, A.; Sayle, T.X.; Sayle, D.C.; von Kalm, L.; Seal, S.; et al. Cellular interaction and toxicity depend on physicochemical properties and surface modification of redox-active nanomaterials. *ACS Nano* **2013**, *7*, 4855–4868. [CrossRef]
36. Pirmohamed, T.; Dowding, J.M.; Singh, S.; Wasserman, B.; Heckert, E.; Karakoti, A.S.; King, J.E.; Seal, S.; Self, W.T. Nanoceria exhibit redox state-dependent catalase mimetic activity. *Chem. Commun.* **2010**, *46*, 2736–2738. [CrossRef] [PubMed]

37. Walkey, C.D.; Olsen, J.B.; Guo, H.; Emili, A.; Chan, W.C. Nanoparticle size and surface chemistry determine serum protein adsorption and macrophage uptake. *J. Am. Chem. Soc.* **2012**, *134*, 2139–2147. [CrossRef] [PubMed]
38. Xia, T.; Kovochich, M.; Liong, M.; Madler, L.; Gilbert, B.; Shi, H.; Yeh, J.I.; Zink, J.I.; Nel, A.E. Comparison of the mechanism of toxicity of zinc oxide and cerium oxide nanoparticles based on dissolution and oxidative stress properties. *ACS Nano* **2008**, *2*, 2121–2134. [CrossRef] [PubMed]
39. Pulido-Reyes, G.; Rodea-Palomares, I.; Das, S.; Sakthivel, T.S.; Leganes, F.; Rosal, R.; Seal, S.; Fernández-Piñas, F. Untangling the biological effects of cerium oxide nanoparticles: The role of surface valence states. *Sci. Rep.* **2015**, *5*, 15613. [CrossRef]
40. Celardo, I.; Pedersen, J.Z.; Traversa, E.; Ghibelli, L. Pharmacological potential of cerium oxide nanoparticles. *Nanoscale* **2011**, *3*, 1411–1420. [CrossRef]
41. Ryss, I.G.; Slutskaya, M.M. Report on the Platinum Sector. *Akademia Nauka SSSR* **1951**, *26*, 216.
42. Jubelin, C.; Muñoz-Garcia, J.; Griscom, L.; Cochonneau, D.; Ollivier, E.; Heymann, M.F.; Vallette, F.M.; Oliver, L.; Heymann, D. Three-dimensional in vitro culture models in oncology research. *Cell Biosci.* **2022**, *12*, 155. [CrossRef]
43. Guillaume, L.; Rigal, L.; Fehrenbach, J.; Severac, C.; Ducommun, B.; Lobjois, V. Characterization of the physical properties of tumor-derived spheroids reveals critical insights for pre-clinical studies. *Sci. Rep.* **2019**, *9*, 6597. [CrossRef]

International Journal of *Molecular Sciences*

MDPI

Article

Silver Nanoparticles as a Novel Tissue Preservative: A Comparative Study with 10% Neutral Buffered Formalin

Safa Taha [1], Amina Ismaeel [2], Muna Aljishi [1], Samvel Selvam [2], Angeleena Esther [3] and Khaled Greish [1,3,*]

1 Department of Molecular Medicine, College of Medicine and Health Sciences, Princess Al Jawhara Center for Molecular Medicine, Genetics and Inherited Diseases, Arabian Gulf University, Manama 329, Bahrain; safat@agu.edu.bh (S.T.); munajma@agu.edu.bh (M.A.)
2 Department of Pathology, College of Medicine and Health Sciences, Arabian Gulf University, Manama 329, Bahrain; aminaaya@agu.edu.bh (A.I.); samvels@agu.edu.bh (S.S.)
3 Clinical Research Center, College of Medicine and Health Sciences, Arabian Gulf University, Manama 329, Bahrain; angeleenae@agu.edu.bh
* Correspondence: khaledfg@agu.edu.bh; Tel.: +973-34358697

Abstract: Tissue preservation plays an essential role in biomedical research and histopathological applications. Traditional methods, despite their efficiency, are associated with compromised long-term tissue integrity and probable ecotoxicities. This study explores the application of silver nanoparticles (AgNPs), known for their antimicrobial properties, as a potential tissue preservative. In this work, AgNPs were synthesized via a chemical reduction method. Heart, liver, and kidney tissues were obtained from BALB/c mice and preserved using 10% neutral buffered formalin (NBF) and AgNPs solution for 72 h. Preservation efficiency was assessed by quantifying and measuring DNA and RNA integrity, evaluating protein stability, and conducting histopathological examinations. This study aimed to compare the performance of AgNPs against 10% NBF across these parameters to determine their suitability as an alternative fixative. Our results showed that AgNPs solution maintained consistent DNA, RNA, and protein concentrations/quality across all tissues over 72 h, whereas formalin treatment led to degradation over time. Conversely, 10% NBF demonstrated better preservation of tissue morphology. These results highlighted the differential strengths of each fixative, with AgNPs excelling in molecular preservation and NBF in structural integrity. Overall, AgNPs exhibited superior qualitative and quantitative preservation of nucleic acids and intracellular proteins, indicating their potential as an alternative to formalin for molecular testing. Despite their demonstrated efficacy in biomolecular preservation, further studies are needed to optimize tissue morphology preservation.

Keywords: silver nanoparticles (AgNPs); BALB/c mice; 10% neutral buffered formalin (10% NBF); GAPDH; nucleic acid extraction; protein extraction; Western blot; RT-PCR

1. Introduction

Tissue preservation and fixation play a vital role in research and have significant implications in medicine, particularly in transplantation, regenerative medicine, forensics, diagnostic pathology, and surgery. Choosing the proper tissue fixative significantly impacts the quality of tissue sections, tissue morphology, protein integrity, and the quality of downstream molecular analyses [1]. Historically, formalin, an aqueous solution of formaldehyde, has been widely employed as a gold standard for tissue preservation due to its efficacy and availability. However, formalin has been classified as a type 1 carcinogen, posing significant health risks to laboratory personnel and environmental hazards due to its toxicity and carcinogenic properties [2,3]. Another significant challenge associated with formalin use is respiratory and mucosal irritation in both clinical and laboratory settings, which can lead to upper respiratory illnesses characterized by symptoms such as cough, dyspnea, and bronchospasm [4]. It can also lead to corrosive injuries to the skin, leading to burns and irritation [5]. Studies on mice have shown that formalin exposure can cause significant damage to vital organs such as the liver and kidneys, including cell deformity, tissue degeneration, and hemorrhages [6,7]. Additionally, formalin fixation leads to significant degradation of nucleic acids due to slow fixation and enzymatic activity, affecting the quality of molecular analyses [8,9]. Moreover, the handling and disposal of formalin also incur legal and environmental costs, prompting the need for safer alternatives [2,3].

Several alternatives to formalin have been explored, including ethanol-based fixatives, glyoxal, and vacuum-based tissue transfer methods, which aim to reduce toxicity while maintaining preservation quality [3,9]. However, these alternatives often compromise either morphological or molecular integrity. Nanotechnology offers novel solutions, with silver nanoparticles (AgNPs) gaining attention for their potent antimicrobial properties and low toxicity at controlled concentrations [10,11]. AgNPs penetrate bacterial cell walls, disrupt membranes, and induce cell death, ensuring sterility in preserved tissues [11,12]. Their physicochemical properties, such as size and surface chemistry, enhance their efficacy [13]. This study aims to evaluate the efficacy of AgNPs in preserving tissue integrity at the molecular and morphological levels, benchmarking their performance against the gold standard, formalin fixation, to assess their potential as a safer, effective alternative.

2. Results

2.1. Silver Synthesis and Characterization

The AgNPs solution was prepared and concentrated as detailed in Section 4.1. The resultant particles showed a maximum visible light absorbance at 424 nm as measured by UV-2600 (Shimadzu-Japan, Kyoto, Japan). The AgNPs showed an average size of 45.3 nm, as measured by Malvern ZEN3600 Zeta Sizer Nano series (Malvern Instruments Inc., Westborough, MA, USA). The prepared nanoparticles had a concentration of 20 μg/mL. AgNPs were further concentrated in a heating oven (50 °C) to acquire the operational concentration in this study of 50 μg/mL, as illustrated in Figure 1.

A.

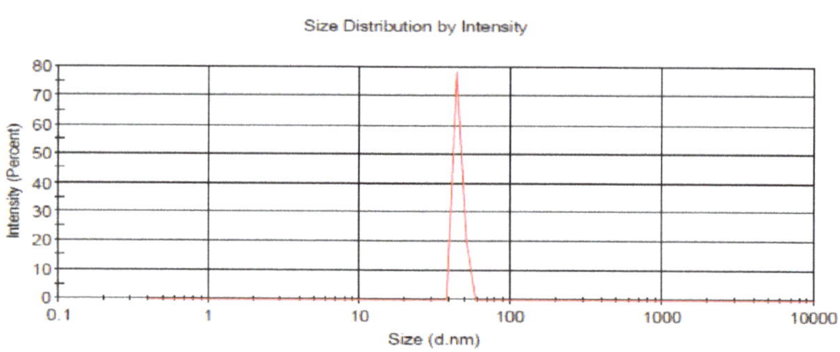

Figure 1. *Cont.*

B.

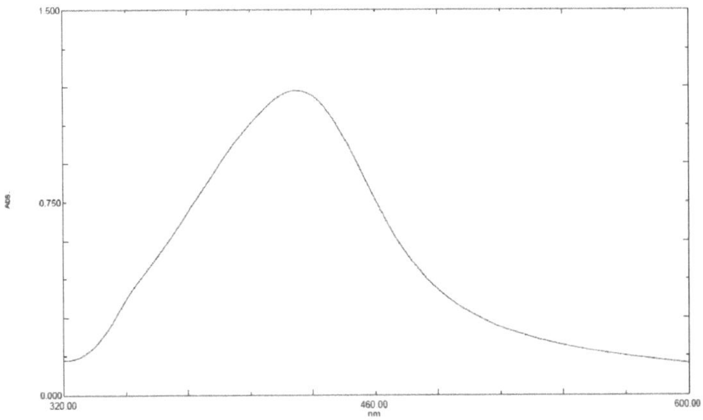

Figure 1. (**A**,**B**): The average size was 45.3 nm, as measured by DLS, with a single peak at 424 nm.

2.2. *DNA Concentration and Purity*

In the liver, the saline control exhibited an initial DNA concentration of 357.00 ± 2.646 ng/µL. AgNPs treatment resulted in a gradual decrease, from 352.00 ± 2.000 ng/µL at 24 h to 309.33 ± 1.528 ng/µL at 72 h, while 10% NBF induced a more pronounced decline, from 250.00 ± 2.000 ng/µL at 24 h to 157.67 ± 2.517 ng/µL at 72 h. These differences were statistically significant at 24, 48, and 72 h ($p < 0.0001$). In the kidney tissue samples, the saline control showed 63.67 ± 1.528 ng/µL at 0 h. AgNPs treatment led to a gradual decrease, from 57.00 ± 2.646 ng/µL at 24 h to 50.33 ± 1.528 ng/µL at 72 h, while 10% NBF showed a more significant decrease, from 49.67 ± 1.528 ng/µL at 24 h to 31.67 ± 2.887 ng/µL at 72 h. Statistical significance was observed at 24 h ($p < 0.01$) and 48 and 72 h ($p < 0.0001$). In the heart, the saline control showed 55.00 ± 1.528 ng/µL at 0 h. AgNPs treatment resulted in a minimal decrease, from 44.00 ± 1.155 ng/µL at 24 h to 43.00 ± 1.528 ng/µL at 72 h, while 10% NBF also showed a decrease, from 35.00 ± 1.528 ng/µL at 24 h to 30.00 ± 1.528 ng/µL at 72 h. These differences were statistically significant at 24, 48, and 72 h ($p < 0.0001$). As illustrated in Figure 2, overall, tissues preserved in 10% NBF consistently showed a substantial reduction in DNA concentration across all tissue types compared to AgNPs solution, indicating greater DNA degradation. This was reflected in the highly significant *p*-values observed, suggesting that the AgNPs solution is comparatively better at DNA preservation over 72 h.

2.3. *Effect of Preservation Time on Nucleic Acid and Protein Integrity*

Initial assessments at 0 h confirmed high-quality DNA, RNA, and protein extraction across all samples, establishing a baseline for evaluating preservation efficacy over time. The data demonstrate that AgNPs-preserved tissues maintained nucleic acid and protein integrity more effectively than formalin-preserved tissues after 24, 48, and 72 h.

DNA purity was measured using the absorbance ratio at 260 nm and 280 nm. Figure 3 illustrates the absorbance ratio from liver, kidney, and heart tissues preserved using either AgNPs or 10% NBF at 0, 24, 48, and 72 h. In the liver, no significant differences in DNA purity were observed between the two preservation methods at 0 and 24 h. However, AgNPs-preserved samples exhibited significantly higher purity at 48 h ($p < 0.05$) and 72 h ($p < 0.01$). For kidney and heart tissues, while no differences were observed at 0 h, AgNP preservation resulted in significantly higher DNA purity at 24, 48, and 72 h ($p < 0.01$ for kidney; $p < 0.001$ for heart). These findings suggest that AgNPs are superior to 10% NBF for maintaining DNA purity, particularly over extended periods, with varying degrees of effectiveness across different tissues.

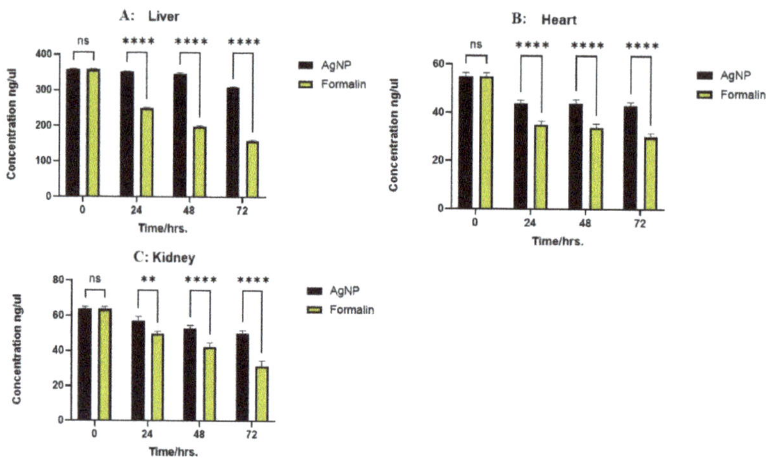

Figure 2. Concentration of extracted DNA (ng/µL) from the liver, kidneys, and heart at different time intervals (0, 24, 48, and 72 h). The bars represent the two preservative treatments: silver nanoparticles (AgNPs) and 10% neutral buffered formalin. The data represent the mean of three biological replicates (n = 3) measured in parallel for each tissue type and time point. Statistical significance between treatments is indicated as follows: (**A**) Liver: No significant difference (ns) is observed at 0 h. Significant differences are observed at 24, 48, and 72 h ($p < 0.0001$ for all time points, indicated by ****). (**B**) Heart: No significant difference (ns) is observed at 0 h. Significant differences are observed at 24, 48, and 72 h ($p < 0.0001$ for all time points, indicated by ****). (**C**) Kidney: No significant difference (ns) is observed at 0 h. Significant differences are observed at 24 h ($p < 0.01$, indicated by **) and at 48 and 72 h ($p < 0.0001$, indicated by ****).

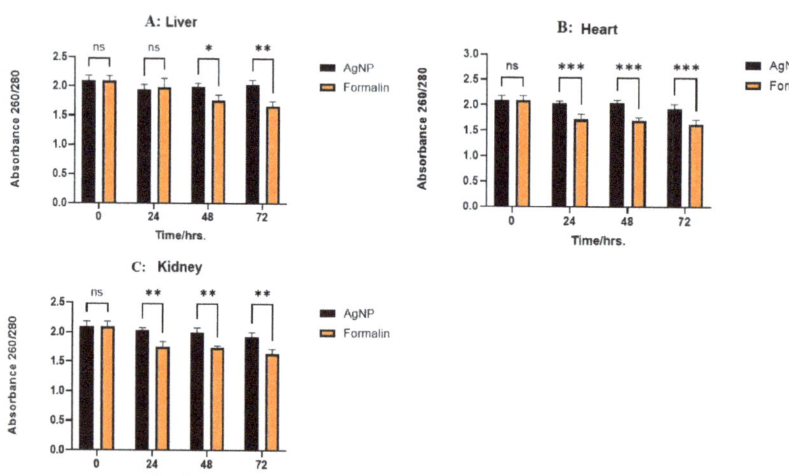

Figure 3. Absorbance ratio (260/280) for liver, heart, and kidney tissues (**A–C**) over time (0, 24, 48, and 72 h) under different preservative conditions (saline, AgNP, and 10% formalin). The data represent the mean of three biological replicates (n = 3) measured in parallel for each tissue type and time point. * $p < 0.05$; ** $p < 0.01$; *** $p < 0.001$. "ns": not significant.

Moreover, the integrity of DNA extracted from liver, heart, and kidney samples preserved in saline (0 time), formalin, and AgNPs was also evaluated using 0.6% agarose gel electrophoresis, as described in Section 4.3.3. Figure 4 illustrates the results, visualized on the gel, revealing distinct banding patterns across the samples. DNA from saline-treated samples (0 time) showed intact, high-molecular-weight bands for all tissues, indicating minimal degradation. In contrast, formalin-preserved samples exhibited significant DNA degradation, with smeared bands observed at 24, 48, and 72 h across all tissues, suggesting that formalin compromises DNA integrity over time. AgNPs-preserved samples displayed better DNA preservation compared to formalin, with clearer bands at 24 and 48 h, though some degradation was evident by 72 h, particularly in kidney samples. These findings highlight AgNPs as a more effective preservative for maintaining DNA integrity than formalin, while saline (0 time) remains the baseline for optimal DNA quality.

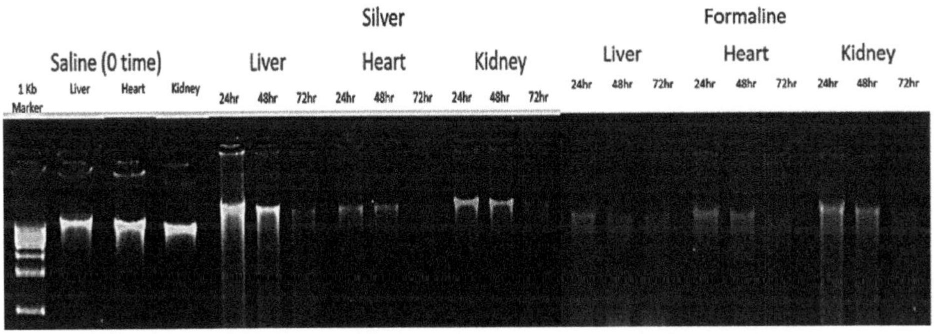

Figure 4. DNA extracted from liver, heart, and kidney samples preserved in saline (0 time), formalin, and AgNPs using 0.6% agarose gel electrophoresis. AgNPs-preserved samples displayed better DNA preservation compared to formalin, with clearer bands at 24 and 48 h, though some degradation was evident by 72 h, particularly in kidney samples.

2.4. Preserving RNA Concentration and Quality: Tissue-Specific Responses to Different Treatments

The effects of the three preservation methods on RNA concentrations across the tissues over a 72-h period were examined. Saline consistently maintained stable RNA concentrations across all three tissues throughout this study (1991.30 $\pm$ 1.310 ng/µL). AgNPs treatment resulted in a gradual, tissue-dependent decline in RNA levels while, in contrast, 10% NBF induced a significant and rapid reduction in RNA concentration across all tissues, with the most pronounced decrease observed in the heart, from 1097 $\pm$ 2.310 ng/µL at 24 h to a mere 864 $\pm$ 2.012 ng/µL at the 72-h mark ($p < 0.001$). The consistent stability observed with saline, the moderate decline with AgNPs, and the substantial reductions with 10% NBF are clearly visualized in Figure 5. These findings suggest that preservation with 10% NBF has a significantly detrimental impact on RNA concentration over time compared to the AgNPs solution. Thus, these insights can inform the selection of appropriate preservative strategies for various tissue types, depending on the specific research needs and the desired preservation outcomes.

The relationships between various tissue types and the impact of different preservative treatments on the quality of RNA nucleic acids extracted were evaluated by analyzing the absorbance ratios at 260 nm and 280 nm. In the liver, AgNPs treatment resulted in a slight absorbance ratio decline from 1.96 to 1.84 over 72 h, while 10% NBF showed a more substantial decrease from 1.86 to 1.73. In the kidney, AgNPs maintained a stable ratio of around 1.95, whereas 10% NBF decreased from 1.94 to 1.83. Similarly, in the heart, the AgNPs ratio remained stable at around 2.0, while the 10% NBF declined from 1.89 to 1.75. Thus, 10% NBF consistently demonstrated a more significant reduction in absorbance ratios, indicating a greater impact on nucleic acid quality compared to AgNPs across all tested tissues. As clearly illustrated in Figure 6, the AgNPs treatment consistently exhibited relatively stable absorbance ratios across all tissues, while the 10% NBF treatment resulted in more pronounced declines. These findings suggest that AgNPs effectively maintained nucleic acid quality. Conversely, 10% NBF demonstrated a detrimental effect, significantly reducing absorbance ratios, particularly in the heart and liver tissues, indicating a greater impact on nucleic acid degradation.

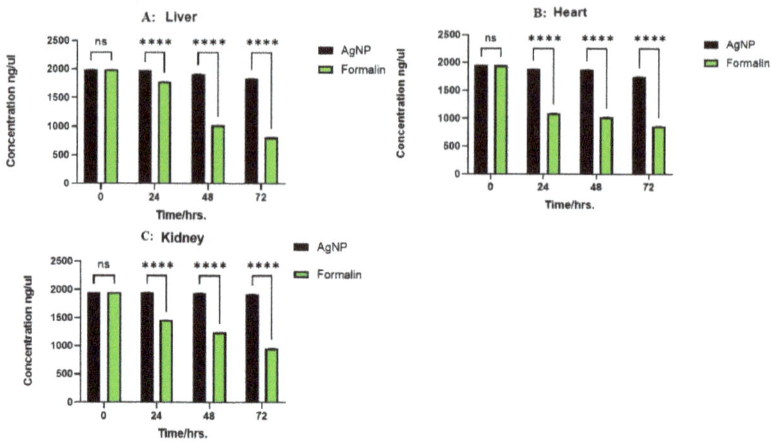

Figure 5. Concentration of extracted RNA (ng/µL) in the liver, heart, and kidney (**A**–**C**) at different time intervals (0, 24, 48, and 72 h). The bars represent preservative treatments, AgNPs and 10% NBF, with statistical significance between groups indicated by **** ($p < 0.0001$). "ns": not significant. The data represent the mean of three biological replicates (n = 3) measured in parallel for each tissue type and time point.

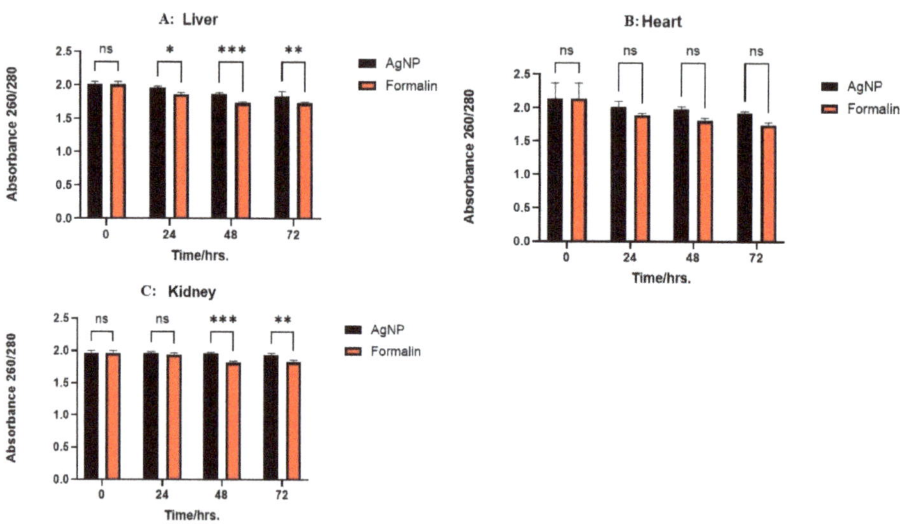

Figure 6. Absorbance ratio (260/280) for liver, heart, and kidney tissues (**A**–**C**) over time (0, 24, 48, and 72 h). The bars represent preservative treatments, AgNPs and 10% NBF, with statistical significance between groups indicated as follows: Liver (**A**): No significant difference (ns) is observed at 0 h. Significant differences are observed at 24 h ($p < 0.05$, indicated by *), 48 h ($p < 0.001$, indicated by ***), and 72 h ($p < 0.01$, indicated by **). Kidney (**C**): No significant difference (ns) is observed at 0 and 24 h. Significant differences are observed at 48 h ($p < 0.001$, indicated by ***) and 72 h ($p < 0.01$, indicated by **). The data represent the mean of three biological replicates (n = 3) measured in parallel for each tissue type and time point.

2.5. GAPDH Expression Determination in Multiple Preservatives

GAPDH, a housekeeping gene, is commonly used as a control gene for normalizing gene expression data [14]. Significant changes in the gene expression levels across different preservatives and time periods were identified by a RT-PCR study of GAPDH expression. The samples maintained in AgNPs demonstrated the highest expression levels of GAPDH across all time points (0, 24, 48, and 72 h). This indicates that AgNPs are effective in preserving the integrity and expression of this housekeeping gene during storage. The samples preserved in saline exhibited significant expression levels of GAPDH, akin to the AgNPs group. The expression levels in the saline group were comparable to, and, in some instances, slightly elevated compared to, the AgNPs group, suggesting that saline is equally effective in maintaining GAPDH expression. Although lower than the other two preservatives, the 10% NBF group exhibited relatively stable expression of the housekeeping gene throughout the assessed time intervals. These findings indicate that both AgNPs and saline are more effective than 10% NBF in maintaining GAPDH expression. Figure 7 depicts the RT-PCR analysis of GAPDH expression across all tissues over a 72-h period.

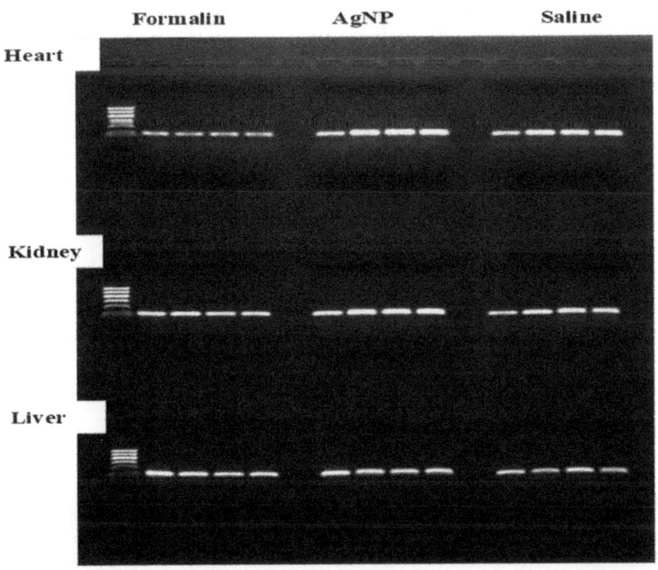

Figure 7. RT-PCR analysis of GAPDH expression across all tissues preserved in the different preservatives. The samples were evaluated at intervals of 0, 24, 48, and 72 h. The bands represent the relative expression levels of GAPDH under different conditions.

GAPDH band intensity was analyzed using the AzurSpot software version 14.2. The heart tissue stored in saline at 0 h demonstrated the highest GAPDH expression level, approximately 1.2 on the intensity scale. The heart tissue stored in AgNPs exhibited consistently high GAPDH expression levels, ranging from approximately 0.9 to 1.0, at the 24-, 48-, and 72-h time points, while heart tissues preserved in 10% NBF showed a gradual decline in GAPDH expression over time, starting at approximately 0.7 at 24 h and decreasing to nearly 0.4 by 72 h. Similarly, kidney tissue treated with saline at 0 h exhibited the highest GAPDH expression level, surpassing 1.1 on the intensity scale. This indicated the peak expression observed in all kidney samples. Kidney samples preserved in silver at 24, 48, and 72 h displayed intensity values between approximately 0.9 and 1.0 on the intensity scale, indicating that AgNPs effectively maintained strong GAPDH expression in kidney samples. The sustained increase in GAPDH levels in kidney tissue during the silver treatment periods is noteworthy, especially compared to the gradual decrease observed in the 10% NBF-preserved kidney samples over the same duration, starting at approximately 0.6 at 24 h and decreasing to about 0.4 by 72 h. The liver tissue stored in saline at 0 h demonstrated the highest GAPDH expression level, surpassing 1.0 on the intensity scale. Liver tissue preserved in AgNPs solution demonstrated consistently high GAPDH expression levels, ranging from approximately 0.8 to 0.9 at the 24, 48, and 72-h time points. In contrast, liver tissue preserved in 10% NBF showed a notable decrease in GAPDH expression over time, starting at approximately 0.6 at 24 h and declining to about 0.3 by 72 h. These findings, illustrated in Figure 8, indicate that saline- and AgNP-based preservation methods were more effective in maintaining GAPDH expression levels than 10% NBF across all three tissue types and time points.

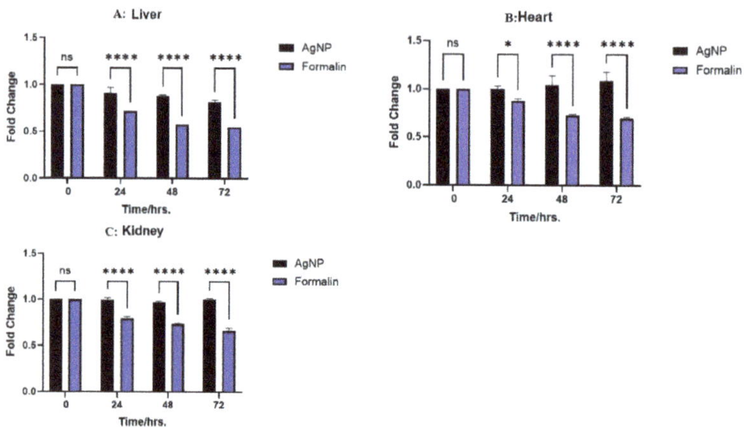

Figure 8. Expression levels of the housekeeping gene GAPDH in the three different tissue types (**A–C**) under 10% NBF and AgNPs preservation treatments and time intervals. Fold change values are shown for AgNPs and 10% NBF. The data represent the mean of three biological replicates (n = 3) measured in parallel for each tissue type and time point. Statistical significance between treatments is indicated as follows: Liver (**A**): No significant difference (ns) is observed at 0 h. Significant differences are observed at 24, 48, and 72 h ($p < 0.0001$ for all time points, indicated by ****). Heart (**B**): No significant difference (ns) is observed at 0 h. Significant differences are observed at 24 h ($p < 0.05$, indicated by *) and at 48 and 72 h ($p < 0.0001$ for both time points, indicated by ****). Kidney (**C**): No significant difference (ns) is observed at 0 h. Significant differences are observed at 24, 48, and 72 h ($p < 0.0001$ for all time points, indicated by ****).

2.6. Protein Extraction and Analysis

The protein profiles of the preserved organs were examined using SDS-PAGE and Western blot analysis. AgNPs and saline-treated tissues yielded clearer, better-resolved protein bands compared to 10% NBF, indicating superior protein preservation. AgNPs-treated tissues maintained protein integrity and GAPDH immunoreactivity over time more effectively than 10% NBF. Consistent GAPDH detection across all treatments validated its use as a reliable loading control and reference protein for comparative protein analysis. These results highlight AgNPs' potential for improved protein preservation, which may benefit downstream proteomic and biomarker discovery studies. Appendix A Figure A1 demonstrates that protein bands from AgNPs-preserved organ samples exhibited greater distinctness and resolution compared to those from 10% NBF-preserved samples.

2.7. Histopathologic Evaluation of Tissue Samples

Gross examination of AgNPs-fixed specimens revealed a softer and paler texture and increased difficulty in microtome sectioning compared to 10% NBF-fixed tissues. Microscopically, AgNPs-fixed paraffin-embedded tissue stained with H&E staining showed pyknotic nuclei, shrinkage of cells, poor cytologic features, blurred disintegrated cytoplasmic membrane contour, and granular cytoplasm with disintegration of tissue morphology (Figures 9–11).

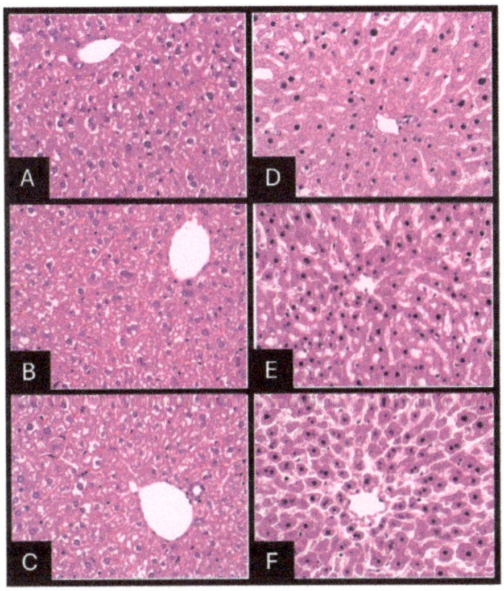

Figure 9. H & E staining of liver (40×). Tissue fixed with 10% NBF (**left panel**) or AgNP solution (**right panel**) for 24 h (**A**,**D**), 48 h (**B**,**E**), and 72 h (**C**,**F**).

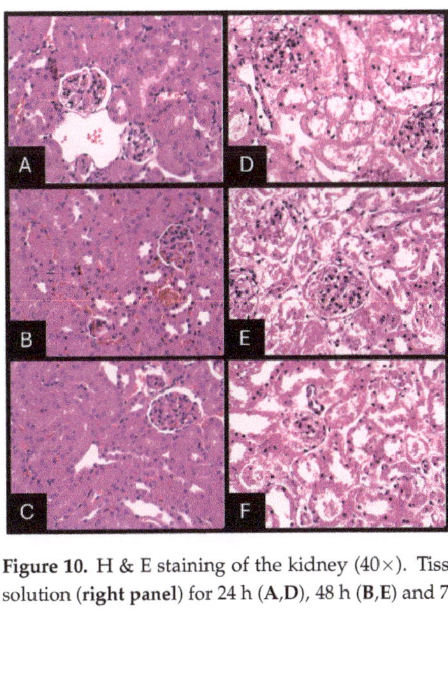

Figure 10. H & E staining of the kidney (40×). Tissue fixed with 10% NBF (**left panel**) or AgNP solution (**right panel**) for 24 h (**A**,**D**), 48 h (**B**,**E**) and 72 h (**C**,**F**).

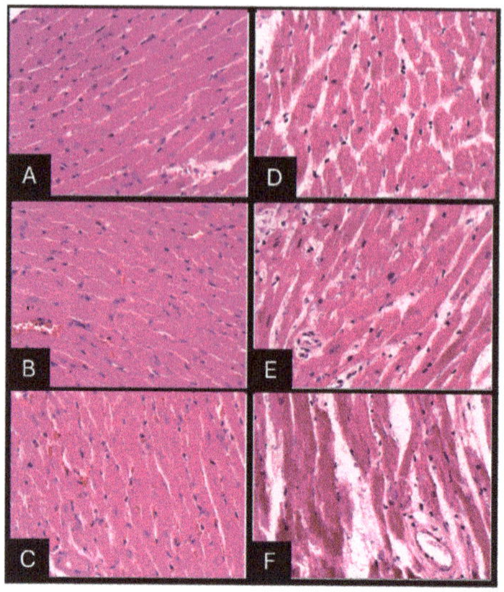

Figure 11. H & E staining of the heart (40×). Tissue fixed with 10% NBF (**left panel**) or AgNP solution (**right panel**) for 24 h (**A**,**D**), 48 h (**B**,**E**) and 72 h (**C**,**F**).

The histopathologic scoring of tissue morphology was calculated and compared to the 10% NBF-fixed paraffin-embedded tissue stained with H&E (Appendix A; Table A1). Scores were found to be inadequate for tissues fixed with an AgNPs solution, in contrast to tissues fixed with 10% NBF, which were adequately preserved at the three time points of fixation for the three organs tested, indicating that AgNPs did not effectively preserve tissue morphology compared to the established 10% NBF standard.

3. Discussion

The application of AgNPs as a tissue preservative remains a relatively under-explored area of research. While numerous studies have investigated the antimicrobial properties of AgNPs and their applications in biomedicine, a notable shortage of research remains in specifically exploring their preservative capabilities in tissues. Consequently, the current investigation aims to elucidate this process, with the findings from this study providing significant insights into the preservation capabilities and limitations of AgNPs compared to the gold standard, formalin.

Our research indicated that AgNPs provided excellent preservation of DNA, RNA, and protein integrity and quantity across all tissue types over a 72-h timeframe, a notable improvement relative to formalin fixation. These results are consistent with the widely documented degradation of nucleic acids by formalin, resulting in a reduced quality of extracted molecules and a subsequent negative impact on sensitive molecular analyses [15,16]. An earlier study on biosynthesized AgNPs for preserving mulberry leaves demonstrated their superior performance in preventing microbial proliferation and maintaining structural integrity [17], suggesting that similar principles may apply to biological tissues. Further research has demonstrated the inhibitory effect of AgNPs on voltage-gated sodium channels [18,19], which may contribute to their preservative effects by stabilizing cellular functions. Another contributing mechanism to the molecular preservation observed with AgNPs was reported by Ekaterina O [20], who demonstrated that AgNPs induce the inactivation of sulfur- and phosphorus-rich enzymes, thereby disrupting cellular metabolism. This contrasts with formalin, which slowly inactivates intracellular enzymes, allowing time for biomolecular degradation by endogenous tissue enzymes. However, AgNPs exhibited suboptimal morphological preservation compared to formalin, as detailed in the limitations.

Compared to formalin, AgNPs offer reduced toxicity and superior biomolecular preservation, which are critical for molecular analyses. However, their suboptimal morphological preservation limits histopathological applications. Alternative fixatives, such as ethanol or glyoxal, also struggle to balance molecular and morphological preservation [9]. Optimizing AgNPs formulations could address these limitations, leveraging their antimicrobial and low-toxicity properties [13]. The size of the AgNPs is likely a critical factor influencing tissue preservation effectiveness, as smaller nanoparticles may penetrate tissues more effectively, enhancing antimicrobial and enzyme-inactivating properties, while larger particles could improve structural stabilization but potentially reduce tissue permeability. Although our study used a standardized AgNP size, variations in nanoparticle size could modulate preservation outcomes, as smaller AgNPs have been shown to exhibit more substantial antimicrobial effects in other applications [21]. Further studies investigating the impact of AgNP size on both molecular and morphological preservation are needed to optimize their utility as a tissue preservative.

Safety concerns associated with formalin further highlight the importance of exploring alternatives like AgNPs for tissue preservation. Formalin's classification as a type 1 carcinogen is associated with significant health risks, including short-term respiratory, nasopharyngeal, ocular, and dermal irritation, as well as long-term metaplastic changes and potential carcinogenesis in the naso- and oropharynx [4,22]. In contrast, AgNPs offer multiple advantages beyond reduced toxicity when used at controlled concentrations. These include superior preservation of DNA, RNA, and protein integrity, as demonstrated in our study (Sections 2.2–2.4), which is critical for downstream molecular analyses such as PCR and Western blotting, where formalin's degradative effects are a known limitation [8,15,16]. Additionally, AgNPs' broad-spectrum antimicrobial activity prevents microbial proliferation, ensuring tissue sterility without additional preservatives [10–12]. Their ability to inactivate sulfur- and phosphorus-rich enzymes further supports biomolecular preservation by halting endogenous degradation processes [19]. While reduced toxicity is a significant driver for adopting AgNPs, these multifaceted benefits—biomolecular preservation, antimicrobial efficacy, and enzyme inactivation—position AgNPs as a promising alternative to formalin. However, challenges in morphological preservation, as outlined in the limitations section, indicate that further optimization is needed to achieve a balance between adequate morphological and superior biomolecular preservation. This could involve modifying AgNPs formulations or combining them with other fixatives to enhance histological outcomes. Future studies optimizing these aspects, as outlined in the final paragraph, could further enhance AgNPs' potential.

The potential of AgNPs as a tissue preservative is further supported by their growing applications in biomedical fields, such as drug delivery, wound healing, and antinociceptive therapies [10,18]. These applications underscore AgNPs' versatility and their capacity to address the limitations of traditional fixatives like formalin. Future studies should focus on optimizing AgNPs-based preservation protocols, including concentration gradients and delivery methods, to improve tissue morphology while maintaining their superior biomolecular preservation. Additionally, exploring the long-term stability of AgNPs-preserved tissues and their compatibility with advanced molecular techniques, such as next-generation sequencing, could further validate their utility. By addressing these challenges, AgNPs could offer a safer, more effective alternative to formalin, mitigating health and environmental risks while enhancing the quality of tissue preservation for research and clinical applications.

Limitations Section

This study provides significant insights into the potential of AgNPs as a tissue preservative, but several limitations warrant consideration. A primary limitation was using a single operational concentration of AgNPs (50 μg/mL), selected based on preliminary experiments and literature indicating effective antimicrobial activity and low toxicity at this level [13]. While this concentration excelled in preserving DNA, RNA, and protein integrity, it resulted in suboptimal preservation of tissue morphology compared to formalin, as evidenced by pyknotic nuclei, cell shrinkage, and poor cytologic features in histological examinations (Section 2.7). Testing a range of AgNsP concentrations (e.g., 10, 25, 75, and 100 μg/mL) could optimize preservative efficacy, particularly for improving tissue morphology by balancing antimicrobial activity, enzyme inactivation, and tissue penetration. Additionally, this study evaluated preservation over a 72-h period, limiting insights into the long-term stability of AgNPs-preserved tissues. The compatibility of AgNPs-preserved tissues with advanced molecular techniques, such as next-generation sequencing, was not assessed, representing another constraint. Future studies addressing these limitations, including concentration optimization, long-term preservation assessments, and compatibility testing with advanced techniques, will enhance AgNPs' applicability as a formalin alternative.

4. Materials and Methods

4.1. *Synthesis and Characterization of AgNPs*

Silver nanoparticles were synthesized via a chemical reduction method utilizing an aqueous solution of silver nitrate ($AgNO_3$) as the precursor for silver ions (Ag^+) [23]. The reduction of Ag^+ to metallic silver (Ag^0) was accomplished by trisodium citrate ($Na_3C_6H_5O_7$), which functioned as both a reducing and capping agent [18]. Briefly, 1 mM of $AgNO_3$ solution was prepared in deionized water (DW) and heated to 95 °C, following which $Na_3C_6H_5O_7$ was added. This solution was stirred vigorously for 7 min. The resulting nanoparticles were collected by ultracentrifugation at 27,000× *g* for 15 min using a Beckman J2-MC/JA-20 centrifuge. The supernatant was decanted, replaced with DW, and then centrifuged three times at 31,000× *g* for 20 min using a Beckman L-80/NVT-65 centrifuge. The size and concentration were measured using a Malvern ZEN3600 Zetasizer Nano series (Malvern Instruments Inc., Westborough, MA, USA).

4.2. *Preparation of Tissue Samples*

Tissue samples, specifically liver, heart, and kidney, were obtained from adult BALB/c mice that had reached their natural endpoints. The current study was approved by the institutional research and ethics committee Ref No: E9-PI-11-24. Following CO_2 euthanasia, in accordance with institutional ethical guidelines, liver, heart, and kidney tissues were extracted and weighed to ensure uniformity. One set of samples was preserved in 10% neutral buffered formalin (10% NBF) and the other in AgNPs solution (50 ug/mL) for up to 72 h. Tissues were also preserved in saline and used as a control at 0 h. DNA, RNA, and proteins were retrieved from all tissues under different treatment conditions at 24 h intervals, and the efficiency of each preservative was compared in terms of quantity and quality and for histological examination.

4.3. *Nucleic Acid Extraction*

4.3.1. DNA Extraction

Genomic DNA was extracted from the samples using the Qiagen DNA Mini Kit (catalog number 51304), according to the manufacturer's instructions [24]. Initially, samples were homogenized in 180 μL buffer ATL, followed by the addition of 20 μL proteinase K, and incubated at 56 °C for 30 min to facilitate cell lysis. Subsequently, 200 μL of buffer AL was added, and the mixture was vortexed before the addition of 600 μL of ethanol (96–100%). The resulting solution was transferred to a mini-spin column and centrifuged at 8000 rpm for 1 min. The column was washed with 500 μL buffer AW1 and buffer AW2, and DNA was eluted with 100 μL buffer AE. Extracted DNA was stored at −20 °C for further analysis.

4.3.2. RNA Extraction

Total RNA was isolated using TRIzol reagent (catalog number 15596026, Thermo Fisher Scientific Inc.) following the manufacturer's protocol [25]. The samples were initially homogenized in 1 mL TRIzol reagent. After 5 min of incubation at room temperature, 200 μL of chloroform was added, and the mixture was shaken vigorously for 15 s before incubating for 2–3 min. The mixture was then centrifuged at 12,000 rpm for 15 min at 4 °C, resulting in clear separation of the aqueous phase containing RNA. The aqueous phase was carefully transferred to a new tube, and RNA was precipitated by adding 500 μL of isopropanol. After a 10-min incubation at room temperature, the samples were centrifuged at 12,000 rpm for 10 min at 4 °C. The RNA pellet was washed with 75% ethanol, air-dried, and then dissolved in 30 μL of RNase-free water. The extracted RNA was stored at −80 °C for future assays.

4.3.3. Quality Assessment

RNA and DNA quality and quantity were assessed using a spectrophotometer, which measures the amount of UV light absorbed by a sample. The ratio of absorbance at 260 nm and 280 nm is used to assess the purity of DNA and RNA. A ratio of ~1.8 is generally accepted as "pure" for DNA; a ratio of ~2.0 is generally accepted as "pure" for RNA. If the A260/A280 ratio is appreciably lower in either case, it may indicate the presence of protein, phenol, or other contaminants that absorb strongly at or near 280 nm [26]. Moreover, DNA purity was assessed via agarose gel electrophoresis, employing a 0.6% agarose gel prepared and solidified with wells for sample loading. Following mixture with a loading dye, DNA samples were introduced into these wells. Electrophoresis was performed in a buffer-filled chamber under an electric current, facilitating the separation of negatively charged DNA fragments based on size. Subsequently, the gel was stained with ethidium bromide and visualized under UV light [27].

4.3.4. cDNA Synthesis and PCR

cDNA synthesis was performed using the Applied Biosystems High-Capacity cDNA Reverse Transcription Kit (catalog number 4368814-Thermo Fisher Scientific Inc., Waltham, MA, USA). In a PCR tube, 10 μL of 2× Reverse Transcription Master Mix was combined with 10 μL of the RNA sample (1 μg) and RNase-free water to achieve a total volume of 20 μL [28]. The reaction was incubated in a thermal cycler at 25 °C for 10 min, followed by incubation at 37 °C for 120 min, 85 °C for 5 min, and then held at 4 °C. After cycling, the synthesized cDNA was stored at −20 °C for subsequent PCR analysis. PCR amplification of GAPDH was conducted using SYBR Green Master Mix and an Applied Biosystems 7500 Fast Real-Time PCR System. The reaction mix consisted of 5 μL of master mix, 0.5 μL each of forward and reverse primers (5 μM) as shown in Table 1, 2 μL of cDNA template, and 2 μL of nuclease-free water [28]. The thermal cycling conditions included an initial denaturation at 95 °C for 10 min, followed by 40 cycles of denaturation at 95 °C for 15 s, annealing at 61 °C for 30 s, extension at 72 °C for 30 s, and a final extension at 72 °C for 5 min.

Table 1. The primer sequences.

Gene Name	Product Size	Primers
GAPDH	120 bp	F- AGG TCG GTG TGA ACC GAT TTG R- TGT AGA CCA TGT AGT TGA GGT CA

4.4. Protein Extraction and Analysis

To extract proteins from 10% NBF-fixed tissues, AgNPs, and saline, samples were collected and cut into small pieces (1–2 mm) using sterile scissors or a scalpel. The tissue pieces were placed into microcentrifuge tubes, and 200 µL of RIPA lysis buffer containing protease inhibitors was added to maintain protein integrity throughout the process. RIPA lysis buffer containing 50 mM Tris-HCl (pH 7.4), 150 mM NaCl, 1% Nonidet P-40, 0.5% sodium deoxycholate, 0.1% SDS, and protease inhibitors was thoroughly mixed with the tissue. The lysate was then incubated on ice for 30 min with occasional vortexing to enhance the lysis. After incubation, the lysate was centrifuged at 12,000× *g* for 10–15 min at 4 °C, and the supernatant containing soluble proteins was carefully collected [29]. For protein quantification, a series of BSA standards (0, 100, 200, 400, 800, and 1000 µg/mL) were prepared by diluting BSA in the same buffer used for extraction. The protein extracts were diluted in RIPA buffer (1:5, or as needed) for optimal measurement. To each standard and sample in a microplate or cuvette, 200 µL Bradford reagent was added, mixed gently, and incubated at room temperature for 5–10 min. The absorbance was measured at 595 nm using a spectrophotometer, and a standard curve was generated from the BSA standards to calculate the protein concentration in the samples. Finally, the extracted protein samples were stored at −80 °C for long-term storage or used immediately for downstream applications.

SDS-PAGE and Western Blot

The protein samples were mixed with the sample buffer and heated at 95 °C for 5 min. The samples were loaded into 12% gradient gel wells and subjected to electrophoresis at 120 V until the dye front reached the gel bottom. Following electrophoresis, Coomassie staining was performed, and the gels were transferred to a nitrocellulose membrane using the wet transfer method for Western blot detection. The membrane was blocked with 5% non-fat milk for 1 h at room temperature, followed by overnight incubation with GAPDH antibody (1:500) at 4 °C. After washing, the membrane was incubated with a secondary anti-mouse antibody (1:1000) for 1 h at room temperature. Detection was performed using WesternSure PREMIUM (Catalog No.50-489-552—LI COR BIOTECH LLC 92695000-Thermo Fisher Scientific Inc.) [30]. Chemiluminescent Substrate reagents were used to visualize protein bands.

4.5. Tissue Processing and Staining

The mouse liver, kidney, and heart specimens were fixed in 10% NBF or AgNPs solution for 24 h, 48 h, and 72 h at room temperature (18–23 °C). The tissues were then processed as usual. In brief, the tissues were dehydrated through an ethanol series, cleared with xylene, and infiltrated with paraffin. The tissues were then embedded in paraffin and sectioned at 4 µm thickness using a microtome. Hematoxylin and eosin (H&E) staining was performed for each tissue sample according to standard protocol to examine the histomorphologic features. The stained sections were digitized and examined utilizing (Aperio LV1, Leica Biosystems, United States) at 40× objective magnification. The tissues were then scored based on previously published histopathological criteria, which are nuclear staining, cytoplasmic staining, tissue morphology, clarity of staining, and uniformity of staining [31]. Each of the variables was given a score of 1 if it was acceptable/present and a score of zero if it was unacceptable/absent. When the total score was ≤ 2, it was graded as inadequate fixation, and when the total score was 3–5, it was graded as adequate fixation.

4.6. Statistical Analysis

Statistical analysis was performed using the Prism 10 software (GraphPad Software, San Diego, CA, USA). The data are presented as mean $\pm$ standard deviation (SD) for each treatment group at each time point. The statistical significance of differences between the AgNPs and formalin treatments was assessed using an unpaired, two-tailed Student's t-test. A p-value of less than 0.05 ($p < 0.05$) was considered to indicate statistical significance.

5. Conclusions

Silver nanoparticles offer a promising alternative to formalin fixation, particularly due to their demonstrated efficacy in preserving DNA, RNA, and protein quantity and integrity. However, their limitations in morphological preservation pose a challenge, necessitating the need for further research and optimization. Given the established health and environmental concerns associated with formalin, pursuing alternative methods such as AgNPs is warranted.

Author Contributions: Conceptualization, K.G.; methodology, K.G., S.T., M.A., A.I., A.E. and S.S.; data curation and data analysis, K.G., S.T., A.E, M.A., A.I. and S.S.; writing—original draft preparation A.I., A.E. and K.G.; writing—review, K.G., A.E. and A.I.; and project administration, K.G. All authors have read and agreed to the published version of the manuscript.

Funding: This research was funded by the Arabian Gulf University research fund to A.I.

Institutional Review Board Statement: This study was approved by the Institutional Research and Ethics Committee (Ref No: E9-PI-11-24).

Informed Consent Statement: Not applicable.

Data Availability Statement: Data supporting the findings of this study are available on request from the corresponding author.

Acknowledgments: All authors would like to acknowledge the animal house facility at AGU for their cooperation in this project.

Conflicts of Interest: The authors declare that they have no conflicts of interest.

Abbreviations

The following abbreviations are used in this manuscript:

AgNPs	Silver nanoparticles
NBF	Neutral buffered formalin
DLS	Dynamic Light Scattering
DNA	Deoxyribonucleic acid
RNA	Ribonucleic acid
GAPDH	Glyceraldehyde-3-phosphate dehydrogenase
SDS-PAGE	Sodium Dodecyl Sulfate-Polyacrylamide Gel Electrophoresis

Appendix A

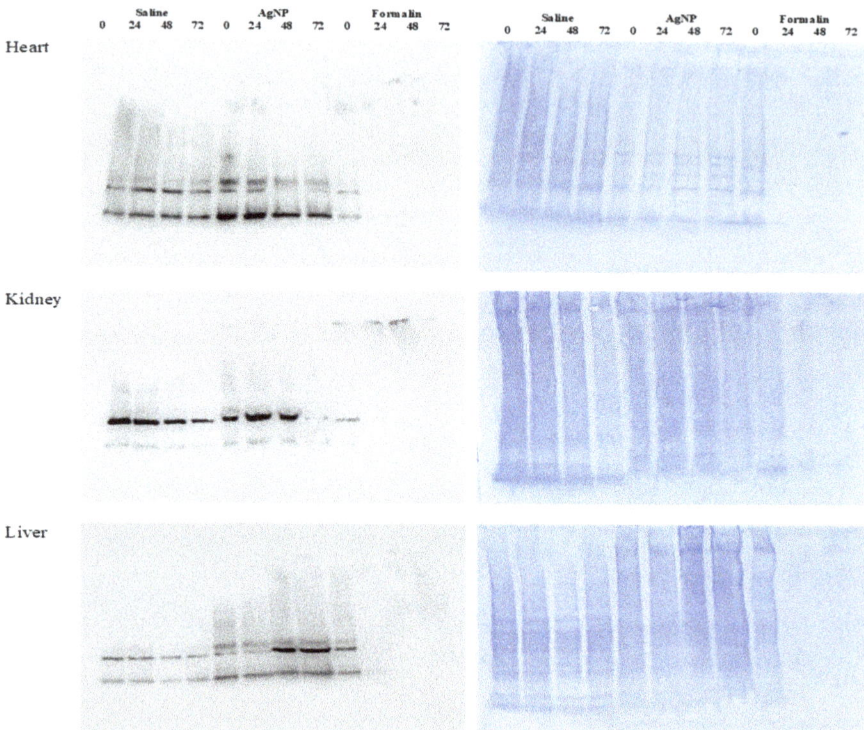

Figure A1. The image displays the results of a Western blot test that used an antibody to find the GAPDH protein in heart, kidney, and liver samples. The Western blot showed clear GAPDH protein bands in the organ samples, showing that this important protein was expressed consistently. The strength and clarity of the GAPDH bands changed based on the preservation method applied. The GAPDH bands in the samples preserved with AgNPs were clear and strong in the heart and kidney. This suggests that this preservation method is the best at keeping the structure and responsiveness of GAPDH intact. The GAPDH bands in the samples treated with 10% neutral buffered formalin (10% NBF) looked a bit less clear. This suggests that these preservation methods may have affected the GAPDH protein structure or how well the antibody could recognize it.

Table A1. Histopathologic scoring of the liver, kidney, and heart specimens after fixation with 10% NBF or AgNP for 24 h, 48 h, and 72 h.

Preservative	Organ	Time (Hours)	Nuclear Staining	Cytoplasmic Staining	Tissue Morphology	Clarity of Staining	Uniformity of Staining	Total Score	Interpretation
10% NBF	Liver	24	1	1	1	1	1	5	Adequate
		48	1	1	1	1	1	5	Adequate
		72	1	1	1	1	1	5	Adequate
	Kidney	24	1	1	1	1	1	5	Adequate
		48	1	1	1	1	1	5	Adequate
		72	1	1	1	1	1	5	Adequate
	Heart	24	1	1	1	1	1	5	Adequate
		48	1	1	1	1	1	5	Adequate
		72	1	1	1	1	1	5	Adequate

Table A1. *Cont.*

Preservative	Organ	Time (Hours)	Nuclear Staining	Cytoplasmic Staining	Tissue Morphology	Clarity of Staining	Uniformity of Staining	Total Score	Interpretation
AgNPs solution	Liver	24	0	0	0	1	1	2	Inadequate
		48	0	0	0	1	1	2	Inadequate
		72	0	0	0	1	1	2	Inadequate
	Kidney	24	0	0	0	0	1	1	Inadequate
		48	0	0	0	0	1	1	Inadequate
		72	0	0	0	0	1	1	Inadequate
	Heart	24	0	0	0	1	1	2	Inadequate
		48	0	0	0	1	1	2	Inadequate
		72	0	0	0	1	1	2	Inadequate

References

1. Kmiec, Z. J.A. Kiernan. Histological and Histochemical Methods: Theory and Practice. 5th Edition, Scion Publishing, 2015, 571 pp. *Folia Histochem. Cytobiol.* **2016**, *54*, 58–59. Available online: https://journals.viamedica.pl/folia_histochemica_cytobiologica/article/view/46908 (accessed on 22 March 2025). [CrossRef]
2. Vucevic, D.D.; Seidman, M.A.; Mesaki, K.; Guan, Z.; Juvet, S.; Liu, M.; Keshavjee, S.; Murphy, K. A Novel Tissue Preservation and Transport Solution as a Substitute for Formalin. *Lab. Investig.* **2023**, *103*, 100198. [CrossRef]
3. Bussolati, G.; Chiusa, L.; Cimino, A.; D'armento, G. Tissue transfer to pathology labs. Under vacuum is the safe alternative to formalin. *Virchows Arch.* **2008**, *452*, 229–231. [CrossRef]
4. Shetty, J.K.; Babu, H.F.; Laxminarayana, K.P.H. Histomorphological Assessment of Formalin versus Nonformalin Fixatives in Diagnostic Surgical Pathology. *J. Lab. Physicians* **2020**, *12*, 271–275. [CrossRef] [PubMed]
5. Swami, P.C.; Raval, R.; Kaur, M.; Kaur, J. Accidental intraoral injection of formalin during extraction: Case report. *Br. J. Oral Maxillofac. Surg.* **2016**, *54*, 351–352. [CrossRef] [PubMed]
6. Mamun, M.M.A.; Rahman, M.A.; Zaman, M.K.; Ferdousi, Z.; Abu Reza, M. Toxicological effect of formalin as food preservative on kidney and liver tissues in mice model. *IOSR J. Environ. Sci. Toxicol. Food Technol.* **2014**, *8*, 47–51. [CrossRef]
7. Islam, M.T.; Akter, S.; Hasan, M.T.; Roy, D.; Sadi, R.S.; Harun, A.B.; Habiba, U.; Khatun, F.; Das, Z.C.; Haider, M.G. Elucidating the effects of formalin as food preservative on gross and histoarchitecture of vital organs of swiss albino mice. *Ecol. J.* **2024**, *6*, 17–28. [CrossRef]
8. Chu, W.-S.; Furusato, B.; Wong, K.; Sesterhenn, I.A.; Mostofi, F.K.; Wei, M.Q.; Zhu, Z.; Abbondanzo, S.L.; Liang, Q. Ultrasound-accelerated formalin fixation of tissue improves morphology, antigen and mRNA preservation. *Mod. Pathol.* **2005**, *18*, 850–863. [CrossRef]
9. Berrino, E.; Annaratone, L.; Detillo, P.; Grassini, D.; Bragoni, A.; Sapino, A.; Bussolati, B.; Bussolati, G.; Marchiò, C. Tissue Fixation with a Formic Acid-Deprived Formalin Better Preserves DNA Integrity over Time. *Pathobiology* **2022**, *90*, 155–165. [CrossRef]
10. Greish, K.; Alqahtani, A.A.; Alotaibi, A.F.; Abdulla, A.M.; Bukelly, A.T.; Alsobyani, F.M.; Alharbi, G.H.; Alkiyumi, I.S.; Aldawish, M.M.; Alshahrani, T.F.; et al. The Effect of Silver Nanoparticles on Learning, Memory and Social Interaction in BALB/C Mice. *Int. J. Environ. Res. Public Health* **2019**, *16*, 148. [CrossRef]
11. Almatroudi, A. Unlocking the Potential of Silver Nanoparticles: From Synthesis to Versatile Bio-Applications. *Pharmaceutics* **2024**, *16*, 1232. [CrossRef] [PubMed]
12. Burdușel, A.-C.; Gherasim, O.; Grumezescu, A.M.; Mogoantă, L.; Ficai, A.; Andronescu, E. Biomedical Applications of Silver Nanoparticles: An Up-to-Date Overview. *Nanomaterials* **2018**, *8*, 681. [CrossRef]
13. Franci, G.; Falanga, A.; Galdiero, S.; Palomba, L.; Rai, M.; Morelli, G.; Galdiero, M. Silver Nanoparticles as Potential Antibacterial Agents. *Molecules* **2015**, *20*, 8856–8874. [CrossRef]
14. Barber, R.; Harmer, D.; Coleman, R.A.; Clark, B.J. GAPDH as a housekeeping gene: Analysis of GAPDH mRNA expression in a panel of 72 human tissues. *Physiol. Genom.* **2005**, *21*, 389–395. [CrossRef]
15. Jones, W.; Greytak, S.; Odeh, H.; Guan, P.; Powers, J.; Bavarva, J.; Moore, H.M. Deleterious Effects of Formalin-Fixation and Delays to Fixation on RNA and miRNA-Seq Profiles. *Sci. Rep.* **2019**, *9*, 6980. [CrossRef]
16. Niemantsverdriet, M.; Knol, A.; van der Starre-Gaal, J.; Hoogland, A.M. Formalin-Free Tissue Embedding Is Less Hazardous and Results in Better DNA Quality. *PLoS ONE* **2024**, *19*, e0316107. [CrossRef]
17. Das, D.; Ghosh, D.; Mandal, P. Preservative Potential of Biosynthesized Silver Nanoparticles: Prevention of Xylem Occlusion and Microbial Proliferation at Postharvest Stage of Preservation. *Environ. Sci. Pollut. Res. Int.* **2021**, *28*, 22038–22063. [CrossRef]

18. Morsi, S.; Pittala, V.; Alqudah, M.; Haider, M.; Greish, K. In Vivo Evaluation of Anti-Nociceptive Effects of Silver Nanoparticles. *Molecules* **2022**, *27*, 7259. [CrossRef]
19. Liu, Z.; Ren, G.; Zhang, T.; Yang, Z. Action Potential Changes Associated with the Inhibitory Effects on Voltage-Gated Sodium Current of Hippocampal CA1 Neurons by Silver Nanoparticles. *Toxicology* **2009**, *264*, 179–184. [CrossRef]
20. Mikhailova, E.O. Green Silver Nanoparticles: An Antibacterial Mechanism. *Antibiotics* **2024**, *14*, 5. [CrossRef]
21. Rai, M.; Yadav, A.; Gade, A. Silver Nanoparticles as a New Generation of Antimicrobials. *Biotechnol. Adv.* **2009**, *27*, 76–83. Available online: https://www.sciencedirect.com/science/article/pii/S0734975008000918 (accessed on 8 May 2025). [CrossRef]
22. Ocimum Scientific Publishers Pty Ltd. Formalin Hazards and Health Insurance Coverage Needs among Exposed Personnel. *Res. Adv. Biomed. Sci. Technol.* **2019**, *1*, 18–21. [CrossRef]
23. Iravani, S.; Korbekandi, H.; Mirmohammadi, S.V.; Zolfaghari, B. Synthesis of Silver Nanoparticles: Chemical, Physical and Biological Methods. *Res. Pharm. Sci.* **2014**, *9*, 385–406.
24. QIAamp DNA Mini Kit and QIAamp DNA Blood Mini Kit Handbook. Available online: https://www.qiagen.com/br/resources/download.aspx?id=bf602afb-cc1e-4032-b762-6e364adf39ff&lang=en (accessed on 20 March 2025).
25. TRIzolTM Reagent. Available online: https://www.thermofisher.cn/order/catalog/product/15596026 (accessed on 20 March 2025).
26. Thermo Scientific. T042-TECHNICAL BULLETIN NanoDrop Spectrophotometers. Available online: https://www.scirp.org/reference/referencespapers?referenceid=3666598 (accessed on 20 March 2025).
27. Lee, P.Y.; Costumbrado, J.; Hsu, C.Y.; Kim, Y.H. Agarose Gel Electrophoresis for the Separation of DNA Fragments. *J. Vis. Exp.* **2012**, *62*, 3923. Available online: https://pmc.ncbi.nlm.nih.gov/articles/PMC4846332/ (accessed on 20 March 2025).
28. High-Capacity cDNA Reverse Transcription Kit. Available online: https://www.thermofisher.com/order/catalog/product/4368814 (accessed on 22 March 2025).
29. Cell Lysis Buffers-BH. Available online: https://www.thermofisher.com/sa/en/home/life-science/protein-biology/protein-purification-isolation/cell-lysis-fractionation/cell-lysis-total-protein-extraction.html (accessed on 22 March 2025).
30. LI COR BIOTECH LLC WesternSure PREMIUM Chemiluminescent Substrate (100 mL Kit). Available online: https://www.fishersci.com/shop/products/westernsure-premium-substrate/50489552 (accessed on 22 March 2025).
31. Alwahaibi, N.; Al Dhahli, B.; Al Issaei, H.; Al Wahaibi, L.; Al Sinawi, S. Effectiveness of Neutral Honey as a Tissue Fixative in Histopathology. *F1000Research* **2022**, *11*, 1014. [CrossRef]

International Journal of *Molecular Sciences*

MDPI

Article

A Complete Approach for circRNA Therapeutics from Purification to Lyophilized Delivery Using Novel Ionizable Lipids

Esther Broset †, Ana Larraga †, Verónica Lampaya, Víctor Navarro, Alexandre López-Gavín, Diego de Miguel, Álvaro Peña, Juan Martínez-Oliván * and Diego Casabona *

Certest Pharma, Certest Biotec Ltd., Polígono Industrial Río Gallego II, Calle J, 1, 50840 San Mateo de Gállego, Spain; ebroset@certest.es (E.B.); alopez@certest.es (A.L.-G.); dmiguel@certest.es (D.d.M.); apena@certest.es (Á.P.)

* Correspondence: jmartinez@certest.es (J.M.-O.); dcasabona@certest.es (D.C.)

† These authors contributed equally to this work.

Abstract: Circular RNA (circRNA) has gained significant attention as a potential therapeutic tool due to its remarkable stability and resistance to degradation by exonucleases. However, scalable and efficient methods for purification and delivery remain critical challenges that must be addressed. In this study, we developed and evaluated an optimized affinity chromatography method using Oligo (dT) columns for the purification of circRNA, achieving high yield and purity compared with high-performance liquid chromatography. Additionally, we investigated the in vivo efficacy of circRNA-Oligo (dT) encapsulated in lipid nanoparticles (LNPs) formulated with emerging ionizable lipids, including CP-LC-0867 and CP-LC-0729. Our results showed that LNPs formulated with CP-LC-0867 consistently produced higher protein expression compared to SM-102, with sustained luciferase activity observed over a 14-day period. Furthermore, we assessed the lyophilization potential of LNP-circRNA-Oligo (dT) using CP-LC-0729 to extend the shelf life and eliminate the need for ultra-low-temperature storage. Remarkably, the lyophilized LNPs exhibited no significant differences in protein expression compared to their non-lyophilized counterparts, demonstrating that lyophilization is a viable strategy for extending the storage and transport of circRNA therapies. These findings underscore the potential of optimized new ionizable lipids, improved purification strategies, and lyophilization techniques to enhance the scalability, stability, and practical application of circRNA therapies.

Keywords: LNP; circRNA; ionizable lipids; Lyophilization; scalable purification

1. Introduction

Circular RNA (circRNA) has emerged as a novel and versatile class of RNA molecules, distinguished by their covalently closed loop structure. This topology grants circRNA remarkable stability, as it lacks the 5′ and 3′ ends that make linear RNA susceptible to exonuclease-mediated degradation. Furthermore, the use of the internal ribosome entry site (IRES) in circRNAs allows them to be translated [1], which has increased the attention they have received in various fields, including gene therapy [2], vaccine development [3,4], and synthetic biology [5]. Its ability to sustain prolonged protein expression makes circRNA an attractive alternative to traditional linear RNA [6], particularly in therapeutic applications where long-lasting effects are desired without the risks associated with genomic integration.

The production of circRNA typically involves in vitro transcription (IVT) with a subsequent circularization and a complex purification, a process to separate circular from linear RNA species [7]. The circularization process has recently evolve from the use of low effective T4 ligase-mediated reactions [8] to the utilization of the highly efficient engineered permuted intron–exon (PIE) self-splicing strategy to promote the circularization of RNA molecules [9,10]. However, the efficient purification of circRNA from precursor, introns, and nicked RNA remains a significant challenge. Those RNA contaminants can compromise the immunity and functionality of circRNA [11,12], which is particularly problematic in applications that require high fidelity and minimal interference, such as in RNA vaccines and gene therapies [13]. Traditional methods for RNA purification, such as gel electrophoresis [5] and high-performance liquid chromatography (HPLC) [9], have been used to isolate circRNA. Nevertheless, RNA yields from these methods are typically less than 1% of the input RNA [2,9], making large-scale circRNA production economically challenging.

The lipid nanoparticle (LNP) system is one of the more effective for in vivo delivery of circRNA-based therapies. LNPs are typically composed of four key components: ionizable lipids, phospholipids, cholesterol, and polyethylene glycol (PEG)-lipids. Among these, ionizable lipids are crucial for the efficient release of RNA cargo into the cytoplasm. These ionizable lipids acquire a positive charge in acidic environments, such as within endosomes, facilitating the release of RNA into the cytoplasm. Despite their pivotal role, current ionizable lipids present several limitations that hinder their broader application. Firstly, endosomal escape is highly inefficient, with only 1–2% of internalized LNPs successfully releasing their RNA cargo into the cytoplasm [14]. This represents a major bottleneck in the intracellular delivery process, significantly limiting the therapeutic potency of RNA-based interventions [15]. Additionally, the vast majority of ionizable lipid screenings have been conducted using mRNA as the model cargo, resulting in limited data on the performance and functionality of LNPs when encapsulating alternative RNA modalities, such as self-amplifying RNA or circRNA [16]. Toxicity is another concern; some ionizable lipids have been associated with elevated inflammatory responses [17]. Moreover, the limited number of clinically approved ionizable lipids, such as MC3, SM-102, and ALC-0315, poses challenges related to complex synthesis process [18] and licensing restrictions and high costs, thereby complicating access and increasing the expense of RNA-based treatments [19]. In this context, the development of novel lipid formulations that are accessible at an affordable cost is not only scientifically necessary but also critical from a public health and equity standpoint [20].

Additionally to the lipid components, the physical and chemical stability of LNP formulations represent another significant challenge for their widespread application [21]. Current RNA-LNP therapies, including the approved mRNA vaccines Comirnaty and Spikevax, require cryogenic preservation and transportation between −80 °C and −60 °C for Comirnaty and at −20 °C for Spikevax [22]. The stringent storage requirements are primarily due to the intrinsic sensitivity of mRNA to environmental factors such as oxygen, enzymatic degradation, and fluctuations in pH. Additionally, complex interactions between the lipid components and degradation by-products from oxidation and hydrolysis further compromise the stability of mRNA. As water, oxygen, and lipids are included in mRNA-LNP formulations, ensuring high stability in liquid mRNA-LNP systems has historically presented some difficulties [23].

Enhancing the stability of mRNA-LNP therapies has shown promise through lyophilization, a process that removes water by sublimation under vacuum at low temperatures [24,25]. One study reported successful lyophilization of mannose-modified LNPs containing SM-102 as the ionizable lipid, resulting in a durable immune response against rabies and SARS-CoV-2 viruses [26]. While lyophilization has been primarily developed for mRNA-based LNPs, there is limited research on its application to circRNA-LNP formulations. Consequently, developing a circRNA-LNP platform that incorporates novel ionizable lipids capable of withstanding the lyophilization process will be crucial for making these therapies more accessible and practical for public health applications.

In this study, we introduce a complete circRNA platform designed to overcome the economic and industrial challenges associated with circRNA-LNP-based therapies. To address the critical issue of low purification yields, we propose an innovative purification method utilizing poly(A) affinity chromatography. This approach significantly enhances purification efficiency, achieving yields comparable to those of mRNA, thus rendering the process feasible for industrial-scale production. Furthermore, we address the challenge of ionizable lipid selection, a key component in LNP formulations, by exploring novel lipid candidates' performance with circRNA [27,28]. Finally, we integrate a lyophilization technique tailored to circRNA-LNP formulations [29], ensuring the practicality and accessibility of these therapies for widespread health applications.

2. Results

2.1. Optimization of In Vitro Transcription Conditions for Enhanced CircRna Production

To optimize our circRNA production platform, we utilized the PIE self-splicing circularization system, which is widely recognized for its efficiency [9]. The initial focus was on enhancing in vitro transcription (IVT) conditions, particularly incubation time and Mg^{2+} concentration, as both are critical determinants of yield and RNA quality [30]. For circRNA production, the Mg^{2+} concentration is particularly important, as it is essential not only for the IVT process but also for the PIE-mediated circularization.

Our optimization began by adjusting the incubation time while maintaining a Mg^{2+} concentration of 16.5 mM and an incubation temperature of 37 °C. We observed that a 3 h incubation period resulted in the highest percentage of circRNA isoforms, indicating it as the optimal duration (Figure 1A). We next varied the Mg^{2+} concentration to evaluate its effect on circRNA yield. In line with our previous optimization studies for mRNA IVT [31], we identified 16.5 mM of Mg^{2+} as the optimal concentration for circRNA production (Figure 1B,C). Under these conditions, approximately 60% of the RNA was in the circular isoform, compared to 40% observed with 20 mM and 10 mM of Mg^{2+}. To quantify the proportion of circular isoforms, we performed PAGE electrophoresis coupled with image analysis (Figure 1B) and confirmed the results using capillary electrophoresis, which yielded consistent findings (Figure S1). Notably, reducing the Mg^{2+} concentration from 10 mM to 5 mM resulted in a complete absence of RNA synthesis. This underscores the critical role of Mg^{2+} concentration in the formation of both mRNA and circRNA.

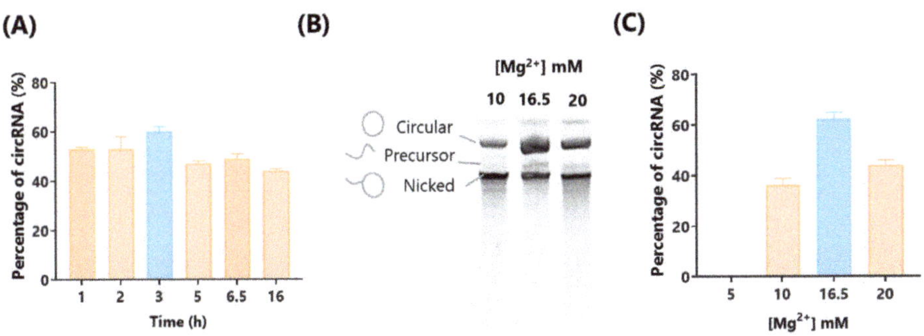

Figure 1. Optimization of incubation time and Mg^{2+} concentration for circRNA production. (**A**) Quantification of circular isoform yield at 1, 2, 3, 5, 6.5, and 16 h of incubation during the in vitro transcription

(IVT) reaction. (**B**) Polyacrylamide gel electrophoresis (PAGE) analysis of circRNA yield at different Mg^{2+} concentrations (10, 16.5, and 20 mM) used on the IVT reaction buffer. (**C**) Percentage of circular RNA isoform based on PAGE results from panel (**B**) quantified by image analysis.

2.2. Scalable circRNA Purification Using Affinity Chromatography

After optimizing the IVT conditions, we shifted our focus to the purification steps. Traditionally, circRNA production involves an IVT step, followed by circularization with the addition of GTP and Mg^{2+}, and a subsequent purification step. The resulting circRNA is then treated with RNase R to eliminate the linear RNA, and then it is further purified. Firstly, to maximize yield, we chose to bypass the RNase R treatment and the subsequent purification step, directly purifying the circRNA after the circularization step. Secondly, to improve purity, we leveraged the polyA tail in the circRNA design [4,9], utilizing affinity chromatography with Oligo (dT) 25 columns (Oligo (dT)) for the purification (Figure S2). This method was compared to a gold-standard approach commonly used, which utilizes silica-based purification. The Oligo (dT) method yielded an average recovery of 18.9 µg of total RNA per µg of IVT input pDNA (Figure 2A). Although this recovery rate is lower than the 55.75 µg per µg of pDNA achieved with the silica-based method (circRNA-kit), it is important to highlight that the Oligo (dT) column resulted in higher purity of the circular isoform. Specifically, the Oligo (dT) method enriched the circRNA isoform to approximately 69.4% of the total RNA, whereas the silica method reached only 51.25%, indicating a markedly better purification performance.

Additionally, after the Oligo (dT) purification, we observed the complete removal of the intron isoform and linear precursors (Figure 2B,C) that are reported to profoundly diminished translation efficiency by inducing immune activation [13]. This efficient precursor removal justifies omitting the RNase R treatment, thereby eliminating the need for an additional, costly enzymatic step. In contrast, when using the silica-based purification method, intron isoforms and linear precursors remained at levels comparable to those observed immediately after the IVT reaction. Furthermore, when evaluating purity, we found that the circRNA/nicked RNA ratio serves as a reliable purity indicator. Oligo (dT) purification achieved a two-fold higher circRNA/nicked RNA ratio compared to the silica method (Figure 2D).

The presence of dsRNA in RNA therapies has been shown to be detrimental to protein production [32,33], as it activates the innate cellular response, leading to the elimination of foreign RNA [34]. HPLC purification has been demonstrated to eliminate immune activation in IVT mRNA samples [32] and circRNA [9,13]; however, there are no data regarding its effect when using Oligo (dT) purification for circRNA. Therefore, to further assess the quality of the purified circRNA, we quantified the percentage of dsRNA% in the samples (Figure 2E). Our analysis demonstrated that affinity chromatography reduced dsRNA levels to 0.05%, significantly below the 0.5% threshold generally considered acceptable for RNA-based therapies. Notably, compared to the circRNA-kit purified using a silica-based method, our circRNA-Oligo (dT) purification resulted in an 18.8-fold reduction in dsRNA content. This substantial decrease emphasizes the capability of Oligo (dT) purification in minimizing dsRNA impurities, which is crucial for enhancing the safety and efficacy of circRNA-based therapies.

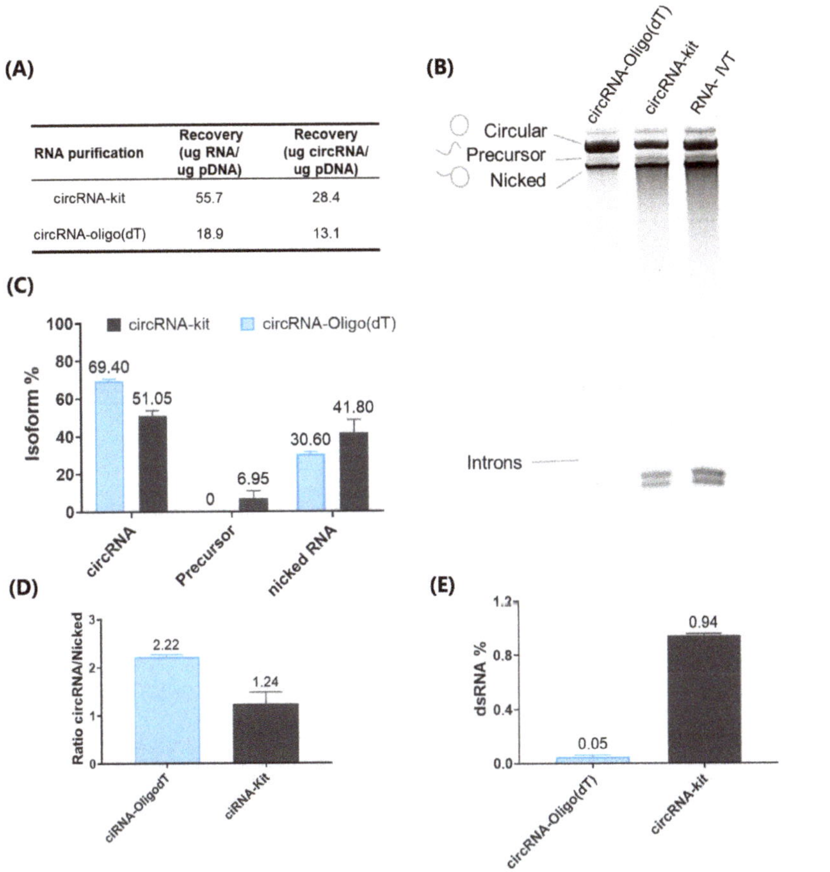

Figure 2. Comparison of circRNA Purification Methods and Quality Assessment. (**A**) Total RNA and circRNA recovery per µg of in vitro transcription (IVT) reaction input pDNA for Oligo (dT) affinity chromatography compared to the silica-based circRNA-kit method. (**B**) Denaturing polyacrylamide gel electrophoresis (PAGE) analysis of RNA before purification (RNA-IVT) and post-purification using either the silica or Oligo (dT) methods. Circular RNA, precursor, nicked, and intronic RNA species are indicated. (**C**) Quantification of the percentage of circular RNA, precursor, and nicked isoforms based on PAGE results in panel (**B**), determined through image analysis. (**D**) circRNA/nicked RNA ratio from panel (**C**). (**E**) Quantification of double-stranded RNA percentage (dsRNA) content through dot-blot analysis, with image-based quantification on samples purified by silica kit (circRNA-kit) or Oligo (dT) methods (circRNA-Oligo (dT).

2.3. *Oligo (dT)-Purified circRNA Exhibits Enhanced In Vitro and In Vivo Translation*

We next assessed the translational capacity of circRNA purified by Oligo (dT) affinity chromatography in vitro, comparing it to circRNA purified by HPLC (circRNA-Ctrl) as a benchmark [35]. Luminescence assays conducted 24 h post-transfection (Figure 3A) showed that circRNA-Oligo (dT) produced 14-fold higher protein levels compared to the commercial circRNA-Ctrl, and 9-fold compared to a circRNA purified by commercial silica-based capture columns (circRNA-kit) in HeLa cells. Additionally, circRNA-Oligo (dT) exhibited a 2-fold increase in protein production compared to conventional *N*1-methylpseudouridine-modified mRNA (Figure 3B). Similar trends were observed in HEK293T (Figure S3) and A549 cells (Figure 3C). Notably, the difference was most pronounced in A549 cells, where circRNA-Oligo (dT) generated 25-fold higher luminescence than circRNA-Ctrl purified using the HPLC method [35] and 9-fold compared to circRNA-kit. This is particularly significant given that A549 cells possess functional Toll-like receptors (TLRs), making them more sensitive to circRNA impurities than HEK293T cells [13]. Furthermore, these transfection results align with our dsRNA impurity data, which showed higher levels of impurities in circRNA-kit than in circRNA-Oligo (dT) (Figure 2).

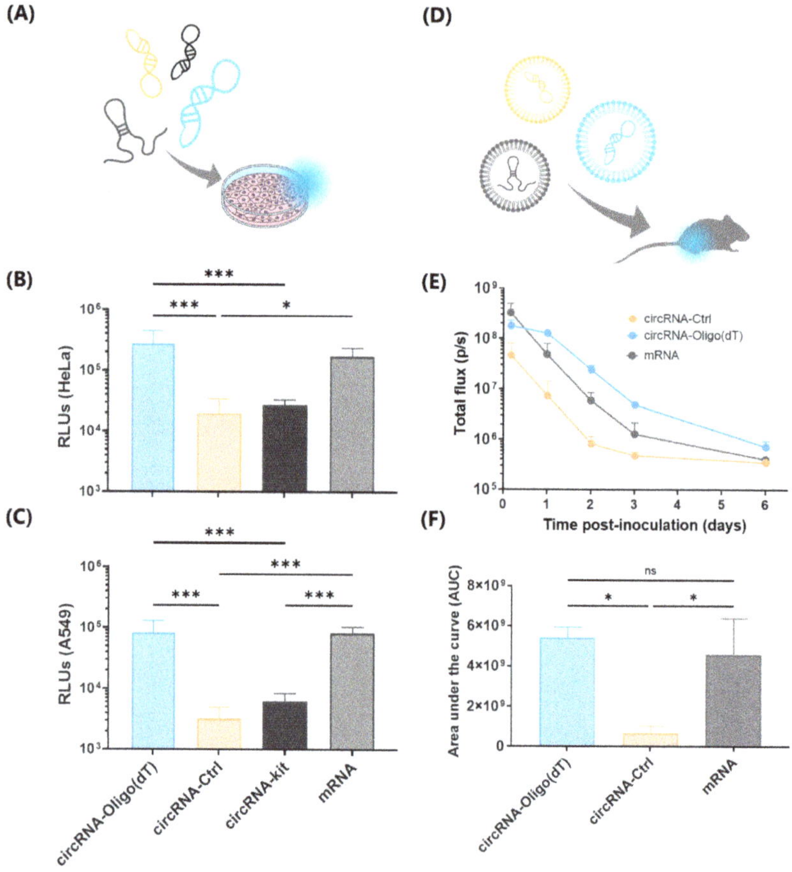

Figure 3. Translation efficacy of circRNA purified by Oligo (dT) affinity chromatography. (**A**) Schematic of in vitro transfection of RNAs using a cationic lipid-based transfection reagent. Oligo (dT)-purified circRNA (circRNA-Oligo (dT) is represented in blue, commercial control circRNA (circRNA-Ctrl) in yellow, circRNA purified by commercial silica-based capture columns (circRNA-kit) is represented in black, and 5-methoxyuridine-modified mRNA (mRNA) in grey. (**B**) HeLa or (**C**) A549 cell lines were transfected with 100 ng/well of circRNA-Oligo (dT), circRNA-Ctrl, circRNA-kit, or mRNA. Luminescence was measured 24 h post-transfection. The x-axis in panel (**B**) is the same as in panel (**C**). (**D**) Scheme of the in vivo procedure to assess translation efficacy of the different RNAs encapsulated into LNPs. (**E**) In vivo luminescence levels monitored over 144 h (6 days) following intramuscular injection of LNPs encapsulating circRNA-Oligo (dT), circRNA-Ctrl, or mRNA in a mouse model. All LNPs were formulated using the commercial ionizable lipid SM-102, maintaining consistent lipid molar ratios across all samples. (**F**) Area under the curve (AUC) analysis of cumulative luciferase expression over the 6-day period represented in **E**. Results are represented as mean ± SD. Statistical significance was determined using one-way ANOVA with Tukey's post hoc test (*: *p*-value < 0.05; ***: *p*-value < 0.001; *ns*: *p*-value > 0.05).

Given that circRNA is intended for use in in vivo therapies, we next assessed the translational capacity of circRNA purified using our new method in a mouse model. We encapsulated circRNA-Oligo (dT), circRNA-Ctrl, and mRNA into conventional LNPs using ionizable lipid SM-102. Following intramuscular administration of RNA-LNPs, luminescence was monitored over 6 days. circRNA-Oligo (dT) showed higher luminescence values than circRNA-Ctrl from the time point at 4 h (Figure 3D,E) and also surpassed those of modified mRNA. Area under the curve (AUC) analysis over the 6-day period indicated that circRNA-Oligo (dT) produced a significantly higher cumulative amount of luciferase protein than circRNA-Ctrl (Figure 3F). Although circRNA-Oligo (dT) produced a similar cumulative amount of protein as mRNA, the persistence of the signal was markedly different. By day 3 (72 h), luminescence from circRNA-Oligo (dT) was 3.7 times higher than that of mRNA, indicating that circRNA-Oligo (dT) may be better suited for therapies requiring sustained protein expression over time.

2.4. CircRNA-Oligo (dT) Encapsulated with Novel Ionizable Lipids Shows Long-Lasting Protein Production In Vivo and Lyophilizability

To further develop an industrially scalable circRNA therapy, it is essential to expand the range of ionizable lipids available for LNP encapsulation. We have developed a broad family of ionizable lipids [27], originally evaluated for mRNA encapsulation and delivery, but with applicability for encapsulating any potential type of nucleic acid. In this study, we focused on evaluating the top three candidates from this family, CP-LC-0729, CP-LC-0743, and CP-LC-1254 (Figure S4), specifically for their efficiency in delivering circRNA-Oligo (dT) in vivo. In addition, we introduced a novel lipid, CP-LC-0867, synthesized using the same platform (Figure 4A), whose structure, although not yet published in the scientific literature, is described in a recent patent [28]. Each LNP formulation was physically characterized (Figure 4C), tested in vitro (Figure S5), and evaluated in a mouse model via intramuscular injection to assess its effectiveness in circRNA delivery and subsequent protein expression (Figure 5A).

Notably, while luminescence was initially planned to be monitored over a 6-day period post-injection, LNPs formulated with CP-LC-0867 exhibited sustained high levels of luciferase expression, prompting us to extend the monitoring period to 14 days. LNPs containing CP-LC-0867 consistently produced higher protein levels compared to those formulated with SM-102, while LNPs with CP-LC-0729 exhibited a similar expression profile to SM-102 (Figure 5B). CP-LC-0743 and CP-LC-1254 also showed similar expression profiles to SM-102, but with a more pronounced decrease in protein levels from 72 h onward. Interestingly, these in vivo results did not align with the in vitro data from HeLa, HEK293T, or A549 cells (Figure S5), where CP-LC-0867 performed similarly to the other lipids. This discrepancy may be due to the shorter time points used in cell-based luminescence assays compared to the extended monitoring in mice. However, it is more likely attributable to the well-documented differences between in vitro and in vivo transfection efficiencies of LNPs, as reported extensively in the literature [27].

When analyzing cumulative luciferase protein expression, as measured by the area under the curve (AUC), CP-LC-0867 generated significantly higher protein levels compared to all other conditions (Figure 5C). Although the AUC analysis showed similar total luciferase production between mRNA and certain circRNA formulations, this was largely due to mRNA producing higher protein levels at early time points, which skews the overall comparison. However, at an intermediate time point of 6 days, the sustained protein production profile became more evident, with CP-LC-0867 exhibiting a 92-fold increase in protein expression and the remaining circRNA formulations showing at least a 6-fold increase compared to mRNA-LNP formulations. These results highlight the long-term stability and effectiveness of circRNA-LNPs, particularly those formulated with CP-LC-0867.

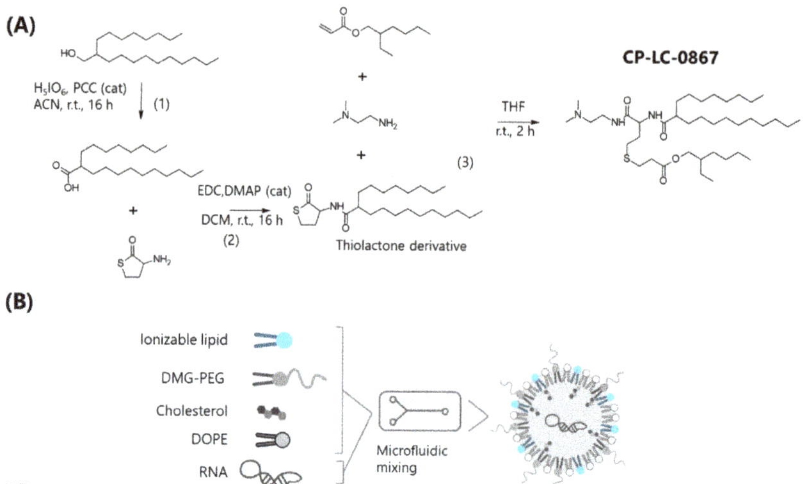

(C)

RNA type	Ionizable lipid	Dimeter [nm]	Polydispersity index	Zeta Potential [mV]	Encapsulation %
mRNA	SM-102	98.57	0.2325	−5.164	90.69
circRNA-oligo(dT)	SM-102	109.9	0.2318	−7.938	93.38
	CP-LC-0729	89.93	0.2912	−11.03	90.47
	CP-LC-0867	113.3	0.2504	−8.672	84.68
	CP-LC-0743	136.9	0.06148	−4.859	83.46
	CP-LC-1254	107.8	0.2943	−9.782	93.98

Figure 4. Synthesis of CP-LC-0867 and physicochemical characterization of circRNA-LNP formulations. (**A**) Schematic representation of the synthetic route to CP-LC-0867: (1) oxidation of the hydrophobic alcohol to a carboxylic acid, (2) amide bond formation between the thiolactone and the carboxylic acid, and (3) a one-pot, multicomponent reaction involving the thiolactone derivative, an acrylate, and a primary amine to yield CP-LC-0867. (**B**) Schematic representation of the encapsulation of circRNA or mRNA into lipid nanoparticles (LNPs) by microfluidic mixing. (**C**) Table summarizing the physical properties of LNP formulations containing circRNA or mRNA and the indicated ionizable lipid. The measured parameters include particle diameter (in nanometers), polydispersity index (PDI) for uniformity, zeta potential (in mV) for surface charge, and encapsulation efficiency (percentage of RNA encapsulated).

We then assessed the lyophilization capacity of LNP-circRNA-Oligo (dT) to extend its storage life and eliminate the need for ultra-low-temperature storage and transport, which represents a significant hurdle to the broader adoption of circRNA therapies. For this study, we selected CP-LC-0729 as the ionizable lipid, as it is one of the most thoroughly characterized lipids among the new ionizable lipids we have tested. Using a previously established lyophilization method [29], we lyophilized LNPs encapsulating circRNA-Oligo (dT) with CP-LC-0729 and evaluated their physical characteristics (Figure S6) and functionality in mice, comparing them with non-lyophilized LNPs (Figure 5D). Lyophilization was also tested with CP-LC-0867, which showed similarly satisfactory results (Figure S7). After intramuscular administration and monitoring luminescence over a 6-day period, we observed no significant differences in the protein production profile (Figure 5E) or cumulative protein expression (Figure 5F) between lyophilized and non-lyophilized LNPs.

These results indicate that lyophilization does not impair the functionality of LNP-circRNA-Oligo (dT) using CP-LC-0729 or CP-LC-0867 as the ionizable lipid, making this new platform a viable option for extending storage life without the need for ultra-low-temperature conditions.

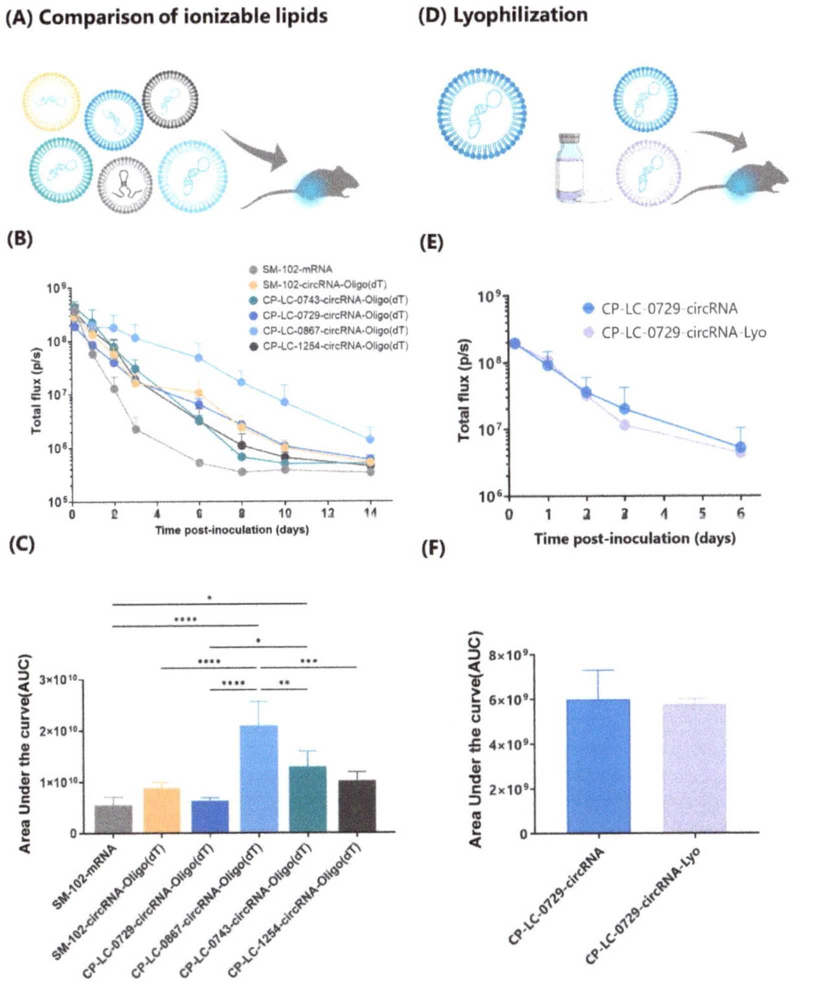

Figure 5. In vivo comparison of different ionizable lipids and lyophilization impact on circRNA-Oligo (dT) LNP. (**A**) Schematic of in vivo comparison of different ionizable lipids used to encapsulate circRNA-Oligo (dT) into LNPs. Control using mRNA as cargo is indicated in light grey and the different ionizable lipids in green, yellow, black, and light or dark blue. (**B**) In vivo luminescence monitoring over a 14-day period following intramuscular injection in mice of LNPs encapsulating circRNA-Oligo (dT) or mRNA. Ionizable lipid used for the LNP formulation is indicated. (**C**) Comparison of cumulative luciferase protein expression (AUC) over the entire monitoring period in panel (**B**). (**D**) Experimental workflow illustrating the lyophilization of LNPs encapsulating circRNA-Oligo (dT) using CP-LC-0729 as the ionizable lipid. Colors in panels C and D represent the different LNP formulations used and are uniformly applied in both panels for direct comparison. (**E**) Luminescence monitoring over a 6-day period following intramuscular injection in mice, comparing lyophilized and non-lyophilized LNP formulations. (**F**) Cumulative luciferase protein expression, calculated via the area under the curve (AUC) of panel (**E**). Results are represented as mean ± SD. Statistical significance was determined using one-way ANOVA with Tukey's post hoc test (*: p-value < 0.05; **: p-value < 0.01; ***: p-value < 0.001; ****: p-value < 0.0001, absence of asterisks: p-value > 0.05).

3. Discussion

Circular RNA has gained significant attention in therapeutic applications due to its exceptional stability and resistance to exonucleases. However, achieving efficient and scalable purification remains a considerable challenge. Traditional methods, such as PAGE electrophoresis and RNase R digestion, although effective, have significant limitations in terms of throughput, scalability, and the ability to achieve high purity [36]. PAGE, commonly employed for RNA separation based on size, conformation, and total molecular charge, provides high resolution but is labor-intensive and prone to contamination, which significantly reduces its practicality for large-scale operations [37]. Likewise, RNase R digestion, which selectively degrades linear RNA while preserving circRNA, often leaves structured linear contaminants and proves cost-prohibitive for large-scale purification. Consequently, while these methods are suitable for initial characterization, they fall short in terms of scalability and widespread industrial application [38].

High-performance liquid chromatography has also been explored for its potential to remove impurities such as dsRNA [9]. However, the HPLC technique reported for circRNA suffer from low recovery yields, as only a small fraction of the total RNA is collected, thus making the method challenging to scale for industrial use [9]. To overcome this scale restrictions, we have developed an innovative affinity chromatography method that for the first time employs Oligo (dT) columns to purify circRNA for therapy purposes. Additionally, we have eliminated the need for RNase R digestion, which provides a cost-effective, scalable alternative to HPLC.

In relation to Mg^{2+} concentration during the IVT process, previous reports largely rely on commercial kits where Mg^{2+} concentrations are undisclosed, making it impossible to modify or optimize [9,11,13]. In this study, we demonstrated that by adjusting the Mg^{2+} concentration, we can significantly increase the proportion of circRNA produced, providing an important optimization strategy for circRNA synthesis.

In terms of in vivo delivery of circRNA, only a limited number of studies have reported successful strategies. Some approaches have employed cationic polymers such as in vivo-jetPEI [11] or O6-stat-N6 CARTs, [3,6] while others have focused on LNPs using ionizable lipids like cKK-E12 [11], MC3 [2,39], SM-102, and ALC-0315 [38]. Nonetheless, several studies that utilize LNPs do not specify the ionizable lipid used [25]. Here, we report the best-performing ionizable lipid described to date, which demonstrated a remarkable 92-fold increase in in vivo protein expression at 6 days post-injection compared to SM-102. Moreover, while previous studies typically employed 5 to 20 μg of RNA for in vivo delivery [2–4,26,39], we achieved robust protein expression with just 1 μg of circRNA. This demonstrates the exceptional efficiency of our novel lipid formulations and their potential to significantly reduce the required therapeutic circRNA dosage.

A critical factor in evaluating circRNA for clinical applications is the duration of gene expression, as circRNA is designed to provide more sustained and stable expression compared to mRNA. To our knowledge, the longest previously reported in vivo signal duration is 42 h using LNPs [13] and 168 h (7 days) using O6-stat-N6 CARTs [6]. In this study, we report an extended signal duration of 336 h (14 days), doubling the 7-day expression window reported in the literature [6]. This prolonged expression suggests that the combination of Oligo (dT) purification and novel ionizable lipids provides superior performance over existing methods, making this approach particularly promising for long-term therapeutic applications.

Among the ionizable lipids evaluated in this study, CP-LC-0867 and CP-LC-0729 demonstrated not only superior in vivo efficacy but also notable advantages in terms of synthetic accessibility and scalability. The previously described synthesis of CP-LC-0729 involves just two steps, with the final step carried out as a straightforward one-pot reaction [27]. Similarly, CP-LC-0867, although requiring three synthetic steps, is produced under mild conditions without the need for stringent parameters [28]. In contrast, clinically established ionizable lipids such as SM-102 [40] and ALC-0315 [41] require four and three synthetic steps, respectively, and typically involve more demanding conditions, including high temperatures, reactions more prone to byproduct formation, and strictly anhydrous environments. These features highlight the practical advantages of CP-LC-0867 and CP-LC-0729 for large-scale and cost-effective manufacturing.

Lyophilization has been extensively studied to enhance the stability of mRNA vaccines [29], but its application to circRNA-LNPs remains largely unexplored, with only one prior study reported to date [26]. However, this study does not specify the ionizable lipid used and focuses on mannose-modified PEG for circRNA-LNP delivery to lymphocytes, without evaluating the impact of lyophilization on LNPs containing conventional, unmodified PEG. To our knowledge, our study provides the first demonstration of successful lyophilization of circRNA-LNPs using novel ionizable lipids [42] and standard PEGylated formulations. By evaluating both the physical integrity and functional preservation of the LNPs post-lyophilization, we offer a more comprehensive and translationally relevant analysis of this strategy's potential to support broader storage and distribution of circRNA therapeutics.

Despite the promising results, our study has some limitations that warrant further investigation. First, while we have demonstrated enhanced circRNA yield and prolonged in vivo expression using Oligo (dT) purification and novel ionizable lipids, the evaluation was based on intramuscular injection. This route of administration is widely used for vaccines and other therapeutic applications, providing valuable insights into the platform's effectiveness. However, it may not fully capture how the system would perform with alternative delivery routes, such as intravenous or subcutaneous injections. Exploring these routes in future studies will offer a broader understanding of the platform's therapeutic potential and expand its applicability across various clinical scenarios. Additionally, we successfully lyophilized circRNA-LNPs, preserving their functionality, and developed an optimized and streamlined process for generating high-quality circRNA. Together, these advancements mark a qualitative breakthrough, setting the stage for more efficient and scalable circRNA therapeutic development. Looking ahead, we aim to explore the long-term stability of lyophilized circRNA-LNPs, building on our previous success with mRNA-LNPs stored for over a year at 4 °C [29], which included the use of CP-LC-0867 and other top-performing lipids. In addition, we will delve deeper into our extensive library of over 1500 ionizable lipids to identify even more effective candidates for circRNA delivery [42]. Furthermore, a thorough evaluation of the safety profile of our novel ionizable lipids will be essential in future studies to determine how they compare with clinically established standards.

In summary, our study provides an important step forward in the scalable production and in vivo delivery of circular RNA (circRNA), utilizing optimized purification strategies and novel ionizable lipids. We demonstrated that the use of Oligo (dT) affinity chromatography enhances the yield and purity of circRNA, allowing for efficient downstream applications without the need for additional enzymatic treatments. Furthermore, we successfully encapsulated circRNA into LNPs formulated with promising ionizable lipids, including CP-LC-0867 and CP-LC-0729, which resulted in superior protein expression profiles compared to conventional lipids such as SM-102. Additionally, our exploration of lyophilization revealed that lyophilized circRNA-LNP formulations retain their functionality, featuring their potential for long-term storage and global distribution. These results demonstrate the substantial potential of this platform to advance the clinical translation of circRNA therapies by enhancing production efficiency, stability, and delivery, which are critical factors for achieving reliable and scalable therapeutic outcomes.

4. Materials and Methods

4.1. Template Plasmid Construction for circRNA

The DNA template for circRNA synthesis was constructed by cloning the Anabaena 3.0 ribozyme, the internal ribosome entry site IRES-CVB3, both as previously described [9], and a firefly luciferase coding sequence optimized for human expression into a pUC-based plasmid under the control of a T7 promoter. Gene synthesis, cloning, and plasmid preparations were outsourced to Genscript (Piscataway, NJ, USA).

Subsequently, the plasmid was linearized with the BspQI restriction enzyme (Hongene, Shanghai, China) at 50 °C for 2 h. The linearized plasmid was then purified using the Wizard® SV gel and PCR clean-up system kit (Promega, Madison, WI, USA) following the manufacturer's instructions.

4.2. Synthesis of RNA and Enrichment of circRNA

Modified linear mRNA and unmodified circRNA were synthesized by in vitro transcription (IVT) using T7 RNA polymerase (Hongene). The transcription reaction was set up with 1 µg of linearized plasmid DNA, 100 U of T7 RNA polymerase, 5 mM rNTPs, and 1× transcription buffer in a total volume of 20 µL. IVT reaction was incubated at 37 °C for 3 h. Following the IVT reaction, the DNA template was digested with 4 U of DNase I (Hongene) at 37 °C for 15 min.

For circRNA synthesis optimization, we systematically varied the IVT incubation times (1, 2, 3, 5, 7, or 16 h) and adjusted the concentration of Mg^{2+} in the transcription buffer to 5, 10, 16.5, or 20 mM.

To enrich the circRNA, the IVT product was denatured at 70° C for 5 min and then immediately placed on ice for 3 min. Subsequently, GTP was added to a final concentration of 2 mM, and the mixture was incubated at 55 °C for 15 min. The reaction was then quenched by the addition of 5 mM EDTA at pH 8 and storage for its subsequent purification.

The circRNA used in this study has an approximate molecular weight of 829.1 kDa, corresponding to 2573 nucleotides. The control circular RNA (circRNA-Ctrl) used in this study was obtained from GenScript (SC8888), where it was purified via high-performance liquid chromatography (HPLC) [35]. This control circRNA contains the same elements as the experimental circRNA constructs produced in our laboratory, ensuring consistency in evaluating the effects of various conditions on circRNA functionality and stability [35].

4.3. Purification and Identification of circRNA

Resulting RNA transcripts were purified by affinity chromatography using POROS Oligo (dT) 25 (ThermoFisher, Bedford, MA, USA). The columns were equilibrated with 5 column volumes of binding buffer (50 mM sodium phosphate, 0.5 M NaCl, 5 mM EDTA, pH = 7.0) before use. The RNA solution was applied to the equilibrated Oligo dT column. The column was washed with 10 column volumes of wash buffer (50 mM sodium phosphate, 5 mM EDTA, pH = 7.0) and was eluted by applying 2 column volumes of double-deionized water. The eluate was collected and immediately placed on ice. A representative chromatogram is presented in Figure S2, illustrating the separation profile achieved during the purification process.

Subsequently, a polishing treatment of the RNAs was performed to enrich the circular isoform relative to the other species. Resulting RNA fractions were visualized on a 4% polyacrylamide gel (PAGE) containing 8 M urea and analyzed using the FIJI software (version Image J2).

The circRNA-Kit was purified using the Monarch® Spin RNA Cleanup Kit (NEB, Ipswich, MA, USA) (50 µg), following the manufacturer's instructions for optimal RNA recovery.

The quality and size of the resulting RNA samples were evaluated using a Bioanalyzer Agilent 2100 (Agilent Technologies, Waldbronn, Germany), with the Agilent RNA 6000 Nano kit. The resulting data were processed and evaluated with 2100 Expert Software (Version: B.02.08.SI648 (SR2)) RNA quality was further evaluated using high-performance liquid chromatography (HPLC). Samples were processed through a CIMac™ SDVB 0.1 mL (BiA Separations, Ajdouscina, Slovenia) analytical reverse-phase column (2 µm channels, Sartorius) connected to an HPLC WATERS e2695. CIMac SDVB column was equilibrated at 65 °C in mobile phase A (50 mM TEAA and 7.5% ACN, pH 7.0). Approximately 0.4 µg RNA was injected on the column. After an initial equilibration step (1 min at 100% MPA), a linear gradient was performed, 0–100% B (50 mM TEAA in 18% CAN pH = 7.0) in 27 min at a flow rate 1 mL min^{-1}. The column was eluted with 100% MPC (50 mM TEAA in 75% CAN pH = 8.2) for 1 min before re-equilibration in 100% MPA.

4.4. Determination of dsRNA Content

The circRNAs were spotted onto positively charged nylon membranes (Nytran SC, Cytiva, München, Germany). The membranes were then blocked with 5% (*w*/*v*) non-fat dried milk in TBS-T buffer (20 mM Tris–HCl, 150 mM NaCl, 0.1% [*v*/*v*] Tween-20, pH 7.4), and incubated with dsRNA-specific mAb J2 (Jena Bioscience, Jena, Germany) at 4 °C overnight. Membranes were washed with TBS-T buffer and incubated with HRP-conjugated goat anti-mouse IgG (Abcam, Cambridge, UK). Following additional wash with TBS-T, the membranes were treated with ECL Plus Western blot detection reagent (Amersham, Chicago, IL, USA). Images were captured using an iBright 750 digital imaging system.

4.5. Synthesis and Characterization of Ionizable Lipids

The ionizable lipids CP-LC-0743, CP-LC-0729, and CP-LC-1254 were synthesized following established protocols based on the Sequential Thiolactone Amine Acrylate Reaction [27,42]. In brief, thiolactone derivatives (0.15 mmoL, 1 equiv.) and acrylate (0.15 mmoL, 1 equiv.) were dissolved in 300 µL of tetrahydrofuran (THF) at room temperature (RT), followed by the addition of the amine (0.15 mmoL, 1 equiv). After two hours, the THF was removed under vacuum. Notably, CP-LC-0743 was previously referred to as A4B2C1, and CP-LC-1254 was previously designated as CP-LC-0729-04 [27].

The synthesis of CP-LC-0867 was carried out as follows: 2-Octyldecanoic acid: 2-octyl-1-dodecanol (1077 mg, 3.5 mmoL) and periodic acid (4144 mg, 18 mmoL) were dissolved in acetonitrile (28 mL). Then, pyridinium chlorochromate (23 mg, 0.030 mmoL) was added to the previous solution. The reaction mixture was stirred 16 h at room temperature. Then, the solvent was evaporated under reduced pressure. Subsequently, the crude was dissolved in AcOEt and first washed twice with water and finally with brine. The organic phase was dried with anhydrous $MgSO_4$, filtered, and evaporated under reduced pressure. The residue was used in the next step without any further purification.

Thiolactone derivative (2-octyl-N-(2-oxotetrahydrothiophen-3-yl) dodecanamide): DL-homocysteine thiolactone hydrochloride (1.8 mmoL) was dissolved in 5 mL of anhydrous dichloromethane at room temperature. Then, triethylamine (1.8 mmoL) was added followed by EDC hydrochloride (1.3 mmoL), 4-(dimethylamino) pyridine (0.20 mmoL), and 2-octyldecanoic acid (1.0 mmoL). The reaction mixture was stirred at room temperature overnight under an argon atmosphere. The organic layer was dried with anhydrous MgSO4, filtered, and evaporated under reduced pressure. The resulting residue was purified by flash chromatography (gradient of hexane/ethyl acetate: 100/0 to 0/100) to afford as a pure product.

MS-QDa: Theorical $[M + H]^+ = 412.32$, experimental $[M + H]^+ = 412.50$.

Thiolactone derivative [27] (0.15 mmoL, 1 equiv.) and 2-ethylhexyl acrylate (0.15 mmoL, 1 equiv.) were dissolved in 300 μL of THF at RT, followed by the addition of N,N-dimethylethylenediamine (0.15 mmoL, 1 equiv.). After two hours, the THF was removed under vacuum, and CP-LC-0867 was purified using a CombiFlash NextGen 300+ with gradient elution, starting from 100% dichloromethane to 50% of an 80/20/1 mixture of DCM/MeOH/NH_4OH (aq).

All synthesized lipids were analyzed by high-performance liquid chromatography (HPLC) equipped with a charged aerosol detector (CAD) and characterized by mass spectrometry (ISQ ThermoFisher, Waltham, MA, USA). The structures of the synthesized lipids are available in Figure S4. ^{1}H-RMN data are shown in Figures S8 and S9 and summarized below:

- CP-LC-0867: MS-QDa: theoretical $[M + H]^+ = 684.57$, experimental $[M + H]^+ = 684.58$; ^{1}H NMR (500 MHz, CDCl3) δ 6.63 (t, J = 5.1 Hz, 1H), 6.25 (d, J = 7.9 Hz, 1H), 4.59 (dt, J= 7.2 Hz J = 7.2 Hz, 1H), 4.07–3.96 (m, 2H), 3.32 (dt, J = 6.4, 5.5 Hz, 2H), 2.80 (t, J = 7.3 Hz, 2H), 2.67–2.49 (m, 4H), 2.40 (t, J = 6.1 Hz, 2H), 2.22 (s, 6H), 2.11–1.99 (m, 2H), 1.94 (m, 1H), 1.57 (m, 3H), 1.45–1.32 (m, 4H), 1.25 (m, 34H), 0.88 (m, 12H).
- CP-LC-0743: MS-QDa: theoretical $[M + H]^+ = 766.25$, experimental $[M + H]^+ = 766.81$; ^{1}H NMR (400 MHz, CDCl3) δ 6.64 (t, J = 4.8 Hz, 1H), 6.26 (d, J = 8.0 Hz, 1H), 5.39–5.29 (m, 2H), 4.59 (dt, J = 7.2 Hz J = 7.2 Hz, 1H), 4.08 (t, J = 6.8 Hz, 2H), 3.32 (dt, J = 5.9 Hz J = 5.9 Hz, 2H), 2.79 (t, J = 7.2 Hz, 2H), 2.67–2.48 (m, 4H), 2.40 (t, J = 6.1 Hz, 2H), 2.22 (s, 6H), 2.11–1.88 (m, 7H), 1.59 (m, 4H), 1.46–1.15 (m, 44H), 0.87 (m, 9H).
- CP-LC-0729: MS-QDa: theoretical $[M + H]^+ = 740.63$, experimental $[M + H]^+ = 740.85$; ^{1}H NMR (400 MHz, CDCl3) δ: 6.76 (m, 1H); 6.37 (d, J = 7.9 Hz, 1H); 4.60 (dt, J = 7.1 Hz J = 7.1 Hz, 1H); 3.98 (d, J = 5.8 Hz, 2H); 3.36 (dt, J = 5.7 Hz J = 5.7 Hz, 2H); 2.78 (t, J = 6.8 Hz, 2H); 2.59 (m, 4H); 2.48 (t, J = 5.9 Hz, 2H); 2.28 (s, 6H); 2.06 (m, 2H); 1.96 (m, 1H); 1.58 (m, 3H); 1.41 (m, 2H); 1.25 (m, 44H); 0.87 (m, 12H).
- CP-LC-1254: MS-QDa: theoretical $[M + H]^+ = 755.63$, experimental $[M + H]^+ = 755.83$; ^{1}H NMR (400 MHz, CDCl3) δ: 6.79 (m, 1H); 4.05 (d, J = 5.8 Hz, 2H); 4.01 (d, J = 5.8 Hz, 2H); 3.45 (t, J = 7.1 Hz, 1H); 3.36 (m, 2H); 2.79 (t, J = 7.6 Hz, 2H); 2.69–2.49 (m, 4H); 2.44 (t, J = 6.0 Hz, 2H); 2.26 (s, 6H); 2.21 (m, 2H); 1.65 (m, 2H); 1.29 (m, 48H); 0.90 (m, 12H).

4.6. Lipid Nanoparticle (LNP) Formulation and Characterization

The synthesis of LPNs was performed using a microfluidic method. An ethanolic lipid solution containing the synthetized CP-LC ionizable lipids or the commercial SM-102 (BocSCI, Shirley, NY, USA), DOPE (Corden Pharma, Plankstadt, Germany), cholesterol (Merk, Rahway, NJ, USA), and DMG-PEG2000 (Cayman, Ann Arbor, MI, USA) at a molar ratio of 50:10:38.5:1.5 was prepared. This lipid mixture was then combined with an aqueous phase containing mRNA or circRNA diluted in 10 mM citrate buffer (pH 4), achieving an ionizable lipid/RNA weight ratio of 10:1. LNPs were formulated using an INano™ microfluidic device set to a total flow rate of 12 mL/min with an aqueous-to-ethanol volume ratio of 3:1. After LNP formation, ethanol was removed by dialysis (Pur-A-Lyzer™ Midi Dialysis Kit, Sigma-Aldrich, St. Louis, MO, USA).

For LNP characterization the average size, polydispersity (PDI), and zeta potential of LNPs were determined using a Malvern Zetasizer Advance Lab Blue Label (Malvern Instruments Ltd., Malvern, UK). Concentration of encapsulated RNA (circRNA or mRNA) was then quantified using the Quant-IT® Ribogreen assay (Invitrogen, Waltham, MA, USA) following the manufacturer's instructions, and LNP encapsulation efficiency were verified by agarose gel electrophoresis.

The final LNP formulation was adjusted to an RNA (circRNA or mRNA) concentration of 100 µg/mL and was subsequently lyophilized and/or stored at 4 °C for subsequent in vivo and in vitro analysis.

4.7. LNP Lyophilization

Lyophilization was performed following previously reported protocols, [29] using a Virtis Genesis Pilot Freeze Dryer (SP, Warminster, PA, USA). In brief, LNPs were lyophilized with 20% maltose as a lyoprotectant in TRIS buffer. The process included three stages: (1) freezing at −50 °C, (2) primary drying at −10 °C under 180 mTorr pressure, and (3) secondary drying at 40 °C, gradually reducing the pressure to 120 mTorr and finally to 50 mTorr. Upon completion, the vials were backfilled with pure nitrogen to maintain an inert atmosphere, sealed, and stored at various temperatures for stability assessments. For reconstitution, 300 µL of RNase-free water was added to each vial, and the contents were gently swirled until a homogenous suspension was obtained.

4.8. Cell Culture and circRNA Transfection

The HeLa (DSMZ GmbH, Braunschweig, Germany), HEK293T (ATCC; American Type Culture Collection, Manassas, VA, USA).) and A549 (NIBSC; (National Institute for Biological Standards and Control, South Mimms, Hertfordshire, UK).) cell lines were maintained in DMEM with high glucose (Merk, Darmstadt, Germany), supplemented with 10% Fetal Bovine Serum (Sigma, Darmstadt, Germany), 1% Penicillin-Streptomycin Solution (GibcoTM, Thermo Fisher Scientific, Waltham, MA, USA) and 2 mM Glutamax (Thermo Fisher Scientific, Waltham, MA, USA). All cell lines were grown in 175 cm^2 flasks.

The day before transfection, cells were detached from culture flasks using trypsinization and subsequently seeded into 96-well plates at a density of 1×10^4 cells per well. Immediately before the transfections, the culture media in each well was replaced with 90 µL of fresh media. For RNA (circRNA or mRNA) transfection using a cationic lipid-based transfection reagent a mixture containing RNA and Lipofectamine MessengerMAX™ (Invitrogen, Thermo Fisher Scientific, Waltham, MA, USA) was pre-incubated in OptiMEM (GibcoTM, Thermo Fisher Scientific, Waltham, MA, USA) media according to the manufacturer's instructions. This RNA-lipofectamine complex was then added to the designated wells in triplicate, achieving a final RNA concentration of 100 ng/well.

For RNA-LNP transfection a suspension of RNA-containing LNPs was added to the designated wells in triplicate to achieve a final RNA concentration of 200,100 and 50 ng/well. The cells were then incubated for 24 h at 37 °C in a 5% CO_2 atmosphere to allow translation of the delivered RNA into luciferase protein.

4.9. Luciferase Activity Quantification In Vitro

At 24 or 48 h post-transfection, the cells were lysed by adding 100 µL of PBS containing 0.1% Triton X-100 to each well and incubating for 10 min at room temperature. This lysis step allows the release of luciferase protein translated from the delivered circRNA or mRNA. Following lysis, 98 µL of the cell lysate was transferred to an opaque 96-well white plate. To each well, 100 µL of a buffered d-Luciferin solution (GoldBio, Olivette, MO, USA) containing 100 mM Tris-HCl (pH 7.8), 5 mM $MgCl_2$, 250 µM CoA, and 150 µM ATP was added, resulting in a final d-Luciferin concentration of 150 µg/mL. Luminescence was measured after a 5 min incubation at room temperature using a FLUOstar Omega (version 5.70) plate reader (BMG LABTECH, Ortenberg, Germany). Cells that had not been treated with any mRNA were used as a negative control to assess background luminescence.

4.10. Administration of LNP-circRNAs in Mice

Female BALB/c mice (Janvier), aged 8–10 weeks and weighing 18–23 g, were acclimatized to the experimental facilities for 3–7 days upon arrival. The mice were housed under controlled conditions, maintaining a room temperature of 20–24 °C, humidity levels between 50 and 70%, and a light intensity of 60 lux with a 12 h light–dark cycle.

For in vivo measurement of Firefly Luciferase activity, LNPs containing 1 μg of the specified RNA in a final volume of 30 μL. were administered via intramuscular injection. At the designated post-injection times, mice were anesthetized with 4% isoflurane administered via a vaporizer for induction, and anesthesia was maintained at 1.5% isoflurane. Following anesthesia, D-luciferin (Quimigen, Madrid, Spain) was administered intraperitoneally at a dose of 150 mg/kg. Ten minutes after D-luciferin administration, luminescence images were captured using the IVIS Lumina XRMS Imaging System (Living Image software version 4.8.2).

Supplementary Materials: The following supporting information can be downloaded at: https://www.mdpi.com/article/10.3390/ijms26115138/s1.

Author Contributions: V.L.: Conceptualization, Data curation, Formal analysis, Investigation, Methodology, Validation, Visualization, Writing—review and editing. A.L.: Conceptualization, Data curation, Formal analysis, Investigation, Methodology, Validation, Visualization, Writing—review and editing. V.N.: Investigation, Methodology. A.L.-G.: Data curation, Formal analysis, Investigation, Methodology, Validation, Visualization, Writing—review and editing. D.d.M.: Conceptualization, Supervision, Writing—review and editing. Á.P.: Investigation, Methodology, Writing—review and editing. E.B.: Conceptualization, Formal analysis, Supervision, Visualization, Writing—original draft. J.M.-O.: Conceptualization, Funding acquisition, Supervision, Writing—review and editing. D.C.: Conceptualization, Project administration, Supervision, Writing—review and editing. All authors have read and agreed to the published version of the manuscript.

Funding: This study was supported by Gobierno de Aragón (Spain) through project IDMF/2023/0003 (Terapias de última generación basadas en ácidos nucleicos nanoencapsulados). Álvaro Peña was supported by the grant DIN2021-011799, funded by the Spanish Ministry of Science and Innovation (MCIN) and the State Research Agency (AEI)/10.13039/501100011033.

Institutional Review Board Statement: The animal study protocol was approved by the Ethics Committee for Animal Experiments of the University of Zaragoza (Project License PI07/23, approved on 07/02/2023). All procedures were conducted in accordance with the Spanish Policy for Animal Protection (RD 53/2013), which aligns with the European Union Directive 2010/63/EU on the protection of animals used for scientific and experimental purposes.

Informed Consent Statement: Not applicable.

Data Availability Statement: The original contributions presented in this study are included in the article/supplementary material. Further inquiries can be directed to the corresponding authors.

Acknowledgments: We would like to extend our sincere gratitude to all the members of Certest Biotec, especially the Certest Pharma group, whose valuable insights and expertise significantly contributed to the success of this research. The authors would like to acknowledge the use of "Servicios Científico Técnicos" at CIBA (IACS-University of Zaragoza). During the preparation of this work, the authors used ChatGPT in order to improve the clarity and accuracy of the English language. After using this tool/service, the authors reviewed and edited the content as needed and take full responsibility for the content of the publication.

Conflicts of Interest: All authors are employees at Certest Pharma Department, Certest Biotec Ltd. J.M.O., A.P., and D.D.M. are inventors on patents related to this publication.

References

1. Pamudurti, N.R.; Bartok, O.; Jens, M.; Ashwal-Fluss, R.; Stottmeister, C.; Ruhe, L.; Hanan, M.; Wyler, E.; Perez-Hernandez, D.; Ramberger, E.; et al. Translation of CircRNAs. *Mol. Cell* **2017**, *66*, 9–21.e7. [CrossRef] [PubMed]
2. Tong, M.; Palmer, N.; Dailamy, A.; Kumar, A.; Khaliq, H.; Han, S.; Finburgh, E.; Wing, M.; Hong, C.; Xiang, Y.; et al. Robust Genome and Cell Engineering via in Vitro and in Situ Circularized RNAs. *Nat. Biomed. Eng.* **2024**, *9*, 109–126. [CrossRef] [PubMed]
3. Amaya, L.; Grigoryan, L.; Li, Z.; Lee, A.; Wender, P.A.; Pulendran, B.; Chang, H.Y. Circular RNA Vaccine Induces Potent T Cell Responses. *Proc. Natl. Acad. Sci. USA* **2023**, *120*, e2302191120. [CrossRef]
4. Qu, L.; Yi, Z.; Shen, Y.; Lin, L.; Chen, F.; Xu, Y.; Wu, Z.; Tang, H.; Zhang, X.; Tian, F.; et al. Circular RNA Vaccines against SARS-CoV-2 and Emerging Variants. *Cell* **2022**, *185*, 1728–1744.e16. [CrossRef]
5. Kameda, S.; Ohno, H.; Saito, H. Synthetic Circular RNA Switches and Circuits That Control Protein Expression in Mammalian Cells. *Nucleic Acids Res.* **2023**, *51*, e24. [CrossRef] [PubMed]
6. Chen, R.; Wang, S.K.; Belk, J.A.; Amaya, L.; Li, Z.; Cardenas, A.; Abe, B.T.; Chen, C.K.; Wender, P.A.; Chang, H.Y. Engineering Circular RNA for Enhanced Protein Production. *Nat. Biotechnol.* **2022**, *41*, 262–272. [CrossRef]
7. Choi, S.-W.; Nam, J.-W. Optimal Design of Synthetic Circular RNAs. *Exp. Mol. Med.* **2024**, *56*, 1281–1292. [CrossRef]
8. Petkovic, S.; Müller, S. RNA Circularization Strategies in Vivo and in Vitro. *Nucleic Acids Res.* **2015**, *43*, 2454–2465. [CrossRef]
9. Wesselhoeft, R.A.; Kowalski, P.S.; Anderson, D.G. Engineering Circular RNA for Potent and Stable Translation in Eukaryotic Cells. *Nat. Commun.* **2018**, *9*, 2692. [CrossRef]
10. Meganck, R.M.; Liu, J.; Hale, A.E.; Simon, K.E.; Fanous, M.M.; Vincent, H.A.; Wilusz, J.E.; Moorman, N.J.; Marzluff, W.F.; Asokan, A. Engineering Highly Efficient Backsplicing and Translation of Synthetic circRNAs. *Mol. Ther. Nucleic Acids* **2021**, *23*, 821–834. [CrossRef]
11. Chen, Y.G.; Chen, R.; Ahmad, S.; Verma, R.; Kasturi, S.P.; Amaya, L.; Broughton, J.P.; Kim, J.; Cadena, C.; Pulendran, B.; et al. N6-Methyladenosine Modification Controls Circular RNA Immunity. *Mol. Cell* **2019**, *76*, 96–109.e9. [CrossRef] [PubMed]
12. Lee, K.H.; Kim, S.; Song, J.; Han, S.R.; Kim, J.H.; Lee, S.-W. Efficient Circular RNA Engineering by End-to-End Self-Targeting and Splicing Reaction Using Tetrahymena Group I Intron Ribozyme. *Mol. Ther. Nucleic Acids* **2023**, *33*, 587–598. [CrossRef] [PubMed]
13. Wesselhoeft, R.A.; Kowalski, P.S.; Parker-Hale, F.C.; Huang, Y.; Bisaria, N.; Anderson, D.G. RNA Circularization Diminishes Immunogenicity and Can Extend Translation Duration In Vivo. *Mol. Cell* **2019**, *74*, 508–520.e4. [CrossRef] [PubMed]
14. Gilleron, J.; Querbes, W.; Zeigerer, A.; Borodovsky, A.; Marsico, G.; Schubert, U.; Manygoats, K.; Seifert, S.; Andree, C.; Stöter, M.; et al. Image-Based Analysis of Lipid Nanoparticle–Mediated siRNA Delivery, Intracellular Trafficking and Endosomal Escape. *Nat. Biotechnol.* **2013**, *31*, 638–646. [CrossRef]
15. Liu, H.; Chen, M.Z.; Payne, T.; Porter, C.J.H.; Pouton, C.W.; Johnston, A.P.R. Beyond the Endosomal Bottleneck: Understanding the Efficiency of mRNA/LNP Delivery. *Adv. Funct. Mater.* **2024**, *34*, 2404510. [CrossRef]
16. Xu, S.; Xu, Y.; Solek, N.C.; Chen, J.; Gong, F.; Varley, A.J.; Golubovic, A.; Pan, A.; Dong, S.; Zheng, G.; et al. Tumor-Tailored Ionizable Lipid Nanoparticles Facilitate IL-12 Circular RNA Delivery for Enhanced Lung Cancer Immunotherapy. *Adv. Mater.* **2024**, *36*, 2400307. [CrossRef] [PubMed]
17. Moghimi, S.M.; Simberg, D. Pro-Inflammatory Concerns with Lipid Nanoparticles. *Mol. Ther.* **2022**, *30*, 2109–2110. [CrossRef]
18. Han, X.; Zhang, H.; Butowska, K.; Swingle, K.L.; Alameh, M.-G.; Weissman, D.; Mitchell, M.J. An Ionizable Lipid Toolbox for RNA Delivery. *Nat. Commun.* **2021**, *12*, 7233. [CrossRef]
19. How to Make Enough Vaccine for the World in One Year. Available online: https://www.citizen.org/article/how-to-make-enough-vaccine-for-the-world-in-one-year/ (accessed on 30 August 2024).
20. Swetha, K.; Kotla, N.G.; Tunki, L.; Jayaraj, A.; Bhargava, S.K.; Hu, H.; Bonam, S.R.; Kurapati, R. Recent Advances in the Lipid Nanoparticle-Mediated Delivery of mRNA Vaccines. *Vaccines* **2023**, *11*, 658. [CrossRef]
21. Kamiya, M.; Matsumoto, M.; Yamashita, K.; Izumi, T.; Kawaguchi, M.; Mizukami, S.; Tsurumaru, M.; Mukai, H.; Kawakami, S. Stability Study of mRNA-Lipid Nanoparticles Exposed to Various Conditions Based on the Evaluation between Physicochemical Properties and Their Relation with Protein Expression Ability. *Pharmaceutics* **2022**, *14*, 2357. [CrossRef]
22. Crommelin, D.J.A.; Anchordoquy, T.J.; Volkin, D.B.; Jiskoot, W.; Mastrobattista, E. Addressing the Cold Reality of mRNA Vaccine Stability. *J. Pharm. Sci.* **2021**, *110*, 997–1001. [CrossRef] [PubMed]
23. Packer, M.; Gyawali, D.; Yerabolu, R.; Schariter, J.; White, P. A Novel Mechanism for the Loss of mRNA Activity in Lipid Nanoparticle Delivery Systems. *Nat. Commun.* **2021**, *12*, 6777. [CrossRef] [PubMed]
24. Ai, L.; Li, Y.; Zhou, L.; Yao, W.; Zhang, H.; Hu, Z.; Han, J.; Wang, W.; Wu, J.; Xu, P.; et al. Lyophilized mRNA-Lipid Nanoparticle Vaccines with Long-Term Stability and High Antigenicity against SARS-CoV-2. *Cell Discov.* **2023**, *9*, 9. [CrossRef]
25. Muramatsu, H.; Lam, K.; Bajusz, C.; Laczkó, D.; Karikó, K.; Schreiner, P.; Martin, A.; Lutwyche, P.; Heyes, J.; Pardi, N. Lyophilization Provides Long-Term Stability for a Lipid Nanoparticle-Formulated, Nucleoside-Modified mRNA Vaccine. *Mol. Ther.* **2022**, *30*, 1941–1951. [CrossRef]

26. Wan, J.; Wang, Z.; Wang, L.; Wu, L.; Zhang, C.; Zhou, M.; Fu, Z.F.; Zhao, L. Circular RNA Vaccines with Long-Term Lymph Node-Targeting Delivery Stability after Lyophilization Induce Potent and Persistent Immune Responses. *mBio* **2024**, *15*, e0177523. [CrossRef]
27. Peña, Á.; Heredero, J.; Blandín, B.; Mata, E.; De Miguel, D.; Toro, A.; Alejo, T.; Casabona, D.; López, A.; Gallego-Lleyda, A.; et al. Multicomponent Thiolactone-Based Ionizable Lipid Screening Platform for Efficient and Tunable mRNA Delivery to the Lungs. *Commun. Chem.* **2025**, *8*, 116. [CrossRef] [PubMed]
28. Giménez-Warren, J.; Heredero García, J.; Matínez Oliván, J.E.; Peña Moreno, Á.; De Miguel Samaniego, D.; Toro Córdoba, A. Ionizable Lipids and Lipid Nanoparticles Containing. Thereof. Patent WO2024110381A1, 30 May 2024.
29. Alejo, T.; Toro-Córdova, A.; Fernández, L.; Rivero, A.; Stoian, A.M.; Pérez, L.; Navarro, V.; Martínez-Oliván, J.; de Miguel, D. Comprehensive Optimization of a Freeze-Drying Process Achieving Enhanced Long-Term Stability and In Vivo Performance of Lyophilized mRNA-LNPs. *Int. J. Mol. Sci.* **2024**, *25*, 10603. [CrossRef]
30. Popova, P.G.; Lagace, M.A.; Tang, G.; Blakney, A.K. Effect of in Vitro Transcription Conditions on Yield of High Quality Messenger and Self-Amplifying RNA. *Eur. J. Pharm. Biopharm.* **2024**, *198*, 114247. [CrossRef]
31. Martínez, J.; Lampaya, V.; Larraga, A.; Magallón, H.; Casabona, D. Purification of Linearized Template Plasmid DNA Decreases Double-Stranded RNA Formation during IVT Reaction. *Front. Mol. Biosci.* **2023**, *10*, 1248511. [CrossRef]
32. Karikó, K.; Muramatsu, H.; Ludwig, J.; Weissman, D. Generating the Optimal mRNA for Therapy: HPLC Purification Eliminates Immune Activation and Improves Translation of Nucleoside-Modified, Protein-Encoding mRNA. *Nucleic Acids Res.* **2011**, *39*, e142. [CrossRef]
33. Camperi, J.; Roper, B.; Freund, E.; Leylek, R.; Nissenbaum, A.; Galan, C.; Caothien, R.; Hu, Z.; Ko, P.; Lee, A.; et al. Exploring the Impact of In Vitro-Transcribed mRNA Impurities on Cellular Responses. *Anal. Chem.* **2024**, *96*, 17789–17799. [CrossRef] [PubMed]
34. Luan, X.; Wang, L.; Song, G.; Zhou, W. Innate Immune Responses to RNA: Sensing and Signaling. *Front. Immunol.* **2024**, *15*, 1287940. [CrossRef] [PubMed]
35. Genscript IVT_RNA_circRNA. Available online: https://www.genscript.com/product/RP_A00021-F_Luc_Circular_RNA_CVB3_.html (accessed on 26 March 2025).
36. Zhang, Z.; Fu, Y.; Ju, X.; Zhang, F.; Zhang, P.; He, M. Advances in Engineering Circular RNA Vaccines. *Pathogens* **2024**, *13*, 692. [CrossRef] [PubMed]
37. Breuer, J.; Rossbach, O. Production and Purification of Artificial Circular RNA Sponges for Application in Molecular Biology and Medicine. *Methods Protoc.* **2020**, *3*, 42. [CrossRef]
38. Bai, Y.; Liu, D.; He, Q.; Liu, J.; Mao, Q.; Liang, Z. Research Progress on Circular RNA Vaccines. *Front. Immunol.* **2023**, *13*, 1091797. [CrossRef] [PubMed]
39. Chen, C.; Wei, H.; Zhang, K.; Li, Z.; Wei, T.; Tang, C.; Yang, Y.; Wang, Z. A Flexible, Efficient, and Scalable Platform to Produce Circular RNAs as New Therapeutics. *BioRxiv* **2022**. [CrossRef]
40. Sabnis, S.; Kumarasinghe, E.S.; Salerno, T.; Mihai, C.; Ketova, T.; Senn, J.J.; Lynn, A.; Bulychev, A.; McFadyen, I.; Chan, J.; et al. A Novel Amino Lipid Series for mRNA Delivery: Improved Endosomal Escape and Sustained Pharmacology and Safety in Non-Human Primates. *Mol. Ther.* **2018**, *26*, 1509–1519. [CrossRef]
41. Hope, M.J.; Mui, B.; Lin, P.J.C.; Barbosa, C.; Madden, T.; Ansell, S.M.; Du, X. Lipid Nanoparticle. Formulations. Patent WO2018081480A1, 3 May 2018.
42. Heredero, J.; Peña, Á.; Broset, E.; Blandín, B.; de Miguel, D.; Alejo, T.; Toro, A.; Mata, E.; López-Gavín, A.; Gallego-Lleyda, A.; et al. Predictive Lung- and Spleen-Targeted mRNA Delivery with Biodegradable Ionizable Lipids in Four-Component LNPs. *Pharmaceutics* **2025**, *17*, 459. [CrossRef]

International Journal of *Molecular Sciences*

MDPI

Article

Chitosan-Coated Silver Nanocomposites: Biosynthesis, Mechanical Properties, and Ag^+ Release in Liquid and Biofilm Forms

Daniel Martínez-Cisterna [1,2,3,4], Lingyun Chen [5,*], Leonardo Bardehle [2,3,6], Edward Hermosilla [2,4], Gonzalo Tortella [2,4], Manuel Chacón-Fuentes [7] and Olga Rubilar [2,4,*]

1 Doctorado en Ciencias de Recursos Naturales, Facultad de Ciencias Químicas y Recursos Naturales, Universidad de La Frontera, Av. Francisco Salazar 01145, Casilla 54-D, Temuco 4811230, Chile; d.martinez11@ufromail.cl
2 Centro de Investigación Biotecnológica Aplicada al Medio Ambiente (CIBAMA), Universidad de La Frontera, Av. Francisco Salazar 01145, Casilla 54-D, Temuco 4811230, Chile; leonardo.bardehle@ufrontera.cl (L.B.); edward.hermosilla@ufrontera.cl (E.H.); gonzalo.tortella@ufrontera.cl (G.T.)
3 Laboratorio de Química Ecológica, Departamento de Ciencias Químicas y Recursos Naturales, Universidad de La Frontera, Av. Francisco Salazar 01145, Casilla 54-D, Temuco 4811230, Chile
4 Department of Chemical Engineering, Universidad de La Frontera, Av. Francisco Salazar 01145, Temuco 4811230, Chile
5 Department of Agriculture, Food and Nutritional Sciences, University of Alberta, Edmonton, AB T6G 2P5, Canada
6 Departamento de Producción Agropecuaria, Facultad de Ciencias Agropecuarias y Medioambiente, Universidad de La Frontera, Av. Francisco Salazar 01145, Casilla 54-D, Temuco 4811230, Chile
7 Agriaquaculture Nutritional Genomic Center, CGNA, Temuco 4780000, Chile; manuel.chacon@cgna.cl
* Correspondence: lingyun1@ualberta.ca (L.C.); olga.rubilar@ufrontera.cl (O.R.)

Abstract: This study explores the biosynthesis, characterization, and evaluation of silver nanoparticles coated with chitosan (AgChNPs) for liquid nanocomposite and biofilm formation in integrated pest management (IPM). AgChNPs were synthesized using *Galega officinalis* leaf extract as a reducing agent, with varying chitosan concentrations (0.5%, 1%, and 2%) and pH levels (3, 4, and 5). Synthesis was optimized based on nanoparticle size, stability, and polydispersity index (PDI) over 21 days. Biofilms incorporating AgChNPs were analyzed for chemical, physical, mechanical, and thermal properties via Ultraviolet-visible spectroscopy (UV-vis), Dynamic Light Scattering (DLS), Zeta Potential Analysis, Fourier Transform Infrared Spectroscopy (FTIR), X-Ray Diffraction (XRD), Transmission Electron Microscopy with Energy Dispersive X-ray Spectroscopy (TEM-EDX), and Inductively Coupled Plasma Optical Emission Spectroscopy (ICP-OES) to quantify silver ionization. TEM confirmed spherical nanoparticles (5.54–61.46 nm), and FTIR validated *G. officinalis* functionalization on chitosan. AgChNPs with 1% chitosan at pH 4 exhibited optimal properties: a size of 207.88 nm, a zeta potential of +42.30 mV, and a PDI of 0.62. Biofilms displayed tunable mechanical strength, with a tensile strength of 3.48 MPa using 5% glycerol and 2% chitosan and an elongation at break of 24.99 mm. TGA showed a two-step degradation process (98.19% mass loss). Ag ionization was 62.57 mg/L in the liquid nanocomposite and 184.07 mg/kg in the biofilms. These findings highlight AgChNPs' potential for controlled-release properties and enhanced mechanical performance, supporting sustainable agricultural applications.

Keywords: biosynthesized nanoparticles; chitosan nanocomposite; silver nanoparticles

1. Introduction

The current research area increasingly focuses on nanomaterials and nanostructures derived from physical, chemical, and biological sources, which play a pivotal role in advancing nanotechnology [1]. Metallic nanoparticles of various compositions have found widespread applications across industry, medicine, and agriculture [2–5]. Among these, silver nanoparticles (AgNPs) are the most extensively utilized nanomaterials, owing to their remarkable antibacterial, antiviral, antifungal, and insecticidal properties [6–10]. Their efficacy has been demonstrated in inducing mortality and developmental disruptions in various insect orders. For instance, AgNPs have shown larvicidal activity against *Spodoptera litura* (Lepidoptera: Noctuidae), strong adulticidal effects on *Callosobruchus maculatus* (Coleoptera: Chrysomelidae), and significant toxicity in *Musca domestica* (Diptera: Muscidae), while also reducing survival in aphids such as *Aphis nerii* (Hemiptera: Aphididae) [11–14]. These effects are primarily attributed to AgNP-induced oxidative stress caused by reactive oxygen species (ROS), membrane disruption, inhibition of key enzymatic systems, and interference with insect development and reproduction [15].

Traditionally, AgNPs are synthesized chemically through the reduction of silver salt solutions using reducing agents such as $NaBH_4$, citrate, or ascorbate. However, these agents are often associated with environmental toxicity and biological hazards [16]. To address these concerns, green synthesis methods have gained prominence, leveraging biological systems and natural sources to eliminate toxic agents and minimize chemical inputs [17]. Plant-based synthesis, in particular, employs extracts, peels, fruits, rhizomes, and leaves containing bioactive compounds such as flavonoids, alkaloids, terpenoids, and polyphenols, which act as both reducing and stabilizing agents, facilitating the formation of AgNPs [18]. In this study, AgNPs were synthesized biologically using *Galega officinalis*, a perennial herbaceous species from the Fabaceae family. This plant was selected for its high total polyphenol and flavonoid content, which enhances its potential for nanoparticle synthesis. This approach offers a cost-effective, scalable, and environmentally friendly alternative, minimizing energy consumption and reducing reliance on toxic substances [19,20].

Stabilizers such as polymers have been extensively utilized to immobilize metal nanoparticles within matrix polymers. These stabilizers act as matrix materials, facilitating the control of nanoparticle growth and ensuring their stabilization [21]. Among these, Ch and its derivatives are natural cationic biopolymers widely recognized for their bioactivity, biocompatibility, and biodegradability [22]. Low-molecular weight Ch has the advantage of producing smaller and more uniform particle sizes [23]. in particular, Ch contains hydroxyl (-OH) and amino ($-NH_2$) groups, which greatly stabilize the surface of AgNPs via electrostatic interactions to prevent their agglomeration [24]. Higher Ch concentrations enhance stability by providing greater coverage, forming a protective layer around AgNPs that controls silver ion release and ensures sustained ion liberation [25]. Therefore, the incorporation of Ch into a matrix of AgNPs has been shown to provide enhanced stability and reduced toxicity while mitigating oxidative corrosion [26,27].

Ch concentration and pH are critical parameters influencing AgNP stabilization and biofilm formation [28]. Moreover, slightly acidic pH conditions further promote denser biofilm formation by influencing Ch's surface charge and its interactions with nanoparticles [29,30]. Glycerol, as a plasticizer in Ch-AgNP matrices, improves biofilm malleability and uniformity while regulating water vapor permeability and physical properties, thereby enhancing their role as protective barriers in food applications [31,32]. Beyond these structural functions, AgChNPs have demonstrated notable insecticidal properties. For instance, AgChNPs showed potent larvicidal and adulticidal activity against *Anopheles stephensi*, *Aedes aegypti*, and *Culex quinquefasciatus* (Diptera: Culicidae), with LC_{50} values ranging from 9.67 to 12.96 ppm depending on the developmental stage [33]. Similarly, AgChNPs tested against *Drosophila suzukii* (Diptera: Drosophilidae) caused mortality rates of up to 96% at 1000 ppm, significantly reduced adult emergence, and induced morphological abnormalities such as deformed wings and cuticle demelanization [34].

The objective of this study was to synthesize AgNPs coated with Ch (AgChNPs) by optimizing pH and Ch concentration to produce a stable colloidal nanocomposite. This optimized liquid nanocomposite was subsequently used to fabricate AgChNPs biofilms (bf-AgChNPs), with pH, glycerol, and Ch concentration as key reaction parameters. The resulting AgChNP biofilms were evaluated in terms of mechanical properties, biodegradability, and Ag^+ release. These advancements aim to integrate AgChNP nanocomposites into sustainable agricultural strategies, particularly within the framework of the integrated pest management (IPM) approach.

2. Results and Discussion

2.1. Characterization of AgChNPs

2.1.1. Visual Observation

The reduction of silver nitrate (Ag^+) to metallic silver (Ag^0) by the reducing agent *G. officinalis* was confirmed by a color change from light yellow to dark brown upon adjusting the pH to 11. This transition indicates the formation of colloidal biosynthesized AgNPs, characterized by the excitation of surface plasmon (SPR) vibrations [35]. Additionally, the color changes observed in AgNPs upon Ch coating further confirmed the formation of the nanocomposite, with variations ranging from light yellow at pH 3, dark brown at pH 4, to light brown at pH 5.

2.1.2. UV-Vis Spectroscopy of AgChNPs

The characterization of AgChNPs using UV-vis spectroscopy was performed to identify the characteristic absorption peaks corresponding to the SPR of the AgChNP nanocomposite (Figure 1). The UV-vis spectrum of biosynthesized AgNPs without Ch exhibited a peak at 410 nm, consistent with findings by Zainab et al. [36], who reported a similar absorption peak for AgNPs synthesized via green methods using *Thermomyces lanuginosus*. For AgChNPs containing 0.5%, 1%, and 2% Ch, the nanocomposite displayed a peak at 420 nm, corresponding to the plasmon absorbance of AgNPs. The red shift observed in the AgNP peak following Ch treatment is attributed to an increase in the refractive index of the medium surrounding the metallic nanoparticles [37]. This observation aligns with reports by Rajeshkumar et al. [38], which identified SPR peaks in the range of 380–460 nm, confirming the successful formation of the AgChNP nanocomposite. Furthermore, Wulandari et al. [39] noted that SPR peaks within the 410–450 nm range are indicative of nanoparticles with a spherical morphology, although additional techniques such as TEM-EDX are required for definitive confirmation.

2.1.3. Dynamic Light Scattering (DLS) and Zeta Potential of AgChNPs

The Z-average particle size (nm) and polydispersity index (PDI) of AgChNPs were determined using DLS measurements, while zeta potential results were used to assess particle stability (Figure 2). AgNPs synthesized using the *Galega officinalis* extract had an average size of 53.79 nm, which increased proportionally with the concentration of Ch used in the treatment. The zeta potential of AgNPs was negative at −25.37 mV, while after all Ch treatments, the AgChNPs turned positive (22.8 to 49.2 mV). The size increase and the change from negative to positive zeta potential were similarly observed in Ch-capped AgNPs synthesized using fungal extracts [40]. In the initial measurement, AgChNPs at 0.5% and pH 3 exhibited the smallest particle size (202.75 nm), followed by AgChNPs at 1% and pH 3, with a size of 224.80 nm. The latter demonstrated the highest zeta potential (+49.20 mV), showing significant differences compared to the control in both size and stability. For PDI, biosynthesized AgChNPs at 0.5% and pH 4 presented the lowest value (0.31), though no significant differences in PDI were observed among treatments or the control.

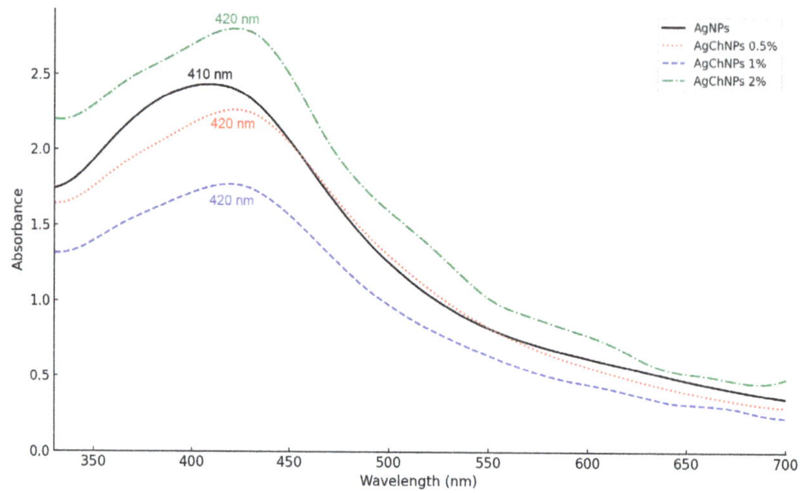

Figure 1. UV-Vis spectra of biosynthesized AgNPs (black), AgChNPs 0.5% (red), 1% (blue), 2% (green).

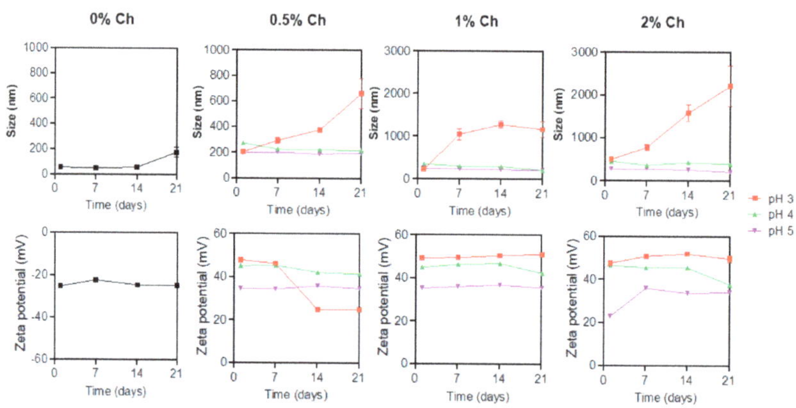

Figure 2. Changes in size and zeta potential of AgChNPs treated with 0.5 to 2% of Ch during 21 days of storage. Black line corresponds to AgNP 0% Ch treatment at pH 12 which was used as control.

Subsequent measurements taken on days 7, 14, and 21 revealed significant increases in size for some AgChNP treatments. AgChNPs at 1% Ch and pH 3 showed a considerable size increase, exceeding the nanometric scale (1–1000 nm) and reaching 1165.65 nm by day 21. Similarly, treatments at 0.5% and 2% Ch and pH 3 exhibited significant size increases by day 21, measuring 658.70 nm and 2209.23 nm, respectively, with the highest values showing significant differences compared to the control.

In contrast, other treatments demonstrated greater stability. AgChNPs at 1% Ch and pH 4 exhibited a decrease in size over time, reaching 207.88 nm on day 21, with values approaching those of the control, which consisted of biosynthesized AgNPs. The control size increased from 53.79 nm to 175.23 nm over the 21 days, with no significant differences observed between the control and AgChNPs at 1% Ch and pH 4. The smallest particle size was observed in AgChNPs at 0.5% Ch and pH 5, measuring 196.33 nm on day 21. However, this treatment exhibited a zeta potential value of +34.73 mV, which indicates good stability and reduced aggregation of the nanoparticles in the dispersion. Generally, a reduction in pH increases the zeta potential of AgChNPs due to the protonation of amino groups in the Ch molecules. Below the pKa of the amino groups (6.0–6.5), these groups become protonated, resulting in a positive charge on the nanoparticles [41].

Regarding zeta potential, after 21 days (Figure 2), AgChNPs at 1% and 2% Ch and pH 3 exhibited the highest values, with +51.03 mV and +39.57 mV, respectively. However, both treatments displayed particle sizes exceeding the nanometric scale (>1000 nm), indicating aggregation and low long-term stability of the AgNPs, which limits their suitability for certain applications. The most favorable treatment in terms of zeta potential and particle size was AgChNPs at 1% Ch and pH 4, with a zeta potential of +42.30 mV.

The effect of reaction parameters on the PDI is summarized in Table 1. After 21 days, the lowest PDI values were observed in AgChNPs at 0.5% and 1% Ch and pH 5, both with a PDI of 0.44. However, no significant differences were noted between these treatments and the control, indicating that the nanoparticles were monodispersed and free from aggregation. Based on the overall findings, the optimal conditions for the biosynthesis of AgChNPs were identified as 1% Ch at pH 4, yielding an average particle size of 207.88 nm, a zeta potential of +42.30 mV, and a PDI of 0.62. This nanocomposite was selected for further characterization via FTIR, XRD, and TEM-EDX analysis.

Table 1. Polydispersity index (PDI) of biosynthesized AgNPs and AgChNPs at 0.5%, 1% and 2%, Ch concentrations at different pH levels over 21 days.

Ch Concentration (%)	pH	PDI			
		Day 1	Day 7	Day 14	Day 21
0.5	3	0.53 ± 0.14	0.39 ± 0.04	0.27 ± 0.08	0.68 ± 0.10
	4	0.31 ± 0.02	0.43 ± 0.09	0.36 ± 0.05	0.45 ± 0.12
	5	0.42 ± 0.10	0.44 ± 0.11	0.43 ± 0.12	0.44 ± 0.11
1	3	0.48 ± 0.12	0.79 ± 0.10	0.69 ± 0.04	0.67 ± 0.15
	4	0.41 ± 0.05	0.39 ± 0.06	0.39 ± 0.06	0.62 ± 0.16
	5	0.38 ± 0.07	0.37 ± 0.06	0.46 ± 0.12	0.44 ± 0.12
2	3	0.46 ± 0.07	0.80 ± 0.07	0.84 ± 0.10	0.75 ± 0.11
	4	0.48 ± 0.09	0.54 ± 0.12	0.40 ± 0.05	0.41 ± 0.07
	5	0.52 ± 0.12	0.50 ± 0.12	0.49 ± 0.12	0.55 ± 0.15
0	12	0.66 ± 0.15	0.68 ± 0.14	0.71 ± 0.13	0.52 ± 0.01

The absence of letters is due to the lack of significant differences.

The optimization of the AgChNP biosynthesis process demonstrated that both pH and Ch concentration significantly influence particle size distribution and zeta potential. These findings are consistent with those of Karuppaiah et al. [42], who reported an increase in nanoparticle size after Ch coating (545 nm), resulting in monodisperse nanoparticles (PDI of 0.26) and a shift in zeta potential from negative (−36.3 mV) for biosynthesized AgNPs to positive (+37.5 mV) upon Ch addition. This shift suggests that the Ch shell imparts a positive surface charge, leading to a positive zeta potential. Similarly, Kanniah et al. [43] highlighted that biopolymer-stabilized metallic nanoparticles are more effective against pH and electrolyte fluctuations compared to chemical stabilizers.

Mbae and Umesha [44] further noted that the degree of deacetylation and concentration of Ch influence the number of ionizable groups on Ch chains coating the nanoparticles, contributing to higher positive zeta potential values (+26 mV). Ch adopts various conformations on the nanoparticle surface, enhancing stabilization. Barbosa et al. [45] reported that medium molecular weight Ch exhibits a zeta potential of approximately +20 mV at neutral pH. However, at lower pH levels (2–4) and concentrations above 0.3% (*w*/*v*), the acidic moieties of the polymeric chains become unionized, increasing hydrophobicity due to interactions with hydronium ions (H_3O^+), which accumulate at the interface between water and hydrophobic media.

This phenomenon aligns with the considerable size increases observed for AgChNPs at pH 3 in this study's DLS measurements. At higher Ch concentrations (0.5%, 1%, and 2%), the hydrophobic effect appears to diminish, likely due to structural and concentration-related changes in Ch conformations, which further influence nanoparticle stability and size.

2.1.4. Fourier-Transform Infrared Spectroscopy (FTIR) of AgChNPs

The FTIR analysis identified biomolecules in *G. officinalis* leaf extract and Ch that contribute to silver ion (Ag^+) reduction, nanoparticle stabilization, and interactions between Ch and biosynthesized AgNPs (Figure 3). The FTIR spectra of AgNPs without Ch exhibited characteristic absorption bands at 3397 cm^{-1} (-OH groups), 2956 cm^{-1} (amine salt stretching), and 2925 cm^{-1} and 2854 cm^{-1} (C-H stretching) [46–48]. Peaks at 1739 cm^{-1}, 1600 cm^{-1}, 1405 cm^{-1}, 1259 cm^{-1}, and 1072 cm^{-1} corresponded to C=O esters/acids, aromatic C=C, C-N aliphatic amines, and C-O stretching vibrations of alcohols [49–51].

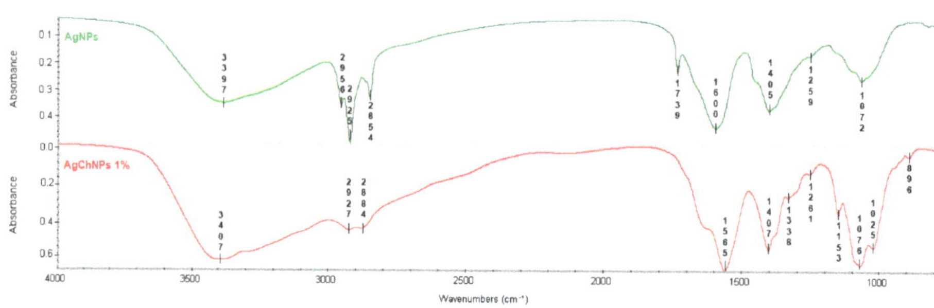

Figure 3. FTIR spectrum of biosynthesized AgNPs (green) and AgChNPs 1% pH 4 (red).

For AgChNPs, the FTIR spectrum revealed prominent bands at 3407 cm^{-1} (O-H/NH_2 stretching), 2956 cm^{-1} (C-H stretching), and 2927 cm^{-1} and 2884 cm^{-1} (CH_2 stretching) [46]. Peaks at 1565 cm^{-1} and 1338 cm^{-1}, along with bands between 1407–1025 cm^{-1}, corresponded to amide I (C=O stretching), amide III (C-N stretching), and CH_3 and C-O-C bending vibrations, indicating the presence of carbohydrate structures [52,53]. Additionally, the absorption band at 896 cm^{-1} confirmed the polysaccharide structure of Ch, consistent with findings by Malakhovska et al. [54], who identified C-H deformation bands as markers of polysaccharide-nanoparticle interactions.

A slight shift in the O-H/NH_2 band from 3396 cm^{-1} in AgNPs to 3407 cm^{-1} in AgChNPs suggests coordination between AgNPs and the functional groups of Ch, contributing to nanocomposite stabilization. This interaction highlights the role of the NH_2 group in metal ion chelation [55]. Moreover, distinct differences were observed in the spectral region between 1000 and 1100 cm^{-1}. While AgNPs exhibited a single peak at 1072 cm^{-1}, AgChNPs displayed two well-resolved bands at 1026 cm^{-1} and 1076 cm^{-1} [25]. These bands are typically attributed to the stretching vibrations of C–O–C and C–O bonds in polysaccharides such as Ch [56]. The presence of two distinct peaks in AgChNPs, compared to the single band in AgNPs, suggests a greater contribution of these vibrational modes due to the Ch coating, providing compelling evidence of its successful integration onto the nanoparticle surface [57]. This chemical modification reflects the formation of a Ch-nanoparticle interface [58]. Therefore, the spectral features between 1000 and 1100 cm^{-1} serve as reliable indicators of the presence of Ch on the AgNP surface, reinforcing its role in enhancing nanoparticle stability and dispersion within the polymer matrix [59].

2.1.5. X-Ray Diffraction (XRD) Analysis of AgChNPs

The XRD analysis was conducted to determine the crystalline structure of silver (Ag) in the biosynthesized nanocomposite. The XRD patterns revealed distinct peaks corresponding to silver crystals, confirming the presence of Ag^0 in the AgChNPs (Figure 4). Specifically, diffraction peaks were observed at 37,83°, 43,51°, 63,73°, and 76,86°, which correspond to the crystallographic planes (111), (200), (220), and (311) of the face-centered cubic (FCC) structure of metallic silver. A broad diffraction hump centered at $2\theta = 20°$ was also observed, which may be attributed to the amorphous scattering of Ch chains, consistent with previously reported patterns for semi-crystalline biopolymers [60,61].

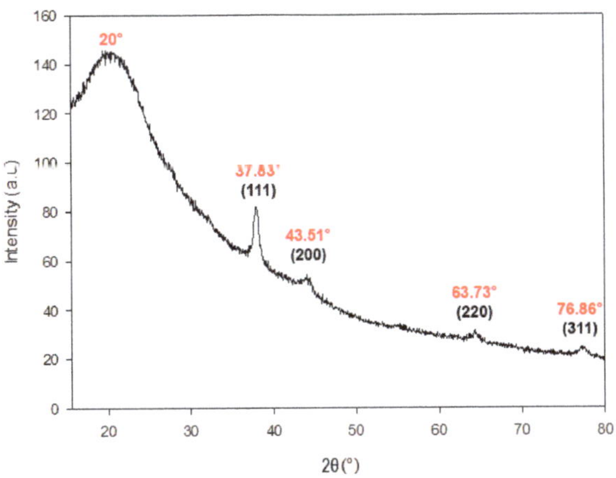

Figure 4. The XRD pattern of biosynthesized AgChNPs (1% Ch, pH 4). The XRD peaks correspond to the crystallographic planes (111), (200), (220), and (311) of the face-centered cubic (FCC) structure of metallic silver.

These results align with Naqvi et al. [62], who reported intense peaks at 38.26°, 44.37°, 65.41°, and 77.29°, indicating the FCC structure of AgNPs synthesized from *Juglans regia* extract. Similarly, Kumari et al. [63] observed peaks at 37.8°, 43.4°, 63.2°, and 76.6° in AgNPs synthesized from *Euphorbia royleana* extract, further supporting the FCC structure of biosynthesized AgNPs.

2.1.6. Transmission Electron Microscopy (TEM-EDX) Analysis of AgChNPs

TEM was employed to assess the morphology and size of biosynthesized AgNPs and AgChNPs. AgNPs reduced by *Galega officinalis* extract (Figure 5a) appeared predominantly spherical, with a narrow size range of 3.07–19.05 nm. These values differed from DLS measurements, which reported sizes between 53.79 and 175.23 nm over 21 days (PDI = 0.52), indicating moderate polydispersity in the colloidal suspension.

Similarly, AgChNPs exhibited spherical morphology in TEM (Figure 5b), with diameters ranging from 5.5 to 61.46 nm, while DLS indicated a larger hydrodynamic diameter of 207.88 nm and a higher PDI of 0.62, suggesting increased polydispersity. These differences reflect the methodological contrast between TEM and DLS. TEM measures dehydrated core sizes, whereas DLS accounts for solvation layers, surface-bound biomolecules, and particle aggregation in solution [64,65]. These findings are consistent with those of Azimi et al. [66], who observed ~25 nm spherical AgNPs using *G. officinalis* via TEM. Variations in synthesis conditions, such as extract preparation, Ag^+ concentration, pH, and temperature, likely explain the size differences between studies [67,68].

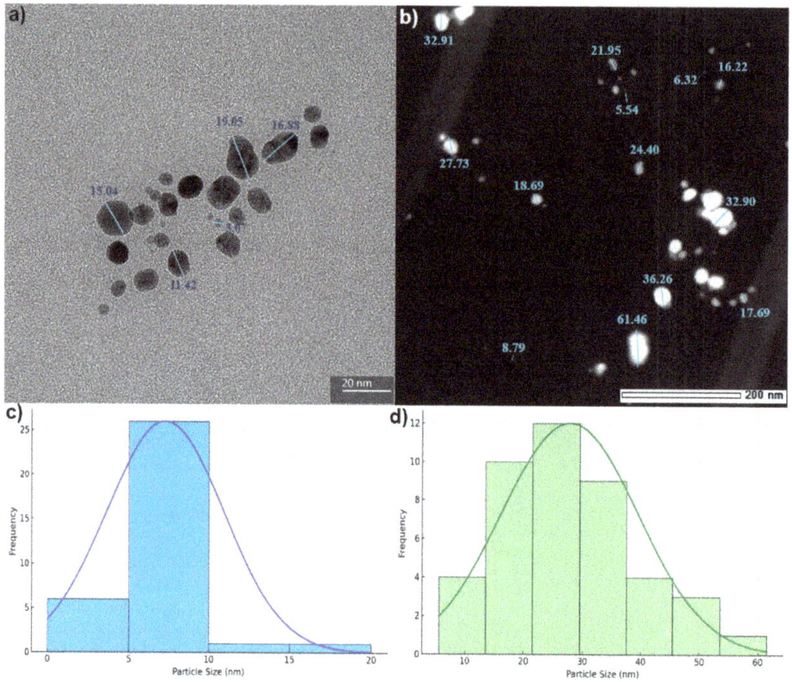

Figure 5. Transmission electron microscopy (TEM) characterization and particle size distribution of silver-based nanomaterials. (**a**) TEM image of biosynthesized AgNPs with sizes ranging from 3.07 to 19.05 nm; (**b**) TEM image of biosynthesized AgChNPs with sizes ranging from 5.54 to 61.46 nm; (**c**) Histogram of AgNP particle size distribution derived from TEM measurements; (**d**) Histogram of AgChNP particle size distribution.

Regarding AgChNPs, Martínez-Cisterna et al. [34] reported a similar trend, with TEM diameters ranging from 2.8 to 65.3 nm and DLS values up to 257.2 nm, highlighting nanoparticle aggregation in suspension. Affes et al. [69] showed that low-molecular weight Ch promotes the formation of ~5 nm AgNPs, supporting the stabilizing role of biopolymers. Wulandari et al. [39] also noted size ranges of 51–255 nm (for AgNPs) and 55–371 nm (for AgChNPs), with DLS consistently reporting larger sizes than SEM due to drying-induced clustering, while better reflecting dispersion behavior.

Energy Dispersive X-ray (EDX) spectroscopy was conducted to validate the elemental composition of the biosynthesized AgChNPs (Figure 6). A strong Ag signal was observed at 2.98 keV, corresponding to a silver mass percentage of 4.71%. This result is consistent with previous studies that report metallic silver nanocrystals exhibiting characteristic absorption spectra within the 2.4–4.0 keV range [70].

In addition to Ag, peaks for elements such as C, N, O, Na, Mg, P, Cl, and Ca were detected. These elements can be attributed to the phytochemicals present in the *G. officinalis* leaf extract and Ch composition. This finding aligns with Wang et al. [71], who also reported the presence of C and O in AgNPs synthesized using a Ch-*Mentha piperita* biocomposite. The detection of these additional elements suggests that *G. officinalis* extract serves a dual role as both a reducing and stabilizing agent during nanoparticle synthesis, enhancing the biocompatibility and functional diversity of AgChNPs.

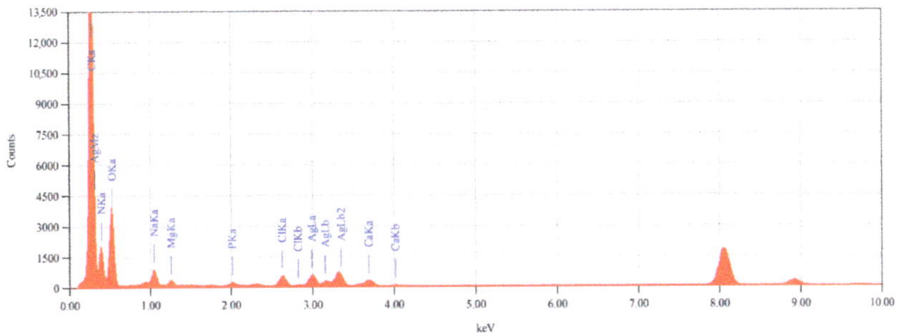

Element	(keV)	Mass%	Counts	Sigma	Atom%	Compound	Mass%	Cation	K
C K (Ref.)	0.277	77.60	98054.83	0.19	86.03				1.0000
N K !	0.392	4.77	8909.00	0.06	4.53				0.6760
O K !	0.525	8.35	20658.32	0.08	6.95				0.5110
Na K	1.041	0.97	4516.52	0.03	0.56				0.2708
Mg K	1.253	0.35	1801.33	0.02	0.19				0.2462
P K	2.013	0.27	1371.48	0.02	0.12				0.2524
Cl K	2.621	0.92	4352.17	0.03	0.35				0.2684
Ca K !	3.690	2.06	2442.71	0.08	0.68				1.0650
Ag L	2.984	4.71	6852.77	0.09	0.58				0.8679
Total		100.00			100.00				

Figure 6. Energy dispersive X ray (EDX) spectrum of biosynthesized AgChNPs (1% Ch, pH 4)

2.2. *Biosynthesis of Ch-Coated Silver Nanoparticle Biofilm (bf-AgChNPs)*

2.2.1. Physical, Structural, and Mechanical Properties Characterization Biofilms

Biofilm thickness measurements (Table 2) showed no significant differences among measurable treatments; however, treatments with low Ch concentrations (0.5% or none) and the control group exhibited insufficient structural integrity. These findings emphasize Ch's role in biofilm stability. For example, Santos et al. [37] observed that Ch concentrations below 1% result in fragile, discontinuous films due to weaker ionic bonding and intermolecular interactions. Similarly, Haghighi et al. [72] demonstrated that 1–2% Ch promotes the formation of robust three-dimensional matrices with cross-linked networks.

Table 2. Thickness, load at tensile strength and extension at break of AgChNP biofilms.

Biofilm Sample				Biofilm Characteristics		
Glycerol (%)	Ch (%)	pH	Assigned Label	Thickness (mm)	Load at Tensile Strength (N)	Extension at Break (mm)
5	0.5	4	bf-glychph-5-05-4	nbf	nbf	nbf
	0.5	5	bf-glychph-5-05-5	nbf	nbf	nbf
	1.0	4	bf-glychph-5-1-4	0.23 ± 0.04 ^{a}	0.80 ± 0.23 ^{a}	6.29 ± 1.54 ^{a}
	1.0	5	bf-glychph-5-1-5	0.28 ± 0.05 ^{ab}	1.41 ± 0.01 ^{ab}	16.05 ± 1.51 ^{ab}
	2.0	4	bf-glychph-5-2-4	0.23 ± 0.02 ^{a}	2.94 ± 0.50 ^{cd}	24.99 ± 3.62 ^{b}
	2.0	5	bf-glychph-5-2-5	0.25 ± 0.05 ^{ab}	3.48 ± 0.06 ^{d}	15.68 ± 1.86 ^{ab}
10	0.5	4	bf-glychph-10-05-4	nbf	nbf	nbf
	0.5	5	bf-glychph-10-05-5	nbf	nbf	nbf
	1.0	4	bf-glychph-10-1-4	0.25 ± 0.03 ^{ab}	0.71 ± 0.23 ^{a}	6.80 ± 1.32 ^{a}
	1.0	5	bf-glychph-10-1-5	0.23 ± 0.04 ^{a}	0.97 ± 0.12 ^{ab}	10.59 ± 2.31 ^{a}
	2.0	4	bf-glychph-10-2-4	0.50 ± 0.07 ^{ab}	2.16 ± 0.10 ^{bc}	11.82 ± 2.16 ^{a}
	2.0	5	bf-glychph-10-2-5	0.28 ± 0.01 ^{ab}	2.75 ± 0.38 ^{cd}	12.65 ± 2.67 ^{a}
15	0.5	4	bf-glychph-15-05-4	nbf	nbf	nbf
	0.5	4	bf-glychph-15-05-5	nbf	nbf	nbf
	1.0	4	bf-glychph-15-1-4	nbf	nbf	nbf
	1.0	5	bf-glychph-15-1-5	nbf	nbf	nbf
	2.0	4	bf-glychph-15-2-4	0.52 ± 0.12 ^{a}	1.09 ± 0.15 ^{ab}	12.73 ± 2.79 ^{a}
	2.0	5	bf-glychph-15-2-5	0.39 ± 0.05 ^{ab}	1.32 ± 0.25 ^{ab}	8.33 ± 097 ^{a}

Within each column, mean ± SD followed by the same letters do not differ significantly (ANOVA, Duncan HSD test, $p \geq 0.05$). nbf: indicates no biofilm formation.

Mechanical property evaluations (Table 2) revealed that the bf-glychph-5-02-05 treatment achieved the highest tensile strength (TS) at 3.48 MPa, while the bf-glychph-5-02-04 treatment exhibited the greatest elongation at break (EAB) at 24.99 mm. These results highlight the synergy between 5% glycerol and 2% Ch, which optimized both TS and EAB. This is consistent with the findings presented by Pava et al. [73], who reported that glycerol enhances EAB by reducing hydrogen bonding while maintaining a balance with TS.

The significant impact of Ch on TS is attributed to its ability to form strong intermolecular interactions [74]. Additionally, the incorporation of AgNPs reinforced the polymer matrix, enhancing mechanical properties and maintaining flexibility through uniform nanoparticle dispersion, contributing to biofilm stability and barrier properties, potentially extending biofilm shelf life [31,75].

2.2.2. Thermal Stability Analysis of AgChNP Biofilms

Thermogravimetric analysis (TGA) was performed to evaluate the thermal stability and decomposition behavior of biofilms containing 15% glycerol and 2% Ch at pH 4 (Figure 7). Two major mass loss events were observed from an initial sample mass of 11.50 mg. The first event, occurring between 40 °C and 190 °C, accounted for a 26.71% loss (−3.07 mg), likely due to the evaporation of residual moisture and volatile compounds. The second event, occurring between 190 °C and 280 °C, resulted in a further 70.94% loss (−8.16 mg), culminating in a total mass loss of 98.19% (−11.30 mg).

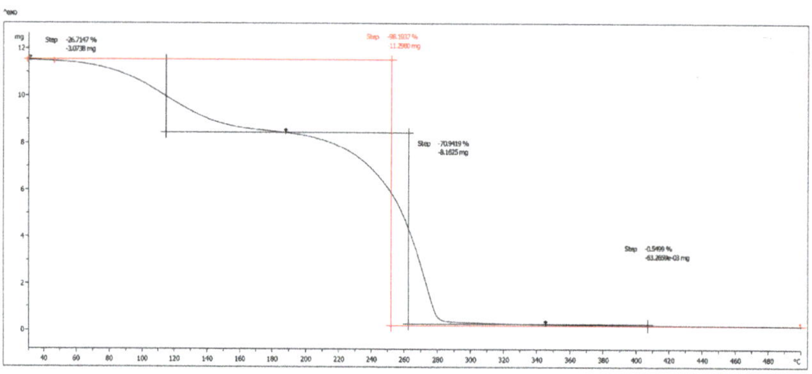

Figure 7. Thermogravimetric analysis (TGA) of AgChNP biofilm at 15% glycerol and 2% Ch concentrations and pH 4.

This two-step degradation pattern is characteristic of polymeric and composite materials, where moisture loss is followed by the thermal decomposition of the polymer matrix at elevated temperatures. Basavegowda and Baek [76] emphasized the environmental benefits and biodegradability of biopolymer-based nanocomposites compared to synthetic polymers. The significant mass loss observed above 240 °C in this study supports the predominance of biodegradable organic components within the biofilms.

These findings highlight the dual advantages of the evaluated nanocomposites: thermal stability suitable for targeted applications and environmentally friendly degradation, making them promising candidates for sustainable end-of-life applications.

2.2.3. Release of Ag^{+} from AgChNP Nanocomposites

The release of metal ions from nanomaterials refers to the process by which ions are liberated from the surface or core of nanoparticles into the surrounding medium. This typically occurs due to chemical or physical interactions, including surface oxidation, dissolution, or nanoparticle degradation [77]. In this study, the concentration of silver ions (Ag^+) released from AgChNP nanocomposites varied significantly depending on the pH level and Ch concentration (Figure 8). The control treatment (AgNPs-ctrl) exhibited the highest Ag^+ concentration in solution (118.60 mg/L), while the AgChNPs 2%, pH 4 treatment displayed the lowest concentration (62.57 mg/L), suggesting enhanced nanoparticle stabilization under high Ch content and acidic conditions.

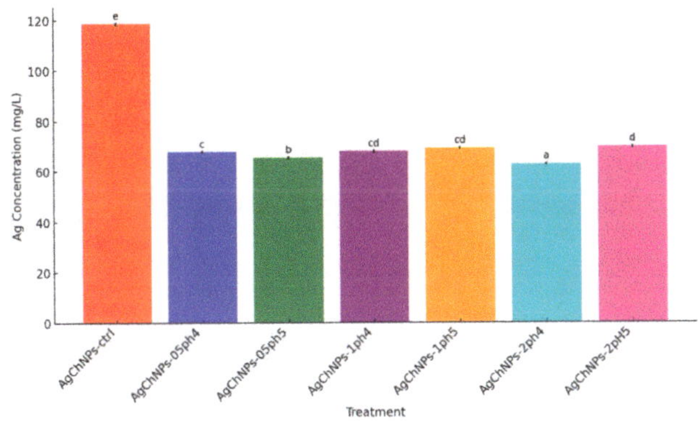

Figure 8. Silver concentration (mg/L) in biosynthesized AgChNP liquid nanocomposites. The figure displays the concentration of silver (Ag) in solution for different AgChNP treatments (0.5%, 1%, and 2% Ch, pH 3, 4 and 5). The letters indicated in the figure identifies statistically significant differences between treatment groups (Duncan's HSD test).

Intermediate Ag^+ levels were observed in AgChNPs 2% at pH 5 (69.60 mg/L), likely due to increased Ag solubility and weakened electrostatic interactions at higher pH values. Similarly, AgChNPs 0.5% at pH 4 and pH 5 showed Ag^+ concentrations of 67.63 mg/L and 65.42 mg/L, respectively, indicating that lower Ch concentrations provide moderate stabilization. Treatments containing 1% Ch exhibited minimal variation between pH conditions (67.79 mg/L at pH 4 and 69.09 mg/L at pH 5), supporting the role of Ch as a stabilizing agent across different environments. These observations are in agreement with findings by Nasef et al. [78], who reported that Ch protonation at acidic pH enhances electrostatic interactions, thereby reducing Ag^+ release.

In the case of biofilm nanocomposites (Figure 9), significant differences in Ag^+ release were also observed. For instance, the sample bf-glychph-5-1-4 exhibited the highest Ag^+ content (1403.00 mg/kg), while bf-glychph-5-2-4 released significantly less (184.07 mg/kg), likely due to pH-dependent interactions between Ch and AgNPs. These results are consistent with previous findings showing enhanced Ag^+ binding to protonated Ch at lower pH, which reduces ion diffusion [79]. Similarly, samples bf-glychph-10-2-4 and bf-glychph-15-2-4 released 304.03 mg/kg and 449.93 mg/kg of Ag^+, respectively, indicating that increasing Ch concentration contributes to matrix robustness and improved film-forming properties.

Compared to conventional AgNPs, which release Ag^+ due to surface oxidation and limited colloidal stability, AgChNPs exhibited a more controlled and sustained release profile under similar conditions. This behavior is primarily attributed to the Ch coating, which forms a semi-permeable polyelectrolyte matrix that modulates ion diffusion and enhances particle stability [80]. The pH-responsive release observed supports the role of protonated Ch in strengthening electrostatic interactions with AgNPs, particularly under acidic conditions, thereby reducing Ag^+ leaching [78]. Furthermore, the Ch matrix acts both as a physical barrier and as an active chelating agent, with amino ($-NH_2$) and hydroxyl (–OH) groups capable of coordinating Ag^+ ions and limiting their mobility [79]. This mechanism is especially relevant in biofilm formulations, where Ch concentrations above 2% enhance structural cohesion and reduce early-phase ion release. Similar findings have been reported in other AgChNP systems, where reduced Ag^{+} release was linked to strong ion binding and slow diffusion through a compact matrix [81].

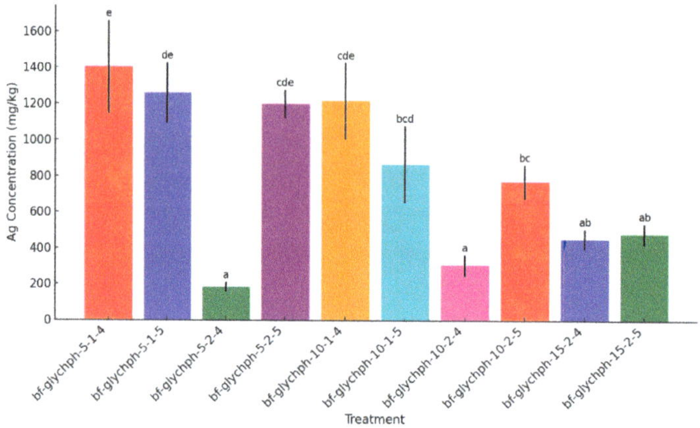

Figure 9. Silver (Ag) concentrations in different AgChNP nanocomposite biofilms. The figure displays the concentration of silver (Ag) in AgChNP biofilm treatments (0.5%, 1%, and 2% Ch, 5%, 10%, 15% glycerol, pH 3, 4 and 5). The letters indicated in the figure identifies statistically significant differences between treatment groups (Duncan's HSD test).

Finally, Ag^{+} release occurred under all tested pH conditions, with the lowest concentrations detected at pH 4, indicating enhanced nanoparticle stability in acidic environments. These features collectively position AgChNPs as promising candidates for pH-sensitive applications, including crop protection films.

3. Materials and Methods

3.1. Plant Leaves and Chemical Obtention

Dried *Galega officinalis* leaves were sourced from Botanic Universe (Burlington, ON, Canada) and processed into small, sifted pieces. Silver nitrate ($AgNO_3$) and low-molecular weight Ch (50–190 kDa; deacetylation degree: > 75%) were purchased from Sigma-Aldrich (Canada Co, Oakville, ON, Canada).

3.2. Biosynthesis of AgNPs (AgNPs)

The AgNPs were synthesized following the methodology of Manosalva et al. [19]. To prepare the extract, 10 g of dried *G. officinalis* leaves were placed in a 250 mL Erlenmeyer flask with 100 mL of ultrapure water (milli-q) and boiled at 100 °C for 5 min. The extract was then cooled to room temperature (25 ± 2 °C), diluted with 10% (v/v) ultrapure water, and filtered using a vacuum pump with Whatman No. 1 filter paper.

A concentrated $AgNO_3$ stock solution (100 mM) was added to the *G. officinalis* extract to achieve a final silver ion concentration of 1.5 mM, and the pH was adjusted to 11 using 1 M NaOH. The reaction was maintained at room temperature (25 ± 2 °C) under continuous stirring for 24 h on a magnetic stirrer. The resulting sample was frozen at −80 °C and lyophilized for characterization and bioassays.

The same AgNP synthesis parameters were consistently applied across all assays.

3.3. Biosynthesis of Ch-Coated AgNPs (AgChNPs)

The biosynthesized AgNPs were treated with Ch following a modified methodology from Wulandari et al. [39]. Briefly, AgNP dispersion and Ch solutions were mixed in a 1:1 volume ratio, with Ch concentrations of 0.5%, 1%, and 2% (prepared in 2% acetic acid). The mixtures were subjected to varying pH conditions (3, 4, and 5), which were adjusted using NaOH and HCl. The resulting AgChNPs were homogenized on a magnetic stirrer at 800 rpm for 24 h. AgNPs without Ch were used as the control.

3.4. Characterization of Ch-Coated AgNPs (AgChNPs)

3.4.1. Dynamic Light Scattering (DLS) and Zeta Potential

The particle size distribution of the AgChNPs was determined using a Zetasizer Nano Zs (Malvern Instruments Ltd., Malvern, UK). Measurements included the average hydrodynamic diameter (nm), PDI, and zeta potential (mV). These parameters were evaluated at 7-day intervals over a 21-day period to assess the size, polydispersity, and stability of the nanoparticles over time. The nanocomposite exhibiting optimal stability and uniformity was selected for further characterization using FTIR, XRD, and TEM-EDX analysis.

3.4.2. UV-Vis Spectroscopy

The formation of biosynthesized AgNPs and their incorporation into AgChNPs were monitored using a SpectraMax UV-vis spectrophotometer (Molecular Devices, San Jose, CA, USA). Absorption spectra were recorded in the range of 300 to 800 nm, with ultrapure water serving as the blank.

3.4.3. FTIR Spectroscopy

The AgChNPs were characterized using FTIR spectroscopy to identify functional groups before and after Ch coating. Spectra were recorded in the range of 500–4000 cm^{-1} using a Nicolet 6700 FT-IR spectrometer (Thermo Fisher Scientific, Waltham, MA, USA).

3.4.4. XRD Analysis

The crystalline structure of the biosynthesized AgChNPs was analyzed using XRD. Measurements were performed with a Bruker D8 Advance X-ray diffractometer (Bruker AXS GmbH, Karlsruhe, Germany) equipped with Cu-Kα radiation (K = 1.5406 Å) at 30 kV and 40 mA, over a 2θ range of 20–100°.

3.4.5. Transmission Electron Microscopy Coupled with Energy Dispersive X-Ray Spectroscopy (TEM-EDX)

Transmission electron microscopy coupled with energy dispersive X-ray spectroscopy (TEM-EDX) was performed using a JEOL JEM-ARM200CF S/TEM equipped with an EDX system (JEOL Ltd., Tokyo, Japan) to confirm the formation, size, and morphology of AgChNPs, as well as to determine the elemental composition and mass percentage within the nanocomposite.

3.5. Biosynthesis of Ch-Coated Silver Nanoparticle Biofilm

The biofilms were prepared using AgChNPs treated with varying Ch concentrations and adjusted to pH levels ranging from 3 to 5, as described in Activity 2.3. These were blended with glycerol at concentrations of 5%, 10%, and 15% (v/v) as a plasticizing agent, following an adaptation of the methodology by Affes et al. [69]. The mixtures were stirred on a magnetic stirrer for 30 min to ensure homogeneity. The resulting biofilm mixtures were then poured into plastic petri dishes (5.2 cm in diameter) and dried at room temperature (25 ± 2 °C) for 48 h to allow complete solvent evaporation. The dried films were carefully peeled from the dish surfaces and stored at 25 °C with a relative humidity of 50%.

3.6. Physical and Structural Characterization of Ch-Coated Silver Nanoparticle Biofilm

3.6.1. Films Thickness

The thickness of the films was measured using a Vogel high-precision external micrometer (DIN 863). Measurements were taken at three random locations on each film sample, and the average thickness was calculated and used for further analysis.

3.6.2. Mechanical Properties

The mechanical properties of nanocomposite biofilms, including tensile strength (TS, MPa) and elongation at break (EAB, %), were assessed using an Instron testing instrument. Rectangular film samples (1.0 × 4.0 cm) were conditioned at 25 °C and 50% RH for seven days prior to testing. Measurements were conducted at a deformation rate of 5 mm/min in accordance with ISO standards (ISO, 1995). Film thickness was measured, and tests were performed at 25 ± 2 °C, with three replicates per formulation.

3.7. Thermal Stability Analysis

The biodegradability of the biofilms was assessed using thermogravimetric analysis (TGA) on a TA Discovery TGA instrument. Samples were heated from 30 °C to 500 °C at a rate of 20 °C/min under a nitrogen atmosphere. Weight loss (Δw), the temperature at the maximum degradation rate (Tmax), and final residue values were evaluated using STARe SW 16.10 software. The analysis focused on AgChNPs formulated with 15% glycerol, 2% Ch, and pH 4, which were selected for their enhanced stability.

3.8. Inductively Coupled Plasma Atomic of Biofilm (ICP-OES)

The Ag^+ concentration in both liquid and biofilm nanocomposites, before and after Ch coating, was quantified using an inductively coupled plasma-optical emission spectrometer (ICP-OES, Thermo iCAP6300 Duo, Thermo Fisher Scientific, Waltham, MA, USA). Samples were atomized in an argon plasma (5500–8000 K), and analyte emissions were detected using yttrium (Y) as an internal standard. Certified multi-element standard solutions ensured calibration, and external reference standards validated the results. Uncoated AgNPs served as controls to establish baseline silver ion release. The method adhered to EPA 200.7 and EPA 6010d standards, ensuring precision and reproducibility.

3.9. Statistical Analysis

All experiments were conducted in triplicate, and the mean values are reported. Statistical analyses were performed using one-way analysis of variance (ANOVA) in SPSS software (version 20.0). Post hoc comparisons among group means were conducted using Duncan's HSD test, with a significance level set at $p < 0.05$.

4. Conclusions

In conclusion, the present study demonstrates that AgChNPs exhibit controlled and pH-dependent Ag^+ ion release, which is enhanced by higher Ch concentrations and acidic conditions. These properties confer distinct advantages over conventional AgNPs, including improved colloidal stability and sustained release kinetics, which are modulated by both matrix density and environmental pH. The dual capacity of Ch to form a mechanical diffusion barrier and to chelate silver ions explains the significant reduction in Ag^+ liberation observed in both colloidal and biofilm formulations.

The incorporation of Ch also significantly improved mechanical properties, such as tensile strength and elongation at break, while enabling controlled silver ion release. In addition, thermogravimetric analysis (TGA) confirmed the thermal stability and biodegradability of these nanocomposites, supporting their suitability for environmentally sustainable applications.

Beyond physicochemical performance, AgChNPs exhibit considerable biological potential. Recent studies have highlighted the efficacy of AgNPs as nanopesticides, with documented toxicity against agriculturally important insect orders. These nanomaterials act through multiple mechanisms, including disruption of enzymatic systems, induction of oxidative stress, and interference with insect development, showing efficacy levels comparable to commercial pesticides. In this context, AgChNPs represent a promising alternative for agricultural implementation, where their pH-responsive release behavior could be harnessed in crop protection matrices (e.g., films, sprays, or coatings) that ensure targeted delivery and reduced environmental impact as part of IPM strategies. Future research should focus on validating these systems under field conditions and evaluating their interactions with crops, pests, and non-target organisms to fully elucidate their benefits and potential ecological risks.

Author Contributions: Conceptualization, D.M.-C., L.C., L.B., E.H., G.T., M.C.-F. and O.R.; software, D.M.-C., L.C., L.B. and O.R. validation, D.M-C., L.C., L.B. and O.R.; resources, D.M-C., L.C., L.B. and O.R.; data curation, D.M.-C., L.C., L.B. and O.R.; writing—original draft preparation, D.M.-C., L.C., L.B., E.H., G.T., M.C.-F. and O.R.; writing—review and editing, D.M.-C., L.C., L.B., E.H., G.T., M.C.-F. and O.R.; visualization, D.M.-C., L.C., L.B., E.H., G.T., M.C.-F. and O.R.; supervision, D.M.-C., L.C., L.B. and O.R.; project administration, D.M.-C., L.C., L.B. and O.R.; funding acquisition, D.M.-C., L.C., L.B. and O.R. All authors have read and agreed to the published version of the manuscript.

Funding: This research was funded by grants FONDECYT 1191089, ANID/FONDAP/1523A0001, ANID 21232172, DIUFRO DI19-0082, UFRO scholarship, the Emerging Leaders in the Americas Program (ELAP), and EduCanada.

Institutional Review Board Statement: Not applicable.

Informed Consent Statement: Not applicable.

Data Availability Statement: Data are contained within the article.

Conflicts of Interest: The authors declare no conflicts of interest. The funders had no role in the design of this study; the collection, analysis, or interpretation of data; the writing of the manuscript; or the decision to publish the results.

References

1. Subramanyam, G.K.; Gaddam, S.A.; Kotakadi, V.S.; Venkata, R.K.C.; Kandasamy, S. Green fabrication of silver nanoparticles by leaf extract of *Byttneria herbacea* Roxb and their promising therapeutic applications and its interesting insightful observations in oral cancer. *Artif. Cells Nanomed Biotechnol.* **2023**, *51*, 83–94. [CrossRef] [PubMed]
2. Smirnov, O.; Kalynovskyi, V.; Zelena, P.; Ivanchenko, S.; Bondarenko, I. Bactericidal activity of Ag nanoparticles biosynthesized from *Capsicum annuum* pericarps against phytopathogenic *Clavibacter michiganensis*. *Sci. Nat.* **2023**, *110*, 15. [CrossRef]
3. Maharkan, A.; Gautam, R.; Lee, G.; Kim, D.; Lee, D.; Acharya, M.; Kim, H.; Heo, Y.; Kim, C. Assessment of skin sensitization potential of zinc oxide, aluminum oxide, manganese oxide, and copper oxide nanoparticles through the local lymph node assay: 5-bromo-deoxyuridine flow cytometry method. *J. Toxicol. Environ. Health A Curr. Issues.* **2025**, *88*, 95–105. [CrossRef] [PubMed]
4. Jiang, Y.Q.; Zhou, P.F.; Zhang, P.; Adeel, M.; Shakoor, N.; Li, Y.B.; Li, M.S.; Guo, M.L.; Zhao, W.C.; Lou, B.Z.; et al. Green synthesis of metal-based nanoparticles for sustainable agriculture. *Environ. Pollut.* **2022**, *309*, 119755. [CrossRef] [PubMed]
5. Meher, A.; Tandi, A.; Moharana, S.; Chakroborty, S.; Mohapatra, S.; Mondal, A.; Dey, S.; Chandra, P. Silver nanoparticle for biomedical applications: A review. *Hybrid Adv.* **2024**, *6*, 100184. [CrossRef]
6. Hamad, A.; Khashan, K.S.; Hadi, A. Silver nanoparticles and silver ions as potential antibacterial agents. *J. Inorg. Organomet. Polym. Mater.* **2020**, *30*, 4811–4828. [CrossRef]
7. Luceri, A.; Francese, R.; Lembo, D.; Ferraris, M.; Balagna, C. Silver nanoparticles.: Review of antiviral properties, mechanism of action and applications. *Microorganisms* **2023**, *11*, 629. [CrossRef]
8. Mussin, K.; Giusiano, G. Biogenic silver nanoparticles as antifungal agents. *Front. Chem.* **2022**, *10*, 1023542. [CrossRef]

9. Wilson, J.J.; Mahendran, S.; Sivakumar, T.; Ponmanickam, P.; Thangaraj, R. Mosquito larvicidal activity of silver nanoparticles synthesized using *Azolla pinnata* against *Culex quinquefasciatus* say (Diptera: Culicidae). *S. Afr. J.* **2023**, *157*, 380–386. [CrossRef]
10. Qu, R.; Chen, M.; Liu, J.; Yu, H.; Liang, J. Blockage of ATPase-mediated energy supply inducing metabolic disturbances in algal cells under silver nanoparticles stress. *J. Environ. Sci.* **2023**, *141*, 141–150. [CrossRef]
11. Manimegalai, T.; Raguvaran, K.; Kalpana, M.; Maheswaran, R. Green synthesis of silver nanoparticle using Leonotis nepetifolia and their toxicity against vector mosquitoes of Aedes aegypti and Culex quinquefasciatus and agricultural pests of Spodoptera litura and Helicoverpa armigera. *Environ. Sci. Pollut. Res.* **2020**, *27*, 43103–43116. [CrossRef] [PubMed]
12. Rouhani, M.; Samih, M.A.; Kalantari, S. Insecticidal effect of silica and silver nanoparticles on the cowpea seed beetle, *Callosobruchus maculatus* F. (Col.: Bruchidae). *J. Entomol. Res.* **2012**, *4*, 297–305. Available online: https://sid.ir/paper/201167/en (accessed on 20 January 2025).
13. Kamaraj, C.; Rajakumar, G.; Rahuman, A.A. Feeding deterrent activity of synthesized silver nanoparticles using *Manilkara zapota* leaf extract against the house fly, *Musca domestica* (Diptera: Muscidae). *Parasitol. Res.* **2012**, *111*, 2439–2448. [CrossRef] [PubMed]
14. Rouhani, M.; Samih, M.A.; Kalantari, S. Insecticide effect of silver and zinc nanoparticles against Aphis nerii Boye de fonscolombe (Hemiptera: Aphididae). *Chil. J. Agric. Res.* **2012**, *72*, 590–594. [CrossRef]
15. Martínez-Cisterna, D.; Rubilar, O.; Tortela, G.; Chen, L.; Chacón-Fuentes, M.; Lizama, M.; Parra, P.; Bardehle, L. Silver nanoparticles as a Potent Nanopesticide: Toxic Effects and Action Mechanisms on Pest Insects of Agricultural Importance—A review. *Molecules* **2024**, *29*, 5520. [CrossRef]
16. Xu, L.; Wang, Y.; Huang, J.; Chen, C.; Wang, H. Silver nanoparticles: Synthesis, medical applications and biosafety. *Theranostics* **2020**, *10*, 8996–9031. [CrossRef]
17. Abdoon, F.; Hasan, H.; Salman, S.; Kareem, F.; Taha, R. Exploiting of green synthesized silver nanoparticles using *Capparis spinosa* L. fruit for spectrophotometric determination of diphenhydramine HCl in pure forms and commercial products. *J. Exp. Nanosci.* **2023**, *18*, 2161525. [CrossRef]
18. Akyüz, G.; Kaymazlar, E.; Ay, H.; Inan, T.; Yildiz, M. Use of silver nanoparticles loaded locust bean gum coatings to extend the shelf-life of fruits. *Biointerface Res. Appl. Chem.* **2023**, *13*, 289. [CrossRef]
19. Manosalva, N.; Tortella, G.; Diez, M.C.; Rubilar, O.; Schalchli, H.; Seabra, A.B. Green synthesis of silver nanoparticles: Effect of synthesis reaction parameters on antimicrobial activity. *World J. Microbiol. Biotechnol.* **2019**, *35*, 88. [CrossRef]
20. Mishra, V.K.; Husen, A.; Rahman, Q. Plant-based fabrication of silver nanoparticles and their application. *Nanomater. Plant Potential* **2019**, *5*, 135–175. [CrossRef]
21. Sharma, A.; Kaur, N.; Nanda, A.; Kaur, M.; Sharma, R.; Sohal, H.S. Recent advances in synthesis and bio-applications of natural stabilizers for metal nanoparticles. *JMMM* **2024**, *34*, 2145. [CrossRef]
22. Hudek, M.; Johnston, K.; Kubiak-Ossowska, K.; Ferro, V.; Mulheran, P. Molecular dynamics study of chitosan adsorption at a silica surface. *J. Phys. Chem. C.* **2024**, *128*, 21531–21538. [CrossRef] [PubMed]
23. Peng, Y.; Song, C.; Yang, C.; Guo, Q.; Yao, N. Low molecular weight chitosan-coated silver nanoparticles are effective for the treatment of MRSA-infected wounds. *Int. J. Nanomed.* **2017**, *112*, 177–188. [CrossRef] [PubMed]
24. Jannoo, K.; Teerapatsakul, C.; Punyanut, A.; Pasanphan, W. Electron beam assisted synthesis of silver nanoparticle in chitosan stabilizer: Preparation, stability and inhibition of building fungi studies. *Radiat. Phys. Chem.* **2015**, *112*, 177–188. [CrossRef]
25. Shinde, S.; Folliero, V.; Chianese, A.; Zannella, C.; De Filippis, A.; Rosati, L.; Franci, G.; Galdiero, M.; Galdiero, M. Synthesis of chitosan-coated silver nanoparticle bioconjugates and their antimicrobial activity against multidrug-resistant bacteria. *Appl. Sci.* **2021**, *11*, 9340. [CrossRef]
26. Poveda, J. Quitina: Compuesto natural y práctico. *Mater. Investig. Joven.* **2018**, *5*, 1–4. Available online: https://revistas.unlp.edu.ar/InvJov/article/view/4308 (accessed on 20 January 2025).
27. González-López, D.; Bustos-Martínez, J.; Hamdam, A.; Schettino-Bermúdez, B.; Chamorro-Ramírez, F. Physical-mechanical and functional properties of Chitosan-based films incorporated with silver nanoparticles. *Mundo Nano* **2024**, *17*, 1e–18e. [CrossRef]
28. Sivera, M.; Kvitek, L.; Soukupova, J.; Panacek, A.; Prucek, R.; Vecerova, R.; Zboril, R. Silver nanoparticles modified by gelatin with extraordinary pH stability and long-term antibacterial activity. *PLoS ONE* **2014**, *9*, e103675. [CrossRef]
29. Farhadi, L.; Mohtashami, M.; Saeidi, J.; Azimi-Nezhad, M.; Taheri, G. Green synthesis of chitosan-coated silver nanoparticle, characterization, antimicrobial activities, and cytotoxicity analysis in cancerous and normal cell lines. *J. Inorg. Organomet. Polym. Mater.* **2022**, *32*, 1637–1649. [CrossRef]
30. Fan, M.; Si, J.; Xu, X.; Chen, L.; Chen, J.; Yang, C.; Zhu, J.W.; Wu, L.; Tian, J.; Chen, X. A versatile chitosan nanogel capable of generating AgNPs in-situ and long-acting slow-release of Ag^+ for highly efficient antibacterial. *Carbohydr. Polym.* **2021**, *257*, 117636. [CrossRef]
31. Samadi, A.; Azandeh, S.; Orazizadeh, M.; Bayati, V.; Rafienia, M.; Karami, M. Fabrication and characterization of glycerol/chitosan/polyvinyl alcohol-based transparent hydrogel films loaded with silver nanoparticles for antibacterial wound dressing applications. *Adv. Biomed. Res.* **2021**, *10*, 4. [CrossRef] [PubMed]

32. Ediyilyam, S.; Lalitha, M.; George, B.; Shankar, S.S.; Waclawek, S.; Cernik, M.; Padil, V. Synthesis, Characterization and Physicochemical properties of Biogenic silver nanoparticle-encapsulated chitosan bionanocomposites. *Polymers* **2022**, *14*, 463. [CrossRef] [PubMed]
33. Mohammed, A.; Al Thabiani, A.; Subrata, T.; Chellasamy, P. Efficacy of chitosan silver nanoparticles from shrimp-shell wastes against major mosquito vectors of public health importance. *Green. Process Synth.* **2020**, *9*, 675–684. [CrossRef]
34. Martínez-Cisterna, D.; Rubilar, O.; Chen, L.; Lizama, M.; Chacón-Fuentes, M.; Quiroz, A.; Parra, P.; Rebolledo, R.; Bardehle, L. Biosynthesized Chitosan-Coated Silver Nanoparticles: Insecticide Activity and Sublethal Effects Against *Drosophila suzukii* (Diptera: Drosophilidae). *Biomolecules* **2025**, *15*, 490. [CrossRef]
35. Suresh, U.; Murugan, K.; Panneerselvam, C.; Madhiyazhagan, P.; Subramaniam, J. *Suaeda maritima*-based herbal coils and green nanoparticles as potential biopesticides against the dengue vector *Aedes aegypti* and the tobacco cutworm *Spodoptera litura*. *Physiol. Mol. Plant Pathol.* **2017**, *101*, 225–235. [CrossRef]
36. Zainab, S.; Jadoon, M.; Sikandar, S.; Hussain, A.; Rehman, R. *Thermomyces lanuginosus*: A prospective thermophilic fungus for green synthesis and stabilization of BioAgNPs through glucoamylase. *Materials* **2023**, *297*, 127442. [CrossRef]
37. Santos, A.F.; Reis, C.A.; Lima, M.S.; Silva, P.A. Effects of low concentrations of chitosan on biofilm formation and mechanical integrity in various pH conditions. *Int. J. Biol. Macromol.* **2019**, *135*, 495–503. [CrossRef]
38. Rajeshkumar, S.; Tharani, M.; Rajeswari, V.; Sivakavinesan, M. Synthesis of greener silver nanoparticle-based chitosan nanocomposites and their potential antimicrobial activity against oral pathogens. *Green Process. Synth.* **2021**, *10*, 658–665. [CrossRef]
39. Wulandari, I.; Pebriatin, B.; Valiana, V.; Indriati, N.; Pratiwi, R. Green synthesis of silver nanoparticles coated by water-soluble chitosan and its potency as non-alcoholic hand sanitizer formulation. *Materials* **2022**, *15*, 4641. [CrossRef]
40. Hermosilla, E.; Diaz, M.; Vera, J.; Contreras, M.J.; Leal, K.; Salazar, R.; Barrientos, L.; Tortella, G.; Rubilar, O. Synthesis of antimicrobial Chitosan-silver nanoparticles mediated by reusable Chitosan fungal beads. *Int. J. Mol. Sci.* **2023**, *24*, 2318. [CrossRef]
41. Kong, M.; Chen, X.; King, K.; Park, H. Antimicrobial properties of chitosan and mode of action: A state of the art review. *Int. J. Food Microbiol.* **2010**, *144*, 51–63. [CrossRef] [PubMed]
42. Karuppaiah, A.; Babu, D.; Servaraj, D.; Ramalingam, G.; Muthusamy, A.; Saravanan, A. Building and behavior of a pH-stimuli responsive chitosan nanoparticles loaded with folic acid conjugated gemcitabine silver colloids in MDA-MB-453 metastatic breast cancer cell line and pharmacokinetics in rats. *Eur. J. Pharm. Sci.* **2021**, *165*, 105938. [CrossRef] [PubMed]
43. Kanniah, P.; Chelliah, P.; Thangapandi, J.; Amalraj, A.; Gopi, S. Green synthesis of antibacterial and cytotoxic silver nanoparticles by *Piper nigrum* seed extract and development of antibacterial silver-based chitosan nanocomposite. *Int. J. Biol. Macromol.* **2021**, *189*, 18–33. [CrossRef] [PubMed]
44. Mbae, K.; Umesha, S. Physicochemical and antimicrobial properties of post-synthesis betanin and chitosan oligosaccharide functionalized silver nanoparticles. *J. Nanopart. Res.* **2020**, *22*, 346. [CrossRef]
45. Barbosa, J.; Abdelsadig, M.; Conway, B.; Batchelor, H. Using zeta potential to study the ionisation behaviour of polymers employed in modified-release dosage forms and estimating their pKa. *Int. J. Pharm.* **2019**, *561*, 100024. [CrossRef]
46. Chowdhury, M.; Hossain, N.; Mostofa, M.D.; Roy, M. Green synthesis and characterization of zirconium nanoparticle for dental implant applications. *Heliyon* **2023**, *9*, e12711. [CrossRef]
47. De Matteis, V.; Cascione, M.; Costa, D.; Rinaldi, R. Aloe vera silver nanoparticles addition in chitosan films: Improvement of physicochemical properties for eco-friendly food packaging material. *J. Mater. Res. Technol.* **2023**, *24*, 1015–1033. [CrossRef]
48. Atrooz, O.; Al-Nadaf, A.; Uysal, H.; El-Gendy, M. Biosynthesis of silver nanoparticles using *Coriandrum sativum* L. extract and evaluation of their antibacterial, anti-inflammatory and antinociceptive activities. *S. Afr. J. Bot.* **2023**, *158*, 219–227. [CrossRef]
49. Pavunraj, M.; Baskar, K.; Duraipandiyan, V.; Al-Dhabi, N.A.; Ignacimuthu, S. Toxicity of Ag nanoparticles synthesized using stearic acid from *Catharanthus roseus* leaf extract against *Earias vittela* and mosquito vectors (*Culex quinquefasciatus* and *Aedes aegypti*). *J. Clust. Sci.* **2017**, *28*, 2477–2492. [CrossRef]
50. Muthukumaran, U.; Govindarajan, M.; Rajeswary, M.; Hoti, S.L. Synthesis and characterization of silver nanoparticles using *Gmelina asiatica* leaf extract against filariasis, dengue, and malaria vector mosquitoes. *Parasitol. Res.* **2015**, *114*, 1817–1827. [CrossRef]
51. Ghramh, H.; Al-Ghamdi, A.; Mahyoub, J.; Mohammed, M.G.; Al-Qahtani, A.; El-Gendy, I.R. Chrysanthemum extract and extract-prepared silver nanoparticles as biocides to control *Aedes aegypti* (L.), the vector of dengue fever. *J. Asia Pac. Entomol.* **2018**, *21*, 205–210. [CrossRef]
52. Akif, H.K.; Olutas, E.; Avcioglu, F.; Kaya, H.; Sahin, B. Quercetin- and caffeic acid-functionalized chitosan-capped colloidal silver nanoparticles: One-pot synthesis, characterization, and anticancer and antibacterial activities. *Beilstein J. Nanotechnol.* **2023**, *14*, 362–376. [CrossRef]
53. Costa, M.; Carreiro, E.; Filho, C.; Sousa, J. Chitosan salts as stabilizing agents for the synthesis of silver nanoparticles (AgNPs). *ChemistrySelect* **2023**, *8*, e202203413. [CrossRef]

54. Malakhovska, T.; Pogodin, A.; Filep, M.; Barannik, A.; Kravets, A. Optical characteristics of silver-based nanocomposites fabricated by an environmentally friendly method. *Semicond. Phys. Quantum. Electron.* **2023**, *26*, 076–083. [CrossRef]
55. Guibal, E. Interactions of metal ions with chitosan-based sorbents: A review. *Sep. Purif. Technol.* **2020**, *38*, 43–74. [CrossRef]
56. Mondéjar-López, M.; López-Jimenez, A.; Ahrazem, O.; Gómez-Gómez, L.; Niza, E. Chitosan coated—Biogenic silver nanoparticles from wheat residues as green antifungal and nanoprimig in wheat seeds. *Int. J. Biol. Macromol.* **2023**, *225*, 964–973. [CrossRef]
57. Shehabeldine, A.; Salem, S.; Ali, O.; Abd-Elsalam, K.; Elkady, F.; Hashem, A. Multifunctional Silver Nanoparticles Based on Chitosan: Antibacterial, Antibiofilm, Antifungal, Antioxidant, and Wound-Healing Activities. *J. Fungi* **2022**, *8*, 612. [CrossRef]
58. Dara, P.K.; Mahadevan, R.; Digita, P.A.; Visnuvinayagam, S.; Kumar, L.; Suseela, M.; Ravishankar, C.N.; Anandan, R. Synthesis and biochemical characterization of silver nanoparticles grafted chitosan (Chi-Ag-NPs): In vitro studies on antioxidant and antibacterial applications. *SN Appl. Sci.* **2020**, *2*, 665. [CrossRef]
59. Abdellatif, A.; Abdelfattah, A.; Aldalaan, S.; Tawfeek, H. Chitosan-capped silver nanoparticles with potent and selective intrinsic activity against the breast cancer cells. *Nanotechnol. Rev.* **2023**, *12*, 20220546. [CrossRef]
60. Reiad, N. Green synthesis of antibacterial chitosan films loaded with silver nanoparticles. *Chin. J. Polym. Sci.* **2013**, *31*, 984–993. [CrossRef]
61. Kumirska, J.; Czerwicka, M.; Kaczynski, Z.; Bychowska, A.; Brzozowski, K.; Thöming, J.; Stepknowski, P. Application of spectroscopic methods for structural analysis of chitin and chitosan. *Mar. Drugs.* **2010**, *8*, 1567–1636. [CrossRef] [PubMed]
62. Naqvi, S.; Kausar, H.; Afzal, A.; Ahmad, F.; Hussain, A. Antifungal activity of *Juglans regia*-mediated silver nanoparticles (AgNPs) against *Aspergillus ochraceus*-induced toxicity in in vitro and in vivo settings. *J. Funct. Biomater.* **2023**, *14*, 221. [CrossRef] [PubMed]
63. Kumari, A.; Naveen; Dhatwalia, J.; Thakur, S.; Radhakrishnan, A.; Chauhan, A.; Chandan, G.; Hyune Choi, B.; Neetika; Nidhi. Antioxidant, antimicrobial, and cytotoxic potential of *Euphorbia royleana* extract-mediated silver and copper oxide nanoparticles. *Chem Pap.* **2023**, *77*, 4643–4657. [CrossRef]
64. Lim, J.; Yeap, S.P.; Che, H.X.; Low, S.C. Characterization of magnetic nanoparticle by dynamic light scattering. *Nanoscale Res. Lett.* **2013**, *9*, 381. [CrossRef]
65. Siddiqi, K.S.; Husen, A.; Rao, R.A.K. A review on biosynthesis of silver nanoparticles and their biocidal properties. *J. Nanobiotechnol.* **2018**, *16*, 14. [CrossRef]
66. Azimi, F.; Mahmoudi, F.; Mahmoudi, F.; Amini, M.M. Synthesis of silver nanoparticles by Galega officinalis and its hypoglycemic effects in type 1 diabetic rats. *Nanomed J.* **2021**, *8*, 255–263. [CrossRef]
67. Singh, R.; Shedbalkar, U.U.; Wadhwani, S.A.; Chopade, B.A. Bacteriagenic silver nanoparticles: Synthesis, mechanism, and applications. *Appl. Microbiol. Biotechnol.* **2015**, *99*, 4579–4593. [CrossRef]
68. Vanlalveni, C.; Lallianrawna, S.; Biswas, A.; Selvaraj, M.; Changmai, B.; Rokhum, S.L. Green synthesis of silver nanoparticles using plant extracts and their antimicrobial activities: A review of recent literature. *RSC Adv.* **2021**, *11*, 2804. [CrossRef]
69. Affes, S.; Maalej, H.; Aranaz, I.; Nasri, M.; Acosta, N. Controlled size green synthesis of bioactive silver nanoparticles assisted by chitosan and its derivatives and their application in biofilm preparation. *Carbohydr. Polym.* **2020**, *236*, 116063. [CrossRef]
70. Cheng, Z.; Tang, S.; Feng, J.; Wu, Y. Biosynthesis and antibacterial activity of silver nanoparticles using *Flos sophorae Immaturus* extract. *Heliyon* **2022**, *8*, e10010. [CrossRef]
71. Wang, W.; Ma, P.; Zhao, Q.; Liu, M.; Zheng, Y. Beneficial properties of the biosynthesized silver/chitosan nanoparticles mediated by *Mentha piperita* in rats with heart failure following myocardial infarction. *Inorg. Chem. Commun.* **2022**, *141*, 109581. [CrossRef]
72. Haghighi, H.; Licciardello, F.; Fava, P.; Siesler, H.H.; Pulvirenti, A. Recent advances on chitosan-based films for sustainable food packaging applications. *Food Packag. Shelf Life* **2020**, *26*, 100551. [CrossRef]
73. Pava, L. Preparation and characterization of plasticized chitosan film: A potential biomaterial in medical field. *Rev. Reto.* **2019**, *7*, 11–24. [CrossRef]
74. Shi, B.; Hao, Z.; Du, Y.; Jia, M.; Xie, S. Mechanical and barrier properties of chitosan-based composite film as food packaging: A review. *Bioresources* **2024**, *19*, 4001–4014. [CrossRef]
75. Rohaeti, E.; Laksono, E.; Rakhmawati, A. Mechanical properties and antibacterial activity of cellulose composite based coconut water with addition glycerol, chitosan, and silver nanoparticle. *Orient. J. Chem.* **2018**, *34*, 3. [CrossRef]
76. Basavegowda, N.; Baek, K.H. Advances in Functional biopolymer-based nanocomposites for active food packaging applications. *Polymers* **2021**, *13*, 4198. [CrossRef]
77. Molleman, B.; Hiemstra, T. Surface structure of silver nanoparticles as a model for understanding the oxidative dissolution of silver ions. *Langmuir* **2015**, *31*, 13361–13372. [CrossRef]
78. Nasef, S.M.; Khozemy, E.E.; Mahmoud, G.A. pH-responsive chitosan/acrylamide/gold/nanocomposite supported with silver nanoparticles for controlled release of anticancer drug. *Sci. Rep.* **2023**, *13*, 7818. [CrossRef]
79. Abd El-Hady, M.; Saeed, S.E. Antibacterial properties and pH-sensitive swelling of in situ formed silver-curcumin nanocomposite based chitosan hydrogel. *Polymers* **2020**, *12*, 2451. [CrossRef]

80. Yan, K.; Xu, F.; Wei, W.; Yang, C.; Wang, D.; Shi, X. Electrochemical synthesis of chitosan/silver nanoparticles multilayer hydrogel coating with pH-dependent controlled release capability and antibacterial property. *Colloids Surf. B Biointerfaces* **2021**, *202*, 111711. [CrossRef]
81. Kumar-Krishnan, S.; Prokhorov, E.; Hernández-Iturriaga, M.; Mota-Morales, J.D.; Vázquez-Lepe, M.; Kovalenko, Y.; Sanchez, I.C.; Luna-Bárcenas, G. Chitosan/silver nanocomposites: Synergistic antibacterial action of silver nanoparticles and silver ions. *Eur. Polym. J.* **2015**, *67*, 242–251. [CrossRef]

International Journal of
Molecular Sciences

MDPI

Article

Tunable Optical Properties of Cu/VSe_2 from the Visible to Terahertz Spectral Range: A First-Principles Study

Elaheh Mohebbi [1], Eleonora Pavoni [1], Pierluigi Stipa [1], Luca Pierantoni [2], Emiliano Laudadio [1,*] and Davide Mencarelli [2]

1 Department of Science and Engineering of Matter, Environment and Urban Planning (SIMAU), Marche Polytechnic University, 60131 Ancona, Italy; e.mohebbi@staff.univpm.it (E.M.); e.pavoni@staff.univpm.it (E.P.); p.stipa@staff.univpm.it (P.S.)

2 Information Engineering Department, Marche Polytechnic University, 60131 Ancona, Italy; l.pierantoni@staff.univpm.it (L.P.); d.mencarelli@staff.univpm.it (D.M.)

* Correspondence: e.laudadio@staff.univpm.it

Abstract: In this study, Density Functional Theory (DFT) and Density Functional Tight-Binding (DFTB) calculations were used to study two different interfaces of Cu/VSe_2 as well as four nanodiodes of VSe_2 bulk including/excluding the Cu layer. We calculated the electronic and optical properties of two systems of two Cu/VSe_2 in which Cu atoms are positioned on the top and at the corner of the VSe_2 monolayer lattice. The electronic band structure calculations revealed that the metallic properties of the VSe_2 monolayer did not change with the interface of Cu atoms; however, the peak around the Fermi level (E_F) in Cu/VSe_2(Top) shifted downward to lower energies. The optical properties showed that in the visible range and the wavelengths related to the interband transition/intraband excitation of Cu atoms, the enhancement of Re(ω) values could be observed for both Cu/VSe_2(Top) and Cu/VSe_2(Corner) nanostructures, while in infrared/terahertz ranges, less/more negative values of Re(ω) were predicted. Through the effect of Cu atoms on the VSe_2 monolayer, the intensity of the peaks in the Im(ω) part of the dielectric constant was increased from 0.2 THz for Cu@VSe_2(Top) and 2.9 THz for Cu@VSe_2(Corner) instead of the zero constant line in the pure system of VSe_2. Refractive index (n) calculations indicated the higher indices at 5.4 and 4.6 for Cu/VSe_2(Top) and Cu@VSe_2(Corner), respectively, in comparison to the value of 2.9 for VSe_2. Finally, DFTB calculations predicted higher current values from I(V) characteristic curves of Au/Cu/VSe_2/Au and Ag/Cu/VSe_2/Ag nanodiodes concerning two other devices without the presence of the Cu layer.

Keywords: VSe_2; copper; electronic property; optical responses; I(V) characteristics

1. Introduction

Two-dimensional material transition-metal dichalcogenides (TMDs), which are built via van der Waals (vdW) epitaxial growth, have been widely investigated due to their remarkable electronic and optical properties [1,2] thanks to their tunable band gap, direct/indirect band gap and efficient optical absorption [3–6]. Moreover, TMDs have attracted enormous research interest due to their other novel physical properties. Strong excitonic effects arise from the Coulombic interactions between carriers caused by the reduced dielectric screening [6–9]. Vanadium diselenide (VSe_2) is a typical TMD that exhibits metallic properties due to the strong electron coupling between all neighboring $V^{4+}-V^{4+}$, and the 1T-VSe_2 polymorph of this material shows an extreme electrical conductivity of 10^6 S/m [10–12]. Importantly, the $3d^1$ odd-electronic configuration of V^{4+} ions also provides rich spin-related information. Dai and co-workers used a theoretical approach to report that a pristine 2D monolayer VSe_2 may exhibit intrinsic magnetic ordering and behave as an unwonted 2D magnetic material [10].

VSe_2 is a particularly attractive candidate system in which the bulk of VSe_2 has a significant structural contribution in the out-of-plane direction (Figure 1A), which has been attributed to a nesting of the strongly three-dimensional Fermi surface at the corresponding wave vector [13]. When reducing the material thickness to a single monolayer (Figure 1B), where the electronic structure must become strictly two-dimensional, a major influence on the charge order in VSe_2 [14] can be expected. Moreover, as mentioned before, first-principles calculations have consistently predicted that a robust ferromagnetic order should emerge for monolayer VSe_2, with a pronounced exchange splitting of the near-E_F electronic states in the 1T polymorph of more than 500 meV [12].

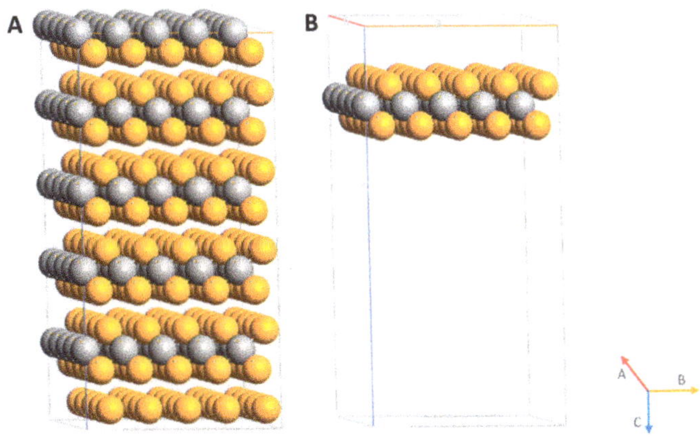

Figure 1. Representation of bulk (**A**) and monolayer (**B**) VSe_2 modeled in this study. Silver and yellow balls presented the V and Se atoms, respectively.

Particular physical properties can provide new ideas for the development of efficient quantum size and localized surface plasmon resonance (LSPR) effects [15,16]. So far, experimental and computational studies have revealed that the LSPR phenomenon of nanomaterial interfaces can extend the light-absorb range of semiconductors from the visible-light to near-infrared-light zone by improving the structures or particle dimensions [17,18] and dielectric surrounding [19,20]. Moreover, the powerful local electric fields produced at the metal–semiconductor interface can remarkably encourage the separation level of photo-generated electron–hole pairs. Elemental metals, i.e., Au, Ag, Cu, Al and Bi, have been explored and found to have LSPR effects, with applications in solar cells [21], photocatalysts [22] and biosensors [23]. The plasmon resonance absorption peak of these metals can be observed around wavelengths of 517 nm (Au), 364 nm (Ag), 675 nm (Cu) and 325 nm (Al). Compared with other noble metals, Cu as a low-cost transition metal has received more attention from developers [24].

Direct energy from light or electricity can be channeled through the decay of surface plasmons, leading to the excitation of hot electron–hole pairs. In metal nanostructures, hot electron generation can take place through intraband excitation within the conduction band edge (CBE) or via interband transitions, which result from transitions from other bands to the unoccupied states of the CBE. It has been observed that in copper (Cu), population decay occurs through the conversion of plasmons into photons (radiation damping) and through non-radiative decay into electron–hole pairs. As illustrated in Figure 2, both intraband excitations and interband transitions in bulk Cu can occur along the x path in the band structure diagram. Due to the requirement that electron energy levels conform to the Fermi–Dirac distribution; incident light excitation leads to the excitation of electrons from occupied states below the Fermi level (E_F) to unoccupied states above E_F. Given Cu's metallic properties, these excited electrons become free carriers that can collectively oscillate in response to alternating external electromagnetic fields. This is the fundamental mechanism behind the LSPR effects produced by Cu [25,26].

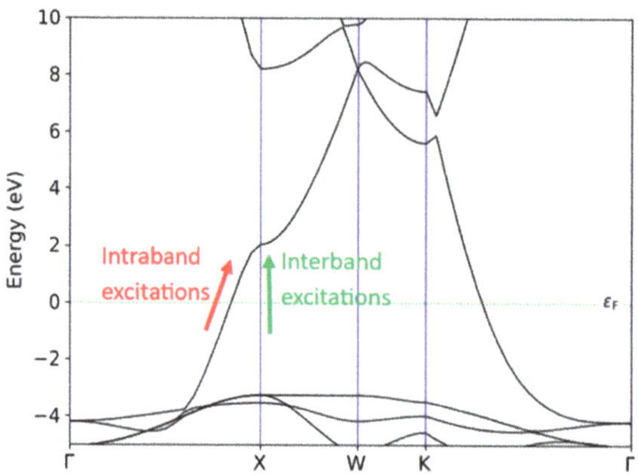

Figure 2. The band structure of metallic Cu computed by the HSE06 functional.

Since most previous studies are related to the morphology, band structure and electromagnetic properties of VSe_2 polymorphs, optical applications are rarely reported for this material. On the other hand, the effects of Cu, Au and Ag metals are mostly considered in the visible region to the infrared zone in the interactions with the semiconductors. In the present work, on the basis of DFT calculations, the regulation rules of Cu/VSe_2 interfaces were investigated systematically and comprehensively for the terahertz (THz) application. We developed two kinds of modeling based on material characterization and devices as follows: (i) the material interfaces linked into an architecture in which we calculated the electronic and optical properties of Cu/VSe_2 (top and corner positions of Cu on the surface of VSe_2) using the accurate Heyd–Scuseria–Ernzerhof (HSE) hybrid methodology; (ii) self-consistent charge Density Functional Tight-Binding (DFTB) calculations carried out to evaluate the electrical features of Cu/VSe_2 nanodevices using different Au and Ag metallic electrodes. Combinations of these methods create new ideas for the constructive application of VSe_2-based and other novel 2D materials in low-frequency THz applications.

2. Results and Discussion

2.1. Structural Parameters and Binding Energy

Bulk VSe_2 has a layered crystal (space group $p6_3mc$) with separate layers stacked along the (001) direction (c-axis direction in Figure 1A). From our DFT calculations, the lattice constants of a = b = 3.42 Å and c = 6.34 Å were predicted for the bulk VSe_2. For the VSe_2 monolayer, the V-Se bonds are 2.51 Å, and the corresponding Se-Se distances are 3.67 Å. The computation results related to structural parameters of the VSe_2 bulk and monolayer in this study are in good agreement with the experimental observations and theoretical calculations [27–29]. According to our previous DFT works based on the adsorption of semiconductor materials of SnSe and $SnSe_2$ on graphene [6], in which the effect of the chemical nature of interfaces on metal–semiconductor hybridization has been described, here, we are focusing on metallic–metallic interaction with Cu adsorption at two different sites over the VSe_2 monolayer. Figure 3 indicates two geometries, where Cu is adsorbed over the VSe_2 surface, near to the V atom (Cu@VSe_2 (Top)), or more simply cornered (Cu@VSe_2 (Corner)), closer to one of the Se atoms on the unit cell of the VSe_2 monolayer. We initially started by considering the geometrical stability levels of different geometries, and the results showed similar binding energies of −0.123 and −0.118 eV for Cu@VSe_2(Top) and Cu@VSe_2(Corner), respectively.

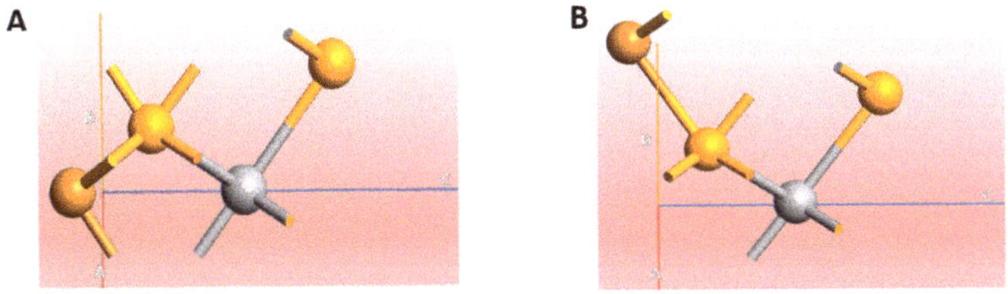

Figure 3. Cu absorption on the VSe_2 monolayer in two different positions (Top (**A**) and Corner (**B**)). Atoms are colored orange, silver and yellow for Cu, V and Se.

2.2. Electronic Structure Calculations

We thus calculated the band structures, projected density of states (PDOS) and Mulliken charges of VSe_2 monolayer and interfaced systems. The outcomes revealed that in both interfaces, we had found a zero-band gap. In a comparison between pure VSe_2 and Cu@VSe_2(Top) (Figure 4), the band structure of pristine VSe_2 indicated that the CBE band crossed the Fermi level and, in M-K paths in BZ, had its maximum/minimum level, respectively, while the top of the valence band edge (VBE) was slightly away from the G point. Moreover, by zooming in the band structures in the zone of G-K, we consistently found a saddle point (SP). The difference in the band structure of the VSe_2 monolayer and Cu/VSe_2(Top) mostly refers to the band dispersions in the zone of SP, in which, after the interaction of Cu with VSe_2, the CBE band shifted to lower energies.

To further validate these results, we carried out PDOS calculations of two systems using the screened hybrid HSE06 functional. The projected PDOS of the VSe_2 monolayer demonstrated that the states between 1 eV and 3.4 eV mainly comprise the V-3*d* orbitals. According to the octahedron crystal field, the V-3*d* orbitals split into the t_2g and the e_g manifolds, which, in the PDOS plot, are contributed at 1–1.4 eV and 1.9–3.4 eV, respectively. Compared with the PDOS of the pristine system in Figure 4A, in the intercalated VSe_2 area with Cu atoms (Figure 4B), the peak around the Fermi level (E_F) shifts downward below the E_F and nearly exhibits a cutoff in this area. That feature indicates a filling of this band.

As confirmed by the Mulliken charge calculations, adsorbed Cu atoms lost charge, and the VSe_2 substrate received charge. These results are in good agreement with the electron-withdrawing character of Cu species, which usually leave the substrate positively charged. The magnitude of the charge transfer between Cu, V and Se along with the absorption can support a more comprehensive description of the Cu/VSe_2 interaction. Mulliken charge calculations reveal a slight transfer of electrons to the $d_z{}^2$ of V and p_x and p_y orbitals of the Se atom near to Cu. The total charges of *p* orbitals of the Se atom lightly changed from 4.125 in the pristine system to 4.552 in the interfaced system. On the other hand, the charge of $d_z{}^2$ orbitals of V increased from 0.480 to 0.898. Since $d_z{}^2$ orbitals are the lowest 3*d* orbital in the CBE band, they can cross the Fermi level and move to lower-energy physics of the VSe_2 monolayer. A similar PDOS description was produced for Cu/VSe_2(Corner), and the results showed a negligible difference between the two different adsorption sites. The PDOS behavior in this study showed similarity with those from relevant experimental works in literature [30].

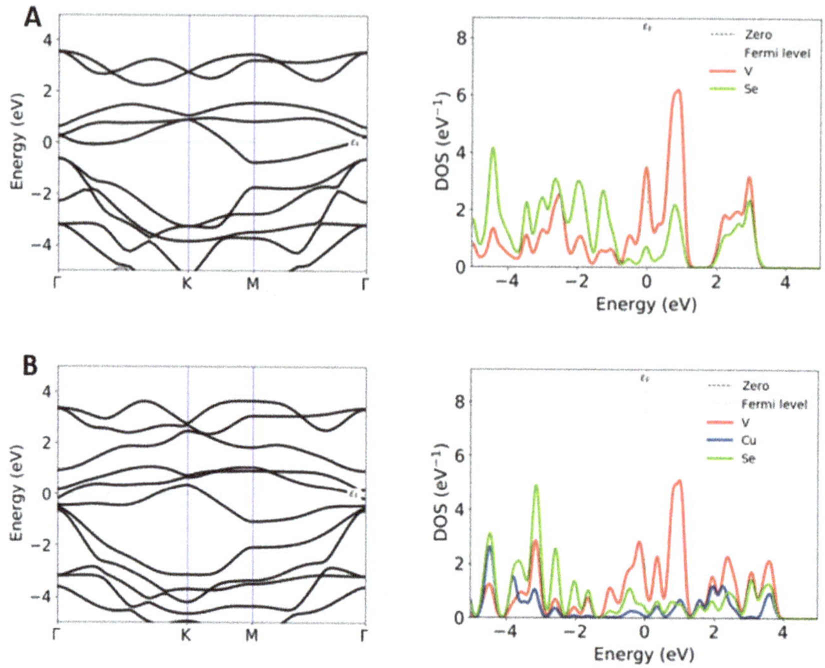

Figure 4. Band structure and PDOS of the VSe_2 monolayer (**A**) in comparison with the Cu@VSe_2(Top) interface (**B**).

2.3. Optical Properties

From a different perspective, Figures 5 and 6 compare the calculated real (Re(ω)) and imaginary (Im(ω)) parts of the dielectric constant of adsorbed systems with respect to pure VSe_2. Compared to the electronic properties, in which the band structure and PDOS of VSe_2 are not drastically affected by the presence of Cu atoms, the Cu/VSe_2 interface can significantly change the optical properties of pristine material. The dielectric functions along the a-, b- and c-axes present in-plane and out-of-plane directions, respectively. The incident in-plane directions are parallel to the symmetrical axis, while the incident out-of-plane direction is perpendicular to the symmetrical axis.

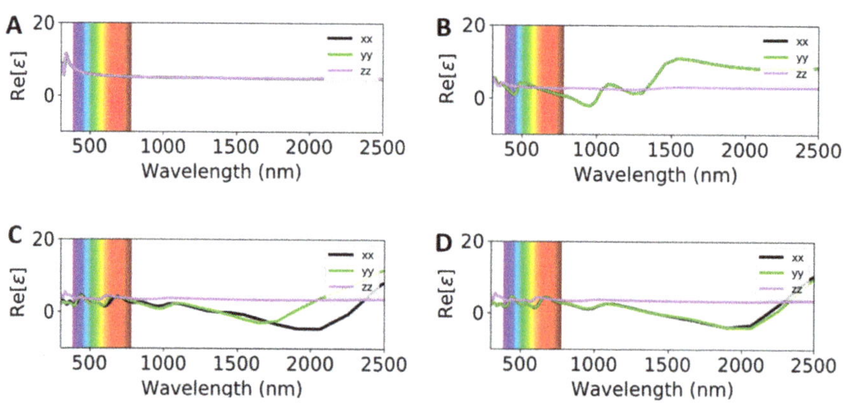

Figure 5. Real of dielectric constants of Cu (**A**), VSe_2 (**B**), Cu/VSe_2(Top) (**C**) and Cu/VSe_2(Corner) (**D**). The color spectrum shows the different wavelengths in the visible range.

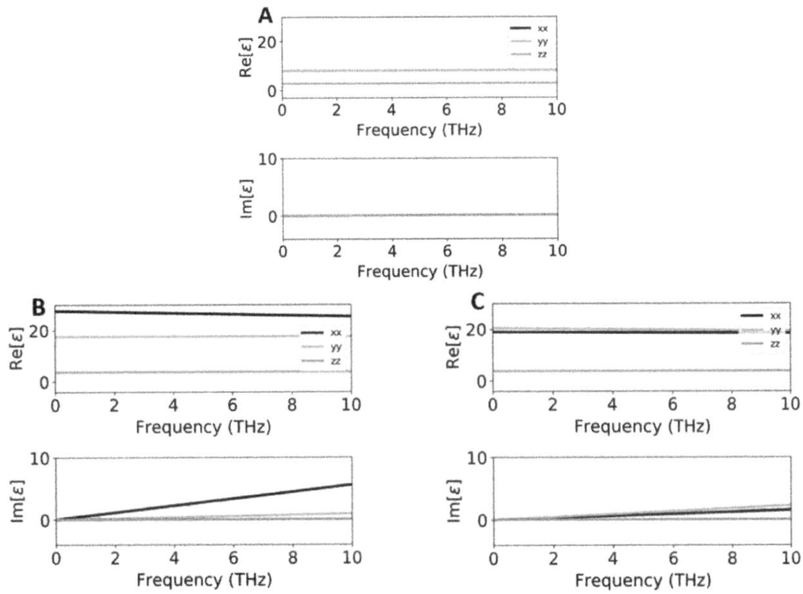

Figure 6. Real and imaginary parts of the dielectric constants of VSe_2 (**A**), Cu/VSe_2(Top) (**B**) and Cu/VSe_2(Corner) (**C**).

The optical properties of Cu bulk show a wavelength of 675 nm where the LSPR of Cu is expected to occur (intraband excitation), while in wavelengths below 550 nm, interband transitions can be observed (Figure 5A) [26]. As illustrated in Figure 5B–D, in the wavelength of the interband transitions for Cu, increasing Re(ω) was seen from 1.16 in pure VSe_2 to 1.96 for Cu/VSe_2(Top) and 1.94 for Cu/VSe_2(Corner). Similarly, Re(ω) values indicated that the low loss of Re(ω) corresponded to Cu/VSe_2(Top) (Figure 5C) and Cu/VSe_2(Corner) (Figure 5D) with respect to pure VSe_2 at the intraband wavelength. At higher wavelengths near to the infrared region at 700–1000 nm, the Re(ω) of adsorbed systems transitioned to less negative values, while at 2100 nm (142 THz), the maximum negative values of −4.72 and −4.00 for Cu/VSe_2(Top) and Cu/VSe_2(Corner) were observed, respectively, in comparison to the positive Re(ω) values in VSe_2.

Since, in this study, considering the optical properties of Cu@VSe_2 in the low-frequency THz range is so important, we report the Re(ω) and Im(ω) parts of the dielectric constant as well as the refractive index (n) of pure and interfaced systems in the range of 0–10 THz. The absorption edge of Im(ω), which is equal to the optical gap, is located at zero along all directions for pure VSe_2 (see Figure 6A). Our HSE results reveal that in both interfaced cases of Cu@VSe_2(Top) and Cu@VSe_2(Corner), the intensity of the peaks started to increase from 0.2 THz for Cu@VSe_2(Top) and 2.9 THz for Cu@VSe_2(Corner). However, the polarization induced by light showed different results, in which the in-plane (a-) axis indicated higher intensities with respect to the in-plane (b-) and out-of-plane c-directions in Cu@VSe_2(Top). About the Cu@VSe_2(Corner) interface, we could not observe significant differences between the a- and b- in-plane axes, while the intensity of b- in-plane was higher than the same axis in the adsorbed system of Cu@VSe_2(Top).

The Re(ω) part of different pure and adsorbed systems (see Figure 6), which is related to the static dielectric constant, showed the positive values of 8.03 and 2.04 along the in-plane and out-of-plane directions, respectively, for the pristine VSe_2 monolayer. These values significantly increased when VSe_2 interacted with Cu atoms by 27.50, 17.45 and

3.69 for Cu@VSe_2(Top) and 19, 20.38 and 3.68 for Cu@VSe_2(Corner) along the in-plane (a-), in-plane (b-) and out-of-plane directions, respectively.

Figure 7A–C show the refractive index (n) of pure VSe_2 (A) in comparison to the interfaced systems of Cu@VSe_2(Top) (B) and Cu@VSe_2(Corner) (C). In Figure 7A, the static refractive indices are 2.9 and 1.8 in the in-plane (a- and b-) and out-of-plane directions, respectively, in the range around 1 THz. Moreover, the corresponding values of the refractive indices are at 5.4 and 4.2 for Cu/VSe_2(Top) and 4.5 and 4.6 for Cu@VSe_2(Corner) along the in-plane (a-) and in-plane (b-) directions, respectively, while a similar value of 1.9 was predicted along the out-of-plane direction, in both adsorbed systems.

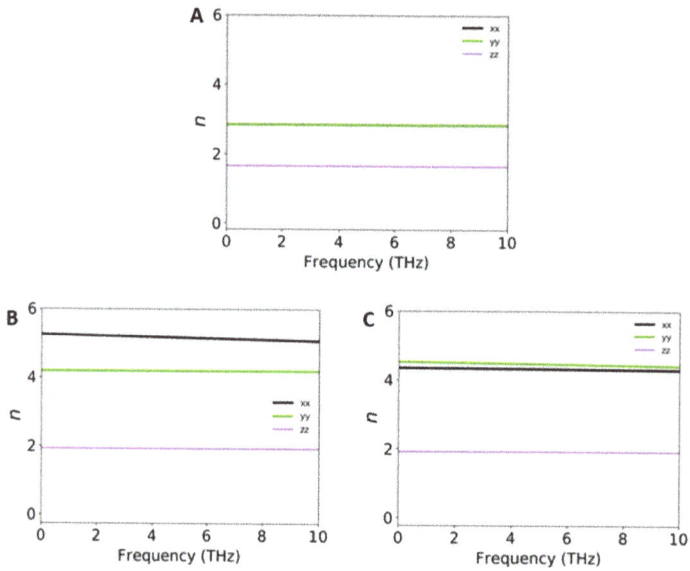

Figure 7. Refractive indexes of VSe_2 (**A**), Cu/VSe_2(Top) (**B**) and Cu/VSe_2(Corner) (**C**).

2.4. I(V) Characteristics

After characterizing the materials of both pure and Cu-adsorbed VSe_2 systems, four distinct nanodiodes were modeled, namely Au/VSe_2/Au, Ag/VSe_2/Ag, Au/Cu/VSe_2/Au and Ag/Cu/VSe_2/Ag. Each diode had identical dimensions of $20 \times 20 \times 83$ Å^3 along the x, y and z axes. These devices were constructed using Au and Ag electrodes, with the inclusion of a Cu layer on top of the VSe_2 material in two of the configurations (refer to Figure 8). The rationale for simulating nanodevices with both Au and Ag lies in the fact that Au possesses a higher work function (5.1 eV) compared to Ag (4.2 eV), which may lead to variations in the electrical properties in diodes. I(V) properties were calculated under a zero-gate ambient voltage in different devices.

The findings illustrated in Figure 9 demonstrate the non-linear I(V) characteristics across four nanodiode configurations. The data indicated that the Au/Cu/VSe_2/Au and Ag/Cu/VSe_2/Ag diodes exhibited significantly higher current values in both positive and negative voltage ranges from -1 V to 1 V when compared to the devices lacking a copper layer. Additionally, the I(V) characteristics of the Ag/VSe_2/Ag diode without the presence of Cu atoms showed some noise because of very high resistance behavior. The resistance fluctuation of noise could be observed for some diode samples, showing the I(V) characteristic dependent on the electric field orientation [31]. Indeed, our theoretical results suggest that the Ag/Cu/VSe_2/Ag device with an appropriate content of Cu possesses abundant electronic traps, which effectively inhibits the recombination of charge carriers and improves the resistivity in I(V) curves of VSe_2. Nevertheless, the difference in work function between Au and Ag did not lead to substantial changes in the current values for the Au/Cu/VSe_2/Au and Ag/Cu/VSe_2/Ag devices, as shown in Figure 9. Only within the voltage range of ±0.3 to ±1 V did the Ag/Cu/VSe_2/Ag diode show marginally higher currents compared to the Au/Cu/VSe_2/Au configuration. Lee et al. [32], through experimental and theoretical investigations, illustrated the effectiveness of photovoltaic devices enhanced by electromagnetic fields. Their research highlighted the differences in hot electron generation resulting from intraband excitation versus interband transition in a plasmonic Cu/TiO_2 nanodiode utilizing Au electrodes, by assessing the current conversion efficiency with a monochromator system.

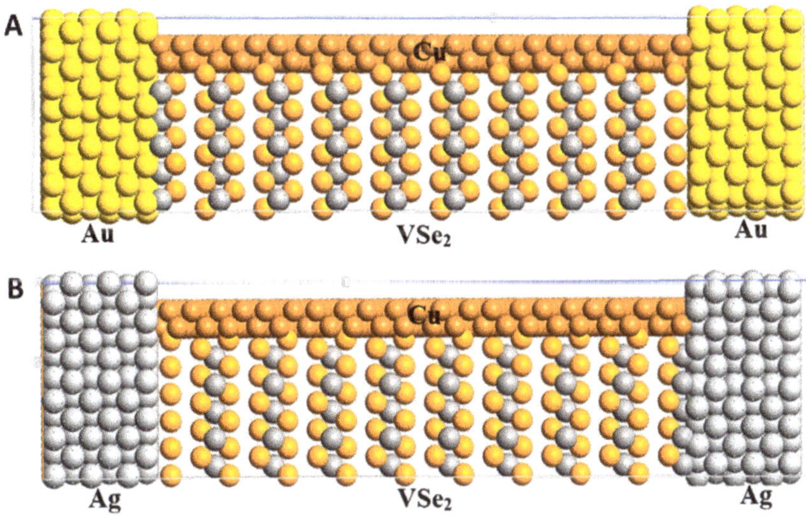

Figure 8. The atomic structure of the Au/Cu/VSe_2/Au (**A**) and Ag/Cu/VSe_2/Ag (**B**) nanodiodes.

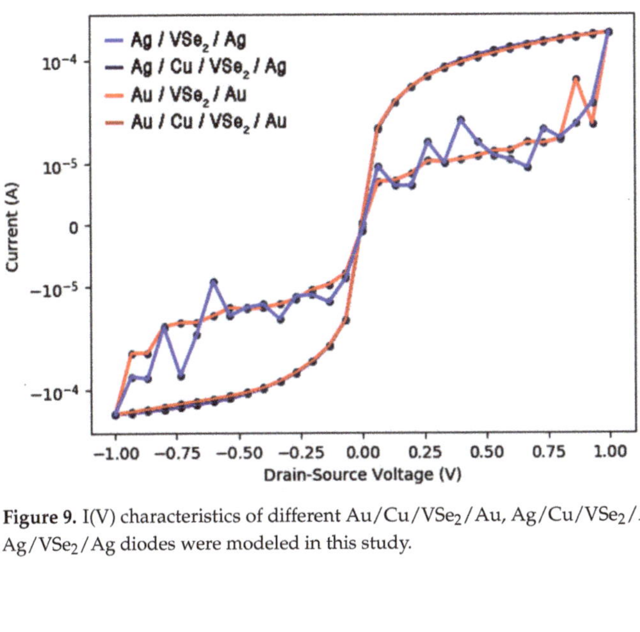

Figure 9. I(V) characteristics of different Au/Cu/VSe_2/Au, Ag/Cu/VSe_2/Ag, Au/VSe_2/Au and Ag/VSe_2/Ag diodes were modeled in this study.

3. Materials and Methods

Density Functional Theory (DFT) [33] calculations were utilized with the Quantum ATK (QATK) code [34]. This software is designed to analyze the chemical and physical properties of various materials and devices [35–37]. The Kohn–Sham (KS) expression [33,38] was addressed using the linear combinations of atomic orbitals (LCAO) basis set [39]. To effectively model the valence orbitals that approximate the core electrons for the elements V, Se, Cu, Au and Ag, norm-conserving PseudoDojo (PDj) pseudopotentials were employed [40]. The exchange-correlation energy was computed using the Perdew, Burke and Ernzerhof (PBE) functional, which is based on generalized gradient approximation (GGA) [41]. For the Brillouin zone sampling, a detailed Monkhorst–Pack k-point grid of 15 × 15 × 15 was used for the optimization of VSe_2 in its bulk form, while a grid of 15 × 15 × 1 was applied for the monolayer configuration. To mitigate the limitations of the DFT method with regard to accurately representing dispersive forces (van der Waals (vdW) interactions), the Grimme (DFT-D3) corrective dispersion term was incorporated [42,43] and corrective dispersion term was added. The band structure, relative band gap and optical properties were assessed using the Heyd–Scuseria–Ernzerhof (HSE) [44] hybrid functionals. Structural optimization was performed until the residual forces dropped below 0.001 eV/Å. The electronic convergence tolerance was established at 10^{-5} eV. A kinetic energy cutoff of 120 Ry was selected for the integration mesh in all calculations. The Gaussian smearing function was chosen to obtain a smooth metallic band for the electronic structure calculations.

The relative dielectric constant ($\epsilon_0(\omega)$) was determined based on Equations (1)–(3) [45]:

$$\epsilon_r(\omega) = (1 + \chi(\omega)), \quad (1)$$

$$\alpha(\omega) = V\epsilon_0\, \chi(\omega), \quad (2)$$

$$\sigma(\omega) = -i\omega\epsilon_0\, \chi(\omega), \quad (3)$$

where χ, V and i are the susceptibility tensor, volume and i-th component (labeling electrons) of the momentum.

Then, we proposed four nanodiodes of the Cu/VSe_2 structure using different electrodes of Au and Ag investigated by DFTB [46]. The DFTB method, using its powered and accurate functionals, is taken into account as a principal tool in this field. The earliest [47,48] formalism of DFTB restricted the computations to non-self-consistent interactions between two atoms; later, a second-order approximation of the Kohn–Sham energy was added to consider the charge self-consistent treatment of the systems [49].

SemiEmpirical (SE) approximation [50] applied by Tight-Binding (TB) calculations was used for the simulations of four nanodiodes: $Au/VSe_2/Au$, $Ag/VSe_2/Ag$, $Au/Cu/VSe_2/Au$ and $Ag/Cu/VSe_2/Ag$. The TB energy $\left(E_0^{TB}\right)$ for an M electron system of N nuclei was resolved with Equation (4):

$$E_0^{TB} = \sum_i^{occ} \left\langle \Psi_i \middle| \hat{H}_0 \middle| \Psi_i \right\rangle + E_{rep} \quad (4)$$

This includes occupied Kohn–Sham eigenstates of Ψ_i, while $\hat{H}_0$ is the Hamiltonian operator, and E_{rep} is rigidly in pairs, repulsive and engaged in short-ranged interactions. To find the solution to the Kohn–Sham equations, the LCAO method is used in an appropriate set of localized atomic orbitals φ_V, the single-particle wave functions Ψ_i, based on Equation (5):

$$\Psi_i(r) = \sum_V C_{Vi}\varphi_V(r - R_a) \quad (5)$$

where R_a denotes the core distance of ion α. In this approach, we utilize confined atomic orbitals represented in a Slater-type form by addressing an adopted Schrödinger equation of a neutral pseudoatom [51].

In the TB Hamiltonian, the non-self-consistent field (nscf) part was parameterized employing the Muller algorithm, where the elements related to the distance-dependent matrix are described with a numerical operator [52]. The electronical transport characteristics were assessed through the non-equilibrium Green's function (NEGF) formalism [49,53,54], while it was assumed that coherent electron transport takes place between the source and drain through the central region.

When the self-consistent density matrix was established, I(V) characteristics were assessed by applying the transmission coefficient (T) with the electron energy (E), as defined by the Landauer equation [55], which is presented in Equation (6):

$$I(V_L T_L V_R T_R) = \frac{e}{h}\sum \int T_\sigma(E)\left[f\left(\frac{E-\mu_R}{K_\beta T_R}\right) - f\left(\frac{E-\mu_L}{K_\beta T_L}\right)\right]dE \quad (6)$$

where f is the Fermi energy, and T_L/T_R are the electron temperatures of the left/right electrodes, where L and R electrodes are the source and the drain, respectively. Moreover, $T_\sigma(E)$ is the transmission coefficient for the spin component σ. Finally, the chemical potentials of the right/left electrodes and V_{bias} can be defined by Equations (7)–(10):

$$\mu_R = E_F^R - eV_R \quad (7)$$

$$\mu_L = E_F^L - eV_L \quad (8)$$

$$\mu_R - \mu_L = e\, V_{bias} \quad (9)$$

$$V_{bias} = V_L - V_R \quad (10)$$

This methodology indicates that it is possible to obtain a quantitative and computationally efficient representation of coherent transport within solid-state physics. This advancement paves the way for enhanced comprehension and manipulation of charge transport characteristics at atomic-scale interfaces under high bias voltages. In the diode simulations, the periodicity was disrupted in the x and y dimensions, while a constant potential was established along the z-axis utilizing Dirichlet boundary conditions. Dirichlet BCs signify that a certain potential (V_0) was fixed at the boundary in the S facet in the simulation cell, following Equation (11):

$$V^H(r) = V_0, r \in S \quad (11)$$

To evaluate the stabilities of interfaced systems, we calculated the binding energy per atom (E_B) using Equation (12):

$$E_B = \frac{E(Interface) - E(Cu) - E(VSe_2)}{N} \quad (12)$$

where $E(Interface)$ refers to the total energy for optimized Cu on the VSe_2 surface, $E(VSe_2)$ and $E(Cu)$ are the total energies of pristine VSe_2 and the optimized phase of the Cu atom, respectively, and N is the number of atoms in the lattice.

4. Conclusions

In summary, a combination of DFT and DFTB computations was carried out to evaluate the material and device features of Cu/VSe_2 by considering the stability and electronic, optical and electrical responses in the VSe_2 monolayer and bulk. Two adsorbed systems of Cu/VSe_2(Top) and Cu/VSe_2(Corner), were preliminarily simulated with the adsorption of Cu atoms on the surface of the VSe_2 monolayer, while Cu atoms were positioned on the top and at the corner of the lattice, respectively. Electronic band structures and PDOS calculations showed the zero-band gap of both adsorbed systems, similar to the pure metallic VSe_2. Optical results indicated that Cu atoms can indeed significantly change the Re(ω) part of the dielectric constant of VSe_2 under visible, infrared and THz ranges. Moreover, the results related to the visible range revealed that in the zone of the interband/intraband of Cu atoms, higher values of Re(ω) were predicted for both Cu/VSe_2(Top) and Cu/VSe_2(Corner), while the negative values of −4.72 (Cu/VSe_2(Top)) and −4.00 (Cu/VSe_2(Corner)) were observed at a higher wavelength of 2100 nm (146 THz). The Im(ω) of the dielectric constant indicated that the peak enhancement was from 0.2 THz for Cu@VSe_2(Top) and 2.9 THz for Cu@VSe_2(Corner). Refractive index (n) calculations showed increasing refractive indices at 5.4 and 4.6 for Cu/VSe_2(Top) and Cu@VSe_2(Corner) along the in-plane (a-) and in-plane (b-) directions, respectively, instead of the corresponding value of 2.9 for the pure system of VSe_2. Then, four nanodiodes—Au/VSe_2/Au, Ag/VSe_2/Ag, Au/Cu/VSe_2/Au and Ag/Cu/VSe_2/Ag—were investigated by the DFTB method. While all simulated nanodevices exhibited non-linear I(V) characteristics, Au/Cu/VSe_2/Au and Ag/Cu/VSe_2/Ag delivered drastically higher current values at positive and negative voltages than the two other nanodiodes, revealing the effect of the copper layer on the electrical properties of VSe_2. Overall, our theoretical study demonstrates the prospects of pure and interfaced VSe_2 nanostructures applied in the THz region.

Author Contributions: Conceptualization, E.M., E.L. and P.S.; methodology, E.M. and E.L.; validation, E.M., E.L., P.S., L.P. and D.M.; investigation, E.M.; data curation, E.M.; writing—original draft preparation, E.M.; writing—review and editing, E.M., E.L., E.P., P.S., L.P. and D.M.; visualization, E.M.; supervision, P.S.; project administration, E.M., E.L., E.P., P.S., L.P. and D.M. All authors have read and agreed to the published version of the manuscript.

Funding: This work is part of the research of HORIZON-EIC-2022-PATHFINDEROPEN-01 "Nanoscale Development of Plasmonic Amplifiers Based on 2D Materials" (PLASNANO) (grant agreement no. 101099552).

Institutional Review Board Statement: Not applicable.

Informed Consent Statement: Not applicable.

Data Availability Statement: No data was used for the research described in the article.

Conflicts of Interest: The authors declare that they have no known competing financial interests or personal relationships that could have appeared to influence the work reported in this paper.

References

1. Chhowalla, M.; Liu, Z.; Zhang, H. Two-Dimensional Transition Metal Dichalcogenide (TMD) Nanosheets. *Chem. Soc. Rev.* **2015**, *44*, 2584–2586. [CrossRef] [PubMed]
2. Choi, W.; Choudhary, N.; Han, G.H.; Park, J.; Akinwande, D.; Lee, Y.H. Recent Development of Two-Dimensional Transition Metal Dichalcogenides and Their Applications. *Mater. Today* **2017**, *20*, 116–130. [CrossRef]

3. Mohebbi, E.; Pavoni, E.; Pierantoni, L.; Stipa, P.; Laudadio, E.; Mencarelli, D. Effect of Different Pseudopotentials on the Phonon Frequencies, Dielectric Constant, and Born Effective Charge of SnSe and $SnSe_2$ Nanostructures: A Density Functional Perturbation Theory Study. *J. Phys. Chem. Solids* **2024**, *185*, 111755. [CrossRef]
4. Li, J.; Jiang, L.; Ma, W.; Wu, T.; Qiu, Q.; Shi, Y.; Zhou, W.; Yao, N.; Huang, Z. VSe_2 Nanosheets for Broadband Photodetection from Visible to Terahertz. *ACS Appl. Nano Mater.* **2022**, *5*, 5158–5167. [CrossRef]
5. Pataniya, P.M.; Patel, V.; Sumesh, C.K. MoS_2/WSe_2 Nanohybrids for Flexible Paper-Based Photodetectors. *Nanotechnology* **2021**, *32*, 315709. [CrossRef] [PubMed]
6. Mohebbi, E.; Pavoni, E.; Pierantoni, L.; Stipa, P.; Zampa, G.M.; Laudadio, E.; Mencarelli, D. Band Gap and THz Optical Adsorption of SnSe and $SnSe_2$ Nanosheets on Graphene: Negative Dielectric Constant of SnSe. *Results Phys.* **2024**, *57*, 107415. [CrossRef]
7. Guan, X.; Yu, X.; Periyanagounder, D.; Benzigar, M.R.; Huang, J.; Lin, C.; Kim, J.; Singh, S.; Hu, L.; Liu, G. Recent Progress in Short-to Long-wave Infrared Photodetection Using 2D Materials and Heterostructures. *Adv. Opt. Mater.* **2021**, *9*, 2001708. [CrossRef]
8. Chowdhury, T.; Sadler, E.C.; Kempa, T.J. Progress and Prospects in Transition-Metal Dichalcogenide Research beyond 2D. *Chem. Rev.* **2020**, *120*, 12563–12591. [CrossRef]
9. Pavoni, E.; Mohebbi, E.; Zampa, G.M.; Stipa, P.; Pierantoni, L.; Laudadio, E.; Mencarelli, D. First Principles Study of WSe_2 and the Effect of V Doping on the Optical and Electronic Properties. *Mater. Adv.* **2024**, *5*, 2230–2237. [CrossRef]
10. Zhang, Z.; Niu, J.; Yang, P.; Gong, Y.; Ji, Q.; Shi, J.; Fang, Q.; Jiang, S.; Li, H.; Zhou, X.; et al. Van Der Waals Epitaxial Growth of 2D Metallic Vanadium Diselenide Single Crystals and Their Extra-high Electrical Conductivity. *Adv. Mater.* **2017**, *29*, 1702359. [CrossRef]
11. Brauer, H.E.; Starnberg, H.I.; Holleboom, L.J.; Strocov, V.N.; Hughes, H.P. Electronic Structure of Pure and Alkali-Metal-Intercalated VSe_2. *Phys. Rev. B* **1998**, *58*, 10031. [CrossRef]
12. Ma, Y.; Dai, Y.; Guo, M.; Niu, C.; Zhu, Y.; Huang, B. Evidence of the Existence of Magnetism in Pristine VX_2 Monolayers (X= S, Se) and Their Strain-Induced Tunable Magnetic Properties. *ACS Nano* **2012**, *6*, 1695–1701. [CrossRef] [PubMed]
13. Feng, J.; Biswas, D.; Rajan, A.; Watson, M.D.; Mazzola, F.; Clark, O.J.; Underwood, K.; Markovic, I.; McLaren, M.; Hunter, A.; et al. Electronic Structure and Enhanced Charge-Density Wave Order of Monolayer VSe_2. *Nano Lett.* **2018**, *18*, 4493–4499. [CrossRef] [PubMed]
14. Strocov, V.N.; Shi, M.; Kobayashi, M.; Monney, C.; Wang, X.; Krempasky, J.; Schmitt, T.; Patthey, L.; Berger, H.; Blaha, P.; et al. Three-Dimensional Electron Realm in VSe_2 by Soft-x-Ray Photoelectron Spectroscopy: Origin of Charge-Density Waves. *Phys. Rev. Lett.* **2012**, *109*, 086401. [CrossRef]
15. Awazu, K.; Fujimaki, M.; Rockstuhl, C.; Tominaga, J.; Murakami, H.; Ohki, Y.; Yoshida, N.; Watanabe, T. A Plasmonic Photocatalyst Consisting of Silver Nanoparticles Embedded in Titanium Dioxide. *J. Am. Chem. Soc.* **2008**, *130*, 1676–1680. [CrossRef]
16. Ramakrishna, G.; Ghosh, H.N. Effect of Particle Size on the Reactivity of Quantum Size ZnO Nanoparticles and Charge-Transfer Dynamics with Adsorbed Catechols. *Langmuir* **2003**, *19*, 3006–3012. [CrossRef]
17. Qin, Y.; Kong, X.-T.; Wang, Z.; Govorov, A.O.; Kortshagen, U.R. Near-Infrared Plasmonic Copper Nanocups Fabricated by Template-Assisted Magnetron Sputtering. *ACS Photonics* **2017**, *4*, 2881–2890. [CrossRef]
18. Grubisic, A.; Schweikhard, V.; Baker, T.A.; Nesbitt, D.J. Multiphoton Photoelectron Emission Microscopy of Single Au Nanorods: Combined Experimental and Theoretical Study of Rod Morphology and Dielectric Environment on Localized Surface Plasmon Resonances. *Phys. Chem. Chem. Phys.* **2013**, *15*, 10616–10627. [CrossRef]
19. Takahata, R.; Yamazoe, S.; Koyasu, K.; Imura, K.; Tsukuda, T. Gold Ultrathin Nanorods with Controlled Aspect Ratios and Surface Modifications: Formation Mechanism and Localized Surface Plasmon Resonance. *J. Am. Chem. Soc.* **2018**, *140*, 6640–6647. [CrossRef]
20. Seh, Z.W.; Liu, S.; Low, M.; Zhang, S.; Liu, Z.; Mlayah, A.; Han, M. Janus Au-TiO_2 Photocatalysts with Strong Localization of Plasmonic Near-fields for Efficient Visible-light Hydrogen Generation. *Adv. Mater.* **2012**, *24*, 2310–2314. [CrossRef]
21. Qiao, L.; Wang, D.; Zuo, L.; Ye, Y.; Qian, J.; Chen, H.; He, S. Localized Surface Plasmon Resonance Enhanced Organic Solar Cell with Gold Nanospheres. *Appl. Energy* **2011**, *88*, 848–852. [CrossRef]
22. Mascaretti, L.; Dutta, A.; Kment, Š.; Shalaev, V.M.; Boltasseva, A.; Zbořil, R.; Naldoni, A. Plasmon-enhanced Photoelectrochemical Water Splitting for Efficient Renewable Energy Storage. *Adv. Mater.* **2019**, *31*, 1805513. [CrossRef] [PubMed]
23. Homola, J. Surface Plasmon Resonance Sensors for Detection of Chemical and Biological Species. *Chem. Rev.* **2008**, *108*, 462–493. [CrossRef]
24. Gawande, M.B.; Goswami, A.; Felpin, F.-X.; Asefa, T.; Huang, X.; Silva, R.; Zou, X.; Zboril, R.; Varma, R.S. Cu and Cu-Based Nanoparticles: Synthesis and Applications in Catalysis. *Chem. Rev.* **2016**, *116*, 3722–3811. [CrossRef]
25. Linic, S.; Aslam, U.; Boerigter, C.; Morabito, M. Photochemical Transformations on Plasmonic Metal Nanoparticles. *Nat. Mater.* **2015**, *14*, 567–576. [CrossRef]

26. Yao, G.-Y.; Zhao, Z.-Y.; Liu, Q.-L.; Dong, X.-D.; Zhao, Q.-M. Theoretical Calculations for Localized Surface Plasmon Resonance Effects of Cu/TiO_2 Nanosphere: Generation, Modulation, and Application in Photocatalysis. *Sol. Energy Mater. Sol. Cells* **2020**, *208*, 110385. [CrossRef]
27. Bayliss, S.C.; Liang, W.Y. Reflectivity and Band Structure of 1T-VSe_2. *J. Phys. C Solid State Phys.* **1984**, *17*, 2193. [CrossRef]
28. Wilson, J.A.; Yoffe, A.D. The Transition Metal Dichalcogenides Discussion and Interpretation of the Observed Optical, Electrical and Structural Properties. *Adv. Phys.* **1969**, *18*, 193–335. [CrossRef]
29. Reshak, A.H.; Auluck, S. Theoretical Investigations of the Electronic and Optical Properties of Pure and Alkali Metal Intercalated 1T-VSe_2. *Phys. B Condens. Matter* **2004**, *349*, 310–315. [CrossRef]
30. Chazarin, U.; Lezoualc'h, M.; Chou, J.; Pai, W.W.; Karn, A.; Sankar, R.; Chacon, C.C.; Girard, Y.; Repain, V.; Bellec, A.; et al. Formation of Monolayer Charge Density Waves and Anomalous Edge Doping in Na Doped Bulk VSe_2. *Adv. Mater. Interfaces* **2023**, *10*, 2201680. [CrossRef]
31. Sita, Z.; Sedlakova, V.; Majzner, J.; Sedlak, P.; Sikula, J.; Grmela, L. Analysis of Noise and Non-Linearity of IV Characteristics of Positive Temperature Coefficient Chip Thermistors. *Metrol. Meas. Syst.* **2013**, *20*, 635–644. [CrossRef]
32. Lee, C.; Park, Y.; Park, J.Y. Hot Electrons Generated by Intraband and Interband Transition Detected Using a Plasmonic Cu/TiO_2 Nanodiode. *RSC Adv.* **2019**, *9*, 18371–18376. [CrossRef] [PubMed]
33. Kohn, W.; Sham, L.J. Self-Consistent Equations Including Exchange and Correlation Effects. *Phys. Rev.* **1965**, *140*, A1133. [CrossRef]
34. Smidstrup, S.; Markussen, T.; Vancraeyveld, P.; Wellendorff, J.; Schneider, J.; Gunst, T.; Verstichel, B.; Stradi, D.; Khomyakov, P.A.; Vej-Hansen, U.G.; et al. QuantumATK: An Integrated Platform of Electronic and Atomic-Scale Modelling Tools. *J. Phys. Condens. Matter* **2019**, *32*, 015901. [CrossRef]
35. Pavoni, E.; Mohebbi, E.; Stipa, P.; Pierantoni, L.; Mencarelli, D.; Dragoman, M.; Aldrigo, M.; Laudadio, E. First-Principles Investigation of Interface Phenomena in Hafnium-Based Metal–Insulator–Metal Diodes. *Nanoscale Adv.* **2023**, *5*, 2748–2755. [CrossRef]
36. Laudadio, E.; Mohebbi, E.; Pavoni, E.; Minnelli, C.; Sabbatini, S.; Stipa, P. Density Functional Theory and Molecular Dynamics Studies on Electrical, Mechanical, and Thermal Properties of TiO2 Nanoparticles Interacting with Poly Lactic-Co-Glycolic Acid. *Colloids Surf. A Physicochem. Eng. Asp.* **2023**, *667*, 131388. [CrossRef]
37. Mohebbi, E.; Pavoni, E.; Pierantoni, L.; Stipa, P.; Laudadio, E.; Mencarelli, D. Insights Into a New Geometric Graphene Diode with Ultrahigh Asymmetry Ratio: A Computational Approach. In Proceedings of the 2023 IEEE 23rd International Conference on Nanotechnology (NANO), Jeju Island, Republic of Korea, 2–5 July 2023; IEEE: Piscataway, NJ, USA, 2023; pp. 238–242.
38. Ernzerhof, M.; Scuseria, G.E. Perspective on "Inhomogeneous Electron Gas" Hohenberg P, Kohn W (1964) Phys Rev 136: B864. *Theor. Chem. Acc.* **2000**, *103*, 259–262. [CrossRef]
39. Sánchez-Portal, D.; Ordejon, P.; Artacho, E.; Soler, J.M. Density-functional Method for Very Large Systems with LCAO Basis Sets. *Int. J. Quantum Chem.* **1997**, *65*, 453–461. [CrossRef]
40. van Setten, M.J.; Giantomassi, M.; Bousquet, E.; Verstraete, M.J.; Hamann, D.R.; Gonze, X.; Rignanese, G.-M. The PseudoDojo: Training and Grading a 85 Element Optimized Norm-Conserving Pseudopotential Table. *Comput. Phys. Commun.* **2018**, *226*, 39–54. [CrossRef]
41. Perdew, J.P.; Burke, K.; Ernzerhof, M. Generalized Gradient Approximation Made Simple. *Phys. Rev. Lett.* **1996**, *77*, 3865. [CrossRef]
42. Ehrlich, S.; Moellmann, J.; Reckien, W.; Bredow, T.; Grimme, S. System-dependent Dispersion Coefficients for the DFT-D3 Treatment of Adsorption Processes on Ionic Surfaces. *ChemPhysChem* **2011**, *12*, 3414–3420. [CrossRef] [PubMed]
43. Sedona, F.; Fakhrabadi, M.M.S.; Carlotto, S.; Mohebbi, E.; De Boni, F.; Casalini, S.; Casarin, M.; Sambi, M. On-Surface Synthesis of Extended Linear Graphyne Molecular Wires by Protecting the Alkynyl Group. *Phys. Chem. Chem. Phys.* **2020**, *22*, 12180–12186. [CrossRef] [PubMed]
44. Henderson, T.M.; Paier, J.; Scuseria, G.E. Accurate Treatment of Solids with the HSE Screened Hybrid. *Phys. Status Solidi* **2011**, *248*, 767–774. [CrossRef]
45. Martin, R.M. *Electronic Structure: Basic Theory and Practical Methods*; Cambridge University Press: Cambridge, UK, 2020; ISBN 1108657478.
46. Mohebbi, E.; Pavoni, E.; Pierantoni, L.; Stipa, P.; Hemmetter, A.; Laudadio, E.; Mencarelli, D. Towards Graphene-Based Asymmetric Diodes: A Density Functional Tight-Binding Study. *Nanoscale Adv.* **2024**, *6*, 1548–1555. [CrossRef]
47. Seifert, G.; Eschrig, H.; Bieger, W. An Approximation Variant of LCAO-X-ALPHA Methods. *Z. Fur Phys. Chem.-Leipz.* **1986**, *267*, 529–539.
48. Foulkes, W.M.C.; Haydock, R. Tight-Binding Models and Density-Functional Theory. *Phys. Rev. B* **1989**, *39*, 12520. [CrossRef]
49. Elstner, M.; Porezag, D.; Jungnickel, G.; Elsner, J.; Haugk, M.; Frauenheim, T.; Suhai, S.; Seifert, G. Self-Consistent-Charge Density-Functional Tight-Binding Method for Simulations of Complex Materials Properties. *Phys. Rev. B* **1998**, *58*, 7260. [CrossRef]

50. Ammeter, J.H.; Bürgi, H.-B.; Thibeault, J.C.; Hoffmann, R. Counterintuitive Orbital Mixing in Semiempirical and Ab Initio Molecular Orbital Calculations. *J. Am. Chem. Soc.* **1978**, *100*, 3686–3692. [CrossRef]
51. Eschrig, H. *Optimized LCAO Method and the Electronic Structure of Extended Systems*; Walter de Gruyter GmbH &, Co., KG: Berlin, Germany, 2022; ISBN 3112483464.
52. Stokbro, K.; Petersen, D.E.; Smidstrup, S.; Blom, A.; Ipsen, M.; Kaasbjerg, K. Semiempirical Model for Nanoscale Device Simulations. *Phys. Rev. B* **2010**, *82*, 075420. [CrossRef]
53. Ozaki, T.; Nishio, K.; Kino, H. Efficient Implementation of the Nonequilibrium Green Function Method for Electronic Transport Calculations. *Phys. Rev. B* **2010**, *81*, 035116. [CrossRef]

54. Zampa, G.M.; Mencarelli, D.; Pierantoni, L. A Full-Wave Time-Dependent Schrödinger Equation Approach for the Modeling of Asymmetric Transport in Geometric Diodes. *Phys. B Condens. Matter* **2023**, *661*, 414917. [CrossRef]
55. Haug, H.; Jauho, A.-P. *Quantum Kinetics in Transport and Optics of Semiconductors*; Springer: Berlin/Heidelberg, Germany, 2008; Volume 2.

International Journal of
Molecular Sciences

MDPI

Article

Point-of-Care Diagnostic Test for Rapid Detection of Infectious Laryngotracheitis Virus by Loop-Mediated Isothermal Amplification and Nanoprobes

Pablo Cea-Callejo [1,2], Claudia Trenado [1], Elías El Mansouri [1], Esperanza Gomez-Lucia [2,3], Ana Doménech [2,3], Mar Biarnés [4], J. Marco Cuenca [5], Christian J. Sánchez-Llatas [1], Ricardo Madrid [1,2,*,†] and Laura Benítez [1,2,*,†]

1 Department of Genetics, Physiology, and Microbiology, School of Biology, Complutense University of Madrid (UCM), 28040 Madrid, Spain; pcea@ucm.es (P.C.-C.); clautren@ucm.es (C.T.); eliaselm@ucm.es (E.E.M.); chsanc01@ucm.es (C.J.S.-L.)
2 Research Group of "Animal Viruses", Complutense University of Madrid, 28040 Madrid, Spain; duato@ucm.es (E.G.-L.); domenech@vet.ucm.es (A.D.)
3 Department of Animal Health, Veterinary Faculty, Complutense University of Madrid (UCM), 28040 Madrid, Spain
4 Centro de Sanidad Avícola de Cataluña y Aragón (CESAC), 43206 Reus, Spain; mbiarnes@cesac.net
5 Department of Physical Chemistry, School of Chemistry, Complutense University of Madrid (UCM), 28040 Madrid, Spain; juliocue@ucm.es
* Correspondence: rimadrid@ucm.es (R.M.); lbenitez@ucm.es (L.B.)
† These authors contributed equally to this work and share last authorship.

Abstract: Infectious laryngotracheitis virus (ILTV), a DNA virus classified as *Gallid alphaherpesvirus 1*, causes a highly contagious respiratory disease in chickens, leading to significant economic losses and health risks for the poultry industry. The rapid detection of ILTV is essential to control its spread and prevent outbreaks. Traditional diagnostic methods like PCR are costly, require specialized personnel, and delay response efforts. To address this, we developed a point-of-care diagnostic test combining loop-mediated isothermal amplification (LAMP) with DNA nanoprobes on respiratory swabs. LAMP targets the ILTV-glycoprotein E (gE) gene, enabling rapid nucleic acid amplification at 65 °C without extraction, making it suitable for on-site detection. DNA nanoprobes provide a colorimetric readout visible to the naked eye. Gold nanoparticles drive this readout, as their red color, based on localized surface plasmon resonance, persists in the presence of ILTV DNA through DNA-DNA hybridization, ensuring reliable detection. The assay achieved 100% sensitivity and specificity for ILTV-gE, with a detection limit of 200 copies per reaction, allowing for the early identification of infections. The results are available within 45 min, enabling prompt measures to control ILTV spread. Cost-effective and user-friendly, this method enhances disease management and biosecurity in poultry farms.

Keywords: infectious laryngotracheitis virus (ILTV); molecular detection; loop-mediated isothermal amplification (LAMP); nanoprobes; point-of-care test (POCT)

1. Introduction

Avian infectious laryngotracheitis (ILT) is a highly contagious respiratory disease in chickens, which poses a considerable economic threat to the poultry industry worldwide, due to severe respiratory outbreaks. Depending on strain virulence, these cause high morbidity (90–100%) and variable mortality [1–3]. It is caused by infectious laryngotracheitis virus (ILTV), a herpesvirus classified as *Iltovirus gallidalpha 1* of the subfamily *Alphaherpesvirinae*, with a DNA genome of around 150 kb encoding 80 predicted viral protein open reading frames (ORFs). The disease usually occurs in acute or subacute forms with mortality rates ranging from 50% to 30%, respectively, with clinical respiratory signs including conjunctivitis and nasal discharge, a decrease in egg production, and weight loss. Its clinical signs resemble those caused by other respiratory viruses, such as infectious bronchitis virus (IBV) and avian metapneumovirus (aMPV). In addition, the existence of asymptomatic carriers makes virus contention more difficult. In severe forms of disease, characteristic hemorrhagic tracheitis is present [1,3,4]. ILTV establishes lifelong latency in the trigeminal ganglia, and birds become long-term carriers, complicating disease control on farms [1,2,5].

ILTV transmission occurs easily in infected chickens, mainly through close contact. This is rapidly enhanced by intensive farming. It is crucial to rapidly detect the virus to implement the most appropriate measures and reduce its economic impact on production. Although conventional laboratory methods remain useful, they are time-consuming and relatively expensive, so they have been replaced by conventional PCR and real-time PCR (qPCR) [6,7]. Rapid and simple on-farm application of these molecular techniques is currently hindered by numerous limitations. Thus, there is growing interest in developing point-of-care tests (POCTs) for on-farm diagnostics to mitigate outbreaks and enhance animal welfare. Ideally, POCTs should be affordable, sensitive, specific, user-friendly, rapid, and robust, requiring only minimal equipment [8,9].

The loop-mediated isothermal amplification (LAMP) technique offers several advantages over PCR, making it an ideal technique for POCTs. Among these, the ability of LAMP to perform amplifications at a constant temperature at 60–72 °C for less than 1 h stands out [10]. In addition, the *Bacillus stearothermophilus* (Bst) enzyme's strand displacement activity eliminates the need for denaturation, simplifying amplification and obviating the need for complex thermal cycling and expensive thermocyclers. Moreover, this enzyme shows remarkable robustness and resistance to inhibitors, making LAMP effective for analyzing a variety of clinical samples, with minimal pre-processing requirements in field conditions [11]. Consequently, developing a POCT-compatible lysis buffer is essential to enable LAMP deployment in resource-limiting settings. The reaction takes place using two primer pairs: the internal primers (FIP-BIP), which are added in a higher concentration so that they bind first to the target, and the external primers (F3-B3), which are present in the reaction at a lower concentration. Each inner primer is designed to target the original and the complementary strand, allowing for the formation of a loop during amplification. The efficiency of the reaction can be enhanced by adding a loop primer pair (LF-LB). The use of four up to six primers to target specific regions makes design more complex than for PCR. However, this complexity yields higher sensitivity and specificity, as LAMP can distinguish eight specific DNA sequences, whereas PCR targets only two [9,12]. The exponential amplification enables rapid results, and POCT-compatible visualization techniques are well-established. Because of these characteristics, LAMP has been successfully applied to the detection of several respiratory viruses in chickens [5], including the detection of ILTV [13–15].

Among the POCT-compatible visualization techniques for LAMP, we find the use of pH indicators, metal indicators (turbidity-based methods), or colorimetric or fluorescent dyes as intercalating agents [5]. These methods, however, lack target specificity. The use of gold nanoparticles (AuNPs), with their localized surface plasmon resonance (LSPR) properties, has emerged as a promising approach to enhance LAMP-based techniques. Typically observed as a distinctive red color around 520 nm and visible to the naked eye, this phenomenon is dependent on the size, shape, and composition of the AuNPs. Consequently, the SPR angle produced on the reflected light wave is specific to a determined surface of the AuNPs [14]. When functionalized with specific oligonucleotides, as DNA nanoprobes, a precise, sensitive, and accurate visualization system for enriched LAMP products is achieved. For instance, under high-salt conditions and in the absence of the specific LAMP product, the red color of DNA nanoprobes changes; it either shifts to longer wavelengths or clears completely, becoming visible to the naked eye. In contrast, the presence of the target LAMP product promotes DNA-DNA hybridization, preserving the red color [16].

This study aimed to validate a highly specific, rapid, affordable, and user-friendly ILTV detection system. This LAMP-based assay, combined with specific DNA nanoprobes targeting the ILTV-gE gene, enables rapid, visual, and in situ ILTV diagnosis, helping to prevent its spread and minimizing associated economic losses on poultry farms.

2. Results

2.1. Optimization of LAMP Assay Targets and Conditions

The three sets of primers, targeting either the ILTV glycoprotein-E (ILTV-gE) or the ILTV thymidine kinase (ILTV-TK) genes, were tested under identical LAMP reaction conditions. Initial validation was performed detecting the ILTV Salisbury 146 vaccine strain as the positive control and a non-template control as the negative control. To optimize the LAMP reaction temperature for each primer set, a temperature gradient from 61 to 69 °C (at 2 °C intervals) was tested. Densitometric analysis of the amplification products (Figure 1) revealed that temperatures between 63 and 67 °C yielded the strongest signals for the gE target. Based on these results, 65 °C was selected as the optimal reaction temperature.

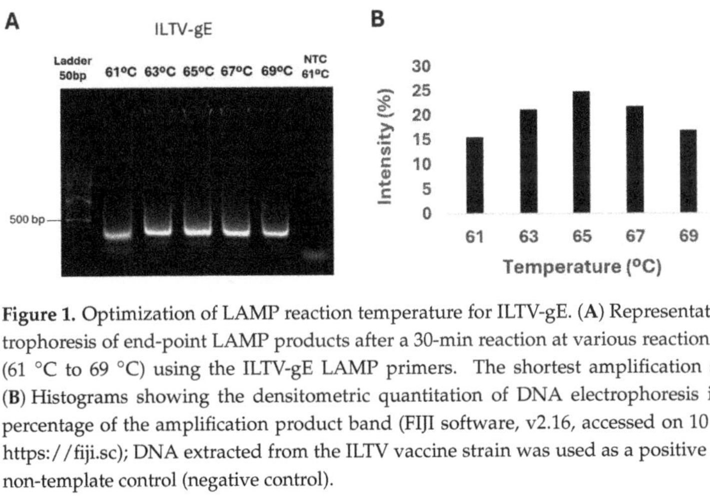

Figure 1. Optimization of LAMP reaction temperature for ILTV-gE. (**A**) Representative DNA electrophoresis of end-point LAMP products after a 30-min reaction at various reaction temperatures (61 °C to 69 °C) using the ILTV-gE LAMP primers. The shortest amplification size is 201 bp. (**B**) Histograms showing the densitometric quantitation of DNA electrophoresis indicating the percentage of the amplification product band (FIJI software, v2.16, accessed on 10 January 2025; https://fiji.sc); DNA extracted from the ILTV vaccine strain was used as a positive sample. NTC, non-template control (negative control).

2.2. LAMP Assays for Sample Detection

Following successful optimization of both gE and TK LAMP amplification, a panel of 32 previously qPCR-validated DNA/RNA extracts (E1 to E32), including 22 ILTV-positive and 10 ILTV-negative samples, was analyzed. Samples were tested in individual reactions using either the ILTV-gE or ILTV-TK primer sets. The ILTV-gE LAMP assay achieved a sensitivity of 100% (32/32), while the ILTV-TK LAMP assay exhibited a lower sensitivity, approximately 84% (21/25) after a 30 min reaction time. Notably, both LAMP assays exhibited a specificity of 100% (10/10) (Figure 2). Based on these results, the ILTV-gE primer set was selected for further validation of the diagnostic system for ILTV with crude samples.

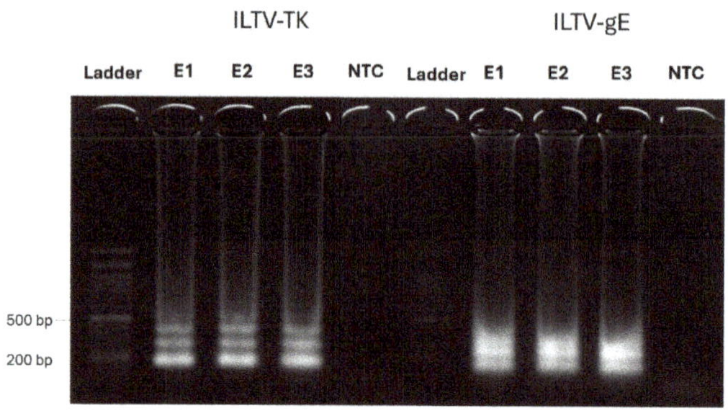

Figure 2. Representative DNA electrophoresis of end-point LAMP products generated using different primer sets for ILTV-TK and ILTV-gE at 65 °C during a 30-min reaction. The shortest amplification size is 219 bp for the ILTV-TK primer set and 201 bp for the ILTV-gE primer set. E1 to E3 correspond to three DNA/RNA extracts of ILTV-positive samples. NTC, non-template control (negative control).

2.3. *Quantitative LAMP (qLAMP) Assay with Crude Samples*

To evaluate the performance of the LAMP assay, a panel of 47 crude samples (S1–S47) and their corresponding nucleic acid extractions (E1–E25) were analyzed. Initially, LAMP assays were carried out using crude samples pretreated with a rapid lysis buffer developed in our laboratory for the detection of enveloped viruses, as described in Section 4. Crude samples were tested with and without a rapid lysis buffer pretreatment designed for virus detection. As shown in Figure 3, the lysis buffer pretreatment did not significantly affect the detection time or efficiency, with comparable results observed between treated crude samples, untreated crude samples, and extracted DNA/RNA.

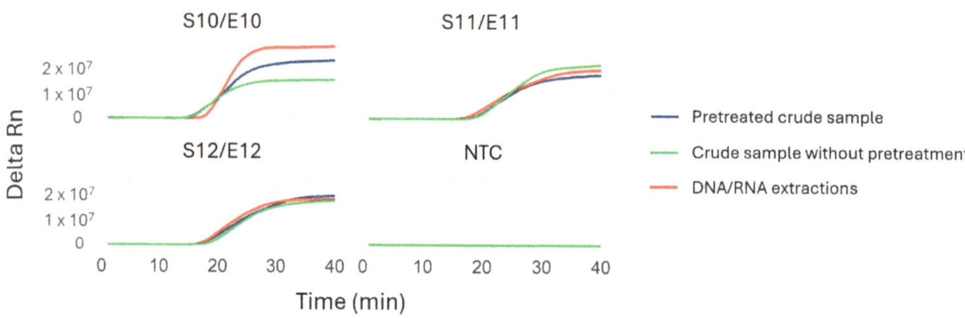

Figure 3. Optimization of the qLAMP protocol using three ILTV-positive samples: crude samples without pretreatment (S10–S12) and their corresponding DNA/RNA extracts (E10–E12), tested using the ILTV-gE primers. Representative amplification plots are shown for: total nucleic acid extractions (E samples, shown in red) and crude samples (S samples) treated with lysis buffer (blue) and untreated (green). Amplification times ranged from 15.1–17.2 min for sample S10/E10, 16.9–17.5 min for S11/E11, and 15.8–18.8 min for S12/E12. Additionally, a non-template control (NTC) was included for comparison alongside these three ILTV-positive samples.

The effect of freeze–thaw cycles on crude sample lysis was investigated using the live attenuated ILTV Salisbury 146 vaccine strain. Duplicate qLAMP assays were performed on the ILTV vaccine and ILTV-positive crude samples, comparing lysis buffer pretreatment, untreated samples, and total nucleic acid extractions. Additionally, to simulate real field conditions, a crude sample negative for respiratory viruses (sample S18) was spiked with the untreated ILTV vaccine strain to assess the efficacy of the lysis buffer pretreatment and potential LAMP inhibition. All samples were amplified successfully within a comparable timeframe, with no significant differences observed between preparation methods (Figure 4).

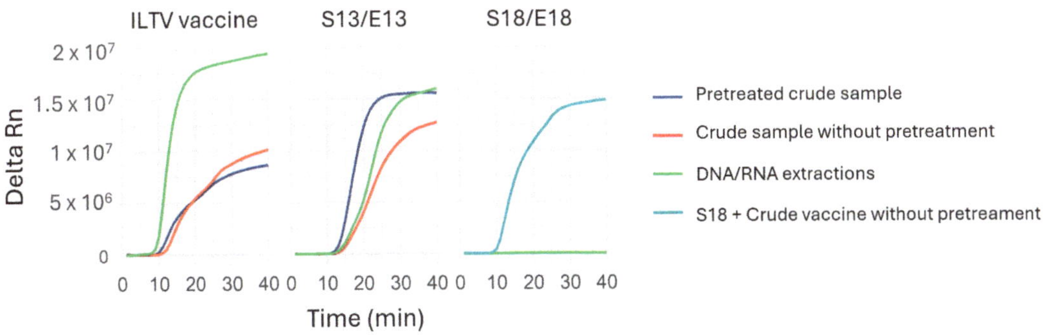

Figure 4. Optimization of qLAMP using ILTV vaccine, ILTV-positive sample (S13 and E13), and ILTV-negative sample (S18 and E18) using ILTV-gE primers. Amplification results are shown for DNA/RNA extractions (green), crude samples (S13 and S18) treated with lysis buffer (dark blue), and untreated (red). The S18 sample and the untreated ILTV vaccine were combined in a 1:1 ratio (light blue). Amplification times ranged from 8.7–10.1 min for the ILTV vaccine and 12.3–14.1 min for sample S13/E13.

Finally, the entire panel of crude samples (S1 to S47) was analyzed without any treatment. All positive samples previously identified by end-point LAMP using their corresponding DNA/RNA extracts (E1 to E25) were also successfully detected in their corresponding crude samples (S1 to S25), maintaining a 100% sensitivity with no false positives (Figure 5).

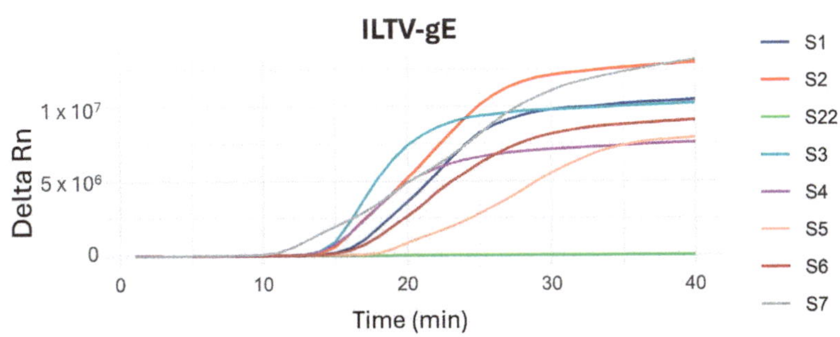

Figure 5. Representative qLAMP amplification plots using ILTV-gE LAMP primers and DNA/RNA extractions from ILTV-positive samples (S1 to S7). S22 was used as negative sample control; this sample was negative for ILTV and positive for aMPV.

2.4. Validation of DNA Nanoprobe Detection System Coupled to LAMP

Specific DNA nanoprobes targeting the ILTV-gE amplicon were employed as a colorimetric readout for viral detection coupled with LAMP reaction. The reaction conditions for the color shift of DNA nanoprobes in the presence or absence of the LAMP product were optimized, allowing for clear differentiation between positive (red color) and negative (greyish-blue color after salt addition) results (Figure 6B). Based on our previous optimization [16], we determined that a 1.3:5 ratio of DNA nanoprobes to LAMP product provides an optimal ratio for efficient detection. To achieve clearer results, we determined that an optimal final volume of 15 μL was required for the binding reaction. This involved adding 8.5 µL of binding buffer and incubating the reaction for 5 min at room temperature. This incubation resulted in a readily observable color difference, which was subsequently confirmed by visible-ultraviolet (VIS-UV) spectroscopy (Figure 6B,C).

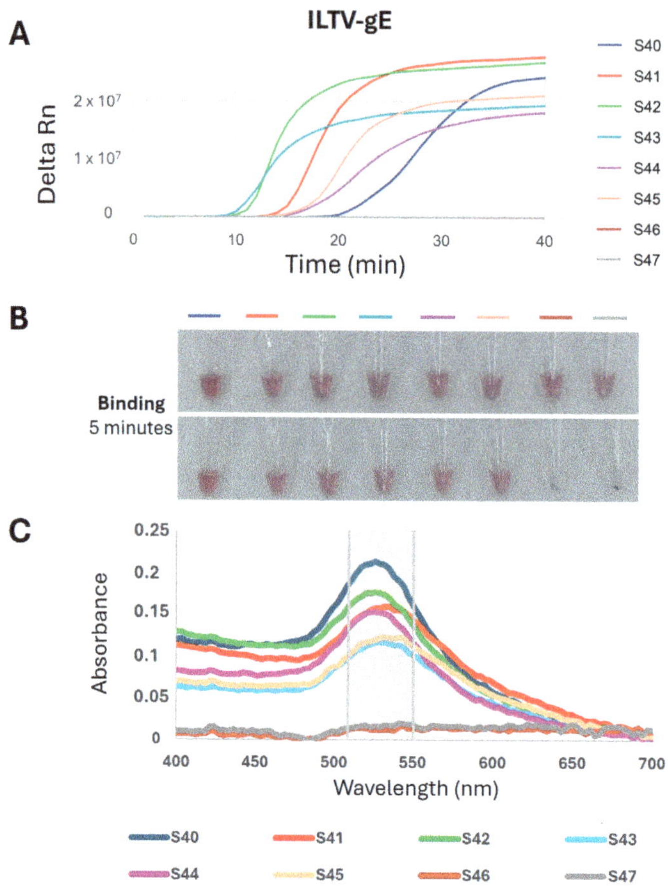

Figure 6. Validation of a two-step LAMP technique coupled with ILTV-gE DNA nanoprobes using crude samples. (**A**) Representative qLAMP amplification plots for ILTV-positive (S40–S45) and ILTV-negative samples (S46–S47) after 40 min of reaction. (**B**) Digital pictures captured after room temperature (RT) incubation of S40 to S47 LAMP products with ILTV-gE DNA nanoprobes, taken at 0 min (upper panel) and 5 min (lower panel). (**C**) VIS-UV spectrum representing the DNA nanoprobes absorbance intensity from 400 nm to 700 nm after 5 min of RT incubation with LAMP products. The region where the maximum absorbance is observed (515–550 nm) is highlighted.

2.5. Limit of Detection (LOD)

The analytical sensitivity, or limit of detection (LOD), of the system combining LAMP with specific ILTV-gE DNA nanoprobes was determined using serial dilutions of the purified TOPO-ILTV-gE plasmid (Figure 7). Based on the results, the LOD of the ILTV detection system was calculated to be 200 copies per reaction.

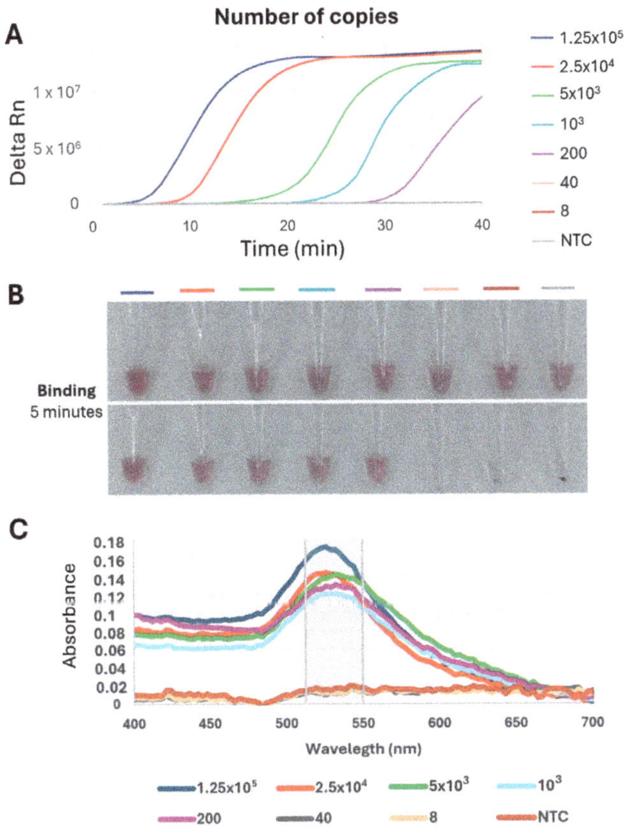

Figure 7. Results of the LOD determination assay using qLAMP coupled with ILTV-gE DNA nanoprobes. (**A**) Amplification plot of qLAMP after 40 min of incubation using serial dilutions of purified TOPO-ILTV-gE plasmid DNA. Each dilution was prepared and analysed in triplicates and amplification plot represents the mean result. (**B**) Digital pictures acquired after room temperature (RT) incubation of the respective LAMP products with ILTV-gE DNA nanoprobes, taken at 0 min (**upper panel**) and 5 min (**lower panel**). (**C**) VIS-UV spectrum representing the DNA nanoprobes absorbance intensity from 400 nm to 700 nm after 5 min of RT incubation with LAMP products. The region where the maximum absorbance is observed (515–550 nm) is highlighted. NTC, non-template control (negative control).

2.6. Optimized Protocol for Molecular Detection of ILTV

We propose a protocol (Figure 8) for the specific and reliable molecular detection of ILTV, using LAMP combined with DNA nanoprobes. Respiratory tract swabs can be used for amplification, either as crude material resuspended in PBS or after being pretreated with a lysis buffer. To validate this protocol, we compared our results with those obtained using DNA/RNA extracts. The amplification reaction includes 12.5 μL of WarmStart MasterMix with dUTP/UDG, 1.6 μM of primers FIP/BIP, 0.2 μM of primers F3/B3, 0.4 μM of primers LF/BF, and 5 μL of sample, for a total volume of 25 μL. Reactions are conducted at 65 °C for 30 min, and results are visualized by incubating 5 μL of RT-LAMP products with 1.5 μL of DNA nanoprobes (50 μM) in a reaction buffer (2 M NaCl, 80 mM $MgCl_2$, and 0.03% Tween-20) for 5 min at room temperature.

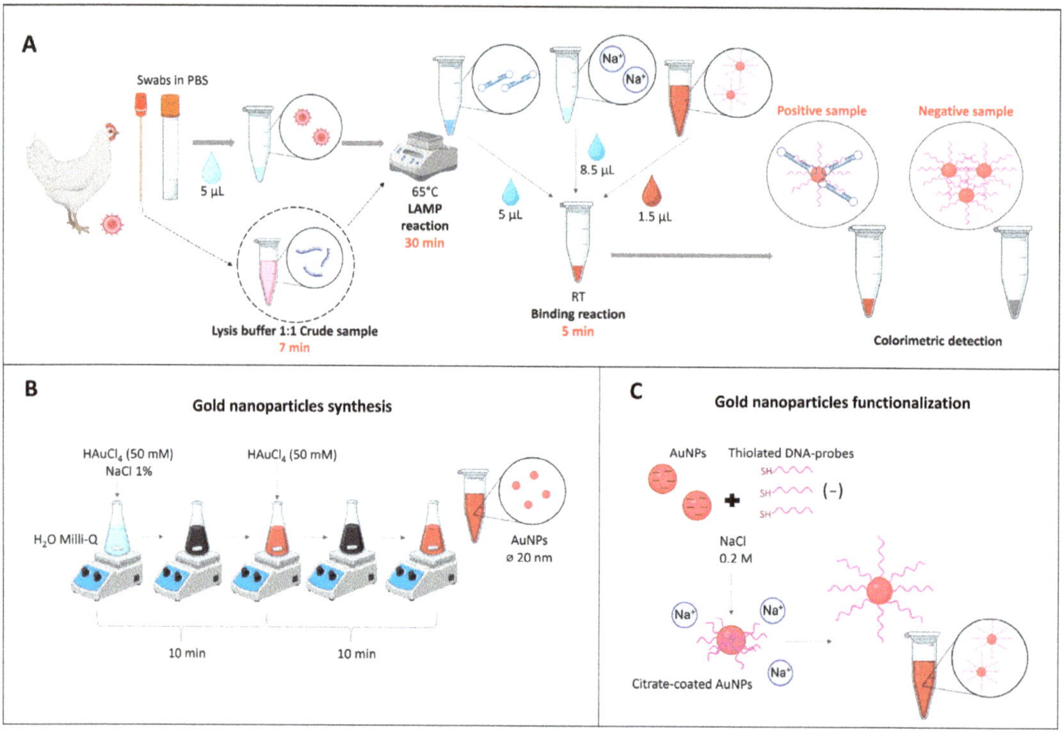

Figure 8. Schematic representation of the experimental procedure. (**A**) LAMP-DNA nanoparticles technique, begins with the sample collection from chickens using respiratory swabs, followed by DNA amplification using LAMP. The detection of amplified LAMP products is then achieved using DNA-functionalized gold nanoparticles, with salt-induced aggregation indicating a positive result. (**B**) Synthesis of 20 nm gold nanoparticles via the Turkevich method. In this process, citrate acts as a reducing and stabilizing agent, conferring a negative charge to the nanoparticles. (**C**) Functionalization of gold nanoparticles (AuNPs) with thiolated DNA probes, although both the AuNPs and the thiolated DNA probes are negatively charged, the addition of salt reduces electrostatic repulsion, allowing for the DNA probes to bind them.

3. Discussion

Accurate and timely diagnosis of viral respiratory infections in poultry is essential for effective outbreak management and control on poultry farms. However, conventional methods often suffer from limitations in sensitivity, specificity, as well as their reliance on complex laboratory equipment and trained personnel. LAMP-based tests offer a promising alternative, providing rapid and specific detection of viral pathogens in the field, eliminating the need for complex thermal cycling. By targeting specific genome sequences, LAMP can facilitate POC diagnostics, which are crucial for quick decision making in farming environments. The integration of colorimetric-based readouts, such as specific DNA nanoprobes, enhances the ease of use and accessibility of these assays.

For on-farm implementation, our diagnostic system notably simplifies testing by obviating nucleic acid purification using a lysis buffer or direct swab processing. Consequently, the elimination of sample transport to a specialized laboratory is achieved. Moreover, the lack of need for complex thermal cycling offers clear time and cost advantages [12]. Unlike our system, reported diagnostic systems for ILTV and other poultry viruses [6,9,11,14] rely on laboratory-based sample processing and, thus, cannot be considered true POC systems.

Notably, without the need for laborious viral genome extraction, our approach achieves ILTV diagnostic sensitivity comparable to standard PCR [17]. Furthermore, this lysis buffer opens up possibilities for multiplexed systems capable of the simultaneous detection of multiple viruses [18].

This study evaluates the performance of LAMP assays targeting the ILTV-gE or -TK genes for a novel on-farm ILTV detection system. While the ILTV-TK gene has been previously assessed [14,19] and is WOAH-recommended for ILTV diagnosis [3], our ILTV-gE gene demonstrated higher sensitivity (100%) compared to the ILTV-TK assay (84%), with both assays exhibiting 100% specificity. To our knowledge, this is the first study to develop LAMP primers targeting the ILTV gE gene, achieving 100% sensitivity and specificity. Thus, the integration of the gE gene target into new on-farm diagnostic systems improves the control and management of the virus directly in poultry production environments.

Additionally, this work presents a LAMP-based assay coupled with colorimetric detection using nanoparticle (AuNP)-based DNA nanoprobes, targeting amplified products with high specificity. This system uses the surface plasmon resonance (SPR) properties of AuNP-based DNA nanoprobes for naked-eye detection, eliminating the need for complex instrumentation. Prior research highlights the potential of AuNP-based systems for pathogen detection, particularly when integrated with LAMP or RT-LAMP and lateral flow biosensors [19–21]. Recently, we successfully applied this approach for aMPV detection, achieving high specificity and sensitivity using DNA nanoprobes targeting the aMPV-F gene [16]. By targeting LAMP-amplified viral DNA sequences, this colorimetric readout minimizes false positives from non-specific amplification or primer dimers, thereby enhancing detection accuracy, specificity, and reliability. This improved specificity addresses those limitations of previous one-step LAMP visualization methods, such as those using hydroxynaphthol blue, turbidimetry, or pH-based systems, which often exhibit reduced sensitivity, especially with field samples [12]. In contrast to fluorescence-based detection, the use of DNA nanoprobes obviates the requirement for specialized instrumentation, resulting in substantial cost savings, and further permits the implementation of multiplexed detection systems employing distinct probes.

ILTV viral loads in respiratory samples often exceed 10^3 genome copies/µL during early infection, enabling reliable detection by most reported systems. While qPCR offers greater sensitivity, isothermal amplification methods like LAMP provide sufficient accuracy for detecting infection at these viral loads, without significant false positives. However, effective early ILTV detection, especially in field settings where sample purity is often a challenge, requires highly sensitive methods. This study focuses on improving the limit of detection (LOD) and optimizing on-farm usability for ILTV diagnostics. Indeed, our combined LAMP-DNA nanoprobe system for ILTV detection achieved an LOD of 200 copies per reaction (40 copies/µL), comparable to previous studies. Similar LODs have been reported for other ILTV targets (100 copies/reaction for the TK gene [13], 200 copies/reaction for a synthetic TK target [14], and 60 copies/µL for the ICP4 gene [22]). However, these methods often lack field validation, a crucial requirement for real-world applicability. Other approaches also present limitations. While a real-time recombinase polymerase amplification (RPA) system reported 100% sensitivity with a 100 copies/reaction LOD using tracheal swabs or lung homogenates [19], its reliance on nucleic acid extraction and fluorescence-based real-time equipment hinders field deployment. Similarly, El-Tholoth et al.'s microfluidic chip-based system targeting the ILTV-polymerase gene [9], though promising for point-of-care (POC) settings, has a high LOD (250 copies/reaction) and reduced sample size, limiting its sensitivity for large-scale diagnostics. The LAMP-DNA nanoprobe system presented in our work addresses these limitations. It significantly improves the LOD for ILTV detection and is optimized for on-farm use, achieving faster detection times, eliminating the need for commercial DNA extraction kits. Moreover, it allows for the visual detection of ILTV within 35 min: 30 min for the LAMP reaction and 5 min for DNA nanoprobe visualization. While a partial lysis buffer slightly increases the assay time, visual detection remains possible within 45 min. These features make this system suitable for point-of-care diagnostics, enabling rapid, simple, cost-effective, and on-site screening for ILTV infection in farm settings.

4. Materials and Methods

4.1. Clinical Samples and Commercial Vaccine

In this study, a comprehensive analysis was conducted involving a total of 54 chickens obtained from farms in Spain, provided by the Centro de Sanidad Avícola de Cataluña y Aragón (CESAC) and Cobb España. Initially, respiratory tract swab samples from 25 chickens (labelled S1 to S25) and their corresponding total nucleic acid extractions (labelled E1 to E25) were processed. These samples had each been previously validated by qPCR or RT-qPCR at the CESAC. Of these, 15 samples tested positive for ILTV (S1–S15), while 10 (S16–S25) tested positive for other respiratory viruses, either aMPV or IBV. Additionally, seven total nucleic acid extracts (E26 to E32) from other chickens, all of which were positive for ILTV by qPCR at the CESAC, were analyzed. Finally, 22 additional respiratory tract swab samples (S26–S47) were evaluated. Of these, 14 were confirmed positive for ILTV by qPCR at CESAC, while 8 were negative for respiratory viruses.

In total, 79 samples were analyzed, including 47 respiratory tract swabs (44 pharyngeal and 3 choanal swabs) along with a panel of 32 total nucleic acid extractions, hereafter referred to as DNA/RNA extractions (Table 1). All respiratory tract swabs were resuspended in 1 mL of phosphate-buffered saline (PBS) and frozen at −20 °C upon arrival at the laboratory. The panel of DNA/RNA extractions was stored at −80 °C.

Table 1. Panel of samples.

Number of Chickens	Samples	Sample Type	Result
25	S1–S25	Respiratory tract swabs [1] (crude samples)	15 positive for ILTV 5 positive for IBV
	E1–E25	DNA/RNA extractions	5 positive for aMPV
7	E26–E32	DNA/RNA extractions	7 positive for ILTV
22	S26–S47	Respiratory tract swabs [1] (crude samples)	14 positive for ILTV 8 negative for respiratory viruses
Total: 54	Total: 79	Total: 47 Respiratory tract swabs [1] (crude samples) 32 DNA/RNA extractions	

[1] Respiratory tract samples correspond to pharyngeal swabs, except for sample S4 (respiratory tract swab), which was positive for ILTV but lacks specific information on the tract of origin, and samples S16, S17, and S18 (choanal swabs), which were positives for IBV.

Live attenuated ILT vaccine from Poulvac®, Salsbury 146 strain (Zoetis Spain S.L., Madrid, Spain) provided in vials with a lyophilized suspension at $\geq 10^{2.5}$ EID50 per doses (0.03 mL), was kindly supplied by the company and used as a positive control. The vials were reconstituted in 1 mL of PBS, and 200 μL total nucleic acids was purified using Purelink viral RNA/DNA Mini Kit (Thermo Fisher Scientific, Waltham, MA, USA), following the manufacturer's recommendations.

4.2. LAMP Primers and Oligonucleotide Probes Design

LAMP primers and oligonucleotide probes for the ILTV envelope glycoprotein E (ILTV-gE) and thymidine kinase (ILTV-TK) coding regions were designed with PrimerExplorerV5 (https://primerexplorer.jp/e/, accessed 10 June 2024) using default parameters. This ensured optimal performance by considering GC content (40–65%), secondary structure formation (free energy below −2.5) and melting temperature (Table 2). Primer sets, each comprising four primers (FIP, F3, BIP, and B3), were designed based on target sequence information. The FIP (BIP) primer consists of sequences from the F1c (B1c) and F2 (B2) regions. The F1, F2, and F3 primers are approximately 20 base pairs long and are selected from the target gene, while B1, B2, and B3 are of similar length (20 bp long) and derived from the complementary strand. The regions F1c and F1, as well as B1 and B1c, are complementary to each other. To design the assay, 34 ILTV-gE gene sequences and 127 ILTV-TK gene sequences published in GenBank (https://www.ncbi.nlm.nih.gov/nuccore, accessed 7 June 2024) were compared using Clustal Omega (https://www.ebi.ac.uk/Tools/msa/clustalo/, accessed 7 June 2024). LAMP primers, including loop primers, were selected in the more conserved regions within the gene, as well as in regions with a higher potential for primer sets. Oligonucleotide LAMP primers used in this study were synthesized by Macrogen-Inc. (Seoul, Republic of Korea). The ILTV-gE oligonucleotide probes, used for functionalization of AuNPs, were designed to target a conserved 17-nucleotide sequence within the LAMP product of the viral gE gene (Table 2). Their secondary structure was further evaluated using the RNAfold server from the Vienna RNA Web Service [23]. Oligonucleotide probes with relatively higher Gibbs free energy (ΔG0) values were selected for synthesis. All oligonucleotide probes utilized in this study were synthesized by Sigma-Aldrich (Merck, Darmstadt, Germany) and featured a 5′-thiol modification for enhanced functionality.

Table 2. Primer sets for LAMP targeting ILTV-gE and ILTV-TK gene sequences, along with corresponding oligonucleotide probe for colorimetric detection.

Primer	Sequence 5′-3′
ILTV-gE-F3	GGCCATGGAAACTACATCGT
ILTV-gE-B3	GGTCGGTGGGAAGACTRT
ILTV-gE-FIP	CCTCTGACGGCTTGATGGTGTCGCGTTCTTTCGCACGTAGA
ILTV-gE-BIP	CCACGCGCACGTGGAATTACATAGGTGGGGCTGTTGTCG
ILTV-gE-Floop	ATCTCCACCTCGTGCGGTG
ILTV-gE-Bloop	TGCTGCCGTTTCATGAACTCA
ILTV-TK-F3	GAAATACAAACCCTAAAGGCT
ILTV-TK-B3	GGAAACACAAACATGCCG
ILTV-TK-FIP	AGCAATAGTCATCTGAACTTCCGTCCGGAAAACTTGAATGTCG
ILTV-TK-BIP	TTGGGATCTGAACGCTGCTACCTATCTACGAGGAATAAGACA
ILTV-TK-Floop	TACGCGACGAGACGCCT
ILTV-TK-Bloop	CAGCTGCATCCGGACCAGA
ILTV-gE-Probe	TTTTTTTTTTATGTGACCCCTGTTCTTA

4.3. LAMP and Quantitative LAMP (qLAMP) Reaction Settings

LAMP assays (end-point and qLAMP) were performed using either crude samples or DNA/RNA extractions with WarmStart® DNA Polymerase or WarmStart® Fluorescent LAMP/RT-LAMP Kit (New England Biolabs, MA, USA), both containing uracil-DNA glycosylase (UDG). To achieve a complete POC diagnostic system, we developed a rapid partial lysis buffer [24] supplemented with Proteinase-K (1U) (NZYTech, Lisbon, Portugal). Swabs from the respiratory tract of chickens, already resuspended in PBS 1X and stored at −20 °C, were used to prepare the reaction by mixing 5 µL of the PBS 1X suspension with an equal volume (1:1) of partial lysis buffer, producing a final volume of 10 µL. This 10 µL mixture was incubated at 56 °C for 5 min for optimal activity of proteinase-K, followed by incubation at 95 °C over 2 min for enzyme inactivation.

The end-point LAMP reaction was performed in a 25 µL volume, consisting of 12.5 µL of WarmStart MasterMix, 1 µL of each primer. The final optimized concentrations were 0.16 µM for internal FIP/BIP primers, 0.1 µM for outer F3/B3 primers, and 0.2 µM for loop LF/LB primers. For DNA/RNA extractions (E1–E32) and vaccines, 2.5 µL of sample was added after 1:1 dilution in Milli-Q water. For crude samples or vaccines pretreated with rapid partial lysis buffer, 5 µL of pretreated sample was added. For crude samples and vaccines without pretreatment, 2.5 µL was used. Milli-Q water was added to adjust the reaction volume to 25 µL. The reaction mixture was prepared at room temperature to allow for UDG activity, followed by incubation at 61–69 °C for 30 min with Bst 2.0 amplification and subsequent enzyme inactivation at 80 °C for 5 min in a Mastercycler® nexus GX2 thermocycler (Eppendorf, Hamburg, Germany). LAMP reactions were analyzed using 2% TAE-agarose gel electrophoresis for visualization.

For qLAMP, 0.5 µL of LAMP fluorescent dye was added to the reaction mix. The fluorescence signal was measured using a QuantStudio™ 5 Real-Time PCR System (Thermo Fisher Scientific), which includes an end-point melt curve step for specificity validation.

To ensure the reproducibility, each purified sample was analyzed in duplicate on separate days by different operators. Crude samples were tested with and without lysis buffer treatment to confirm the robustness of the results.

4.4. Gold Nanoparticle Synthesis and Functionalization

The 20 nm AuNPs were synthesized following the Turkevich method [25]. Briefly, 50 mL of Milli-Q water was heated to boiling under gentle agitation in a 50 mL Erlenmeyer flask. Once boiling, the agitation speed was increased to 1100 rpm, and 400 µL of HAuCl4 (50 mM) and 2 mL of NaCl dihydrate (1%) were injected into the center of the vortex created by the stirring magnet. The solution progressed from clear to black and finally to deep red. A 5 cm diameter Petri dish was placed on top of the flask to minimize water evaporation. Heating and refluxing under agitation continued for 10 min to produce 16 nm AuNPs. To achieve the target particle size of 20 nm, an additional 400 µL of 50 mM HAuCl4 solution was injected while keeping the flask on the heat source, and the conditions were maintained for an additional 10 min. The reaction was then stopped and cooled at 23 °C (Figure 8B). The resulting 20 nm AuNPs were centrifuged at 8500 rpm for 15 min using a Minispin (Eppendorf), and the pellet was redispersed in Milli-Q water. This process was repeated until the UV-visible spectrum showed an absorbance of 0.1 at the maximum absorption wavelength of 520 nm.

To functionalize AuNPs with synthetic oligonucleotides, a previously described protocol was followed [16], incorporating an initial dithiothreitol (DTT) nanoprobe reduction and a modified salt-aging process adapted from Hurst et al. [26] (Figure 8C). Following a final rinse to remove unbound probe-oligonucleotides, the resulting DNA nanoprobes were resuspended in 10 mM PBS and stored at 4 °C until use.

4.5. Detection by DNA Nanoprobes

To optimize colorimetric detection, 1.5 µL of ILTV-gE or ILTV-TK DNA nanoprobes was incubated with 5 µL of the corresponding end-point LAMP product in a reaction solution buffered with 25 mM Tris at pH 7.5. The reaction buffer was supplemented with binding buffer, containing 2 M NaCl, 80 mM $MgCl_2$, and 0.03% Tween-20. The reaction was carried out in a final volume of 15 µL. Nuclease-free water was used as a non-template control (NTC). Colorimetric detection was performed at room temperature for 5 min. Detection parameters were adjusted based on absorbance readings at a wavelength of 400–700 nm, focusing on the maximum absorbance at 540 nm using a NanoDrop 2000 Spectrophotometer (Thermo Fisher Scientific). All binding experiments were performed in duplicate.

4.6. Plasmid Constructs

ILTV-gE coding region was amplified using 10 µM of external LAMP primer (F3 and B3) by conventional PCR with AmpliTaq DNA Polymerase (Thermo Fisher Scientific). The PCR product was then cloned using TOPO™ TA Cloning™ Kit and transformed into TOP10 Chemically Competent *E. coli* (Thermo Fisher Scientific). The TOPO-ILTV-gE plasmid was confirmed by automated Sanger sequencing.

4.7. Determination of the Limit of Detection (LOD)

To determine the LOD, five-fold dilutions of a plasmid DNA stock at 20 ng/µL were performed. These dilutions, ranging from 250,000 copies to 8 copies, were then used for qLAMP amplification with ILTV-gE primers. Each dilution was tested in triplicate, and the mean values are presented in an amplification plot (Figure 7). The corresponding LAMP products were assayed for DNA nanoprobe detection in duplicate experiments. The plasmid and fragment sizes were used to calculate the number of plasmid copies in each dilution for the LOD calculation. The assay was conducted on two separate days to assess the consistency and reproducibility of the results.

5. Conclusions

In conclusion, this study underscores the crucial advancements in diagnosing viral respiratory infections in poultry using LAMP-based assays. The innovative integration of colorimetric readouts with DNA nanoprobes significantly enhances the ease of use and accessibility of these tests. Our research demonstrates that targeting the ILTV-gE gene with LAMP assays achieves 100% sensitivity and specificity, outperforming the ILTV-TK gene assay. This marks the first study to develop primers for the ILTV-gE gene, paving the way for new on-farm diagnostic systems that improve virus control and management in poultry production environments.

The direct processing of swabs in this work eliminates the need for nucleic acid purification, facilitating true point-of-care diagnosis. With this system, we obtained results in only 35 min (30 min for LAMP and 5 min for detection). Optionally, a partial lysis buffer can be used, which slightly increases the total time but still allows for detection in under 45 min, while enabling multiplex amplification of different viruses. These approaches save time and reduce costs associated with sample transportation and laboratory analysis. They provide visible results to the naked eye, without requiring sophisticated equipment, supporting early, rapid, simple, and cost-effective in situ screening for ILTV infections on farms. Although further field trials will be necessary for its validation as a POC system, the high specificity, sensitivity, and ease of use of this test support its immediate applicability on farms.

Author Contributions: Conceptualization, P.C.-C., R.M. and L.B.; formal analysis, P.C.-C., C.T. and R.M.; investigation, P.C.-C., C.T., E.E.M., M.B., J.M.C. and C.J.S.-L.; methodology P.C.-C., R.M. and L.B.; writing—original draft, P.C.-C., A.D., R.M. and L.B.; writing—review and editing, P.C.-C., C.T., E.G.-L., A.D., R.M. and L.B.; resources, E.G.-L., A.D., M.B. and J.M.C.; funding and acquisition, A.D. and L.B.; supervision, A.D., R.M. and L.B.; project administration, R.M. and L.B. All authors have read and agreed to the published version of the manuscript.

Funding: This research was funded by the State Research Agency of the Spanish Government, grant PID2020-114956GB-100. P.C.C. was supported by PAIT08/21-02/2021-08 contract, the Investigo Program, grant CT19/23-INVM-19 from MICINN; and the Industrial Doctorate Program, grant IND2019/BIO-17124 from Autonomous Community of Madrid (Spain).

Institutional Review Board Statement: Not Applicable.

Informed Consent Statement: Not Applicable.

Data Availability Statement: The original contributions presented in this study are included in the article. Further inquiries can be directed to the corresponding authors.

Acknowledgments: We greatly acknowledge the Center for Avian Health of Cataluña and Aragón (CESAC) and COBB España for their strong support and assistance in accessing diverse resources used in the study.

Conflicts of Interest: The authors declare that the research was conducted in the absence of any commercial or financial relationships that could be construed as a potential conflict of interest.

Abbreviations

The following abbreviations are used in this manuscript:

ILTV	Infectious laryngotracheitis virus
ORFs	Open reading frames
IBV	Infectious bronchitis virus
aMPV	Avian metapneumovirus
LAMP	Loop-mediated isothermal amplification
qPCR	Real time PCR
qLAMP	Quantitative LAMP
POCT	Point-of-care test
AuNPs	Gold nanoparticles
LSPR	Localized surface plasmon resonance
ILTV-gE	ILTV glycoprotein-E
ILTV-TK	ILTV thymidine kinase
NTC	Non-template control
UV	Ultraviolet
LOD	Limit of detection
WHOA	World health organization of animals
ICP4	Infected cell protein 4
aMPV-F	aMPV fusion protein
Bst	*Bacillus stearothermophilus*

References

1. García, M.; Spatz, S. Infectious Laryngotracheitis. In *Diseases of Poultry*; John Wiley & Sons, Inc.: Hoboken, NJ, USA, 2020; pp. 189–209, ISBN 978-1119371168.
2. Gowthaman, V.; Kumar, S.; Koul, M.; Dave, U.; Murthy, T.R.G.K.; Munuswamy, P.; Tiwari, R.; Karthik, K.; Dhama, K.; Michalak, I.; et al. Infectious Laryngotracheitis: Etiology, Epidemiology, Pathobiology, and Advances in Diagnosis and Control—A Comprehensive Review. *Vet. Q.* **2020**, *40*, 140–161. [CrossRef] [PubMed]
3. WOAH. Avian Infectious Laryngotracheitis. Available online: https://www.woah.org/en/what-we-do/standards/codes-and-manuals/terrestrial-manual-online-access/ (accessed on 11 December 2024).

4. Menendez, K.R.; García, M.; Spatz, S.; Tablante, N.L. Molecular Epidemiology of Infectious Laryngotracheitis: A Review. *Avian Pathol.* **2014**, *43*, 108–117. [CrossRef] [PubMed]
5. Padzil, F.; Mariatulqabtiah, A.R.; Tan, W.S.; Ho, K.L.; Isa, N.M.; Lau, H.Y.; Abu, J.; Chuang, K.-P. Loop-Mediated Isothermal Amplification (LAMP) as a Promising Point-of-Care Diagnostic Strategy in Avian Virus Research. *Animals* **2021**, *12*, 76. [CrossRef] [PubMed]
6. Kirkpatrick, N.C.; Mahmoudian, A.; O'Rourke, D.; Noormohammadi, A.H. Differentiation of Infectious Laryngotracheitis Virus Isolates by Restriction Fragment Length Polymorphic Analysis of Polymerase Chain Reaction Products Amplified from Multiple Genes. *Avian Dis.* **2006**, *50*, 28–34. [CrossRef] [PubMed]
7. Oldoni, I.; García, M. Characterization of Infectious Laryngotracheitis Virus Isolates from the US by Polymerase Chain Reaction and Restriction Fragment Length Polymorphism of Multiple Genome Regions. *Avian Pathol.* **2007**, *36*, 167–176. [CrossRef]
8. Urdea, M.; Penny, L.A.; Olmsted, S.S.; Giovanni, M.Y.; Kaspar, P.; Shepherd, A.; Wilson, P.; Dahl, C.A.; Buchsbaum, S.; Moeller, G.; et al. Requirements for High Impact Diagnostics in the Developing World. *Nature* **2006**, *444* (Suppl. S1), 73–79. [CrossRef]
9. El-Tholoth, M.; Bai, H.; Mauk, M.G.; Anis, E.; Bau, H.H. Molecular Detection of Infectious Laryngotracheitis Virus in Chickens with a Microfluidic Chip. *Animals* **2021**, *11*, 3203. [CrossRef]
10. Notomi, T.; Okayama, H.; Masubuchi, H.; Yonekawa, T.; Watanabe, K.; Amino, N.; Hase, T. Loop-Mediated Isothermal Amplification of DNA. *Nucleic Acids Res.* **2000**, *28*, E63. [CrossRef]
11. Karanis, P.; Ongerth, J. LAMP—A Powerful and Flexible Tool for Monitoring Microbial Pathogens. *Trends Parasitol.* **2009**, *25*, 498–499. [CrossRef]
12. Atceken, N.; Munzer Alseed, M.; Dabbagh, S.R.; Yetisen, A.K.; Tasoglu, S. Point-of-Care Diagnostic Platforms for Loop-Mediated Isothermal Amplification. *Adv. Eng. Mater.* **2023**, *25*, 2201174. [CrossRef]
13. Xie, Q.; Ji, J.; Pickens, T.T.; Du, L.; Cao, Y.; Li, H.; Wang, L.; Ma, J.; Bi, Y. Rapid Detection of Infectious Laryngotracheitis Virus Isolates by Loop-Mediated Isothermal Amplification. *J. Virol. Methods* **2010**, *165*, 71–75. [CrossRef] [PubMed]
14. Yu, J.; Xie, J.; Cao, Y.; Kang, R.; Lin, Y.; Wei, Y.; Liao, D.; Li, X.; Ye, Y.; Pan, M.; et al. A Modified Loop-Mediated Isothermal Amplification Method for Detecting Avian Infectious Laryngotracheitis Virus. *Anim. Biotechnol.* **2021**, *32*, 766–773. [CrossRef] [PubMed]
15. El-Tholoth, M.; Anis, E.; Bau, H.H. Two Stage, Nested Isothermal Amplification in a Single Tube. *Analyst* **2021**, *146*, 1311–1319. [CrossRef] [PubMed]
16. Cea-Callejo, P.; Arca-Lafuente, S.; Gomez-Lucia, E.; Doménech, A.; Biarnés, M.; Blanco, A.; Benítez, L.; Madrid, R. An Affordable Detection System Based on RT-LAMP and DNA-Nanoprobes for Avian Metapneumovirus. *Appl. Microbiol. Biotechnol.* **2024**, *108*, 414. [CrossRef]
17. Mahmoudian, A.; Kirkpatrick, N.C.; Coppo, M.; Lee, S.-W.; Devlin, J.M.; Markham, P.F.; Browning, G.F.; Noormohammadi, A.H. Development of a SYBR Green Quantitative Polymerase Chain Reaction Assay for Rapid Detection and Quantification of Infectious Laryngotracheitis Virus. *Avian Pathol. J. WVPA* **2011**, *40*, 237–242. [CrossRef]
18. Oscorbin, I.; Filipenko, M. Bst Polymerase—a Humble Relative of Taq Polymerase. *Comput. Struct. Biotechnol. J.* **2023**, *21*, 4519–4535. [CrossRef] [PubMed]
19. Zhu, Y.; Zeng, F.; Sun, J.; Liu, X.; Wu, M.; Huang, B.; Lian, Y.; Xiao, L.; Ma, L.; Zhang, S.; et al. Application of Recombinase Polymerase Amplification Method for Rapid Detection of Infectious Laryngotracheitis Virus. *Mol. Cell. Probes* **2020**, *54*, 101646. [CrossRef]
20. Ou, S.-C.; Giambrone, J.J. Infectious Laryngotracheitis Virus in Chickens. *World J. Virol.* **2012**, *1*, 142–149. [CrossRef] [PubMed]
21. Akalın, P.; Yazgan-Karataş, A. Development of a Nucleic Acid-Based Lateral Flow Device as a Reliable Diagnostic Tool for Respiratory Viral Infections. *MethodsX* **2023**, *11*, 102372. [CrossRef]
22. Chen, X.; Zhou, Q.; Li, S.; Yan, H.; Chang, B.; Wang, Y.; Dong, S. Rapid and Visual Detection of SARS-CoV-2 Using Multiplex Reverse Transcription Loop-Mediated Isothermal Amplification Linked With Gold Nanoparticle-Based Lateral Flow Biosensor. *Front. Cell. Infect. Microbiol.* **2021**, *11*, 581239. [CrossRef]
23. Gruber, A.R.; Lorenz, R.; Bernhart, S.H.; Neuböck, R.; Hofacker, I.L. The Vienna RNA Websuite. *Nucleic Acids Res.* **2008**, *36*, W70–W74. [CrossRef] [PubMed]
24. Arca-Lafuente, S.; Yépez-Notario, C.; Cea-Callejo, P.; Lara-Aguilar, V.; Crespo-Bermejo, C.; Martín-Carbonero, L.; de los Santos, I.; Briz, V.; Madrid, R. Development and Validation of a New and Rapid Molecular Diagnostic Tool Based on RT-LAMP for Hepatitis C Virus Detection at Point-of-Care. *Methods* **2024**, *232*, 43–51. [CrossRef] [PubMed]
25. Turkevich, J.; Stevenson, P.C.; Hillier, J. A Study of the Nucleation and Growth Processes in the Synthesis of Colloidal Gold. *Discuss. Faraday Soc.* **1951**, *11*, 55. [CrossRef]
26. Hurst, M.R.H.; Young, S.D.; O'Callaghan, M. Development of a Speciesspecific Probe for Detection of Serratia Entomophila in Soil. *N. Z. Plant Prot.* **2008**, *61*, 222–228. [CrossRef]

International Journal of *Molecular Sciences*

MDPI

Review

Quantum Dot-Based Luminescent Sensors: Review from Analytical Perspective

Alissa Loskutova, Ansar Seitkali, Dinmukhamed Aliyev and Rostislav Bukasov *

Department of Chemistry, Nazarbayev University, Kabanbay Batyr Ave. 53, Astana 010000, Kazakhstan
* Correspondence: rostislav.bukasov@nu.edu.kz

Abstract: Quantum Dots (QDs) are small semiconductor nanoparticles (<10 nm) with strong, relatively stable, and tunable luminescent properties, which are increasingly applied in the sensing and detection of various analytes, including metal ions, biomarkers, explosives, proteins, RNA/DNA fragments, pesticides, drugs, and pollutants. In this review, we critically assess recent developments and advancements in luminescent QD-based sensors from an analytical perspective. We collected, tabulated, and analyzed relevant data reported in 124 peer-reviewed articles. The key analytical figures of merit, including the limit of detection (LOD), excitation and emission wavelengths, and size of the particles were extracted, tabulated, and analyzed with graphical representations. We calculated the geometric mean and median LODs from those tabulated publications. We found the following geometric mean LODs: 38 nM for QD-fluorescent-based sensors, 26 nM for QD-phosphorescent-based sensors, and an impressively low 0.109 pM for QD-chemiluminescent-based sensors, which demonstrate by far the best sensitivity in QD-based detection. Moreover, AI-based sensing methods, including the ATTBeadNet model, optimized principal component analysis(OPCA) model, and Support Vector Machine (SVM)-based system, were reviewed as they enhance the analytical performance of the detection. Despite these advances, there are still challenges that include improvements in recovery values, biocompatibility, stability, and overall performance. This review highlights trends to guide the future design of robust, high-performance, QD-based luminescent sensors.

Keywords: quantum dots; applications of nanomaterials; LOD; sensitivity; luminescence; chemiluminescence; fluorescence; phosphorescence

1. Introduction

Quantum dots are nanoparticles with semiconducting properties whose size is, at most, twice as large as the size of the Bohr radius of corresponding exciton and whose electronic properties differ from that of the bulk material they were made of due to a higher surface-to-volume ratio and consequent high reactivity [1–3]. They are usually composed of the elements of the groups II A to VI A or III A to V A; however, there are also variations, with IV A–VI A, I B–III A–VI A, or IV A group elements [1,4]. The size of QDs varies from a few nanometers to a few tens of nanometers, which allows for a gradual increase of 1–2 eV in the energy of the electronic transition compared with the bulk material [2,3]. Their unique photophysical characteristics, such as wide absorption spectra, adjustable and small emission profiles, and strong resistance towards photobleaching, allow them to have many applications in biosensing and imaging [5,6]. QDs have discrete energy levels, and by changing their size or alloying their core, one can modulate their bandgap [7–9].

QDs also exhibit one of the most important properties of quantum confinement, which makes them useful in the production of light-emitting diodes (LEDs), transistors, and solar cells [3].

Quantum dots have attracted the interest of scientists since the 1980s. They emerged as a result of advancements in chemical processes for production in the doping of multicomponent glasses, the synthesis of nanoparticles by colloidal chemistry, and the epitaxial growth of nanostructures and thin films [10]. In 1981, Alexey Ekimov and Alexey Onushchenko designed the first QDs made of CuCl [11]. Microcrystals grown in the transparent dielectric matrix have demonstrated the short wave shift (up to 0.1 eV) of exciton absorption bands due to the quantum size effect. They have used multicomponent silicate glasses with copper and chlorine added as their compounds in a concentration of about 1%. The treatment of such glasses with high temperatures created the characteristic spectra of the exciton absorption of CuCl crystals in the transparent regions of the matrix, which points out the formation of a solid supersaturated solution due to phase decomposition. The term "quantum dots" was first introduced by M. A. Reed et al. in their 1986 publication about spatial quantization in GaAs–AlGaAs multiple quantum dots [12]. Over the next fifteen years, the quantum size effect in the absorption spectra was studied for quantum dots fabricated from copper halides and cadmium chalcogenides, such as CuCl, CuBr, CdSe, CdS, and CdSSe [13–20].

The most common synthetic methods of synthesizing QDs include colloidal and hydrothermal synthesis. Colloidal synthesis, especially the high-temperature hot injection method, allows for the production of high-quality QDs with controlled size and composition by creating conditions for rapid nucleation and growth [21]. That method results in the high purity, uniformity, and photochemical stability of QDs, which is crucial for optoelectronic applications [3]. Recent advancements, including the use of new ligands and environmentally friendly, aqueous-based methods, have improved control over QD properties [3]. Generally, the aqueous synthesis of QDs requires the presence of three starting components: a metal precursor, which is typically a water-soluble salt; a chalcogenide, such as S, Se, or Te; and a stabilizer [10]. Chalcogenizers, such as NaHTe or NaHSe, are being made in situ from Se or Te powder and $NaBH_4$ as a reducing agent; however, sometimes potassium salts of chalcogens, such as K_2TeO_3, are being used instead of pure elements [22–24]. Stabilizers are meant to provide control over nucleation at early stages and restrict particle growth, which allows precise size control [10]. Additionally, machine learning is increasingly used to optimize synthesis conditions, enhancing efficiency and performance prediction [25–27]. The hydrothermal method is a cost-effective and easy approach for synthesizing quantum dots with a controlled size and shape. It involves heating a precursor solution containing metal ions and ligands in a sealed, high-pressure vessel. As the temperature and pressure increase, the solution becomes supersaturated, leading to QD nucleation. Parameters like the temperature, pressure, and reaction time can be adjusted to tune the QD properties. Compared to other techniques, hydrothermal synthesis operates at lower temperatures and is more accessible [3].

Apart from traditional II–VI group QDs, silicon QDs (SiQDs) have gained the attention of researchers due to their simple synthesis, low cost, excellent optical characteristics, low biotoxicity, and environmental friendliness [28]. Reports have been made about the application of SiQDs for the detection of L-cysteine, hydrogen sulfide, vitamin B12, nitrite, hydroquinone, and other types of compounds [29–33]. Another new direction in the development of QDs is focused on perovskite QDs (Per-QDs). Their adjustable redox characteristics, thermal and chemical durability, active electrical structure, electronic and ionic conduction, oxygen sorption capacity, and high oxygen mobility make them solid candidates for optical sensors to detect trace amounts of both organic and inorganic compounds, as well as biomolecules [34]. Among synthetic methods, the sol-gel approach appears to be particularly advantageous for Per-QDs since it allows for precise control over stoichiometry and homogeneity of the final product [35].

In order to stabilize QDs and achieve desirable surface characteristics, scientists utilize ligand engineering. For example, Zhang and coworkers have developed fluorophenethyl ammonium bromide (FPEABr) ligands to modify the surface of $CsPbBr_3$ Per-QDs. Their results have shown successful adsorption of ligands onto the QDs' surface and a decrease in bromine vacancy defects. The application of FPEABr allowed for a strong quantum yield of over 90% and the good stability of the material, which have led to an improved maximum luminance of Per-QD light-emitting diodes (QLEDs) [36]. Jiang et al. have utilized phenethylamine (PEA) to modify the InP QDs used in QLEDs, where some of the long-chain oleylamine ligands were replaced by short-chain PEA ligands. PEA-engineered QDs have exhibited a higher quantum yield (71.0% vs. 85.5%) and a higher maximum external quantum efficiency (1.9% vs. 3.5%) [37].

Machine learning (ML) currently is also widely used to design and adjust the synthetic procedure. It determines the required conditions for synthesizing QDs with desirable properties, thus optimizing the process. In contrast, it is nearly impossible to consider all the possible parameter combinations manually [38]. For example, Chen et al. have demonstrated the use of ML tools to analyze the various conditions required for the preparation of CsPbX3 (where X stands for halogen) Per-QDs. The authors report that ML not only assisted in the elucidation of the nucleation-growth process but also allowed for the synthesis of Per-QDs with an exact emission wavelength and a full width at half the maximum (FWHM) of the photoluminescence spectrum. With ML, they have analyzed different values of parameters such as the Pb:Cs molar ratio, reaction halogen ratio, reaction temperature, and flow rate. As a result, the ML model has satisfactorily predicted the synthesis result, depending on the given conditions. Moreover, the developed tool allows for applying other restrictions apart from the wavelength and the FWHM, including, for instance, the shortest time, lower temperature, or minimum lead amount [38]. Abdel-Latif et al. have utilized AI technology similarly; however, the number of adjustable parameters was expanded to 10, while the accessible parameter space exceeded 2×10^7 [39]. Such opportunities offered by AI development play a crucial role in designing new sensor devices with improved performance.

Before proceeding to our review, we also searched for previously published review studies on the topic of luminescent quantum dot-based sensing. Most of the review papers that we found were published in the last 4 years. Atabaev's 2018 mini-review of heteroatom-doped carbon dots (CDs) focuses on synthesis methods and their adjustable photoluminescence for sensing and bioimaging applications [40]. Ding et al. discuss the quantum dot–based biosensors for antibiotic residues, detailing QD types (Cd-based, carbon, graphene), functionalization strategies, and optical transduction mechanisms (quenching, energy transfer) in complex samples [41]. Sabzehmeidani et al. provide insights into various QD platforms (metal, carbon, g-C_3N_4) and their fluorescent/electrochemical sensing mechanisms to detect antibiotics, evaluating detection metrics across luminescent and electrochemical methods [42]. Mehta and colleagues review electrochemiluminescent (ECL) QDs, discussing synthesis, surface engineering, ECL mechanisms, and their integration into sensitive, low-background analytical devices [43]. Finally, the latest 2023 review by Sargazi et al. offers a broad comparison of fluorescent nanomaterials, not limited to QDs, for antibiotic sensing in food and environmental matrices, emphasizing design principles and fluorescence pathways [44]. Among those reviews published previously, we found no exhaustive comparison of analytical performance parameters, such as the limit of detection (LOD), analytical range, excitation/emission wavelength, relative standard deviation of luminescence intensity signal (RSD), and others. Here we tabulate the results from the papers reporting these parameters, calculate the average and median values for three major luminescence methods, including fluorescence, phosphorescence, and chemiluminescence, and finally discuss the differences between these methods, as well as their strong and weak sides.

2. Sensing Applications Based on the Fluorescence of QDs

Fluorescence, along with phosphorescence, is one of the two types of photoluminescence, which can be described as the emission of light resulting from the absorption of electromagnetic radiation, usually in the form of light. For example, when excited by UV radiation, fluorescent material emits light within the visible range. The absorption of a photon with a certain wavelength excites the molecule, making it reach a higher energy level. Then, after a certain time, the molecule relaxes back to the ground state and emits a photon. For fluorescence, this transition occurs from the excited singlet state to the ground singlet state, where the spin of the molecule does not change, and thus the fluorescence lifetime is short (in the ns range) [28].

Quantum dots possess properties different from bulk materials due to quantum confinement and are frequently used as fluorescent nanoprobes. QDs emit fluorescence due to the recombination of electrons and holes (excitons), and this fluorescence can be influenced by modifications to their surface or ligands, which affect their exciton recombination efficiency. By changing these surface characteristics, QDs can be tuned for application in fluorescent sensors that detect target analytes through their either direct or indirect interactions with QDs. The analyte concentration is determined based on the linear relationship between the fluorescence intensity and the analyte concentration. Functional QD-based sensors are typically developed by adjusting surface ligands to enable specific interactions—such as adsorption or chelation—with target molecules [45]. The fluorescent properties of QDs can be tuned by adjusting their size: bigger particles will have a smaller band gap between valence and conduction bands and thus will emit light with longer wavelengths (red shift), while smaller particles appear to emit light with shorter wavelengths (blue shift) due to the larger band gap. The fluorescence lifetime, ability to control the particle size, narrow emission, and wide excitation spectra of QDs allow for their wide application in biosensing, bioimaging, nanomedicine, LED lamps, and solar cells [46–48]. One of the first applications of QDs in fluorescence sensor design was demonstrated by Depu Chen's group in 2001 for the detection of human Immunoglobulin G (IgG) with ZnS-coated CdSe QDs attached to the detection antibody of a sandwich-type immunocomplex with an LOD of 1 nM [49].

There are multiple applications of surface-enhanced fluorescence in biosensing, including in surface-enhanced fluorescence (SEF), that apply QDs and CDs [50]. For instance, there are applications of commercially available QDs and organic CDs in SEF at the surface of bacteria on various substrates including gold, and Al film, which reported high SEF enhancement factors (EF) of up to several 100s and a high contrast of the same order of magnitude up to several 100s [51–53]. It was also found that SEF EF had a strong negative correlation with the measured toxicity of QDs, at least for the observation of QD-labeled bacteria, or, in other words, the highest fluorescent enhancement was demonstrated by the least toxic QDs [51]. Therefore, the cytotoxicity of QDs should be minimized for the optimal/effective imaging of QD-labeled bacteria.

Oftentimes, QD-based sensing devices rely on fluorescence quenching strategies such as FRET. Fluorescence (or Förster) resonance energy transfer (FRET) is a quantum mechanical process of resonance between transition dipoles. It requires donor (fluorophore) and acceptor (chromophore) particles, and the fluorescence spectrum of the donor particle should overlap with the absorption spectrum of the acceptor particle. FRET is very sensitive to the distance between particles, which should be kept at 5–10 nm [54]. Their wide absorption and tunable emission bands, which allow for the controllable overlap of emission and absorption spectra, as well as the large Stokes shift to suppress the direct excitation from acceptors, make QDs excellent FRET donors [55].

In addition to FRET, other mechanisms, such as the inner filter effect (IFE), static quenching, and dynamic quenching can also influence fluorescence. IFE leads to a decrease in fluorescence intensity without affecting the fluorophore's lifetime, since it is created by the absorption of excitation or the emission of light by other sample components. Oppositely, static quenching involves the formation of a non-fluorescent complex between the fluorophore and quencher in the ground state, while dynamic quenching results from collisional interactions in the excited state. Both static and dynamic quenching reduce the fluorescence intensity, but they impact the fluorescence lifetime differently. These quenching mechanisms are often identified by analyzing spectral overlaps and lifetime changes [34].

QDs have attracted a lot of attention due to their outstanding optical properties, including high brightness, photostability, and tunable emission spectra, which make them effective fluorescent probes for sensing devices developed to detect a wide range of analytes in biomedical, environmental, and industrial contexts. However, a systematic comparison of QD-based fluorescence sensors' performance across different analyte types remains limited in the literature. To fill in this void in QD-related literature, we collected, tabulated, and analyzed data from 42 peer-reviewed analytical papers, just for QD fluorescence-based sensing. The attention was mainly driven to such parameters as the limit of detection (LOD), linear range, and relative standard deviation (RSD). Other parameters, including the excitation and emission wavelength, quantum yield, and recovery percent were also recorded. Table 1 demonstrates the data collected from those papers.

As shown in Table 2, QD-based fluorescence sensors generally exhibit high sensitivity, with most of the values being in the nM range, and the geometric mean and median being equal to 38 nM and 23 nM, respectively. The analysis of the linear range width has shown that currently papers often lack performance according to this parameter since the median is somewhat lower than two orders of magnitude, and the average just slightly exceeds it. Although there were some examples of extremely wide linear ranges spanning 8.00 orders of magnitude, the majority of the values lie within 1.00–3.00 orders of magnitude, which underlines the necessity for improvement.

All the analytes detected in 42 analytical papers were divided into seven groups: explosives, metal ions, proteins, drugs, pesticides and herbicides, small biological molecules, and environmental and industrial pollutants. Then, the geometric mean and median of the LOD and the average and the median of the logarithmic range were calculated for each group and compared. Figure 1 highlights the areas where QD-based fluorescent sensors perform well, as well as the areas where improvements are awaited.

Table 1. Analytical performance in QD-based fluorescent sensors.

First Name	Nanostructure	Size	Sensing Molecule	Excitation and Emission Wavelength	Quenching Mechanism	Analyte	LOD	Range	Log(Range)	RSD (Average, Range)	Other FoM	Preparation
Hong 2012, [56]	CdTe QDs@TGA-Eu^{3+}	N/A	N/A	ex.: 360 nm; em.: 540 nm	PET	nucleoside triphos-phates	2.0 nM	70–150 nM	0.33	N/A	N/A	reflux synthesis, $CdCl_2$, NaHTe, TGA
Ban 2015, [57]	amine-capped Si QDs	2.1 nm	APTMS	ex.: 350 nm; em.: 465 nm	FRET	TNT	1.0 nM	5–500 nM	2.00	2.63%; 2.3–2.9%	QY: 22%	hydrothermal; APTMS, trisodium citrate
Peveler 2015, [58]	red ZnS-shelled CdSe QDs	N/A	OMe	ex.: 365 nm, em.: 608 nm	ET	DNT	0.38 µM	15–35 µM	0.75	N/A	N/A	decomposition; CdO, TOP-Se, Zn-DTCA, surface ligands CD, CX, OH, OMe)
Peveler 2015, [58]	green ZnS-shelled CdSe QDs	N/A	CX	ex.: 365 nm, em.: 544 nm	ET	TNT	0.44 µM	15–35 µM	0.75	N/A	Quenching %: 73%	decomposition; CdO, TOP-Se, Zn-DTCA, surface ligands CD, CX, OH, OMe)
Peveler 2015, [58]	red ZnS-shelled CdSe QDs	N/A	OMe	ex.: 365 nm, em.: 608 nm	ET	tetryl	0.73 µM	15–85 µM	0.75	N/A	Quenching %: 47%	decomposition; CdO, TOP-Se, Zn-DTCA, surface ligands CD, CX, OH, OMe)
Peveler 2015, [58]	red ZnS-shelled CdSe QDs	N/A	OMe	ex.: 365 nm, em.: 608 nm	ET	RDX	2.5 µM	15–85 µM	0.75	N/A	N/A	decomposition; CdO, TOP-Se, Zn-DTCA, surface ligands CD, CX, OH, OMe)
Peveler 2015, [58]	blue ZnS-shelled CdSe QDs	N/A	OH	ex.: 365 nm, em.: 516 nm	ET	PETN	0.35 µM	15–85 µM	0.75	N/A	N/A	decomposition; CdO, TOP-Se, Zn-DTCA, surface ligands CD, CX, OH, OMe)
Zhang 2015, [59]	carboxylated CdT QDs	10 nm	MIPs	ex.: 397 nm, em.: 540 nm	ET	phycocyanin	5.9 nM	0.02–0.8 µM	1.60	3.20%	Recovery %: 94.0–105.0%	hot injection method in aqueous phase; $Cd(NO_3)_2 \cdot 4H_2O$, Te powder, NaBH4
Li 2016, [60]	CdTe QDs	3.08 nm	aptamer	ex.: 400 nm, em.: 543 nm	IFE	bisphenol A	8.2 nM	10–80 ng/mL	0.90	1.62%; 1.08–1.91%	Recovery %: 95.3–102%	microwave-assisted; N2-saturated $CdCl_2$, Te powder, $NaBH_4$, TGA
Qian 2016, [61]	3-MPA-CdTe gQDs@SiO2@ Lcys-CdTe rQDs	3.18 nm	L-cysteine (Lcys)	ex.: 365 nm, em.: 625 nm	static quenching	TNT	3.3 nM	10–8000 nM	2.90	5.28%; 3.5–8.7%	Recovery %: 95.5–108.5%	colloidal; Te powder, $NaBH_4$, $CdCl_2$
Singh 2016, [62]	ZnO QDs	N/A	N/A	ex.: 320 nm, em.: 525 nm	static and dynamic quenching	free chlorine	41 nM	0.05–0.7 µM	1.15	N/A	N/A	ultrasonication; $Zn(OAc)_2$, ethanol, APTES

Table 1. *Cont.*

First Name	Nanostructure	Size	Sensing Molecule	Excitation and Emission Wavelength	Quenching Mechanism	Analyte	LOD	Range	Log(Range)	RSD (Average, Range)	Other FoM	Preparation
Chang 2017, [63]	Mn-doped MBA-capped ZnS QDs	3.5 nm	MBA	ex.: 310 nm, em.: 610 nm	no quenching	transferrin (TRF)	5.7 nM	0.1–10 μM	2.00	0.43%; 0.2–0.8%	Recovery %: 86.9–97.5%	wet-chemical precipitation method; $ZnSO_4 \times 7H_2O$, $MnSO_4 \times 4H_2O$
Qian 2017, [64]	rQDs@SiO2@gQDs	2.34 nm (gQDs); 3.62 nm (rQDs)	1,10-phenanthroline (phen)	ex.: 365 nm, em.: 640 nm	static quenching	Cd^{2+}	0.17 nM	0.5–2000 nM	3.60	5.37%; 4.2–6.7%	Recovery %: 96.4–101.4%	colloidal; Te powder, $NaBH_4$, $CdCl_2$, MPA
Tang 2017, [65]	aN QDs (amino-nitrogen)	5.0 nm	cysteine	ex.: 320 nm, em.: 419 nm	FRET	cysteine	0.10 μM	0.3–3 μM	1.00	2.83%; 1.5–3.7%	QY: 34%; Recovery %: 90.0–106.7%	microwave-assisted; 2-azido imidazole, ammonia
Yu 2017, [66]	Polymer CdTe/CdS QDs	10 nm	CCP	ex.: 380 nm, em.: 420 nm	FRET	hydrogen peroxide	0.10 mM	0.2–4 mM	1.30	N/A	N/A	colloidal; Na_2TeO_3, $NaBH_4$, $CdCl_2$, MPA
Yu 2017, [66]	Polymer CdTe/CdS QDs	10 nm	CCP	ex.: 380 nm, em.: 420 nm	FRET	glucose	50 μM	0.1–5 mM	1.70	N/A	Recovery %: 94.93–105.89%	colloidal; Na_2TeO_3, $NaBH_4$, $CdCl_2$, MPA
Zhou 2017, [67]	Eu-ZnO QDs	5 nm	DPA	ex.: 360 nm, em.: 530 nm	no quenching	CaDPA	3.0 nM	0.004–4 μM	3.00	N/A	QY: 1.89% without DPA, 10.69% with DPA	sol-gel; $Zn(OAc)_2$, KOH, APTES, $Eu(NO_3)_3$
Liu 2018, [33]	Si QDs	6.87 nm	N/A	ex.: 410 nm, em.: 512 nm	ET	hydroquinone	2.6 μM	6–100 μM	1.22	2.17%; 1.242–3.464%	Recovery %: 92.5–105.1%	hydrothermal; DAMO
Pourghobadi 2018, [68]	TGA-CdTe QDs	3 nm	TGA	ex.: 360 nm, em.: 580 nm	ET	dopamine	0.35 μM	0.5–10 μM	1.30	4.92%; 2.5–6.28%	Recovery %: 92–106%	aqueous; Te powder, $NaBH_4$, $CdCl_2$, TGA
Xing 2018, [69]	MAA-ZnTe QDs	12 nm	N/A	ex.: 289 nm, em.: 551 nm	static quenching	Fe^{3+}	4.9 μM	2–100 μM	1.70	3.88%	N/A	colloidal; $Zn(OAc)_2$, Na_2TeO_3, MAA
Zhao 2018, [70]	WS_2 QDs	8 nm	N/A	ex.: 360 nm, em.: 445 nm	FRET	dopamine	3.3 μM	5–50 μM	1.00	N/A	QY: 21.75%	liquid exfoliation of bulk crystals; WS_2, 1-methyl-pyrrolidinone
Chen 2019, [23]	CdTe QDs	2.25 nm	TGA	ex.: 330 nm, em.: 602 nm	dynamic quenching	Ag^+	5.0 nM	5–200 nM	1.60	5.40%; 2.1–9.7%	QY: 56% compared with Rhodamine B in ethanol; Recovery %: 94.5–112.3%	aqueous; K_2TeO_3, $NaBH_4$, $Cd(OAc)_2$, TGA

Table 1. *Cont.*

First Name	Nanostructure	Size	Sensing Molecule	Excitation and Emission Wavelength	Quenching Mechanism	Analyte	LOD	Range	Log(Range)	RSD (Average, Range)	Other FoM	Preparation
Feng 2019, [71]	MIPs layer coated on CdTe QDs	N/A	MIPs	ex.: 350 nm, em.: 570 nm	static quenching	tetrabromob-isphenol-A	0.55 nM	1.0–60.0 ng/mL	1.78	3.11%; 0.9–6.2%	QY of QDs: 64%; QY of MIP-QDs: 35%; Recovery %: 89.6–107.9%	aqueous; $CdCl_2$, K_2TeO_3
Li 2019, [72]	SiQDs	2.5 nm	N/A	ex.: 390 nm, em.: 460 nm	IFE	bovine hemoglobin	12 nM	0.01–10 µM	3.00	N/A	QY: 19.47%; Recovery %: 93.7–109.0%	hydrothermal; APTES, TSIM, L-AA
Najafi 2019, [73]	Pd-doped CdTe QDs	3 nm	TGA	ex.: 340 nm, em.: 529 nm	static quenching	diazinon	3.3 nM	2.3–100 µM	1.64	1.73%; 1.4–2.0%	Recovery %: 95.8–102.4%	hydrothermal; Te powder, $NaBH_4$, $CdCl_2$, $Pd(OAc)_2$, TGA
Safari 2019, [74]	MPA-capped Ni-doped CdTe QDs	2 nm	MPA	ex.: 330 nm, em.: 540 nm	static quenching	pyrazinamide	0.50 µM	2–100 µM	1.70	3.13%; 2.9–3.4%	Recovery %: 97.5–101.0%	hydrothermal; Te powder, $NaBH_4$, $CdCl_2$, $Ni(NO_3)_2$
Wang 2019, [75]	CA-CdS QDs	8.2 nm	CA	ex.: 420 nm, em.: 570 nm	static quenching	Cu^{2+}	9.2 nM	10 nM–50 µM	3.70	3.05%; 2.9–3.2%	QY: 18.82%; Recovery %: 95.80–99.70%	hydrothermal; $CdCl_2$, thioacetamide, citric acid
Zhang 2019, [76]	CdTe QDs	N/A	MIPs	ex.: 350 nm, em.: 566 nm	ET	pesticide 2,4-D	90 nM	0.83–100 µM	2.08	4.92%; 4.7–5.9%	Recovery %: 94.2–107.0%	hot injection method in aqueous phase; $Cd(NO_3)_2{\cdot}4H_2O$, Te powder, $NaBH_4$
Wang 2020, [24]	CdTe QDs	3.5 nm	TGA	ex.: 350 nm, em.: 560 nm	static quenching	propafenone	23 nM	0.07615–20.50 µM (0.026–7.0 µg/mL)	2.43	3.79%; 2.85–4.31%	Recovery %: 95.3–102.4%	aqueous; Te powder, $NaBH_4$, $CdCl_2$
Zhang 2020, [77]	S-doped Si QDs	4.77 nm	$-NH_2$	ex.: 345 nm, em.: 425 nm	FRET	Fe^{3+}	0.21 µM	1–20 µM	1.30	0.73%; 0.00–2.14%	QY (ref quinine): 66%; Recovery %: 90–106%	hydrothermal; APTES, trisodium citrate
Gao 2021, [78]	AuNPs-CdTe QDs@MPA	N/A	cysteamine	ex.: 365 nm, em.: 540 nm	FRET	TNT	0.24 nM	1 nM–5 µM	3.70	8.64%; 5.9–10.2%	Recovery %: 86.70–112.6%	hydrothermal; $CdCl_2$, Te powder, $NaBH_4$, MPA
Liu 2021, [79]	Ti3C2 MQDs (MXene QDs)	2 nm	N/A	ex.: 330 nm, em.: 430 nm	FRET	curcumin	0.20 µM	0.05–10 µM	2.30	N/A	N/A	microwave-assisted; Ti_3AlC_2, HF
Liu 2021, [79]	Ti3C2 MQDs (MXene QDs)	2 nm	N/A	ex.: 330 nm, em.: 430 nm	FRET	hypochlorite	5.0 µM	25–150 µM; 150–275 µM	1.04	N/A	N/A	microwave-assisted; Ti_3AlC_2, HF

Table 1. *Cont.*

First Name	Nanostructure	Size	Sensing Molecule	Excitation and Emission Wavelength	Quenching Mechanism	Analyte	LOD	Range	Log(Range)	RSD (Average, Range)	Other FoM	Preparation
Yang 2021, [80]	ZnCdS QDs@MIP; CdTeS QDs@SiO2	11 nm	APTES	ex.: 380 nm, em.: 530 nm	ET	ascorbic acid	0.78 μM	1–500 μM	2.70	1.23%; 1.044–1.663%	Recovery %: 96.0–99.0%	ZnCdS: aqueous refluxing method; $Zn(OAc)_2$, $CdCl_2$, MPA, Na_2S; CdTeS: two-step procedure; Te powder, $NaBH_4$, $CdCl_2$, thiourea
Yi 2021, [81]	β-CD-MoS2 QDs	3 nm	β-CD	ex.: 295 nm, em.: 435 nm	PET	parathion-methyl	13 nM	0.01–18 mg/L (37.99 nM–68.39 μM)	3.26	3.50%; 2.67–4.42%	Recovery %: 93–105.6%	hydrothermal; $Na_2MoO_4{\cdot}2H_2O$, glutathione
Zhang 2021, [82]	CdTe QDs	5 nm	TGA	ex.: 365 nm, em.: 550 nm	static quenching	salbutamol	42 nM	62.7–209 nM	0.52	6.17%; 5.15–7.21%	Recovery %: 81.1–89.3%	hydrothermal; Te powder, $NaBH_4$, $CdCl_2$, TGA
Zhao 2021, [83]	NALC-CdS QDs (N-acetyl-L-cysteines)	2.03 nm	NALC	ex.: 360 nm, em.: 453 nm	ET	Cu^{2+}	0.48 μM	1–25 μM	1.40	3.2%; 2.3–4.1%	QY: 34.31% (w.r.t. Rhodamine 6G in absolute ethanol); Recovery %: 99.6–101.6%	one-pot low-temperature hydrothermal route; $CdCl_2$, NALC, thioaceamide
Aznar-Gadea 2022, [84]	green and red CdSe QDs	2.9 nm	N/A	ex.: 404 nm, em.: 560 nm	ET, FRET	3-nitrotoluene	10 pM	10 pM–1 mM	8.00	N/A	PL QY: 30%	hot injection method; CdO, Se solution
Aznar-Gadea 2022, [84]	green and red CdSe QDs	2.9 nm	N/A	ex.: 404 nm, em.: 560 nm	ET, FRET	4-nitrotoluene	0.50 nM	0.5 nM–1 mM	6.30	N/A	PL QY: 30%	hot injection method; CdO, Se solution
Aznar-Gadea 2022, [84]	green and red CdSe QDs	2.9 nm	N/A	ex.: 404 nm, em.: 560 nm	ET, FRET	2,3-dimethyl-2,3-dinitrobutane	0.50 nM	0.1 nM–0.1 mM	6.00	N/A	PL QY: 30%	hot injection method; CdO, Se solution
Aznar-Gadea 2022, [84]	green and red CdSe QDs	2.9 nm	N/A	ex.: 404 nm, em.: 560 nm	ET, FRET	picric acid	0.10 μM	100 nM–1 M	7.00	N/A	PL QY: 30%	hot injection method; CdO, Se solution
Wang 2022, [85]	WxOy QDs	3.35 nm	N/A	ex.: 320 nm, em.: 383 nm	IFE, FRET, PET	Tetracycline	19 nM	5–50 μM	1.00	1.82%; 0.9–3.77%	Recovery %: 97.94–109.35%	one-pot ethanol–thermal method; WS_2, H_2O_2
Narasimhappa 2023, [86]	CdS QDs	N/A	N/A	ex.: 365 nm, em:. 440 nm	FRET	tetracycline	23 nM	10–100 μM	1.00	1.82%; 0.90–3.77%	QY: 55.8%; Recovery %: 80.1–106.7%	extracellular synthesis; Citrobacter freundii, cysteine, $CdCl_2$

Table 1. *Cont.*

First Name	Nanostructure	Size	Sensing Molecule	Excitation and Emission Wavelength	Quenching Mechanism	Analyte	LOD	Range	Log(Range)	RSD (Average, Range)	Other FoM	Preparation
Zhong 2023, [87]	G-MoS2 QDs (glutathione)	2 nm	GSH	ex.: 360 nm, em.: 430 nm	IFE	hypochlorite	12 nM	1–30 μM	1.48	2.14%; 0.72–4.41%	QY: 6.81%; Recovery %: 99.5–102.96%	hydrothermal; $Na_2MoO_4{\cdot}2H_2O$, glutathione, MilliQ water
Singh 2024, [88]	Zn_3N_2 QDs	N/A	N/A	ex.: 320 nm, em.: 408 nm	dynamic quenching	Cu^{2+}	22 nM	2.5–50 μM	1.30	2%	QY: 29.56%	hydrothermal; $Zn(NO_3)_2 \times 6H_2O$, NH_3
Singh 2024, [88]	Zn_3N_2 QDs	N/A	N/A	ex.: 320 nm, em.: 408 nm	dynamic quenching	Mn^{2+}	64 nM	0.05–5 μM	2.00	2%	QY: 29.56%	hydrothermal; $Zn(NO_3)_2 \times 6H_2O$, NH_3
Velamala 2024, [89]	$CsPbBr_3$@D-TA Per-QDs	13.1 nm	D-TA	ex.: 380 nm, em.: 522 nm	static quenching (aggregation)	superoxide anion	40 nM	0.125–25 μM	2.30	1.58%; 0.99–1.99%	QY: 29.8%; Recovery %: 98.43–99.81%	in-situ precipitation method; $PbBr_2$, CsBr, D-TA
Zhang 2024, [90]	CdTe-MIP/SiO_2	N/A	MIPs	ex.: 365 nm, em.: 609 nm	FRET	malachite green	3.7 nM	0.01–20 μM	3.30	1.40%; 0.5–2.5%	QY: 19.7%; Recovery %: 98.4–101.5%	hydrothermal; Te powder, $NaBH_4$, $CdCl_2$, MPA
Kailasa 2025, [91]	Malt@$MAPbBr_3$ QDs	7.25 nm	D-maltose	ex.: 430 nm, em.: 535 nm	static quenching	γ-aminobutyric acid (GABA)	8.4 nM	0.05–10 μM	2.30	0.96%; 0.23–1.26%	QY: 23.74%; Quenching efficiency: >90%; Recovery %: 98.13–99.96%	ligand-assisted reprecipitation; MABr, $PbBr_2$, 1-octadecene, oleyamine, maltose
Makwana 2025, [34]	$LaSrO_3$ Per-QDs	4.45 nm	N/A	ex.: 336 nm, em.: 421 nm	FRET	bilirubin (BR)	10 nM	0.025–25 μM	3.00	0.70%; 0.48–0.86%	QY: 36.91%; Recovery %: 98.90–100.01%	microwave-assisted sol-gel method; $La(NO_3)_3{\cdot}6H_2O$, $Sr(NO_3)_2$, L-serine
Makwana 2025,[34]	$LaSrO_3$ Per-QDs	4.45 nm	N/A	ex.: 336 nm, em.: 421 nm	IFE	epinephrine (EP)	17 nM	0.05–10 μM	2.30	0.48%; 0.30–0.77%	QY: 36.91%: Recovery %: 99.24–99.98%	microwave-assisted sol-gel method; $La(NO_3)_3{\cdot}6H_2O$, $Sr(NO_3)_2$, L-serine
Zhang L. 2025, [92]	MoS_2 QDs	N/A	APBA	ex.: 320 nm, em.: 375 nm	dynamic and static quenching, IFE	glutathione (GSH)	0.48 μM	10–500 μM	1.70	3.74%; 2.45–5.52%	Assay time: 1 min; Recovery %: 98.73–103.34%	hydrothermal; $Na_2MoO_4{\cdot}2H_2O$ and cysteine

Table 1. *Cont.*

First Name	Nanostructure	Size	Sensing Molecule	Excitation and Emission Wavelength	Quenching Mechanism	Analyte	LOD	Range	Log(Range)	RSD (Average, Range)	Other FoM	Preparation
Zhang L. 2025, [92]	MoS2 QDs	N/A	APBA	ex.: 320 nm, em.: 375 nm	dynamic and static quenching, IFE	ascorbic acid	0.19 μM	10–100 μM	1.00	2.71%; 0.48–6.91%	Assay time: 1 min; Recovery %: 97.44–104.7%	hydrothermal; $Na_2MoO_4 \cdot 2H_2O$ and cysteine
Zhang Y. 2025, [28]	Si QDs	3 nm	N/A	ex.: 367 nm, em.: 464 nm	ET, IFE	Hg^{2+}	3.0 nM	0.5–5 μM	1.00	12.27%; 4.11–18.8%	QY: 29.4%; Recovery %: 80.3–109%	microwave-assisted; trisodium citrate dehydrate, DAMO, glycerol

Abbreviations: N/A—not available; APTMS —3-aminopropyltrimethoxysilane; TOP—trioctylphosphine; CD—cyclodextrin; CX—calix [4] arene; DTCA—dithiocarbamate; FoM—figure of merit; TGA—thioglycolic acid; MPA—mercaptopropionic acid; MBA—mercaptophenylboronic acid; CCP—cationic conjugated polymers; CaDPA—calcium dipicolinate; APTES—3-aminopropyltriethoxysilane; CA—citric acid; L-AA—L-ascorbic acid; TSIM—trimethylsilylimidazole; DAMO—*N*-[3-(trimethoxysilyl)propyl]-ethylenediamine; MAA—mercaptoacetic acid; NALC—N-acetyl-L-cysteine; GSH—glutathione; rQDs—red quantum dots; gQDs—green quantum dots; QY—quantum yield; TNT—2;4;6-trinitrotoluene; DNT—2;4-dinitrotoluene; PETN—pentaerythritol tetranitrate; RDX—cyclotrimethylenetrinitramine; APBA—3-aminophenylboronic acid; D-TA—D-tartaric acid; Per-QDs—perovskite QDs; MIP—molecularly imprinted polymers; FRET—fluorescence (or Förster) resonance energy transfer; ET—electron transfer; PET—photoinduced electron transfer; IFE—inner filter effect, NPs—nanoparticles.

Table 2. The average LOD, log(range), and RSD were calculated from all the data shown in Table 1. (The number in brackets indicates the number of averaged data points).

Geometric mean (LOD)	38.3 nM (54)
Median (LOD)	23.0 nM (54)
Average (log(range))	2.14 (54)
Median (log(range))	1.70 (54)
Average (RSD)	3.20% (36)

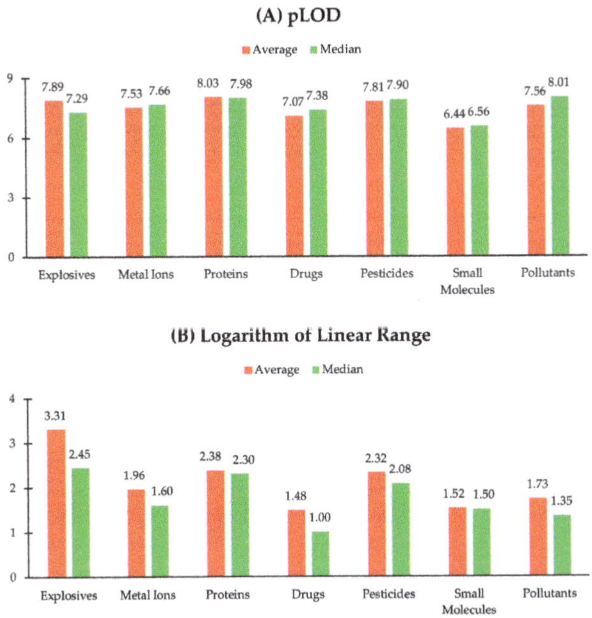

Figure 1. Column charts showing the analytical performance of 42 sensors evaluated based on the (**A**) geometric mean and median of negative logarithm of LOD (pLOD) and (**B**) on the width of the linear range (average and median) for 7 categories of analytes: explosives [57,58,61,78,84], metal ions [23,28,64,69,75,77,83,88], proteins [34,59,63,72], drugs [24,74,80,82,85,86,92], pesticides [73,76,81], small biological molecules [56,65,66,68,70,79,89,91,92], and pollutants [33,60,62,67,71,79,87,90].

The analysis of the LOD, shown in Figure 1A, has shown that proteins were detected most sensitively when the geometric mean of the LOD was equal to 8.15 nM (pLOD = 8.03). Articles detecting proteins (BR, EP, bovine hemoglobin, phycocyanin, and TRF) span the period from 2015 to 2025, with no significant change in LOD values, which indicates that there have not been many improvements made since 2015 in QD-based protein detection. However, for example, for BR detection, the sensitivity level of the system proposed by Makwana is sufficient in correlation with the normal plasma levels of this protein, which range from 3.42 to 20.52 μM [34,93]. Explosives have performed a bit weaker; however, their geometric mean was 12.9 nM, which is not much higher than the LOD for proteins. The lowest median among all the groups of analytes was calculated for environmental and industrial pollutants. This fact underlines that these three categories of analytes require highly sensitive devices to detect even trace amounts of them due to safety concerns. Small biological molecules have shown both the highest geometric mean and the highest median of the LOD, with the values being 0.360 μM and 0.275 μM, respectively. This may point out that in living organisms, these molecules are typically present at a high level, so there is no need for ultrasensitive detection. The difference between the median and the geometric mean values of the pLOD, which can be observed in Figure 1A for explosives and pollutants, suggests that there was skewness of the data, meaning there were outperformers: either positive, as with explosives, or negative, as for pollutants.

The analysis of the linear range (expressed in log scale) across the analyte categories shows a variation in sensor performance, as can be observed from Figure 1B. Explosives, proteins, and pesticides exhibit the widest working ranges, each with median log (linear range) values exceeding two orders of magnitude in width. Explosives demonstrate the broadest linear range, with an average of 3.31 and a median of 2.45. This nearly one-order-of-magnitude difference suggests the presence of outstanding sensors that skew the average upward. In contrast, drugs have the narrowest linear range, with a median of one order of magnitude (1.00), pointing out a potential limitation in the applicability of current drug sensors. This narrow range highlights the need for the further optimization of sensors designed for pharmaceutical detection. Overall, the linear range across all the analytes spans from 0.33 to 8.00 orders of magnitude [82,84], illustrating the difference in the performance of fluorescent QD-based sensors and pointing toward areas where further enhancement is required. Additionally, analyte groups, such as drugs and small biological molecules, had narrower average and median log(range) values, potentially indicating lower variability in the sensitivity of the detection methods used for these categories. On the other hand, explosives and proteins had broader log(range) values, which may reflect a wider diversity in sensor performance or detection strategies across different studies. Thus, for the system designed for the detection of explosives proposed by Aznar-Gadea and coworkers, logarithmic ranges are in the range of six (2,3-dimethyl-2,3-dinitrobutane) to eight (3-nitrotoluene) orders of magnitude, which are the widest ranges among all the analyzed papers [84].

It is also worth noting that the distinction between the geometric mean and median LOD values within each category can hint at the distribution skewness—groups with large discrepancies between these values, such as explosives, where there is a ten-fold difference, contain a few highly sensitive detection methods that significantly lower the mean.

Overall, these trends can guide future sensor development priorities, indicating areas where improved sensitivity is still needed (e.g., small biomolecules) and where reproducibility across platforms should be enhanced (e.g., explosives).

Gao and coworkers have developed a sensor for TNT detection and have tested three detection methods, which include colorimetric detection, fluorescence, and Raman spectroscopic detection. In the sensor design process, they utilized cysteamine-modified AuNPs with a 532 nm absorption wavelength and MPA-capped CdTe quantum dots with a 540 nm emission wavelength. As shown in Figure 2A, at first, AuNPs-QD assemblies are formed due to electrostatic interactions between QDs@MPA and AuNPs@Cys. The Fluorescence of QDs is quenched via the FRET mechanism. After TNT is added, the QDs-AuNPs composite is disassembled, which turns on fluorescence and colorimetric signals. An increase in the fluorescence intensity due to an increase in the TNT concentration can be observed with the naked eye, as shown in Figure 2A. Apart from that, the strong affinity of TNT to Cys leads to the formation of the Meisenheimer complex and the subsequent self-aggregation of AuNPs, which generates a strong Raman enhancement and allows for the sensitive detection of TNT in complex matrices. The LOD for the fluorescence detection of TNT achieved by Li and Ren's groups was 0.24 nM, and the linear range of the sensor spans 3.70 orders of magnitude, from 1 to 5 μM [78].

Wang and coworkers have developed a CdTe QD-based system for the detection of propafenone, with an LOD of 23.2 nM and a linear range of 76.15 to 20.5 μM. Figure 2B demonstrates the suggested binding model between propafenone and CdTe QDs in acidic media, where a connection happens due to electrostatic attraction and hydrogen bonds, which results in the formation of larger ionic association complexes. This leads to the quenching of the fluorescence of QDs by hindering the electron transfer in the semiconductor nanocrystal from an excited to a ground state.

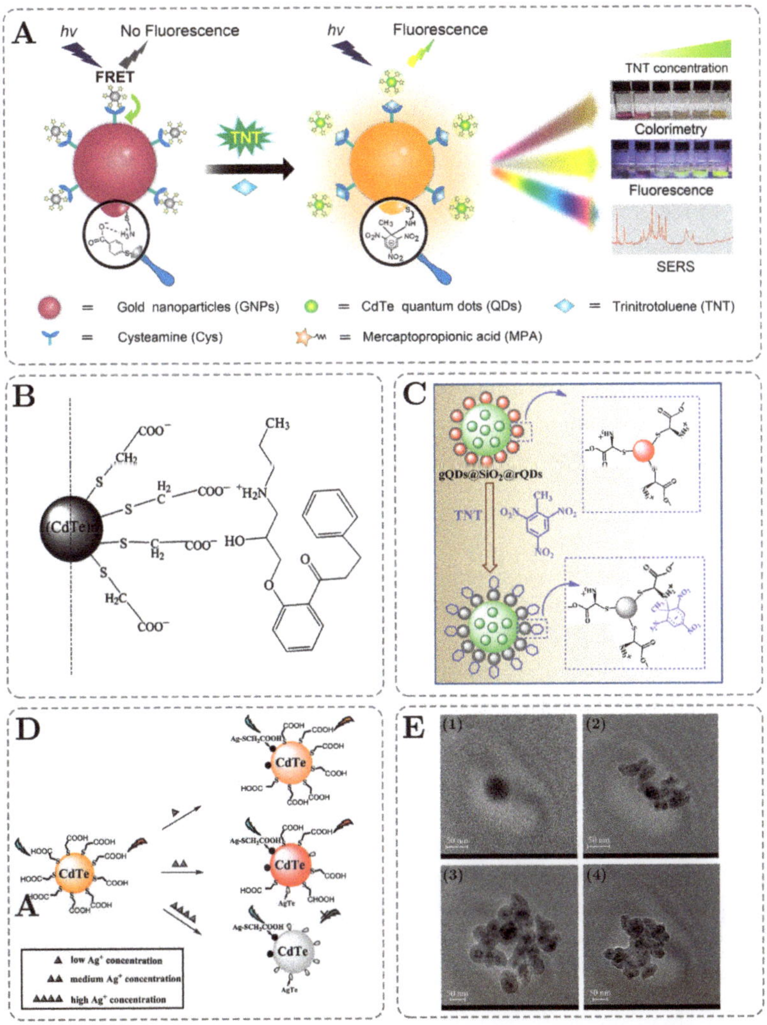

Figure 2. Schemes, graphs, and photographs adapted from academic papers describing QD-based fluorescence sensor mechanisms and morphology. (**A**) A graphical representation of the detection of TNT with AuNPs-CdTe QDs by fluorescence, colorimetry, and Raman spectroscopy. Adapted with permission from Gao et al. [78]. Copyright © 2021, Elsevier. B.V. (**B**) A graphical illustration of the suggested linking mode between thioglycolic acid-modified CdTe QDs and propafenone. Adapted with permission from Wang et al. [24]. Copyright © 2020, Elsevier. B.V. (**C**) A schematic representation of a TNT sensing mechanism with MPA-capped green CdTe QDs embedded into the SiO_2 sphere and red-emitting CdTe QDs conjugated onto the SiO_2 surface. Adapted with permission from Qian et al. [61]. Copyright © 2016, Elsevier. B.V. (**D**) A schematic representation of the working mechanism of a CdTe QD-based sensor for Ag^+ detection in lake water. Adapted with permission from Chen et al. [23]. Copyright © 2019, Elsevier. B.V. (**E**) Transmission electron microscopy (TEM)

photographs of (1) CdTe QDs, (2) CdTe QDs@SiO_2, (3) MIP-CdTe QDs, and (4) non-imprinted particles (NIP)-CdTe QDs designed for tetrabromobisphenol-A detection. Adapted with permission from Feng et al. [71].

Qian and coworkers have developed a TNT sensor based on MPA-capped CdTe gQDs and Lcys-capped CdTe rQDs whose mechanism of operation is shown in Figure 2C. Due to the electron deficiency of TNT and the electron richness of Lcys, Meisenheimer complexes were formed between TNT and Lcys ligands through the acid–base pairing interactions, hydrogen bonding, and electrostatic co-interactions. The quenching of the fluorescence emission of rQDs at 625 nm is induced by the electron transfer from rQDs to the aromatic ring of TNT, while the emission of interior gQDs at 508 nm stayed almost unaffected. The variation in the intensity ratio of the dual emissions led to a significant shift in the fluorescence color, enabling the visual detection of TNT on-site. The sensor has exhibited an LOD of 3.3 nM and a wide linear range of 10 to 8 μM [61].

Chen and coworkers have designed a sensor for the selective and sensitive detection of Ag^+ ions in lake water. The device is based on TGA-capped CdTe QDs, which were synthesized via a one-pot aqueous method. Green, orange, and red QDs were obtained from potassium tellurite, cadmium acetate, and thioglycolic acid by changing the reaction time and temperature. Experimental results have shown that green-emitting QDs demonstrate the highest sensitivity, with their LOD being 5 nM, while red-emitting QDs exhibit the widest linear detection range, from 5 to 200 μM. As can be seen from Figure 2D, the interaction between Ag^+ and QDs is based on the formation of an Ag–S bond with TGA, and the sensor has different responses for different silver ion concentrations. At a low Ag^+ concentration, there was no significant change in the emission wavelength of QDs; however, the fluorescence intensity was slightly weakened. Medium Ag^+ concentrations have led to a red shift in the emission wavelength of orange QDs, while at a high Ag^+ content, the fluorescence of QDs was completely quenched [23].

Feng and coworkers have developed a sensor for tetrabromobisphenol-A detection in e-wastes based on MIP-coated CdTe QDs. Figure 2E demonstrates the morphologies of bare QDs (1), QDs@SiO2 (2), MIP-QDs (3), and NIP)-QDs (4) through the TEM images. The increase of the MIP-coated particle diameter compared to the bare QDs can be observed from Figure 2E(1,2) to Figure 2E(3), demonstrating that the MIP-QDs possessed a big surface area with well-defined imprinted sites for binding template molecules. From the images, it can be also seen that the particles were spherical and were dispersed uniformly. Figure 2E(3,4) illustrate that the MIP-QDs and the NIP-QDs exhibited similar morphologies, with their shape and size becoming slightly rough and irregular due to sol-gel polymerization throughout the synthesis process. The designed sensor has shown a LOD of 0.552 nM (0.3 ng/mL) and a linear range of 1.0 to 60.0 ng/mL [71].

Some other representative examples of fluorescence-based sensing with QDs are shown in Figure 3 below.

The best-performing sensor was developed by Aznar-Gadea and coworkers, and its mechanism of operation is schematically shown in Figure 3A. At first, red- and green-emitting CdSe QDs were embedded into polycaprolactone as a host polymer matrix. Authors have determined that the signals of green and red QDs towards the same compounds are different, which allows for the two-dimensional (2D) mapping of experimental results where the intensity of red QDs is plotted against the intensity of green QDs, as depicted in Figure 3A. Such 2D maps allow for building fingerprints for each compound, which leads to a decreased number of false positives and increases the assay performance towards multiple types of explosives. Figure 3B demonstrates the calibration curves of four tested analytes, namely 3-NT, 4-NT, DMDNB, and PA. All the compounds, except picric acid, can be detected at a subnanomolar concentration, with the lowest LOD achieved for 3-NT at 10 pM. From the calibration curves, it can also be concluded that linear ranges for the different types of analytes span from six to eight orders of magnitude, with the widest range being achieved for 3-NT [84].

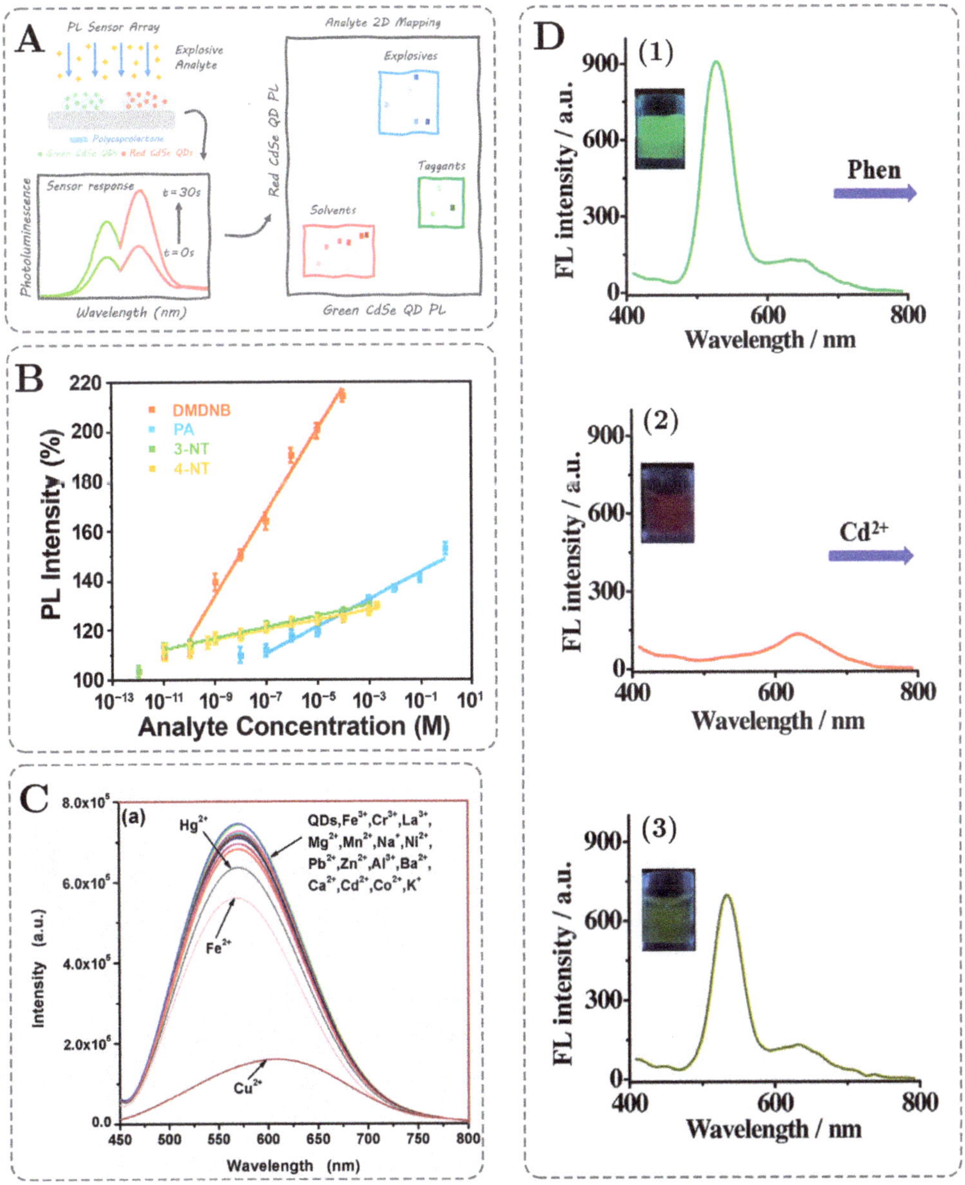

Figure 3. Examples of efficient fluorescence-based sensing with the application of QDs: (**A**) A schematic mechanism of a CdSe QD-based sensor for the detection of four explosives (3-nitrotoluene (3-NT), 4-nitrotoluene (4-NT), 2,3-dimethyl-2,3-dinitrobutane (DMDNB), and picric acid (PA)). (**B**) Calibration plots for the detection of four explosives (3-NT, 4-NT, DMDNB, and PA). Adopted under Creative Common CC BY license from Aznar-Gadea et al. [84]. Copyright © 2022, American Chemical Society. (**C**) Fluorescence spectra illustrating the impact of different metal ions on CA-CdTe QD fluorescence during the Cu^{2+} sensing. Adopted under Creative Common CC BY license from

Wang et al. [75]. Copyright © 2019, MDPI, Basel, Switzerland. (**D**) Photographs and fluorescence spectra of (1) rQDs@SiO_2@gQDs, (2) rQDs@SiO_2@gQDs-Phen (20 μM), and (3) rQDs@SiO_2@gQDs-Phen-Cd^{2+} (2 μM) produced by the nanosensor for Cd^{2+} detection. The excitation wavelength was 365 nm. Adapted with permission from Qian et al. [64]. Copyright © 2017, Elsevier. B.V.

Wang and coworkers proposed a sensor for the selective detection of Cu^{2+} ions by the citric acid-capped CdS QDs. The authors managed to achieve a high performance, with the LOD of the sensor being 9.2 nM and the linear range spanning 3.70 orders of magnitude, from 10 nM to 50 μM. Figure 3C illustrates the selectivity test for Cu^{2+} ions with 17 different heavy transition metals, from which it can be observed that only copper ions quench the fluorescence of the QDs, while other metals do not have a significant effect on the fluorescent signal. Also, the red shift of the fluorescence signal towards Cu^{2+} can be noticed. The red shift suggests that there is a formation of a new energy level below the conduction band of the CdS, which can be attributed to the photochemical reduction of adsorbed copper ions by CA-CdS QDs. From an XPS analysis, which is not shown here, the authors have proved the presence of both Cu^{2+} and Cu^{+} and concluded that the quenching of fluorescence takes place due to the deposition of CuS on the surface of the QDs [75].

Qian and coworkers have developed a Cd^{2+} nanosensor based on CdTe red- and green-emitting QDs (rQDs and gQDs) in the form of rQDs@SiO2@ to detect cadmium ions at such low concentrations as 0.17 nM. In their study, 1,10-Phenanthroline (phen) was used as the sensing molecule. Figure 3D illustrates the fluorescence spectra of the as-obtained rQDs@SiO2@gQDs (Figure 3D(1)), after the addition of phen (Figure 3D(2)), and after the addition of Cd^{2+} ions (Figure 3D(3)). Upon excitation with a 365 nm wavelength, the as-obtained rQDs@SiO2@gQDs showed two highly resolved emission peaks at 525 and 635 nm, while the fluorescence intensity of the gQDs loaded on the surface of the hybrid spheres was significantly quenched. The addition of the Cd^{2+} ions to the system resulted in a noticeable recovery of the fluorescence because of the removal of phen ligands from the surface of gQDs due to the strong chelating of phen with Cd^{2+}. Thus, it is possible to use cadmium cations as the fluorescence activation switch, disrupting the photo-induced hole transfer process and restoring the fluorescence of the gQDs. Apart from changes in the spectra, there is a color change following each step described above, as can be observed from Figure 3D. Since the intensity of the fluorescence of the rQDs stayed unaffected all the time, it can be used as a reference signal, thus providing an in-system correction for environmental effects. Meanwhile, the fluorescence of the gQDs changed, and the quenching efficiency in the pH range of 6 to 9 took the values from 94.33 to 97.79% [64].

3. Chemiluminescence of QDs: Applications in Sensing and Detection

Chemiluminescence in nanomaterials is a process where light is emitted from the nanostructure, like in the QDs in this review article, during the chemical reaction, rather than from the remission of the absorbed light. It has raised the interest of scientists as a promising tool for sensing and detection techniques. The classification of the types of chemiluminescence varies, but mostly, the following three types are classified: general chemiluminescence (CL), bioluminescence (BL), and the most common in sensing, electrochemiluminescence (ECL).

General chemiluminescence involves the emission of light due to chemical reactions that are not related to biological or electrochemiluminescent triggers. For instance, in forensic science, the oxidation of luminol can be used for the detection of blood. Typically involving enzyme-catalyzed reactions, bioluminescence is found in living organisms, such as fireflies, glowworms, and some marine animals. Finally, electrochemiluminescence is triggered by an electrochemical reaction, where the periodical applied potential triggers the light emission. It is commonly used in analytical chemistry for sensitive analyses, like immunoassays [94].

While fluorescence was and remains the most common technique for sensing purposes, during the 2000s, researchers demonstrated the capabilities of chemiluminescence for detection. For instance, in 2007, Heyou Han's research group developed a method that allowed the determination of hydrogen peroxide with an LOD of 60 nM by electrochemiluminescence from thiol-capped Cd-Te QDs. This method was novel in terms of its analytical performance compared to previous reports for the analyte. Another well-recognized paper published a few years later by Lin et al. focused on the nitrite sensing by CL of fluorescent carbon dots induced with peroxynitrous acid, which acquired a detection limit of 53 nM. Together, these and other papers demonstrated the potential of CL quantum dots in analytical chemistry [95,96].

There are also some review papers published in the last several years that investigate the chemiluminescence of QDs, mostly from qualitative sight. The newest one by Yang et al. mainly explores the diverse roles of dots in electrochemiluminescent bioanalysis, while also presenting some quantitative comparisons. In contrast, the 2021 review by Tzani et al. investigates the mechanisms and applications of direct and indirect chemiluminescence. It highlights the recent improvements in the enhancement of CL efficiency with various chemical structures. The last one, a review paper by the Yi Lv group outlines the recent advancements of liquid-phase CL systems, where quantum dots are primarily used as a final emitting species via direct redox reactions of chemiluminescent resonance energy transfer, emphasizing their direct analytical applications aside from roles as catalysts or enhancers [94,97,98].

Recent advancements in this field involve the application of new materials, such as perovskite and MXene quantum dots, for the development of highly sensitive platforms for biosensing, as well as the exploration of multiplex detection using multicolor techniques. Table 3 provides detailed information regarding the chemiluminescence–QD-based sensors developed for sensing purposes. Specific nanostructures, brief preparation methods, analyte molecules, and important analytical parameters, such as the LOD, QDs' size, emission wavelength, and linear parameters, are all covered to help identify the trends and observe the common elements of sensing [99–101].

The information gathered in Table 3 spans the past 13 years of quantitative analytical research. The largest class of analytes found in the reviewed papers consists of clinical biomarkers, while other analytes include small molecules, toxins, and contaminants, as well as pesticides and herbicides. Some of the abbreviations used are **Ab** for antibody, **black-P** for black phosphorus, **FL** and **PL** for the wavelengths reported for fluorescence or photoluminescence exclusively, **IgG** for immunoglobulin G, **LR** for leucine–arginine, **MIL** for materials of institute Lavoisier (adapted from French), **N/D** for values that are not present explicitly in the articles, and **RET** for resonance energy transfer.

Table 3. Analytical performance of QD–chemiluminescence-based sensors.

Year, Last Name	Nanostructure	Preparation (RS: Reflux Synthesis. TGA: Thioglycolic Acid. 3-MPA: 3-Mercaptopropionic Acid)	Analyte	LOD	Range	Size (nm)	Emission λ (nm)	RSD	R^2	Recovery
Han 2007, [95]	thiol-capped CdTe QDs	RS: $CdCl_2$, TGA, NaHTe	hydrogen peroxide	60 nM	$(2.0 \times 10^{-7}–1.0 \times 10^{-5})$ M	3.32	620	4.80%	0.999	N/D
Wang 2012, [102]	SiO_2-QD-Ab2	Solution aging: $CdCl_2$, methiopropamine, NaHTe	human IgG (HIgG)	0.58 fM	$(1.0 \times 10^{-10}–1.0 \times 10^{-5})$ g/L	N/D	705	4.60%	0.996	83.8–120.6%
Fang 2012, [22]	CdSe QDs	RS: $CdCl_2$, TGA, NaHSe	α-fetoprotein	70 fM	$(5.0 \times 10^{-9}–1.0 \times 10^{-4})$ g/L	2.5	575 (PL)	1.50%	N/D	N/D
Liu 2014, [103]	CdSe QDs	RS; $CdCl_2$, HMP, MPA, Na_2SeO_3, N_2H_4	dopamine	3.0 nM	$(3.0 \times 10^{-9}–1.0 \times 10^{-5})$ M	N/D	541	4.30%	N/D	104.5–106.9%
Dong 2014, [104]	CdSe@ZnS QDs	RS: $CdCl_2$, TGA, NaHTe	thrombin	1.4 fM	$(1.0 \times 10^{-14}–1.0 \times 10^{-10})$ M	4.5	550	2.3% to 4.1%	0.995	100.8–102.4%
Zhang 2015, [105]	CdS QDs	RS: $CdCl_2$, methiopropamine, thioacetamide	microcystin-LR	2.8 pM	$(1.0 \times 10^{-8}–5.0 \times 10^{-5})$ g/L	3.5	646 (PL)	3.36%	0.999	97.7–101%
Wang 2015, [106]	CdTe/CdS cores-mall/shellthick QDs	RS: $CdCl_2$, methiopropamine, NaHTe	Cu^{2+}	20 nM	$(1.0 \times 10^{-7}–1.0 \times 10^{-5})$ M	4.8	710	3.3%	0.998	93.3–104.6%
Dong 2017, [107]	CdTe QDs (QDs)-embedded mesoporous silica nanospheres	N/D	carcinoembryonic antigen	1.7 fM	$(1.0 \times 10^{-9}–8.0 \times 10^{-5})$ g/L	3.45	676	5.93%	0.996	96.24–105.26%
Zhao 2017, [108]	MoS_2-QDs	RS; MoS_2, Pd-Au CHs	lipopolysaccharide	0.07 fg/mL	$(1.0 \times 10^{-13}–5.0 \times 10^{-5})$ g/L	4.2	625	1.53%	0.998	N/D
Wu 2017, [109]	Ag_2S:Mn QDs	RS; 3-MPA, acetic acid, $AgNO_3$, Na_2S, $Mn(Ac)_2$	laminin	3.6 fM	$(1.0 \times 10^{-8}–1.0 \times 10^{-4})$ g/L	4	626	N/D	0.997	96.08–105.56%
Dong 2017, [110]	Si QDs	RS: aminopropyl trimethoxysilane, trisodium citrate	target DNA	16 aM	$(1.0 \times 10^{-16}–1.0 \times 10^{-12})$ M	6.72	475	3.50%	0.994	N/D
Yang 2018, [111]	CdTe QDs	RS; $CdCl_2$, trisodium citrate dehydrate, Na_2TeO_3, MPA, $NaBH_4$	cardiac troponin-I antigen	19 aM	$(1.1 \times 10^{-12}–1.1 \times 10^{-5})$ g/L	5	650	0.92%	0.999	93.6–109.7%
Hu 2018, [112]	CdSe QDs (CdSe QDs)	RS: sodium sulfite, selenium powder, $CdCl_2$, 3-MPA, N_2H_4	hydrogen peroxide	0.50 μM	$(5.0 \times 10^{-7}–5.0 \times 10^{-4})$ M	8	470, 610	4.90%	0.972	87.6–104.3%
Zhang 2018, [113]	Black-P QDs	Sonication: black phosphorus powder, N-Methyl-2-pyrrolidone	dopamine	22 pM	$(1.0 \times 10^{-10}–5.0 \times 10^{-8})$ M	8.2	610	1.30%	0.994	103 to 105%
Wang 2018, [114]	CdTe QDs	RS: NaHTe, $CdCl_2$, TGA	clopyralid	4.1 pM	$(2.0 \times 10^{-11}–3.5 \times 10^{-8})$ M	8	N/D	1.78–2.42%	0.997	93.3–105.7%
Lei 2018, [115]	SnS_2 QDs	Hydrothermal method: $SnCl_4$, L-cysteine	anti-CMV pp65	0.33 fM	$(1.0 \times 10^{-15}–1.0 \times 10^{-7})$ M	5	665	N/D	0.996	97.78–108.20%

Table 3. *Cont.*

Year, Last Name	Nanostructure	Preparation (RS: Reflux Synthesis. TGA: Thioglycolic Acid. 3-MPA: 3-Mercaptopropionic Acid)	Analyte	LOD	Range	Size (nm)	Emission λ (nm)	RSD	R^2	Recovery
Gao 2019, [116]	CdTe QDs	RS; $CdCl_2$, Na_2TeO_3, 3-MPA, $NaBH_4$	ochratoxin A	0.42 pM	$(5.0 \times 10^{-10}–5.0 \times 10^{-5})$ g/L	3	585	2.90%	0.999	96.1–100.7%
Liu 2019, [117]	Black-P QDs	Sonication: black phosphorus powder, N-Methyl-2-pyrrolidone	lysozyme	2.0 fM	$(1.0 \times 10^{-10}–1.0 \times 10^{-7})$ g/L	8.2	N/D	0.56%	0.991	98–102%
Du 2020, [118]	MOF-5-wrapped CdS quantum dot	RS: $CdCl_2$, Na_2S, mercaptopropionic acid	cardiac Troponin I	0.21 fM	$(1.0 \times 10^{-11}–1.0 \times 10^{-6})$ g/L	5.4	685475	9.25%	0.993	98.0–104.7%
Chen 2020, [119]	CdTe QDs@NH_2-MIL-88(Fe)	RS: $CdCl_2$, Na_2TeO_3, trisodium citrate, $NaBH_4$, 3-MPA	malathion	0.91 fM	$(1.0 \times 10^{-12}–1.0 \times 10^{-6})$ g/L	7.5	689 (1.8 eV)	7.00%	0.993	99.2–110%
Xu 2020, [120]	Multiwall, carbon-nanotube-enhanced, molecularly imprinted QDs	N/D	cyfluthrin	0.11 nM	$(2.0 \times 10^{-7}–1.0 \times 10^{-3})$ g/L	N/D	N/D	2.70%	0.998	86.0–98.6%
Liu 2021, [121]	CdS QDs	RS: 3-MPA, $CdCl_2$, NaOH, Na_2S, ethanol.	17β-estradiol	0.11 pM	$(1.0 \times 10^{-11}–1.0 \times 10^{-8})$ g/L	N/D	N/D	1.1–2.5%	0.99	98.9–118.4%
Feng 2021, [122]	MIL-53(Al)@CdS QDs	N/D	trichlorfon	5.1 pM	$(1.0 \times 10^{-11}–1.0 \times 10^{-4})$ M	<10	517	2.50%	0.997	97–105%
Rasoulzadeh 2021, [123]	$AgInS_2$ QDs	RS; $AgNO_3$, $InCl_3$, sodium citrate, TGA, thiourea, deionized water, ethanol.	glutathione	0.28 nM	$(1.0 \times 10^{-9}–5.0 \times 10^{-6})$ M	2	680	3.10%	0.994	98–101%
Zhao 2021, [124]	Black-P-CdTe QDs	RS; $CdCl_2$, BP-nanosheets, Na_2TeO_3, 3-MPA, $NaBH_4$	miRNA-126	29 aM	N/D	N/D	709	1.19%, 2.85%	0.998	98.8–104%
Yang 2021, [99]	Ir NRs@CdS QDs	RS: $CdCl_2$, Na_2S, L-cysteine	ethyl paraoxon	1.7 pM	$(5.0 \times 10^{-12}–5.0 \times 10^{-8})$ M	N/D	543	2.38%, 2.68%	0.995	91.9–108.0%
Jia 2022, [125]	CdSe@CdS QDs	N/D	ochratoxin A	2.2 nM	$(1.0 \times 10^{-6}–1.0 \times 10^{-4})$ g/L	15	634	2.1–6.5%	0.994	97.3–105.6%
Li 2022, [101]	$CH_3NH_3PbBr_3$ QDs@SiO_2	Ligand-assisted reprecipitation method; $PbBr_2$, CH_3NH_3Br, APTES, toluene	aflatoxin B1	27 fM	$(1.0 \times 10^{-11}–1.0 \times 10^{-5})$ g/L	N/D	566	1.2-2.8%	0.997	101.7–106.7%
Liu 2022, [126]	luminol/MoS_2 QDs@zeolitic imidazolate framework-8	RS; MoS_2, luminol, $ZnNO_3$ solution	miRNA21	15 aM	N/D	2	N/D	N/D	0.998	N/D

Table 3. *Cont.*

Year, Last Name	Nanostructure	Preparation (RS: Reflux Synthesis. TGA: Thioglycolic Acid. 3-MPA: 3-Mercaptopropionic Acid)	Analyte	LOD	Range	Size (nm)	Emission λ (nm)	RSD	R^2	Recovery
Yang 2023, [127]	ECL-RET sensor with TGA-capped CdS QDs	RS: $CdCl_2$, TGA, Na_2S	EGFR T790M ctDNA	3.4 aM	$(1.0 \times 10^{-17}$–$1.0 \times 10^{-13})$ M	5.7	497 (FL)	2.6%	0.991	N/D
Yang 2023, [127]	ECL-RET sensor with TGA-capped CdS QDs	RS: $CdCl_2$, TGA, Na_2S	EGFR T790M ctDNA	8.1 aM	$(5.0 \times 10^{-17}$–$1.0 \times 10^{-13})$ M	5.7	497 (FL)	2.6%	0.994	N/D
Li 2023, [100]	Mo_2TiC_2 QDs	Acid etching, alkaline treatment, and microwave-assisted synthesis; Mo_2TiAlC_2 powder, NH_4HF_2, TMAOH, ultrapure water.	miRNA-27a-3p	1.0 fM	$(1.0 \times 10^{-15}$–$1.0 \times 10^{-8})$ M	2.7	594	2.00–2.59%	0.992	89.1–104.2%
Liu 2024, [128]	CdSe@ZnS/ MXene@NaAsc	N/D	uric acid	18 pM	$(1.0 \times 10^{-10}$–$1.0 \times 10^{-4})$ M	4	N/D	2.81%	0.987	88.40–94.65%
Gong 2025, [129]	AgInZnS QDs	RS: methiopropamine, $AgNO_3$, $In(Ac)_3$, $Zn(Ac)_2$	carcinoembryonic antigen	16 fM	$(5.0 \times 10^{-14}$–$1.0 \times 10^{-8})$ M	N/D	620	1.13%	0.994	95.5–103.3%

The geometric average of the detection limits is 1.1×10^{-13} M or 109 fM, which is noticeably lower compared to 26 and 38 nM for phosphorescence and fluorescence, respectively, which share approximately the same geometric mean value. This is attributed to the absence of excitation light in CL: since chemiluminescence is produced via chemical reaction, background interference, and autofluorescence are significantly minimized, leading to a greater signal-to-noise ratio. Moreover, half of the detections had an LOD better than 27.2 fM, with the lowest included in this review being 3.4 aM, for the sensing of cancer mutations on a linear DNA analyte with the application of CdS dots in 2023 by Yang et al. [127].

The arithmetic average for the width of the linear range, calculated as $\log_{10}(c_{max}/c_{min})$, results in 4.27 orders of magnitude, with the best-reviewed being 8.7 orders of magnitude for the detection of lipopolysaccharide by Zhao et al. [108], with MoS_2 dots in 2017. There is some positive correlation between the pLOD and the width of the linearity, meaning that for a lower LOD, there is an expected wider linear range; however, the correlation is not so strong, with a Pearson coefficient of +0.58.

The central values for the emission wavelength and the size of the quantum dots are 615 nm and 4.8 nm, respectively. Compared to fluorescence and phosphorescence, 615 nm for chemiluminescence is almost no different than phosphorescence, but around one standard deviation higher than the same value for fluorescence. The size of the particles is also greater for CL than for the other two luminescence types, but its variance is too big for a conclusion based on the given sample size.

Regarding the reproducibility of the methods, the mean value of the relative standard deviation for the measurements and the linearity of the calibration curve is equal to 5.1% and 0.9945, respectively. For the recoveries, the absolute average deviation from 100%, in percent recovery, is 5.05%.

It is noted that almost all sensors, with only a few exceptions, have utilized electrically induced CL with electrodes, rather than direct chemiluminescence from spontaneous reactions like luminol and hydrogen peroxide. This selection is based on ECL's capacity to precisely regulate the timing and localization of light generation, which enhances the precision of the measurements. It also deserves to be mentioned that ECL offers a wider dynamic range and higher sensitivity while preserving the simplicity and stability of traditional CL sensors due to less dependence on experimental conditions for reactions. Moreover, ECL-QDs can be used not only as emitter material but also as coreactants or ECL-RET components, which enables them to be used in new sensing mechanisms unavailable for conventional methods [97,130]

Aside from the analytical applications of the nanomaterials described, we could not find articles that utilize chemiluminescent quantum dots for clinical applications.

The articles featuring chemiluminescent sensors with outstanding analytical performances and intuitive visual representations are compiled in Figures 4 and 5 below to summarize key trends and regularities in chemiluminescent sensing using quantum dots.

Figure 4A presents a synthetic pathway for the fabrication of a highly sensitive electrochemiluminescent biosensor based on CdTe QDs enriched with a metal–organic framework to enhance the detection of a key biomarker for myocardial infarction–cardiac troponin-I (cTnI). The novelty of the sensor lies in the signal amplification method, where an isoreticular metal–organic framework-3 is integrated with CdTe dots to utilize it not only as a high-capacity carrier but also as a coreactant accelerator, which is crucial for ECL. The synthesis involved the usage of 2-amino terephthalic acid as the organic ligand to encapsulate CdTe within IRMOF-3, which significantly increased the packing of QDs and improved the emission intensity by making possible the conversion of persulfate into highly reactive sulfate radicals. Then, using this composite as a labeling probe in a sandwich immunoassay, an ultra-low LOD of 19.2 aM and a wide linear range of seven orders of magnitude were acquired. The mechanism applied not only enhanced the quantum dots' ECL performance but also ensured superior stability and selectivity, which is required for broader applications [111].

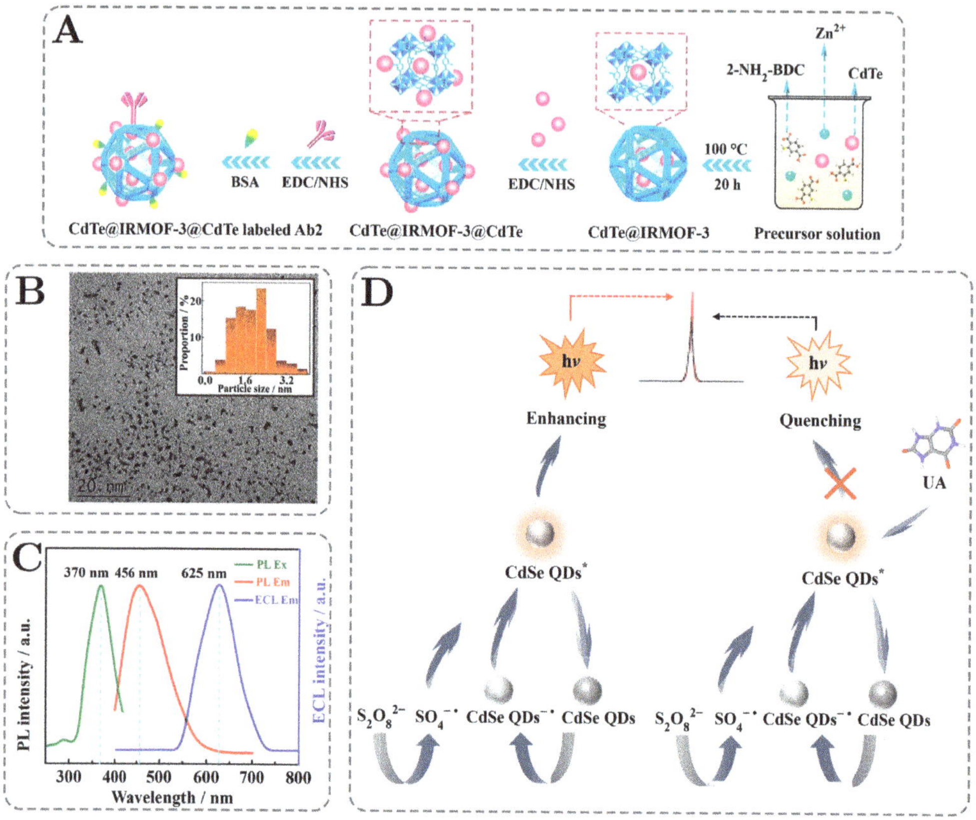

Figure 4. Analytical papers based on luminescence sensing techniques. (**A**) The synthesis of an ECL immunosensor with dramatically enhanced CdTe-QD intensity based on isoreticular metal–organic framework-3 (IRMOF-3) with a narrow RSD and a low LOD. Adapted with permission from Yang et al. [111]. Copyright © 2018 American Chemical Society. (**B**) A TEM image with a size distribution insert of MoS_2 QDs used in a biosensor for the atto-molar detection of microRNA with a 12.7-fold increase in ECL efficiency. Adapted with permission from Liu et al. [126]. Copyright © 2022 American Chemical Society. (**C**) Normalized PL excitation, PL emission, and ECL spectra of MoS_2 quantum dots applied in the eco-friendly, sensitive detection of lipopolysaccharide, achieving the widest linear range, of 8.7 orders of magnitude, among all the reviewed articles. Adapted with permission from Zhao et al. [108]. Copyright © 2017, American Chemical Society. (**D**) A possible, and common for ECL, mechanism of the molecularly imprinted electrochemiluminescence sensor used for the sensing of uric acid with a ratiometric analysis. Adapted with permission from Liu et al. [128], © 2023 Elsevier B.V. All rights reserved.

Figure 4B provides a transmission electron microscopy image of the molybdenum disulfide quantum dots that were used in the fabrication of the luminol/MoS_2 QDs@Zeolitic imidazolate framework-8 for the sensitive detection of microRNA-21. A high level of this RNA is a marker of cancer or cardiac disease. As seen on the inset, the distribution of particle sizes is not uniform and is clustered in a wide range between 0.8 nm and 2.8 nm.

While variability in the dots' diameter is not considered to be a desirable phenomenon as it may cause batch-to-batch variability and inhomogeneous ECL signal, for the given case, the detection was very sensitive, with a limit of 14.6 aM. Typically, luminol is used in a pair with hydrogen peroxide to produce the chemiluminescent signal; however, as was shown in this paper, the application of MoS_2 dots as a coreactant to luminol enhances the efficiency of the signal up to 12.7 times [126].

Continuing with MoS_2 dots, Figure 4C demonstrates the spectra for this type of QDs in terms of their excitation by photoluminescence and emission by both photoluminescence and chemiluminescence. The PL excitation peak is the most blue-shifted compared to any emission peak; however, ECL is more red-shifted than the PL emission peak. Under electrochemical excitation, charges can localize on or near QD surface states more readily than under optical excitation. Emissions from these lower-energy states or excitons partially "trapped" at the surface tend to appear red-shifted relative to the band-edge emission that dominates photoluminescence. In their paper, Zhao et al. demonstrated the synthesis of non-toxic MoS_2 dots that are used in sensing lipopolysaccharide, which is considered a contaminant in biotechnology and research. The achieved linearity range of Pd-Au, convex, hexoctahedron-immobilized QDs is 8.7 orders of magnitude, which is the largest found for this review value. The acquired limit of detection is 0.07 fg/mL [108,131].

Figure 4D demonstrates the mechanism of the common ECL from the work of Liu et al., who have synthesized an MXene@NaAsc-enhanced sensor based on CdSe@ZnS for the detection of uric acid, with an achieved LOD of 18.13 pM. This method, like many others, utilizes coreactants with QDs to achieve an enhanced ECL signal. This is possible due to the generation of reactive intermediates and energy transfer facilitation caused by coreactants. Hydrogen peroxide, persulfates, oxalates, and tripropylamine are typical examples of coreactants. In the given case, persulfate ion is used. Under applied potential CdSe@ZnS QDs and persulfate ions, both gain an electron from a reduced radical species:

$$\text{CdSe@ZnS QDs} + e^- = \text{CdSe@ZnS QDs}^{-\bullet}$$

$$S_2O_8 + e^- = SO_4^{-\bullet} + SO_4^{2-}$$

The reduced QDs then react with a sulfate radical anion to initiate an excited state in the QDs (denoted as QDs*), which then relaxes back to the ground state with the emission of a photon:

$$\text{CdSe@ZnS QDs}^{-\bullet} + SO_4^{-\bullet} = \text{CdSe@ZnS QDs}^* + SO_4^{2-}$$

$$\text{CdSe@ZnS QDs}^* = \text{CdSe@ZnS QDs} + h\nu$$

However, in the presence of uric acid, a portion of the quantum dot radicals is oxidized, preventing the formation of their excited state; consequently, a reduction of ECL signal is observed:

$$\text{CdSe@ZnS QDs}^{-\bullet} + \text{Uric Acid} = \text{CdSe@ZnS QDs} + \text{Uric Acid}^{-\bullet}$$

The concentration of the analyte is then found by the linear fitting of the ratio of the ECL signal to the applied current and the decimal logarithm of uric acid concentration [128].

Figure 5A,B present an electrochemiluminescent biosensor based on cadmium sulfide quantum dots, which have applications in detecting cancer-related mutations in circulating tumor DNA (ctDNA). The sensor is designed to identify the epidermal growth factor receptor (EGFR) T790M mutation, a critical marker in non-small-cell lung carcinoma. Typical methods, like polymerase chain reaction (PCR), require a significant number of extra nucleotides on both sides of the region of interest for successful detection, whereas the developed sensor can work with short segments (18–100 nucleotides) of DNA with terminal mutations. It consists of a glass carbon electrode with immobilized thioglycolic acid-CdS QDs and AuNP-labeled hairpin DNA (AuNP-haiDNA).

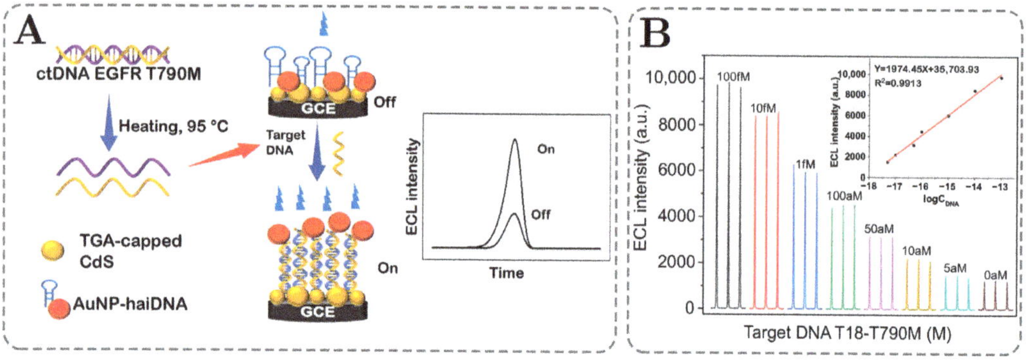

Figure 5. (**A**,**B**) ECL-resonance energy transfer system allowing detection of cancer mutations on linear DNA analytes with atto-molar sensitivity. Adopted with permission from Yang et al. [127]. Copyright © 2023 American Chemical Society.

As shown in Figure 5A, during the "off state" of the sensor, Au nanoparticles cause the quenching of ECL emission from QDs due to high proximity; however, when the target DNA is introduced into the system, it selectively binds to complementary hairpins on the nanoparticles, which forces structural changes, decreasing the proximity of AuNP-haiDNA to the glassy carbon electrode (GCE), causing an increase in the emission signal. The performance of the sensor depends on the mutation position: when the terminal position is analyzed, an LOD of 3.4 aM with a linear range of 50 aM to 100 fM is achieved; in comparison, a higher detection limit of 8.1 aM but a wider linear range of 10 aM to 100 fM is achieved for the middle mutation position. Since complementary DNA is used in this method, it shows excellent selectivity. As a result, the method could contribute to highly sensitive, rapid, and cost-effective ctDNA mutation detection for cancer diagnostics.

Figure 5B depicts a common setup for chemiluminescent measurements and calibration. As seen, the vertical axis represents the ECL emission intensity in arbitrary units for varied concentrations of the analyte, whereas the horizontal axis indicates the time. The inset graph provides the semilogarithmic calibration plot of the emission intensity versus the log of the concentration of a standard solution. The short-term emission is induced by forcing an electrochemiluminescent reaction with an applied potential, which is then recorded. To reduce the variability of the signal, the signals from a few stimulations are averaged. The same steps are taken on each of the standard solutions. Finally, the average maximum ECL emission intensity is plotted against the concentration on the semilogarithmic plot to achieve a linear relationship for calibration [127].

4. Phosphorescence of QDs: Applications in Sensing and Detection

Because of their unique optical properties and long-lived excited states, phosphorescent quantum dots (PQDs) became one of the prominent molecular sensing tools. Another reason might be the obstacles that are caused by fluorescent quantum dots and yet are solved by PQDs, such as delayed fluorescence, since it has the ability to undergo intersystem crossing from singlet to triplet states [132].

In addition to that, fluorescent sensor arrays have several disadvantages, such as autofluorescence and scattering light originating from endogenous biomolecules. PQDs solve these issues, particularly Mn-doped ZnS QDs, by emitting orange phosphorescence (~590 nm) via energy transfers from the ZnS host lattice to the Mn^{2+} dopants, followed by a spin-forbidden transition from the excited triplet state ($^4T_1$) to the ground state ($^6A_1$), which achieves a longer emission lifetime. A longer emission lifetime allows us to effectively filter out short-lived background fluorescence and Rayleigh scattering. Furthermore, the difference between the excitation and emission wavelengths is significant because of Mn-doped ZnS QDs' large Stokes shifts, which in turn reduce the spectral overlap and enhance the signal clarity. These differences can be seen in Table 4: the excitation occurs at 295–304 nm, while the emission stays around 590 nm. Because of all of these advantages, PQDs demonstrate great results in detecting hyaluronic acid, melamine, heparin, and other biomolecules [132–135].

For a long time, fluorescent quantum dots dominated the molecular sensing field. However, He et al. showed the benefits of PQDs for the detection of enoxacin in 2008 by demonstrating a 58.6 nM detection limit [136]. In the following years, researchers started discovering the potential of PQDs in molecular sensing during metal ion detection and environmental pollutant detection [137–139]. Most PQDs are ZnS-based because of their wide band gap and surface passivation ability. Moreover, the protection of the quantum dot core by ZnS from the environment plays a significant role in the reduction of photobleaching and an increased emission time, leading to better sensing abilities [140].

In addition to ZnS properties, the diameter of PQDs also should be considered since it affects various properties, such as the emission wavelength, functionalization surface area, and photostability. Therefore, we analyzed various PQDs in Table 4, and the most common radius is 3.5 nm, which completely agrees with the theoretical calculations, according to which, the smaller the QDs' size, the more intense the emissions for the detections of analytes with low concentrations.

According to 54 studies on PQDs, we found the following analytical conclusions. The geometric average LOD is 26 nM, with a median of 38 nM. The average PQD diameter is 4.6 nm (median 3.7 nm). The values for the average and the median RSD are 3.9% and 3.4%, respectively. This data shows how efficient and reproducible PQDs are. In addition, R^2 (0.996 for average and median) demonstrates the consistency of PQDs.

Another important thing to consider is analyzing the best-performing PQDs, along with the worst ones, to understand their strengths and weaknesses. Therefore, we took PQDs corresponding to these parameters and examined their synthesis/functioning mechanisms.

The best-performing PQDs, demonstrating the lowest LOD (79 fM), utilized Mn-ZnS QDs capped with mercaptopropionic acid (MPA). To positively charge QD nanohybrids, these PQDs were functionalized with poly (diallyldimethylammonium chloride) (PDAD), a cationic polymer, as illustrated in Figure 6A. The purpose of the positive charging was to improve their room-temperature phosphorescence and minimize background interference by utilizing strong electrostatic interactions after the addition of the analyte, HA, since it is negatively charged [134]. In Figure 6B, we analyzed the optical properties of these PQDs. According to the figure, these PQDs have a maximum absorption peak at 295 nm (curve a) and an emission peak at 590 nm (curve b), where the transition of electrons from the triplet excited state $^4T_1$ to the $^6A_1$ ground state of Mn^{2+} is responsible for the emission. It is also important to note that PDAD does not change the absorption/emission properties of Mn-ZnS QDs since PDAD-functionalized Mn–ZnS QDs have the same absorption/emission peak at 295/590 nm as unmodified Mn–ZnS QD [134].

Table 4. Analytical performance of sensing based on phosphorescence quantum dots (PQDs).

Year, Family Name	Nanostructure	Preparation	Analyte	LOD	Range	Diameter (nm)	RSD (%)	R^2	Excitation Wavelength (nm)	Emission Wavelength (nm)	Binding Molecule
Wu 2010, [132]	Mn-doped ZnS QDs	CS	Glucose	3.0 μM	10 μM–1 mM	3	3.2	0.9855	290	595	1-ethyl-3-(3-dimethylaminopropy)carbodiimide (EDC)/N-hydroxysuccinimide (NHS)
Wang 2010, [141]	Mn-doped ZnS QDs	CS	ascorbic acid	9.0 nM	0.05–0.8 μM	4.2	4.8	0.9913	337	595	Sodium Tripolyphosphate
Wang 2011, [135]	ZnS QDs	Combination of colloidal and template-assisted synthesis	2,4,6-trinitrotoluene	50 nM	0.05–1.8 μM	5	3.5	0.9902	316	580	3-Aminopropyltriethoxysilane-functionalized manganese
Yu 2011, [142]	Mn-doped ZnS QDs	CS	DNA	0.10 μM	0.08–12 mg L^{-1}	3.5	3.7	N/D	316	590	methyl violet
Zhang 2013, [143]	Mn-doped ZnS QDs	N/D	DNA	27 pM	0 to 45 nM	N/D	3.73	0.9991	N/D	581	N/D
Wu 2013, [138]	Mn-doped ZnS QDs	Combination of colloidal and template-assisted synthesis	trypsin	40 nM	0.1–1.2 μM	3.8	N/D	0.9917	280	590	bovine serum albumin (BSA)
Dan 2013, [144]	Mn-doped ZnS QDs	N/D	domoic acid (DA)	67 nM	0.25–3.5 μM	N/D	0.65	0.99	300	590	Polyethyleneimine
Tan 2013, [145]	Mn-doped ZnS QDs	CS	bovine hemoglobin	38 nM	1.0×10^{-7}–5.0×10^{-6} mol L^{-1}	7	1.8	0.998	340	585	3-Mercaptopropyltriethoxysilane (MPTS)
Wang 2013, [146]	Mn-doped ZnS QDs	CS	catechol	53 nM	0.5–5 μM	N/D	3.2	0.9962	316	595	Sodium Tripolyphosphate
Bian 2013, [147]	Mn-doped ZnS QDs	CS	L-ascorbic acid	0.72 μM	2.5–37.5 μM	9	1.4	0.9983	315	583	N-acetyl-L-cysteine
				14 μM	2.5–47.5 μM	9	1.8	0.995	306	580	L-cysteine
Bian 2014, [148]	Mn-doped ZnS QDs	CS	Co^{2+}	60 nM	1.25×10^{-6}–3.25×10^{-5} M	10	2.3	0.9989	315	583	N-acetyl-L-cysteine
Zhu 2014, [149]	Mn-doped ZnS QDs	Microwave-assisted synthesis	indapamide	0.89 μM	1.5–80 μM	4.2	3.4	0.995	310	585	MPA
Gong 2014, [150]	Mn-doped ZnS QDs	CS	clenbuterol	12 nM	5–1000 ng·mL^{-1}	3.5	2.9	0.9985	295	590	MPA
Zhang 2015, [151]	Mn-doped ZnS QDs	CS	protamine	33 nM	0.2–3.0 μg mL^{-1}	3.5	2.76		295	590	MPA

Table 4. *Cont.*

Year, Family Name	Nanostructure	Preparation	Analyte	LOD	Range	Diameter (nm)	RSD (%)	R^2	Excitation Wavelength (nm)	Emission Wavelength (nm)	Binding Molecule
Bi 2015, [152]	Mn-doped ZnS QDs	CS	DNA	0.14 μM	0.2–20 mg L^{-1}	2.5	1.7	0.998	295	590	MPA
Gong 2015, [153]	Mn-doped ZnS QDs	CS	phosphopeptides	N/D	1.6–2800 ng mL^{-1}	3.5	2.5	0.9969	295	590	MPA
Zhang 2015, [154]	Mn-doped ZnS QDs	CS	quercetin	0.16 μM	0.1–6.0 mg mL^{-1}	3.5	4.6	0.996	295	590	MPA
Ertas 2015, [155]	Mn-doped ZnS QDs	CS	double stranded DNA/ idarubicin	0.48 μM	2.0–20.0 μM	3.5	4.35	0.9943	290	590	L-cysteine
Bian 2015, [156]	ZnS QDs	CS	histidine	0.74 μM	1.25–30 μM	10	0.65	0.997	315	589	Co^{2+}-adsorbed N-acetylL-cysteine (NAC)
Zhang 2015, [157]	Mn-doped ZnS QDs	CS	glucose	7.0 μM	20–800 μM	2.7	0.5	0.996	310	597	N/D
Chang 2016, [133]	Mn-doped ZnS QDs	CS	human serum albumin	10 nM	0.02–1.0 μM	N/D	N/D	0.9799	301	602	Rhodamine B (Rh B)
Chen 2016, [158]	Mn-doped ZnS QDs	CS	Pb^{2+} ions	2.2 nM	1-100 μg L^{-1}	3	0.54	0.9962	280	590	glutathione
Gong 2016, [159]	Mn-doped ZnS QDs	CS	DNA	46 pM	15 μg L^{-1}– 40 mg L^{-1}	3	1.9	0.998	295	590	MPA
Liu 2017, [160]	Mn-doped ZnS QDs	CS	trypsin	1.8 nM	0.88–15.6 μg mL^{-1}	N/D	5	0.993	316	598	Cytochrome c (Cyt c)
Lv 2017, [161]	Mn-doped ZnS QDs	CS	miRNA-21	1.6 nM	8–80 nM	3.5	3.4	0.989	295	590	ROX-DNA
Zhang 2017, [162]	Mn-doped ZnS QDs	N/D	alkaline phosphatase	N/D	0.001–0.1 U/L	N/D	N/D	0.999	312	600	PNPP (p-nitrophenylphosphate)
Zhang 2017, [163]	Mn-doped ZnS QDs	Hydrothermal/CS	thiram	25 nM	50 nM–2.5 μM	5	N/D	0.9975	312	590	mercaptosuccinic acid (MSA)
Pacheco 2017, [164]	Mn-doped ZnS QDs	CS	warfarin	4.7 μM	1.07×10^{-5} M– 4.50×10^{-5} M	2.7	N/D	0.997	276	584	l-cysteine (L-cys)
Zhang 2017, [165]	Mn-doped ZnS QDs	CS	patulin	0.32 μM	0.43-6.50 μmol L^{-1}	10.25	4.2	0.9945	324	585	3-Mercaptopropyltriethoxysilane (MPTS)
Deng 2017, [166]	Mn-doped ZnS QDs	CS	permanganate anions (MnO_4^-)	0.24 μM	0.5–100 μM	7.75	2.95	0.9981	315	585	L-cysteine
Lv 2017, [167]	Mn-doped ZnS QDs	CS	transgenic 35S promoter DNA	4.0 nM	12–300 nM	5	7.9	0.997	295	590	DNA

Table 4. *Cont.*

Year, Family Name	Nanostructure	Preparation	Analyte	LOD	Range	Diameter (nm)	RSD (%)	R^2	Excitation Wavelength (nm)	Emission Wavelength (nm)	Binding Molecule
Li 2018, [134]	Mn-doped ZnS QDs	CS	alkaline phosphatase		0.15–18 U L^{-1}	4	4.2	0.994	295	590	Eu^{3+}
Li 2018, [168]	Mn-doped ZnS QDs	CS	Micrococcal nuclease base		2 × 10−3–8.0 × 10−2 U mL^{-1}	3.5	5.2	0.993	295	595	DNA-ROX
Wei 2018, [169]	Mn-doped ZnS QDs	CS	2,4,6-trichlorophenol	35 nM	0.1–30 μmol L^{-1}	N/D	4	0.99	320	594	magnetite (Fe_3O_4)
Liu 2018, [170]	Mn-doped ZnS QDs	CS	adriamycin	0.45 μM	0.5–64.0 μM	4.5	N/D	0.9932	316	585	poly(diallyldimethylammonium chloride) (PDDA)
Li 2018, [134]	Mn-doped ZnS QDs	CS	hyaluronic acid (HA)	79 fM	0.08–2.8 μg mL^{-1}	4	2.1	0.995	295	590	MPA
Zou 2018, [171]	Mn-doped ZnS QDs	CS	copper(II)	6.0 nM	0.01–12 μM	3.6	3.2	0.994	316	590	alginate
Qin 2018, [172]	Mn-doped ZnS QDs	CS	resveratrol	10 nM	0.03–14 μM	3.5	2.7	N/D	295	595	MPA
Luo 2019, [173]	Mn-doped ZnS QDs	CS	4-nitrophenol	60 nM	0.1–100 μM	4	4.6	0.998	295	590	N/D
Chen 2019, [174]	Mn-doped ZnS QDs	CS	cephalexin	2.3 nM	2.5–50 μg·L^{-1}		5	0.9985	295	590	thioglycolic acid (TGA)
Zhao 2019, [175]	Mn-doped ZnS QDs	CS	picric acid	6.1 nM	2.0–180 ng mL^{-1}	5.8	2.2	0.9985	316	600	Melamine (MA)
Miao 2019, [137]	Mn-doped ZnS QDs	CS	melamine	1.6 μM	0.005–6 mM	4	2.6	0.997	297	590	ssDNA
Liu 2020, [176]	Mn-doped ZnS QDs	CS	tetracyclines	8.6 nM	50–1.5 × 10^5 nM	3.8	6	0.995	289	583	L-Cysteine
Jayasinghe 2020, [177]	Mn-doped ZnS QDs	CS	aflatoxins	11 pM	2–20 μg L^{-1}	2.3	20	0.9947	290	594	Polyethylene glycol (PEG)
Jinadasa 2020, [178]	Mn-doped ZnS QDs	CS	As(III), As(V)	0.12 nM	0–20 μg L^{-1}	N/D	10	N/D	289	595	(3-aminopropyl) triethoxysilane and an As(III) ionic
Lv 2020, [179]	Mn-doped ZnS QDs	CS	phenol	2.1 μM	5.0 to 55 μmol L^{-1}	5.5	3.7	0.9984	330	600	3-mercaptopropyltriethoxysilane (MPTS)
Liu 2021, [180]	Mn-doped ZnS QDs	CS	thyroxine	2.0 nM	4.85 nmol/L–1.59 μmol/L	3.5	N/D	0.999	295	590	MPA
			carbamazepine	3.4 nM	7.9 nmol/L–1.555 μmol/L	3.5	N/D	0.995	295	590	

Table 4. *Cont.*

Year, Family Name	Nanostructure	Preparation	Analyte	LOD	Range	Diameter (nm)	RSD (%)	R^2	Excitation Wavelength (nm)	Emission Wavelength (nm)	Binding Molecule
Qin 2021, [181]	Mn-doped ZnS QDs	CS	alkaline phosphatase		0.2–10 U/L	3.65	N/D	0.993	310	680	pyrophosphate
Chen 2021, [182]	Mn-doped ZnS QDs	CS with sol-gel synthesis	norfloxacin	2.5 nM	1–90 μg L^{-1}	5.5	7	0.9993	300	590	magnetite (Fe_3O_4)
Fan 2021, [139]	Mn-doped ZnS QDs	CS	chlorpyrifos	0.89 μM	0–80 μM	N/D	1	0.99	275	600	N/D
Kong 2023, [183]	Mn–ZnS QDs@g-C_3N_4	N/D	2,4,6-trinitrotoluene	0.56 μM	0–12 μM	3.6	6	0.997	318	582	mercaptoethylamine (MEA)
Yang 2025, [184]	Mn-doped ZnS QDs	CS	lead (II)	2.6 nM	5×10^{-6}–100 mM	2.12		0.9979	360	596	1-thioglycerol
Summary of Average/Median Values in Table 4											
Median LOD (M)	3.8×10^{-8}	Median Diameter (nm)	3.7	Median RSD (%)	3.4	Median R^2	0.996	Median Excitation wave-length (nm)	300	Median Emission wavelength (nm)	590
Geometric Average LOD (M)	2.6×10^{-8}	Average Diameter (nm)	4.6	Average RSD (%)	3.9	Average R^2	0.996	Average Excitation wave-length (nm)	304	Average Emission wavelength (nm)	590

The abbreviations used are CS for colloidal synthesis, DNA for deoxyribonucleic acid, MPA for 3-mercaptopropionic acid, ROX for 6-carboxy-x-rhodamine, and N/D for the values that are not explicitly mentioned in the papers.

The PQDs with the worst performance (the LOD is 13.8 μM) were synthesized using the wet-chemical approach, where $ZnSO_4 * 7H_2O$, $MnCl_2 * 4H_2O$, and N-Acetyl-L-cysteine (NAC) were mixed in a 50 mL reaction volume. According to the article, the optimal pH for PQD synthesis was 11, which was regulated by 1.0 M NaOH. Using nitrogen bubbling, dissolved oxygen was removed so that the $Na_2S * 9H_2O$ solution could be injected into the mixture to start the synthesis of NAC-Mn/ZnS QDs. After purification, centrifugation, and vacuum drying, the desired PQDs were obtained. This synthesis mechanism is briefly described in Figure 6C. However, in that case, they replaced NAC with cysteine as the stabilizing agent. It seems like the most important factor for the LOD in this case is the substitution of NAC by cysteine since the LOD significantly increases from 0.72 to 13.8 μM [147].

The PQDs with the smallest diameter (2.12 nm) demonstrate absorption at 300–350 nm and emission at 596 nm, according to Figure 6D. Because of the strong coupling between the $3d^5$ electrons of Mn^{2+} and the s-p electrons of the ZnS host, the electron transition occurs from $^4T_1$ to $^6A_1$ of the Mn^{2+} ions doped within the ZnS lattice, which makes the emission 100% pure dopant [184].

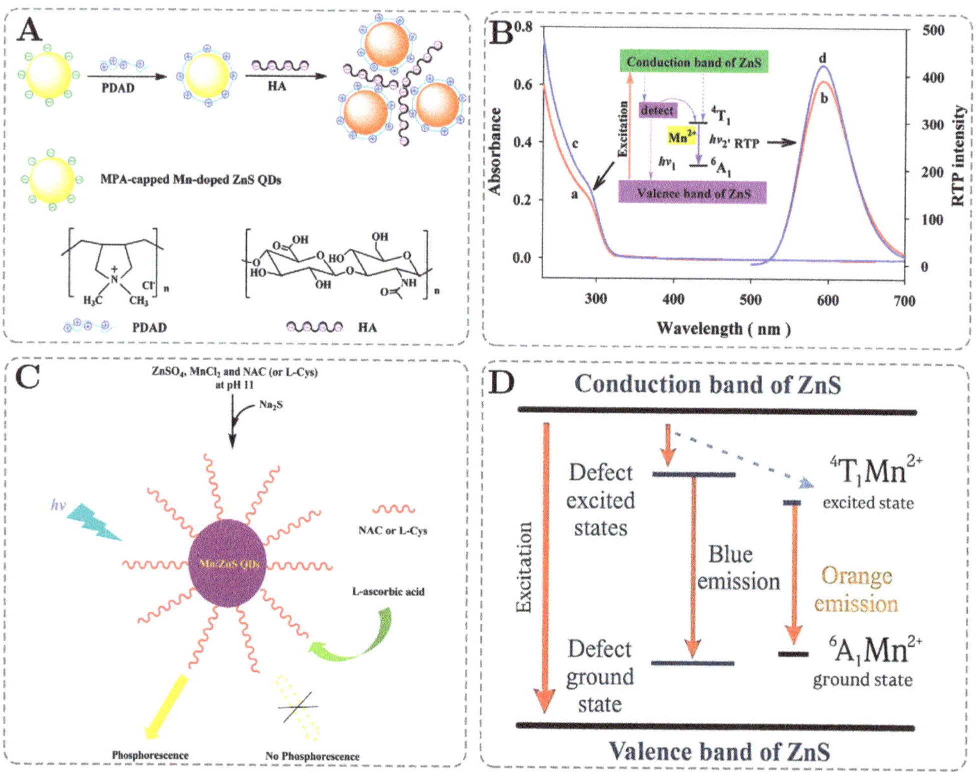

Figure 6. (**A**) Schematic illustration of fabrication of PDAD–Mn–ZnS QDs for HA detection. Adopted under Creative Common CC BY license from Li et al. [134]. Copyright © 2018. (**B**) UV/vis spectra (curves a and c) and RTP emission spectra (curves b and d) of Mn–ZnS QDs (10 mg mL^{-1}; curves a and b) and PDAD–Mn–ZnS QDs (10 mg mL^{-1}; curves c and d). Adopted under Creative Common CC BY license from Li et al. [134]. Copyright © 2018. (**C**) Schematic illustration of sensor design. Adapted with permission from Bian et al. [147]. Copyright © 2013, Elsevier. B.V. (**D**) Schematic of emission mechanism of TG-Mn-ZnS QDs.

It can be concluded that the PQDs with the lowest LOD utilized electrostatic interaction as one of their techniques to achieve very high sensitivity by using PDAD, while the performance of the PQDs with the highest (worst) LOD significantly suffered from the substitution of the binding molecule joining NAC to cysteine. Therefore, it can be clearly seen how modifications and ligand choices play a significant role in the detection performance of PQDs.

5. Quantitative Analysis of Relationships Between Sensitivity and Other QD Parameters in QD-Based Sensing

In order to understand which QD parameters are likely to impact/improve the QDs' sensitivity (LOD) in analysis, we have represented the relationships between the pLOD (M) and the major parameters of quantum dots reported in the literature, including excitation wavelengths, emission wavelengths, and QDs' average diameter or size. We graphically represent those relationships in Figure 7. We have also calculated the linearity coefficients displayed on the plots, as well as the Pearson correlation coefficients displayed in Table 5.

Table 5. Pearson correlation coefficients for sensitivity and major QD parameters.

Parameter		Size			Excitation λ			Emission λ		
Type		FL	P	CL	FL	P	CL	FL	P	CL
Correl	LOD	0.384	0.181	0.181	0.115	−0.043	-	−0.227	−0.163	−0.210
	pLOD	−0.466	−0.299	−0.211	0.179	−0.126	-	0.313	0.133	−0.024
Sample Size		38	35	25	54	49	-	54	50	30

As can be concluded from the data, the wavelength of excitation light has little influence on the pLOD (hereinafter, detection limit) for both fluorescence- and phosphorescence-based quantum dots. However, there is a significant difference between the typical values for the excitation wavelengths for these two methods. For fluorescence, the values are generally gathered around 359 ± 31 nm, with many values exceeding 350 nm. In contrast, the phosphorescent excitation wavelengths are typically equal to 304 ± 17 nm, with almost no data points above 350 nm, which is noticeably shorter than those of FL.

As a matter of observation, the phosphorescent systems require higher energy excitation wavelengths compared to fluorescent ones by almost 50 nm, which seemingly populates triplet excited states and defect states more efficiently, since P is based on them. However, we were unable to find sources that explicitly explain why P-based QDs have lower excitation wavelengths than FL-QDs. The variation of values is narrower for phosphorescent systems, with distinct clusters around 295 nm and 315 nm that are attributed to well-known classes of phosphorescent nanomaterials, like Mn^{2+}-doped ZnS quantum dots. It is also worth noting that the best detection limits for P-based sensors were achieved at higher energy excitation light.

There is a small but noticeable positive correlation between sensitivity and the emission wavelength both in fluorescence and phosphorescence, though to a smaller extent in the latter one, with Pearson correlation coefficients of +0.31 and +0.13 for FL and P, respectively. This may be related to a reduced autofluorescence from biological samples at lower energy emission light levels. Additionally, QDs that emit red light tend to be bigger and, therefore, brighter. For the fluorescence, longer emission wavelengths also reduce the interference from the excitation light, which can be close in wavelength. In phosphorescence, a longer lifetime of the excited triplet state helps the separation of the signal from short-lived noise, which is particularly useful for the detection of trace analytes. The mean value for the emission wavelength resulted in 509 ± 74 nm for fluorescence-based sensing and 590 ± 5 nm for phosphorescent-based sensing. Similar to the emission light, the excitation light's wavelengths are more clustered for phosphorescence, with distinct groups around 304 ± 17 nm for Mn-doped ZnS dots due to less available nanomaterials with tunable phosphorescence. Chemiluminescent QDs show no meaningful correlation with r = −0.02 and data points that are scattered across the plot. For chemiluminescence, sensitivity is more dependent on the efficiency of the chemical or the electrochemical reaction, rather than on the emission wavelength. The average emission wavelength for CL is 602 ± 75 nm.

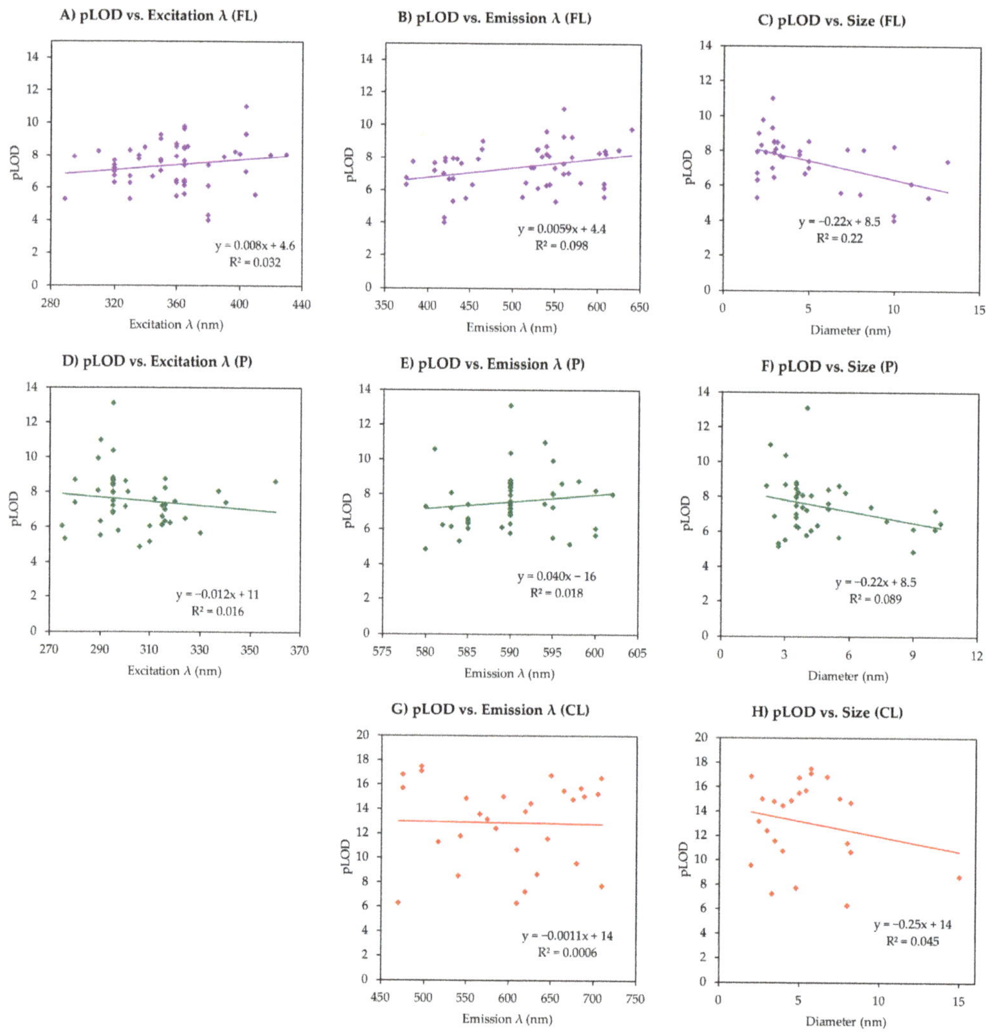

Figure 7. Scatter plots presenting the relationships between the negative decimal logarithm of the detection limit and the emission wavelength, excitation wavelength, and quantum dots' size for the three types of QDs considered for analysis in this review paper. Parts (**A–C**) display fluorescent QDs (FL), parts (**D–F**) phosphorescent (P), and parts (**G,H**) chemiluminescent (CL).

As expected for quantum dots as nanomaterials with tunable properties, the strongest correlation is observed between the detection limit and quantum dots' size. In each of the three types of luminescence, there is a negative correlation, showing that higher sensitivity is achieved with smaller QDs. The correlation is the strongest for FL, with r = −0.47, weaker for P, with r = −0.30, and minimal for CL, with r = −0.21. For fluorescent quantum dots, a smaller size leads to a higher surface-to-volume ratio, making QDs more responsive towards surface interactions, such as analyte binding, pH levels, or ion concentrations. Moreover, smaller dots have stronger confinement, which increases the energy gap and promotes radiative energy relaxations, enhancing their sensitivity to the environment. This size correlation is not so firm for phosphorescence and chemiluminescence as these systems are more dependent on triplet state dynamics or reaction kinetics than on particle size. The average QD sizes obtained are 4.9 $\pm$ 3.2 nm for FL, 4.6 $\pm$ 2.2 nm for P, and 5.3 $\pm$ 2.8 nm for CL.

Table 6 below presents the arithmetic mean and median values for the pLOD, emission wavelength, excitation wavelength, and particle diameter.

Table 6. Average and median values of essential QD parameters for FL, P, and CL (fluorescence, phosphorescence, and chemiluminescence).

Average	FL	P	CL	Median	FL	P	CL
pLOD	7.4	7.6	13.0	pLOD	7.6	7.4	13.6
Excitation λ	359	304	-	Excitation λ	360	299	-
Emission λ	509	590	602	Emission λ	530	590	615
Size	4.9	4.6	5.3	Size	3.3	3.8	4.8

As seen, the pLOD values for fluorescence (FL) and phosphorescence (P) are nearly identical, with averages of 7.4 and 7.6 and medians of 7.6 and 7.4, respectively. This indicates that FL has slightly more high-sensitivity results than low-sensitivity ones, while in the case of P, the average is elevated by a few exceptionally low LOD values reported in the literature. As mentioned above, the comparable sensitivity of phosphorescent QDs, despite their lower emission intensity, is explained by their delayed emission, which allows for the elimination of the main drawbacks of fluorescence—namely, autofluorescence, Rayleigh scattering, and stray light—through techniques such as time-resolved phosphorescence [133,185]. The same explanation goes to chemiluminescent nanomaterials, which do not have any excitation light at all, which allows measurements with reduced background noise.

The average, 304 nm, and median, 299 nm, values for the excitation wavelength for phosphorescence are lower compared to these of fluorescence, which are 359 and 360 nm, respectively, whereas the emission value is greater. The emission wavelengths for chemiluminescence, which are 602 nm on average and 615 for the median are pretty much comparable to those of phosphorescence, since both methods experience lower energy electron transitions, though for slightly different reasons: triplet transition for P, and a lower energy intake for CL. As a matter of observation, the size of chemiluminescent quantum dots is slightly larger than that of fluorescent or phosphorescent quantum dots. This is related to the fact that larger CL-QDs possess better performance, making them more desirable in research. With the increased particle size, the degree of matching between the energy produced during the chemical reaction and the energy gap of the QDs is maximized, which means that the production of excited state QDs might be more efficient. Furthermore, since the quantum yield of the luminescence in QDs is inversely proportional to the confinement energy, and the confinement energy decreases with the increasing size, it must be the case that luminescence efficiency should correspondingly increase [186]. Other papers also verify that for larger QDs, the emission intensity of the ECL is greater [187].

6. Perspectives of Improvements in Sensing

Quantum dots exhibit excellent luminescent properties; however, due to surface trap sites, they are prone to the deterioration of their properties, such as the quantum yield and stability, which is still considered a challenge for researchers [188]. For perovskite QDs, limited stability hinders their characterization and slows down their technological progress [189,190]. Some of the common methods for QD stabilizations, as described by Sanjayan et al., include ligand exchange/doping, coating silica shells, and developing polymer encapsulations [191].

Yu et al. have utilized a selenium coating to stabilize colloidal HgTe QDs by tuning precursor reactivity. Se stabilization allowed for enhanced colloidal stability, surface passivation, and a variety of possibilities for further doping. As a result, the researchers have constructed adjusted *p-i-n* HgTe colloidal QD infrared photodetectors that exhibit an ultra-low dark current of 3.26×10^{-6} A cm^{-2} at -0.4 V and a room-temperature-specific detectivity of 5.17×10^{11} Jones at wavelength ≈ 2 µm, which leads to almost one order of magnitude improvement [192].

A group working with Mattoussi has developed a strategy to stabilize CsPbBr3 perovskite QDs (Per-QDs) by using coatings of polyzwitterion polymers, which were prepared via a nucleophilic addition reaction between amine-modified sulfobetaine anchors and solubilizing motifs of poly(isobutylene-*alt*-maleic anhydride). By their nature, Per-QDs tend to degenerate upon storage, processing, or testing, as well as lose their colloidal properties, absorption, and emission features after exposure to such polar media as alcohols or water. Typically used capping molecules include oleamine and oleic acid, which bind weakly to the Per-QDs and desorb quickly. That also contributes to the instability of Per-QDs. Authors list several methods of Per-QD stabilization, which include using additional ligand species, like alkyl-zwitterion, dimethylammonium, or octylphosphoric acid; ligand exchange with such small molecules as dimethylammonium bromide or 2,2′-iminodibenzoic acid; and embedding Per-QDs with a protective macroscale coating, including a polystyrene matrix. The hydrophilicity of Per-QDs was promoted by PEG blocks or zwitterion motifs, and a variety of prepared multifunction polymers allowed for a range of colloidal QDs, plasmonic nanomaterials, and magnetic NPs that demonstrate excellent photophysical properties and outstanding stability [189].

Apart from the stabilization of QDs, coating and/or ligand doping also enhances their biocompatibility [188,193]. Some of the reports of the stabilization of semiconductor QDs to improve their biocompatibility include works presented by Wang et al. and Nie et al.; however, both have utilized toxic synthetic components and media such as TOPO (trioctylphosphine oxide) or HDA (hexadecylamine) [194,195]. In contrast, Parani et al. have synthesized gelatin-stabilized, semiconductor (CdSe/CdS/ZnS)-core/double-shell quantum dots. In their study, they used the described QDs as fluorescent probes for *in vitro* HeLa cell imaging, and the experimental results showed that gelatin stabilization decreased the cytotoxicity by 50%. Moreover, gelatin stabilization and the CdS/ZnS double shell increased the photoluminescence quantum yield from around 20% (just CdSe QDs) to almost 60% (CdTe/CdS/ZnS/gelatin). The properties of the prepared QDs were preserved for a year [193].

The stabilization and reduction of cytotoxicity for any biosensing materials remain a hot topic in research, especially in QD-based sensors as they tend to use toxic components (such as transitional metals) that are released upon the degradation of the dots. However, there are reported attempts to improve the situation. For instance, in their study, Ali et al. demonstrate that the encapsulation of MPA-CdTe QDs with polyethylene glycol (PEGylation) shifts the half maximal inhibitory concentration (IC_{50}) from less than 0.12 nM to more than 0.21 nM after 24 h, leading to the possibility of applying almost twice the concentration of QDs. Furthermore, the cell viability was detected to be 65% for the PEGylated sensor compared to 45% for the bare MPA-CdTe at the same 0.2 nM concentration. Although the specific sensing results are not mentioned, they reported the acquisition of a brighter luminescence signal from the nanomaterial, and, therefore, a lower LOD, while still preserving the viability of a cell. Such improvement is attributed to the ability of poly(ethylene furanoate (PEF) to shield surface-bound Cd^{2+} release and to neutralize the surface charge, almost doubling the tolerable concentration [196]. Another interesting observation was presented by the research group of K. Kim, who conducted a cytotoxicity assessment of CdSe/ZnS QDs. The diameter of the yellow dots' core was 5.2 nm, while for green QDs it was 2.2 nm. They have demonstrated that bigger, yellow QDs, despite having dimmer light emission per particle compared to green ones, can remain non-toxic for the cells at concentrations up to 179 µg/mL, whereas green dots are lethal for the cells at 28 µg/mL. However, the usage of yellow dots decreased the signal-to-noise ratio by 33%, which means that in practical sensing, scientists must balance the choice: brighter (green) QDs might deliver a better LOD at the cost of a narrow cytotoxicity window [197]. Finally, for cytotoxicity, the study by Chen et al. focused on the cellular intake, intracellular dynamics, and short-term toxicity of PEG-coated quantum dots. They found that branched 6-armed PEG-amine/mPEG QDs have almost 10 times lower cellular uptake compared to linear PEG-QDs, which remarkably reduces the short-term cytotoxic effects; however, they still have long-term delayed effects related to the accumulation of radical oxygen species in their intracellular space, leaving the potential space for future modifications to resolve that issue [198]. Hence, while there is still a lack of direct quantitative comparison studies between raw and modified QDs, it can be seen that decreased cytotoxicity can lead to sensing improvements by increasing the maximum allowed biological sensing concentration of luminophore, leading to a brighter luminescence response. Future works might focus on combining these approaches to further limit the release of the toxic degradation products of QDs, which is also directly linked to the stability of a nanomaterial, as discussed above.

7. Application of Machine Learning in QD-Based Luminescent Sensors

The implementation of machine learning (ML) techniques into QD-based sensing platforms allows for achieving the enhancement of sensor performance, data interpretation, and automation. ML and deep learning-based sensing systems have a promising potential for biomarker detection [199,200]. For example, a research group of Qi and He have developed a QD- and magnetic bead (MB)-based ultrasensitive device for the detection of protein biomarkers, which was tested for interleukin-6 (IL-6). The authors have developed a novel deep learning model, ATTBeadNet, based on the UNet3+ architecture, which was used to count the magnetic beads on the images with a higher accuracy than previous AI models, such as ImageJ (code can be found at: https://github.com/foodszhang/ATTBeadNet; accessed on 9 July 2025). ATTBeadNet enables the accurate counting of fluorescent and total magnetic beads by processing paired fluorescence (FL) and "reference field" (RF) images. This bead counting is crucial for determining protein biomarkers, such as IL-6, since one Ab1-bound, streptavidin-coated capture MB binds one IL-6. The model identifies beads involved in a newly formed sandwich immunocomplex, which consists of antibody-functionalized MBs, IL-6, and horse-radish peroxidase (HRP)-labeled secondary antibodies. CdS QDs are utilized as signal reporters due to their narrow spectral width and high optical stability, which allow for signal clarity enhancement and strongly benefit FL microscopy. The use of RF images instead of traditional bright-field (BF) images improves signal-to-background ratios, which enables sensitive and automated detection in the assay. The designed system allows for the ultrasensitive quantification of IL-6 with an outstanding LOD of 3.1 fM and a linear range of 5 to 100 fM [199].

In another research article, Saren's group applied an optimized machine learning approach, the OPCA model. It enhances principal component analysis (PCA) by incorporating a neighborhood rough set algorithm to improve gene feature selection for gastrointestinal tumor classification. This model has yielded high classification precision (over 99%) and accuracy (above 94%) for both colon and gastric cancer datasets. The selected gene features were then applied to design a multicolor CdSe/ZnS QD-based immunobiosensor for the simultaneous detection of four tumor markers, namely α-fetoprotein (AFP), cancer embryonic antigen (CEA), cancer antigen 19-9 (CA19-9), and carbohydrate antigen 125 (CA125). Each marker was conjugated to a different color-emitting CdSe/ZnS QD, which allowed for multiplexed detection with a minimized spectral overlap. The OPCA model has led to a more efficient feature extraction, which improved the performance of the biosensor in tumor classification. The sensor was successfully tested, and the working ranges for all four biomarkers are from 2.0 to 51.5 ng/mL. Such integration of ML with QD-based biosensing demonstrates a promising direction for the high-throughput, accurate detection of cancer biomarkers and the clinical diagnosis of gastrointestinal tumors [200].

A research group of Liu and Wang have applied ML to a $CsPbBr_3$ perovskite QD (Per-QD)-based device for the rapid and accurate detection and sterilization of food pathogens, such as *E. coli*, *S. aureus*, *S. typhimurium*, *Listeria monocytogenes* (*L. monocytogenes*), and *P. aeruginosa*. The ML algorithm, namely a Support Vector Machine (SVM), was utilized for the analysis of relative signal intensity changes. The concentration ranges at which pathogens can be detected by the proposed system span four orders of magnitude from 1×10^3 to 1×10^7 CFU/mL. The LODs were also reasonably low for each of the five bacteria: *E. coli*—94 CFU/mL, *S. aureus*—117 CFU/mL, *S. Typhimurium*—93 CFU/mL, *L. monocytogenes*—136 CFU/mL, and *P. aeruginosa*—100 CFU/mL. The interaction of the QDs with different bacteria suspensions and various concentrations caused different degrees of fluorescence quenching to occur due to the aggregation of the Per-QDs, which in turn led to changes in color. After this, relative signal intensity changes, ΔRGB (red-green-blue), were captured by the Color Grab application on a smartphone and analyzed with SVM, which allowed for the precise identification and sensitive quantification of five bacterium types. Studies have also revealed that the designed systems possess strong bactericidal properties, with the deactivation efficiency of *E. coli* and *S. aureus* being more than 99% at 30 min after detection [201].

8. Conclusions

Sensing and bio-sensing methods based on QD applications demonstrated high (typically nanomolar or lower LOD molarity) sensitivity in the detection of a variety of analytes. The luminescence of QDs measured in sensing applications can employ fluorescence, phosphorescence, and chemiluminescence. QD-based methods using chemiluminescence demonstrated, on average, the highest sensitivity, typically in the sub-picomole range of molarity, which is five or six orders of magnitude higher with regard to sensitivity or lower with regard to the LOD in comparison with the average sensitivity/LOD of the QD-based sensing methods, where fluorescence or phosphorescence are measured.

Generally, although the number of publications utilizing machine learning tools for sensor development is considerably low, this approach has gained popularity recently, with the majority of the above-described works being published in 2025. As these reports have demonstrated, the implementation of ML strategies allows for enhanced sensitivity, fast and precise data analysis, and a decrease in the operations performed manually, i.e., the automatization of the detection process, which increases both the accuracy and the precision of the assays.

Author Contributions: Conceptualization, R.B.; methodology, R.B., A.L., D.A. and A.S.; formal analysis, R.B., A.L., D.A. and A.S.; resources and funding acquisition, R.B.; writing—original draft preparation, R.B., A.L., D.A. and A.S.; writing—review and editing, R.B., A.L., D.A. and A.S.; figure visualization, R.B., A.L., D.A. and A.S.; supervision, R.B. All authors have read and agreed to the published version of the manuscript.

Funding: This research was funded by the Nazarbayev University Faculty Development Competitive Research Grant Program 2023–2025, grant number 20122022FD4140.

Conflicts of Interest: The authors declare no conflicts of interest.

References

1. Zhang, Y.; Liu, B.; Liu, Z.; Li, J. Research progress in the synthesis and biological application of quantum dots. *New J. Chem.* **2022**, *46*, 20515–20539. [CrossRef]
2. Brichkin, S.B.; Razumov, V.F. Colloidal quantum dots: Synthesis, properties and applications. *Russ. Chem. Rev.* **2016**, *85*, 1297. [CrossRef]
3. Agarwal, K.; Rai, H.; Mondal, S. Quantum dots: An overview of synthesis, properties, and applications. *Mater. Res. Express* **2023**, *10*, 062001. [CrossRef]
4. Valizadeh, A.; Mikaeili, H.; Samiei, M.; Farkhani, S.M.; Zarghami, N.; Kouhi, M.; Akbarzadeh, A.; Davaran, S. Quantum dots: Synthesis, bioapplications, and toxicity. *Nanoscale Res. Lett.* **2012**, *7*, 480. [CrossRef]
5. Das, A.; Snee, P.T. Synthetic Developments of Nontoxic Quantum Dots. *ChemPhysChem* **2016**, *17*, 598–617. [CrossRef]
6. Bawendi, M.G.; Wilson, W.L.; Rothberg, L.; Carroll, P.J.; Jedju, T.M.; Steigerwald, M.L.; Brus, L.E. Electronic structure and photoexcited-carrier dynamics in nanometer-size CdSe clusters. *Phys. Rev. Lett.* **1990**, *65*, 1623–1626. [CrossRef]
7. Klimov, V.I. Spectral and Dynamical Properties of Multiexcitons in Semiconductor Nanocrystals. *Annu. Rev. Phys. Chem.* **2007**, *58*, 635–673. [CrossRef]
8. Steckel, J.S.; Snee, P.; Coe-Sullivan, S.; Zimmer, J.P.; Halpert, J.E.; Anikeeva, P.; Kim, L.-A.; Bulovic, V.; Bawendi, M.G. Color-Saturated Green-Emitting QD-LEDs. *Angew. Chem. Int. Ed.* **2006**, *45*, 5796–5799. [CrossRef] [PubMed]
9. Jang, E.; Jun, S.; Pu, L. High quality CdSeS nanocrystals synthesized by facile single injection process and their electroluminescence. *Chem. Commun.* **2003**, *24*, 2964–2965. [CrossRef]
10. Rempel, A.A.; Ovchinnikov, O.V.; Weinstein, I.A.; Rempel, S.V.; Kuznetsova, Y.V.; Naumov, A.V.; Smirnov, M.S.; Eremchev, I.Y.; Vokhmintsev, A.S.; Savchenko, S.S. Quantum dots: Modern methods of synthesis and optical properties. *Uspekhi Khimii* **2024**, *93*, 1–62. [CrossRef]
11. Ekimov, A.I.; Onushchenko, A.A. Quantum size effect in three-dimensional microscopic semiconductor crystals. *Lett. JETP* **1981**, *34*, 363. [CrossRef]
12. Reed, M.A.; Bate, R.T.; Bradshaw, K.; Duncan, W.M.; Frensley, W.R.; Lee, J.W.; Shih, H.D. Spatial quantization in GaAs–AlGaAs multiple quantum dots. *J. Vac. Sci. Technol. B Microelectron. Process. Phenom.* **1986**, *4*, 358–360. [CrossRef]
13. Ekimov, A.I.; Onushchenko, A.A. Size quantization of the electron energy spectrum in a microscopic semiconductor crystal. *Lett. JETP* **1984**, *40*, 337. [CrossRef]
14. Ekimov, A.I.; Efros, A.L.; Onushchenko, A.A. Quantum size effect in semiconductor microcrystals. *Solid State Commun.* **1985**, *56*, 921–924. [CrossRef]
15. Vandyshev, Y.V.; Dneprovskiĭ, V.S.; Ekimov, A.I.; Okorokov, D.K.; Popova, L.B.; Efros, A.L. Nonlinear optical properties of semiconductor microcrystals. *Sov. J. Exp. Theor. Phys. Lett.* **1987**, *46*, 495. [CrossRef]
16. Ekimov, A.I.; Efros, A.L.; Ivanov, M.G.; Onushchenko, A.A.; Shumilov, S.K. Donor-like exciton in zero-dimension semiconductor structures. *Solid State Commun.* **1989**, *69*, 565–568. [CrossRef]
17. Ekimov, A.I.; Efros, A.L.; Shubina, T.V.; Skvortsov, A.P. Quantum-size stark effect in semiconductor microcrystals. *J. Lumin.* **1990**, *46*, 97–100. [CrossRef]
18. Ekimov, A.I.; Hache, F.; Schanne-Klein, M.C.; Ricard, D.; Flytzanis, C.; Kudryavtsev, I.A.; Yazeva, T.V.; Rodina, A.V.; Efros, A.L. Absorption and intensity-dependent photoluminescence measurements on CdSe quantum dots: Assignment of the first electronic transitions. *JOSA B* **1993**, *10*, 100–107. [CrossRef]
19. Itoh, T.; Nishijima, M.; Ekimov, A.I.; Gourdon, C.; Efros, A.L.; Rosen, M. Polaron and exciton-phonon complexes in CuCl nanocrystals. *Phys. Rev. Lett.* **1995**, *74*, 1645. [CrossRef]
20. Ekimov, A. Growth and optical properties of semiconductor nanocrystals in a glass matrix. *J. Lumin.* **1996**, *70*, 1–20. [CrossRef]
21. Liu, B.; Altintas, Y.; Wang, L.; Shendre, S.; Sharma, M.; Sun, H.; Mutlugun, E.; Demir, H.V. Record High External Quantum Efficiency of 19.2% Achieved in Light-Emitting Diodes of Colloidal Quantum Wells Enabled by Hot-Injection Shell Growth. *Adv. Mater.* **2020**, *32*, 1905824. [CrossRef] [PubMed]
22. Fang, X.; Han, M.; Lu, G.; Tu, W.; Dai, Z. Electrochemiluminescence of CdSe quantum dots for highly sensitive competitive immunosensing. *Sens. Actuators B Chem.* **2012**, *168*, 271–276. [CrossRef]
23. Chen, B.; Liu, J.; Yang, T.; Chen, L.; Hou, J.; Feng, C.; Huang, C.Z. Development of a portable device for Ag+ sensing using CdTe QDs as fluorescence probe via an electron transfer process. *Talanta* **2019**, *191*, 357–363. [CrossRef] [PubMed]

24. Wang, X.; Liu, Z.; Gao, P.; Li, Y.; Qu, X. Surface functionalized quantum dots as biosensor for highly selective and sensitive detection of ppb level of propafenone. *Spectrochim. Acta Part A Mol. Biomol. Spectrosc.* **2020**, *227*, 117709. [CrossRef]
25. Yang, W.; Fidelis, T.T.; Sun, W.-H. Machine Learning in Catalysis, From Proposal to Practicing. *ACS Omega* **2020**, *5*, 83–88. [CrossRef]
26. Mater, A.C.; Coote, M.L. Deep Learning in Chemistry. *J. Chem. Inf. Model.* **2019**, *59*, 2545–2559. [CrossRef]
27. Krishnadasan, S.; Brown, R.J.C.; deMello, A.J.; deMello, J.C. Intelligent routes to the controlled synthesis of nanoparticles. *Lab A Chip* **2007**, *7*, 1434–1441. [CrossRef]
28. Zhang, Y.; Ren, D.; Shi, Y.; Yuan, R.; Ye, H.; Yin, X.-B.; Chi, H. A smartphone sensing fluorescent detection of mercury ion based on silicon quantum dots in environment water. *Spectrochim. Acta Part A Mol. Biomol. Spectrosc.* **2025**, *325*, 125135. [CrossRef]
29. Ma, Y.; Mei, H.; Li, Y.; Zhou, P.; Mao, G.; Wang, H.; Wang, X. A novel raiometric fluorescence probe based on silicon quantum dots and copper nanoclusters for visual assay of l-cysteine in milks. *Food Chem.* **2022**, *379*, 132155. [CrossRef]
30. Wei, N.; Liang, Z.-Y.; Fang, W.-L.; Guo, X.-F.; Wang, H.; Zhang, H.-X. Facile synthesis of non-modified yellow emission silicon quantum dots and their visualization of hydrogen sulfide in living cells and onion tissues. *J. Colloid Interface Sci.* **2023**, *642*, 145–153. [CrossRef]
31. Zhao, D.; Liu, H.; Zhang, Z.; Xiao, X.; Li, J. Preparation of green luminescent silicon quantum dots by synergistic method for VB12 detection and antimicrobial property research application. *Colloids Surf. B Biointerfaces* **2022**, *220*, 112868. [CrossRef] [PubMed]
32. Wei, N.; Wei, M.-X.; Huang, B.-H.; Guo, X.-F.; Wang, H. One-pot facile synthesis of green-emitting fluorescent silicon quantum dots for the highly selective and sensitive detection of nitrite in food samples. *Dye. Pigment.* **2021**, *184*, 108848. [CrossRef]
33. Liu, Y.; Wang, Q.; Guo, S.; Jia, P.; Shui, Y.; Yao, S.; Huang, C.; Zhang, M.; Wang, L. Highly selective and sensitive fluorescence detection of hydroquinone using novel silicon quantum dots. *Sens. Actuators B Chem.* **2018**, *275*, 415–421. [CrossRef]
34. Makwana, K.P.; Deshpande, M.P.; Malek, N.I.; Park, T.J.; Kailasa, S.K. LaSrO3 perovskite quantum dots as a fluorescent probe for the detection of bilirubin and epinephrine via FRET and IFE mechanisms. *J. Photochem. Photobiol. A Chem.* **2025**, *461*, 116172. [CrossRef]
35. Cernea, M.; Monnereau, O.; Llewellyn, P.; Tortet, L.; Galassi, C. Sol–gel synthesis and characterization of Ce doped-$BaTiO_3$. *J. Eur. Ceram. Soc.* **2006**, *26*, 3241–3246. [CrossRef]
36. Zhang, Q.; Jiang, M.; Yan, G.; Feng, Y.; Zhang, B. Surface ligand engineering involving fluorophenethyl ammonium for stable and strong emission $CsPbBr_3$ quantum dots and high-performance QLEDs. *J. Mater. Chem. C* **2022**, *10*, 5849–5855. [CrossRef]
37. Jiang, W.; Kim, B.; Chae, H. Phenethylamine ligand engineering of red InP quantum dots for improving the efficiency of quantum dot light-emitting diodes. *Opt. Lett.* **2020**, *45*, 5800–5803. [CrossRef]
38. Chen, G.; Zhu, X.; Xing, C.; Wang, Y.; Xu, X.; Bao, J.; Huang, J.; Zhao, Y.; Wang, X.; Zhou, X.; et al. Machine Learning-Assisted Microfluidic Synthesis of Perovskite Quantum Dots. *Adv. Photonics Res.* **2023**, *4*, 2200230. [CrossRef]
39. Abdel-Latif, K.; Epps, R.W.; Bateni, F.; Han, S.; Reyes, K.G.; Abolhasani, M. Self-Driven Multistep Quantum Dot Synthesis Enabled by Autonomous Robotic Experimentation in Flow. *Adv. Intell. Syst.* **2021**, *3*, 2000245. [CrossRef]
40. Atabaev, T.S. Doped Carbon Dots for Sensing and Bioimaging Applications: A Minireview. *Nanomaterials* **2018**, *8*, 342. [CrossRef]
41. Ding, R.; Chen, Y.; Wang, Q.; Wu, Z.; Zhang, X.; Li, B.; Lin, L. Recent advances in quantum dots-based biosensors for antibiotics detection. *J. Pharm. Anal.* **2022**, *12*, 355–364. [CrossRef]
42. Sabzehmeidani, M.M.; Kazemzad, M. Quantum dots based sensitive nanosensors for detection of antibiotics in natural products: A review. *Sci. Total Environ.* **2022**, *810*, 151997. [CrossRef]
43. Mehta, J.; Dilbaghi, N.; Kumar Singhal, N.; Marrazza, G.; Kaushik, A.; Kumar, S. Electrochemiluminescent quantum dots as emerging next generation sensing platforms. *Chem. Eng. J.* **2023**, *477*, 146958. [CrossRef]
44. Sargazi, S.; Fatima, I.; Hassan Kiani, M.; Mohammadzadeh, V.; Arshad, R.; Bilal, M.; Rahdar, A.; Díez-Pascual, A.M.; Behzadmehr, R. Fluorescent-based nanosensors for selective detection of a wide range of biological macromolecules: A comprehensive review. *Int. J. Biol. Macromol.* **2022**, *206*, 115–147. [CrossRef] [PubMed]
45. Wang, Z.; Yao, B.; Xiao, Y.; Tian, X.; Wang, Y. Fluorescent Quantum Dots and Its Composites for Highly Sensitive Detection of Heavy Metal Ions and Pesticide Residues: A Review. *Chemosensors* **2023**, *11*, 405. [CrossRef]
46. Sudewi, S.; Sai Sashank, P.V.; Kamaraj, R.; Zulfajri, M.; Huang, G.G. Understanding Antibiotic Detection with Fluorescence Quantum Dots: A Review. *J. Fluoresc.* **2024**, *35*, 2527–2551. [CrossRef] [PubMed]
47. Zhong, L.; Zhang, L.; Li, Y.; Liang, X.; Kong, L.; Shen, X.; Wu, T. Assessment of the Toxicity of Quantum Dots through Biliometric Analysis. *Int. J. Environ. Res. Public Health* **2021**, *18*, 5768. [CrossRef]
48. Bian, F.; Sun, L.; Cai, L.; Wang, Y.; Zhao, Y. Quantum dots from microfluidics for nanomedical application. *WIREs Nanomed. Nanobiotechnology* **2019**, *11*, e1567. [CrossRef]
49. Sun, B.; Xie, W.; Yi, G.; Chen, D.; Zhou, Y.; Cheng, J. Microminiaturized immunoassays using quantum dots as fluorescent label by laser confocal scanning fluorescence detection. *J. Immunol. Methods* **2001**, *249*, 85–89. [CrossRef]
50. Sultangaziyev, A.; Bukasov, R. Review: Applications of surface-enhanced fluorescence (SEF) spectroscopy in bio-detection and biosensing. *Sens. Bio-Sens. Res.* **2020**, *30*, 100382. [CrossRef]

51. Sultangaziyev, A.; Akhmetova, A.; Kunushpayeva, Z.; Rapikov, A.; Filchakova, O.; Bukasov, R. Aluminum foil as a substrate for metal enhanced fluorescence of bacteria labelled with quantum dots, shows very large enhancement and high contrast. *Sens. Bio-Sens. Res.* **2020**, *28*, 100332. [CrossRef]
52. Bukasov, R.; Kunushpayeva, Z.; Rapikov, A.; Zhunussova, S.; Sultangaziyev, A.; Filchakova, O. High Contrast Surface Enhanced Fluorescence of Carbon Dot Labeled Bacteria Cells on Aluminum Foil. *J. Fluoresc.* **2020**, *30*, 1477–1482. [CrossRef]
53. Bukasov, R.; Filchakova, O.; Gudun, K.; Bouhrara, M. Strong Surface Enhanced Florescence of Carbon Dot Labeled Bacteria Cells Observed with High Contrast on Gold Film. *J. Fluoresc.* **2018**, *28*, 1–4. [CrossRef]
54. Qiushu, C.; Alper, K.; Xudong, F. *An Optofluidic FRET Laser Using Quantum Dots as Donor*; SPIE: San Francisco, CA, USA, 2016; p. 97250I.
55. Shopova, S.I.; Cupps, J.M.; Zhang, P.; Henderson, E.P.; Lacey, S.; Fan, X. Opto-fluidic ring resonator lasers based on highly efficient resonant energy transfer. *Opt. Express* **2007**, *15*, 12735–12742. [CrossRef] [PubMed]
56. Hong, J.; Pei, D.; Guo, X. Quantum dot-Eu3+ conjugate as a luminescence turn-on sensor for ultrasensitive detection of nucleoside triphosphates. *Talanta* **2012**, *99*, 939–943. [CrossRef]
57. Ban, R.; Zheng, F.; Zhang, J. A highly sensitive fluorescence assay for 2,4,6-trinitrotoluene using amine-capped silicon quantum dots as a probe. *Anal. Methods* **2015**, *7*, 1732–1737. [CrossRef]
58. Peveler, W.J.; Roldan, A.; Hollingsworth, N.; Porter, M.J.; Parkin, I.P. Multichannel Detection and Differentiation of Explosives with a Quantum Dot Array. *ACS Nano* **2016**, *10*, 1139–1146. [CrossRef]
59. Zhang, Z.; Li, J.; Wang, X.; Shen, D.; Chen, L. Quantum Dots Based Mesoporous Structured Imprinting Microspheres for the Sensitive Fluorescent Detection of Phycocyanin. *ACS Appl. Mater. Interfaces* **2015**, *7*, 9118–9127. [CrossRef]
60. Li, Y.; Xu, J.; Wang, L.; Huang, Y.; Guo, J.; Cao, X.; Shen, F.; Luo, Y.; Sun, C. Aptamer-based fluorescent detection of bisphenol A using nonconjugated gold nanoparticles and CdTe quantum dots. *Sens. Actuators B Chem.* **2016**, *222*, 815–822. [CrossRef]
61. Qian, J.; Hua, M.; Wang, C.; Wang, K.; Liu, Q.; Hao, N.; Wang, K. Fabrication of l-cysteine-capped CdTe quantum dots based ratiometric fluorescence nanosensor for onsite visual determination of trace TNT explosive. *Anal. Chim. Acta* **2016**, *946*, 80–87. [CrossRef]
62. Singh, K.; Mehta, S.K. Luminescent ZnO quantum dots as an efficient sensor for free chlorine detection in water. *Analyst* **2016**, *141*, 2487–2492. [CrossRef]
63. Chang, L.; He, X.; Chen, L.; Zhang, Y. Mercaptophenylboronic acid-capped Mn-doped ZnS quantum dots for highly selective and sensitive fluorescence detection of glycoproteins. *Sens. Actuators B Chem.* **2017**, *243*, 72–77. [CrossRef]
64. Qian, J.; Wang, K.; Wang, C.; Ren, C.; Liu, Q.; Hao, N.; Wang, K. Ratiometric fluorescence nanosensor for selective and visual detection of cadmium ions using quencher displacement-induced fluorescence recovery of CdTe quantum dots-based hybrid probe. *Sens. Actuators B Chem.* **2017**, *241*, 1153–1160. [CrossRef]
65. Tang, Z.; Lin, Z.; Li, G.; Hu, Y. Amino Nitrogen Quantum Dots-Based Nanoprobe for Fluorescence Detection and Imaging of Cysteine in Biological Samples. *Anal. Chem.* **2017**, *89*, 4238–4245. [CrossRef] [PubMed]
66. Yu, M.; Zhao, K.; Zhu, X.; Tang, S.; Nie, Z.; Huang, Y.; Zhao, P.; Yao, S. Development of near-infrared ratiometric fluorescent probe based on cationic conjugated polymer and CdTe/CdS QDs for label-free determination of glucose in human body fluids. *Biosens. Bioelectron.* **2017**, *95*, 41–47. [CrossRef]
67. Zhou, R.; Zhao, Q.; Liu, K.-K.; Lu, Y.-J.; Dong, L.; Shan, C.-X. Europium-decorated ZnO quantum dots as a fluorescent sensor for the detection of an anthrax biomarker. *J. Mater. Chem. C* **2017**, *5*, 1685–1691. [CrossRef]
68. Pourghobadi, Z.; Mirahmadpour, P.; Zare, H. Fluorescent biosensor for the selective determination of dopamine by TGA-capped CdTe quantum dots in human plasma samples. *Opt. Mater.* **2018**, *84*, 757–762. [CrossRef]
69. Xing, X.; Wang, D.; Chen, Z.; Zheng, B.; Li, B.; Wu, D. ZnTe quantum dots as fluorescence sensors for the detection of iron ions. *J. Mater. Sci. Mater. Electron.* **2018**, *29*, 14192–14199. [CrossRef]
70. Zhao, X.; He, D.; Wang, Y.; Fu, C. Facile fabrication of tungsten disulfide quantum dots (WS2 QDs) as effective probes for fluorescence detection of dopamine (DA). *Mater. Chem. Phys.* **2018**, *207*, 130–134. [CrossRef]
71. Feng, J.; Tao, Y.; Shen, X.; Jin, H.; Zhou, T.; Zhou, Y.; Hu, L.; Luo, D.; Mei, S.; Lee, Y.-I. Highly sensitive and selective fluorescent sensor for tetrabromobisphenol-A in electronic waste samples using molecularly imprinted polymer coated quantum dots. *Microchem. J.* **2019**, *144*, 93–101. [CrossRef]
72. Li, D.; Xu, X.; Zhou, P.; Huang, Y.; Feng, Y.; Gu, Y.; Wang, M.; Liu, Y. A facile synthesis of hybrid silicon quantum dots and fluorescent detection of bovine hemoglobin. *New J. Chem.* **2019**, *43*, 19338–19343. [CrossRef]
73. Najafi, S.; Safari, M.; Amani, S.; Mansouri, K.; Shahlaei, M. Preparation, characterization and cell cytotoxicity of Pd-doped CdTe quantum dots and its application as a sensitive fluorescent nanoprobe. *J. Mater. Sci. Mater. Electron.* **2019**, *30*, 14233–14242. [CrossRef]
74. Safari, M.; Najafi, S.; Arkan, E.; Amani, S.; Shahlaei, M. Facile aqueous synthesis of Ni-doped CdTe quantum dots as fluorescent probes for detecting pyrazinamide in plasma. *Microchem. J.* **2019**, *146*, 293–299. [CrossRef]

75. Wang, Z.; Xiao, X.; Zou, T.; Yang, Y.; Xing, X.; Zhao, R.; Wang, Z.; Wang, Y. Citric Acid Capped CdS Quantum Dots for Fluorescence Detection of Copper Ions (II) in Aqueous Solution. *Nanomaterials* **2019**, *9*, 32. [CrossRef]
76. Zhang, Z.; Ma, X.; Jia, M.; Li, B.; Rong, J.; Yang, X. Deposition of CdTe quantum dots on microfluidic paper chips for rapid fluorescence detection of pesticide 2,4-D. *Analyst* **2019**, *144*, 1282–1291. [CrossRef] [PubMed]
77. Zhang, X.; Li, C.; Zhao, S.; Pang, H.; Han, Y.; Luo, X.; Tang, W.; Li, Z. S doped silicon quantum dots with high quantum yield as a fluorescent sensor for determination of Fe^{3+} in water. *Opt. Mater.* **2020**, *110*, 110461. [CrossRef]
78. Gao, R.; Li, D.; Zhang, Q.; Zheng, S.; Ren, X.; Deng, W. GNPs-QDs core–satellites assembly: Trimodal platform for on-site identification and detection of TNT in complex media. *Sens. Actuators B Chem.* **2021**, *328*, 128960. [CrossRef]
79. Liu, M.; Bai, Y.; He, Y.; Zhou, J.; Ge, Y.; Zhou, J.; Song, G. Facile microwave-assisted synthesis of Ti3C2 MXene quantum dots for ratiometric fluorescence detection of hypochlorite. *Microchim. Acta* **2021**, *188*, 15. [CrossRef]
80. Yang, M.; Wang, C.; Liu, E.; Hu, X.; Hao, H.; Fan, J. A novel ascorbic acid ratiometric fluorescent sensor based on ZnCdS quantum dots embedded molecularly imprinted polymer and silica-coated CdTeS quantum dots. *J. Mol. Liq.* **2021**, *337*, 116438. [CrossRef]
81. Yi, Y.; Zeng, W.; Zhu, G. β-Cyclodextrin functionalized molybdenum disulfide quantum dots as nanoprobe for sensitive fluorescent detection of parathion-methyl. *Talanta* **2021**, *222*, 121703. [CrossRef]
82. Zhang, J.; Wei, Y.; Qiu, S.; Xiong, Y. A highly selective and simple fluorescent probe for salbutamol detection based on thioglycolic acid-capped CdTe quantum dots. *Spectrochim. Acta Part A Mol. Biomol. Spectrosc.* **2021**, *247*, 119107. [CrossRef] [PubMed]
83. Zhao, R.; Wang, Z.; Tian, X.; Shu, H.; Yang, Y.; Xiao, X.; Wang, Y. Excellent fluorescence detection of Cu^{2+} in water system using N-acetyl-L-cysteines modified CdS quantum dots as fluorescence probe. *Nanotechnology* **2021**, *32*, 405707. [CrossRef]
84. Aznar-Gadea, E.; Rodriguez-Canto, P.J.; Sánchez, S.A.; Martínez-Pastor, J.P.; Abargues, R. Luminescent CdSe Quantum Dot Arrays for Rapid Sensing of Explosive Taggants. *ACS Appl. Nano Mater.* **2022**, *5*, 6717–6725. [CrossRef]
85. Wang, X.; Li, L.; Jiang, H.; Zhangsun, H.; Wang, Q.; Sun, X.; Wang, L. Highly selective and sensitive fluorescence detection of tetracyclines based on novel tungsten oxide quantum dots. *Food Chem.* **2022**, *374*, 131774. [CrossRef] [PubMed]
86. Narasimhappa, P.; Singh, S.; Ramamurthy, P.C. Synthesis of water-soluble CdS quantum dots for the fluorescence detection of tetracycline. *Environ. Pollut.* **2023**, *338*, 122682. [CrossRef] [PubMed]
87. Zhong, Y.; Guo, L.; Zou, Y.; Chen, Y.; Lu, Z.; Wang, D. Rapid and ratiometric fluorescent detection of hypochlorite by glutathione functionalized molybdenum disulfide quantum dots. *Spectrochim. Acta Part A Mol. Biomol. Spectrosc.* **2023**, *295*, 122649. [CrossRef] [PubMed]
88. Singh, P.D.D.; Murthy, Z.V.P.; Kailasa, S.K. Zinc nitride quantum dots as an efficient probe for simultaneous fluorescence detection of Cu^{2+} and Mn^{2+} ions in water samples. *Microchim. Acta* **2024**, *191*, 161. [CrossRef]
89. Velamala, L.K.; Patel, M.R.; Deshpande, M.P.; Gul, A.R.; Park, T.J.; Kailasa, S.K. Fluorescence detection of superoxide anion with $CsPbBr_3$ perovskite quantum dots in aqueous media. *J. Mol. Liq.* **2024**, *414*, 126211. [CrossRef]
90. Zhang, Y.; Wang, C.; Wei, G.; Wang, X.; Liu, W.; Yang, G.; Zhang, P.; Li, Q.; Geng, X.; Chen, L.; et al. Facile fluorescence detection of malachite green in fish using molecularly imprinted polymers doped CdTe quantum dots based system. *Food Chem.* **2024**, *442*, 138458. [CrossRef]
91. Kailasa, S.K.; Patel, M.R.; Deshpande, M.P.; Shin, E.; Choi, Y.; Park, T.J. Maltose-functionalized MAPbBr3 fluorescent perovskite quantum dots with strong water resistance for detection of γ-aminobutyric acid as a neurological biomarker. *J. Photochem. Photobiol. A Chem.* **2025**, *463*, 116282. [CrossRef]
92. Zhang, L.; Shao, K.; Zhong, Y.; Guo, L.; Ge, J.; Lu, Z.; Wang, D. Molybdenum disulfide quantum dots for rapid fluorescence detection of glutathione and ascorbic acid. *Spectrochim. Acta Part A Mol. Biomol. Spectrosc.* **2025**, *326*, 125189. [CrossRef] [PubMed]
93. Creeden, J.F.; Gordon, D.M.; Stec, D.E.; Hinds, T.D. Bilirubin as a metabolic hormone: The physiological relevance of low levels. *Am. J. Physiol.-Endocrinol. Metab.* **2020**, *320*, E191–E207. [CrossRef] [PubMed]
94. Tzani, M.A.; Gioftsidou, D.K.; Kallitsakis, M.G.; Pliatsios, N.V.; Kalogiouri, N.P.; Angaridis, P.A.; Lykakis, I.N.; Terzidis, M.A. Direct and Indirect Chemiluminescence: Reactions, Mechanisms and Challenges. *Molecules* **2021**, *26*, 7664. [CrossRef] [PubMed]
95. Han, H.; Sheng, Z.; Liang, J. Electrogenerated chemiluminescence from thiol-capped CdTe quantum dots and its sensing application in aqueous solution. *Anal. Chim. Acta* **2007**, *596*, 73–78. [CrossRef]
96. Lin, Z.; Xue, W.; Chen, H.; Lin, J.-M. Peroxynitrous-Acid-Induced Chemiluminescence of Fluorescent Carbon Dots for Nitrite Sensing. *Anal. Chem.* **2011**, *83*, 8245–8251. [CrossRef]
97. Yang, E.; Zhang, Y.; Shen, Y. Quantum dots for electrochemiluminescence bioanalysis—A review. *Anal. Chim. Acta* **2022**, *1209*, 339140. [CrossRef]
98. Song, H.; Su, Y.; Zhang, L.; Lv, Y. Quantum dots-based chemiluminescence probes: An overview. *Luminescence* **2019**, *34*, 530–543. [CrossRef]
99. Yang, G.; He, Y.; Zhao, J.; Chen, S.; Yuan, R. Ratiometric electrochemiluminescence biosensor based on Ir nanorods and CdS quantum dots for the detection of organophosphorus pesticides. *Sens. Actuators B Chem.* **2021**, *341*, 130008. [CrossRef]
100. Li, M.; Li, Z.; Wang, P.; Ma, Q. A novel bimetallic MXene derivative QD-based ECL sensor for miRNA-27a-3p detection. *Biosens. Bioelectron.* **2023**, *228*, 115225. [CrossRef]

101. Li, J.; Wang, Q.; Xiong, C.; Deng, Q.; Zhang, X.; Wang, S.; Chen, M.-M. An ultrasensitive $CH_3NH_3PbBr_3$ quantum dots@SiO_2-based electrochemiluminescence sensing platform using an organic electrolyte for aflatoxin B1 detection in corn oil. *Food Chem.* **2022**, *390*, 133200. [CrossRef]
102. Wang, J.; Han, H.; Jiang, X.; Huang, L.; Chen, L.; Li, N. Quantum Dot-Based Near-Infrared Electrochemiluminescent Immunosensor with Gold Nanoparticle-Graphene Nanosheet Hybrids and Silica Nanospheres Double-Assisted Signal Amplification. *Anal. Chem.* **2012**, *84*, 4893–4899. [CrossRef] [PubMed]
103. Liu, S.; Zhang, X.; Yu, Y.; Zou, G. A Monochromatic Electrochemiluminescence Sensing Strategy for Dopamine with Dual-Stabilizers-Capped CdSe Quantum Dots as Emitters. *Anal. Chem.* **2014**, *86*, 2784–2788. [CrossRef] [PubMed]
104. Dong, Y.-P.; Gao, T.-T.; Zhou, Y.; Zhu, J.-J. Electrogenerated Chemiluminescence Resonance Energy Transfer between Luminol and CdSe@ZnS Quantum Dots and Its Sensing Application in the Determination of Thrombin. *Anal. Chem.* **2014**, *86*, 11373–11379. [CrossRef]
105. Zhang, J.-J.; Kang, T.-F.; Hao, Y.-C.; Lu, L.-P.; Cheng, S.-Y. Electrochemiluminescent immunosensor based on CdS quantum dots for ultrasensitive detection of microcystin-LR. *Sens. Actuators B Chem.* **2015**, *214*, 117–123. [CrossRef]
106. Wang, J.; Jiang, X. Anodic near-infrared electrochemiluminescence from CdTe/CdS coresmall/shellthick quantum dots and their sensing ability of Cu^{2+}. *Sens. Actuators B Chem.* **2015**, *207*, 552–555. [CrossRef]
107. Dong, H.; Han, T.-T.; Ren, L.-L.; Ding, S.-N. Novel sandwich-structured electrochemiluminescence immunosensing platform via CdTe quantum dots-embedded mesoporous silica nanospheres as enhanced signal labels and Fe_3O_4@SiO_2@PS nanocomposites as magnetic separable carriers. *J. Electroanal. Chem.* **2017**, *806*, 32–40. [CrossRef]
108. Zhao, M.; Chen, A.-Y.; Huang, D.; Chai, Y.-Q.; Zhuo, Y.; Yuan, R. MoS2 Quantum Dots as New Electrochemiluminescence Emitters for Ultrasensitive Bioanalysis of Lipopolysaccharide. *Anal. Chem.* **2017**, *89*, 8335–8342. [CrossRef]
109. Wu, F.-F.; Zhou, Y.; Wang, J.-X.; Zhuo, Y.; Yuan, R.; Chai, Y.-Q. A novel electrochemiluminescence immunosensor based on Mn doped Ag_2S quantum dots probe for laminin detection. *Sens. Actuators B Chem.* **2017**, *243*, 1067–1074. [CrossRef]
110. Dong, Y.-P.; Wang, J.; Peng, Y.; Zhu, J.-J. Electrogenerated chemiluminescence of Si quantum dots in neutral aqueous solution and its biosensing application. *Biosens. Bioelectron.* **2017**, *89*, 1053–1058. [CrossRef]
111. Yang, X.; Yu, Y.-Q.; Peng, L.-Z.; Lei, Y.-M.; Chai, Y.-Q.; Yuan, R.; Zhuo, Y. Strong Electrochemiluminescence from MOF Accelerator Enriched Quantum Dots for Enhanced Sensing of Trace cTnI. *Anal. Chem.* **2018**, *90*, 3995–4002. [CrossRef]
112. Hu, Y.; Chen, C.; Liu, Y.; Wang, S.; Guo, Z.; Hu, Y. Dual-signals electrochemiluminescence ratiometry based the synergic effect between luminol and CdSe quantum dots for direct detection of hydrogen peroxide. *J. Electroanal. Chem.* **2018**, *815*, 61–67. [CrossRef]
113. Zhang, L.; Tian, K.; Dong, Y.; Ding, H.; Wang, C. Electrogenerated chemiluminescence of $Ru(bpy)_3{}^{2+}$ at a black phosphorus quantum dot modified electrode and its sensing application. *Analyst* **2018**, *143*, 304–310. [CrossRef] [PubMed]
114. Wang, Q.; Li, S.; Li, J. A molecularly imprinted sensor with enzymatic enhancement of electrochemiluminescence of quantum dots for ultratrace clopyralid determination. *Anal. Bioanal. Chem.* **2018**, *410*, 5165–5172. [CrossRef]
115. Lei, Y.-M.; Zhou, J.; Chai, Y.-Q.; Zhuo, Y.; Yuan, R. SnS_2 Quantum Dots as New Emitters with Strong Electrochemiluminescence for Ultrasensitive Antibody Detection. *Anal. Chem.* **2018**, *90*, 12270–12277. [CrossRef] [PubMed]
116. Gao, J.; Chen, Z.; Mao, L.; Zhang, W.; Wen, W.; Zhang, X.; Wang, S. Electrochemiluminescent aptasensor based on resonance energy transfer system between CdTe quantum dots and cyanine dyes for the sensitive detection of Ochratoxin A. *Talanta* **2019**, *199*, 178–183. [CrossRef] [PubMed]
117. Liu, H.; Zhang, Y.; Dong, Y.; Chu, X. Electrogenerated chemiluminescence aptasensor for lysozyme based on copolymer nanospheres encapsulated black phosphorus quantum dots. *Talanta* **2019**, *199*, 507–512. [CrossRef]
118. Du, D.; Shu, J.; Guo, M.; Haghighatbin, M.A.; Yang, D.; Bian, Z.; Cui, H. Potential-Resolved Differential Electrochemiluminescence Immunosensor for Cardiac Troponin I Based on MOF-5-Wrapped CdS Quantum Dot Nanoluminophores. *Anal. Chem.* **2020**, *92*, 14113–14121. [CrossRef]
119. Chen, P.; Liu, Z.; Liu, J.; Liu, H.; Bian, W.; Tian, D.; Xia, F.; Zhou, C. A novel electrochemiluminescence aptasensor based CdTe QDs@NH2-MIL-88(Fe) for signal amplification. *Electrochim. Acta* **2020**, *354*, 136644. [CrossRef]
120. Xu, J.; Zhang, R.; Liu, C.; Sun, A.; Chen, J.; Zhang, Z.; Shi, X. Highly Selective Electrochemiluminescence Sensor Based on Molecularly Imprinted-quantum Dots for the Sensitive Detection of Cyfluthrin. *Sensors* **2020**, *20*, 884. [CrossRef]
121. Liu, Y.; Li, B.; Yao, Y.; Yang, B.; Tian, T.; Miao, Y.; Liu, B. An electrochemiluminescence sensor for 17β-estradiol detection based on resonance energy transfer in α-FeOOH@CdS/Ag NCs. *Talanta* **2021**, *221*, 121479. [CrossRef]
122. Feng, D.; Wei, F.; Wu, Y.; Tan, X.; Li, F.; Lu, Y.; Fan, G.; Han, H. A novel signal amplified electrochemiluminescence biosensor based on MIL-53(Al)@CdS QDs and SiO2@AuNPs for trichlorfon detection. *Analyst* **2021**, *146*, 1295–1302. [CrossRef]
123. Rasoulzadeh, F.; Amjadi, M. The chemiluminescence of $AgInS_2$ quantum dots and its application as a sensing platform for glutathione assay. *J. Photochem. Photobiol. A Chem.* **2021**, *420*, 113493. [CrossRef]
124. Zhao, J.; He, Y.; Tan, K.; Yang, J.; Chen, S.; Yuan, R. Novel Ratiometric Electrochemiluminescence Biosensor Based on BP-CdTe QDs with Dual Emission for Detecting MicroRNA-126. *Anal. Chem.* **2021**, *93*, 12400–12408. [CrossRef]

125. Jia, M.; Jia, B.; Liao, X.; Shi, L.; Zhang, Z.; Liu, M.; Zhou, L.; Li, D.; Kong, W. A CdSe@CdS quantum dots based electrochemiluminescence aptasensor for sensitive detection of ochratoxin A. *Chemosphere* **2022**, *287*, 131994. [CrossRef]
126. Liu, W.; Su, M.; Chen, A.; Peng, K.; Chai, Y.; Yuan, R. Highly Efficient Electrochemiluminescence Based on Luminol/MoS_2 Quantum Dots@Zeolitic Imidazolate Framework-8 as an Emitter for Ultrasensitive Detection of MicroRNA. *Anal. Chem.* **2022**, *94*, 9106–9113. [CrossRef] [PubMed]
127. Yang, F.; Gong, J.; Li, M.; Jiang, X.; Zhang, J.; Liao, M.; Zhang, H.; Tremblay, P.-L.; Zhang, T. Electrochemiluminescent CdS Quantum Dots Biosensor for Cancer Mutation Detection at Different Positions on Linear DNA Analytes. *Anal. Chem.* **2023**, *95*, 14016–14024. [CrossRef] [PubMed]
128. Liu, M.; Wang, Y.; Tang, S.; Wang, W.; Liang, A.; Luo, A. A ratiometric molecular imprinted electrochemiluminescence sensor based on enhanced luminescence of CdSe@ZnS quantum dots by MXene@NaAsc for detecting uric acid. *Bioelectrochemistry* **2024**, *156*, 108610. [CrossRef]
129. Gong, Q.; Guo, Y.; Wang, X.; Zhang, L.; Liu, D.; Nie, G. Dual quenching ECL strategy based on AgInZnS quantum dots and a new co-reaction promoter oxygen vacancy-modified P5AIn/TiO_2 for sensitive CEA detection. *Sens. Actuators B Chem.* **2025**, *423*, 136844. [CrossRef]
130. Chinnadayyala, S.R.; Park, J.; Le, H.T.N.; Santhosh, M.; Kadam, A.N.; Cho, S. Recent advances in microfluidic paper-based electrochemiluminescence analytical devices for point-of-care testing applications. *Biosens. Bioelectron.* **2019**, *126*, 68–81. [CrossRef]
131. Veamatahau, A.; Jiang, B.; Seifert, T.; Makuta, S.; Latham, K.; Kanehara, M.; Teranishi, T.; Tachibana, Y. Origin of surface trap states in CdS quantum dots: Relationship between size dependent photoluminescence and sulfur vacancy trap states. *Phys. Chem. Chem. Phys.* **2015**, *17*, 2850–2858. [CrossRef]
132. Wu, P.; He, Y.; Wang, H.-F.; Yan, X.-P. Conjugation of Glucose Oxidase onto Mn-Doped ZnS Quantum Dots for Phosphorescent Sensing of Glucose in Biological Fluids. *Anal. Chem.* **2010**, *82*, 1427–1433. [CrossRef]
133. Chang, N.; Mao, J.; Lu, Y.; Yang, J.; Pu, Y.; Zhang, S.; Liu, Y. Time-resolved phosphorescent sensor array based on quantum dots for recognition of proteins. *Sens. Actuators B Chem.* **2016**, *233*, 17–24. [CrossRef]
134. Li, D.; Qin, J.; Lv, J.; Yang, J.; Yan, G. "Turn on" room-temperature phosphorescent biosensors for detection of hyaluronic acid based on manganese-doped ZnS quantum dots. *RSC Adv.* **2018**, *8*, 2873–2879. [CrossRef]
135. Wang, Y.-Q.; Zou, W.-S. 3-Aminopropyltriethoxysilane-functionalized manganese doped ZnS quantum dots for room-temperature phosphorescence sensing ultratrace 2,4,6-trinitrotoluene in aqueous solution. *Talanta* **2011**, *85*, 469–475. [CrossRef]
136. He, Y.; Wang, H.-F.; Yan, X.-P. Exploring Mn-Doped ZnS Quantum Dots for the Room-Temperature Phosphorescence Detection of Enoxacin in Biological Fluids. *Anal. Chem.* **2008**, *80*, 3832–3837. [CrossRef] [PubMed]
137. Miao, Y.; Wang, R.; Suna, X.; Yan, G. Preparation of DNA functional phosphorescent quantum dots and application in melamine detection in milk. *RSC Adv.* **2019**, *9*, 21147–21154. [CrossRef]
138. Wu, P.; Zhao, T.; Tian, Y.; Wu, L.; Hou, X. Protein-Directed Synthesis of Mn-Doped ZnS Quantum Dots: A Dual-Channel Biosensor for Two Proteins. *Chemistry* **2013**, *19*, 7473–7479. [CrossRef] [PubMed]
139. Fan, M.; Gan, T.; Yin, G.; Cheng, F.; Zhao, N. Molecularly imprinted polymer coated Mn-doped ZnS quantum dots embedded in a metal–organic framework as a probe for selective room temperature phosphorescence detection of chlorpyrifos. *RSC Adv.* **2021**, *11*, 27845–27854. [CrossRef]
140. Verma, N.; Singh, A.K.; Saini, N. Synthesis and characterization of ZnS quantum dots and application for development of arginine biosensor. *Sens. Bio-Sens. Res.* **2017**, *15*, 41–45. [CrossRef]
141. Wang, H.F.; Li, Y.; Wu, Y.Y.; He, Y.; Yan, X.P. Ascorbic acid induced enhancement of room temperature phosphorescence of sodium tripolyphosphate-capped Mn-Doped ZnS quantum dots: Mechanism and bioprobe applications. *Chemistry* **2010**, *16*, 12988–12994. [CrossRef]
142. He, Y.; Yan, X. Mn-doped ZnS quantum dots/methyl violet nanohybrids for room temperature phosphorescence sensing of DNA. *Sci. China Chem.* **2011**, *54*, 1254–1259. [CrossRef]
143. Zhang, L.; Zhang, R.; Cui, P.; Cao, W.; Gao, F. An efficient phosphorescence energy transfer between quantum dots and carbon nanotubes for ultrasensitive turn-on detection of DNA. *Chem. Commun.* **2013**, *49*, 8102–8104. [CrossRef]
144. Dan, L.; Wang, H.-F. Mn-Doped ZnS Quantum Dot Imbedded Two-Fragment Imprinting Silica for Enhanced Room Temperature Phosphorescence Probing of Domoic Acid. *Anal. Chem.* **2013**, *85*, 4844–4848. [CrossRef] [PubMed]
145. Tan, L.; Kang, C.; Xu, S.; Tang, Y. Selective room temperature phosphorescence sensing of target protein using Mn-doped ZnS QDs-embedded molecularly imprinted polymer. *Biosens. Bioelectron.* **2013**, *48*, 216–223. [CrossRef] [PubMed]
146. Wang, H.-F.; Wu, Y.-Y.; Yan, X.-P. Room-Temperature Phosphorescent Discrimination of Catechol from Resorcinol and Hydroquinone Based on Sodium Tripolyphosphate Capped Mn-Doped ZnS Quantum Dots. *Anal. Chem.* **2013**, *85*, 1920–1925. [CrossRef] [PubMed]
147. Bian, W.; Ma, J.; Guo, W.; Lu, D.; Fan, M.; Wei, Y.; Li, Y.; Shuang, S.; Choi, M.M.F. Phosphorescence detection of L-ascorbic acid with surface-attached N-acetyl-L-cysteine and L-cysteine Mn doped ZnS quantum dots. *Talanta* **2013**, *116*, 794–800. [CrossRef]

148. Bian, W.; Ma, J.; Liu, Q.; Wei, Y.; Li, Y.; Dong, C.; Shuang, S. A novel phosphorescence sensor for Co^{2+} ion based on Mn-doped ZnS quantum dots. *Luminescence* **2014**, *29*, 151–157. [CrossRef]
149. Zhu, D.; Li, W.; Wen, H.-M.; Chen, Q.; Ma, L.; Hu, Y. Microwave-assisted aqueous synthesis of Mn-doped ZnS quantum dots and their room-temperature phosphorescence detection of indapamide. *Anal. Methods* **2014**, *6*, 7489–7495. [CrossRef]
150. Gong, Y.; Fan, Z. Melamine-modulated mercaptopropionic acid-capped manganese doped zinc sulfide quantum dots as a room-temperature phosphorescence sensor for detecting clenbuterol in biological fluids. *Sens. Actuators B Chem.* **2014**, *202*, 638–644. [CrossRef]
151. Zhang, Z.; Miao, Y.; Zhang, Q.; Yan, G. Facile and sensitive detection of protamine by enhanced room-temperature phosphorescence of Mn-doped ZnS quantum dots. *Anal. Biochem.* **2015**, *478*, 90–95. [CrossRef]
152. Bi, L.; Yu, Y.-H. Phosphorescent quantum dots/ethidium bromide nanohybrids based on photoinduced electron transfer for DNA detection. *Spectrochim. Acta Part A Mol. Biomol. Spectrosc.* **2015**, *140*, 479–483. [CrossRef] [PubMed]
153. Gong, Y.; Fan, Z. Highly selective manganese-doped zinc sulfide quantum dots based label free phosphorescent sensor for phosphopeptides in presence of zirconium (IV). *Biosens. Bioelectron.* **2015**, *66*, 533–538. [CrossRef]
154. Zhang, Z.; Miao, Y.; Lian, L.; Yan, G. Detection of quercetin based on Al^{3+}-amplified phosphorescence signals of manganese-doped ZnS quantum dots. *Anal. Biochem.* **2015**, *489*, 17–24. [CrossRef]
155. Ertas, N.; Satana Kara, H.E. l-Cysteine capped Mn-doped ZnS quantum dots as a room temperature phosphorescence sensor for in-vitro binding assay of idarubicin and DNA. *Biosens. Bioelectron.* **2015**, *70*, 345–350. [CrossRef] [PubMed]
156. Bian, W.; Wang, F.; Wei, Y.; Wang, L.; Liu, Q.; Dong, W.; Shuang, S.; Choi, M.M.F. Doped zinc sulfide quantum dots based phosphorescence turn-off/on probe for detecting histidine in biological fluid. *Anal. Chim. Acta* **2015**, *856*, 82–89. [CrossRef] [PubMed]
157. Zhang, J.; Zhu, A.; Zhao, T.; Wu, L.; Wu, P.; Hou, X. Glucose oxidase-directed, instant synthesis of Mn-doped ZnS quantum dots in neutral media with retained enzymatic activity: Mechanistic study and biosensing application. *J. Mater. Chem. B* **2015**, *3*, 5942–5950. [CrossRef]
158. Chen, J.; Zhu, Y.; Zhang, Y. Glutathione-capped Mn-doped ZnS quantum dots as a room-temperature phosphorescence sensor for the detection of Pb^{2+} ions. *Spectrochim. Acta Part A Mol. Biomol. Spectrosc.* **2016**, *164*, 98–102. [CrossRef]
159. Gong, Y.; Fan, Z. Room-Temperature Phosphorescence Turn-on Detection of DNA Based on Riboflavin-Modulated Manganese Doped Zinc Sulfide Quantum Dots. *J. Fluoresc.* **2016**, *26*, 385–393. [CrossRef]
160. Liu, W.; Li, H.; Wei, Y.; Dong, C. A label-free phosphorescence sensing platform for trypsin based on Mn-ZnS QDs. *RSC Adv.* **2017**, *7*, 26930–26934. [CrossRef]
161. Lv, J.; Miao, Y.; Yan, G. Detection of tumor marker miRNA21 based on phosphorescent resonance energy transfer of Mn–ZnS QDs. *RSC Adv.* **2017**, *7*, 41063–41069. [CrossRef]
162. Zhang, J.; Lu, X.; Lei, Y.; Hou, X.; Wu, P. Exploring the tunable excitation of QDs to maximize the overlap with the absorber for inner filter effect-based phosphorescence sensing of alkaline phosphatase. *Nanoscale* **2017**, *9*, 15606–15611. [CrossRef] [PubMed]
163. Zhang, C.; Zhang, K.; Zhao, T.; Liu, B.; Wang, Z.; Zhang, Z. Selective phosphorescence sensing of pesticide based on the inhibition of silver(I) quenched ZnS:Mn^{2+} quantum dots. *Sens. Actuators B Chem.* **2017**, *252*, 1083–1088. [CrossRef]
164. Pacheco, M.E.; Castells, C.B.; Bruzzone, L. Mn-doped ZnS phosphorescent quantum dots: Coumarins optical sensors. *Sens. Actuators B Chem.* **2017**, *238*, 660–666. [CrossRef]
165. Zhang, W.; Han, Y.; Chen, X.; Luo, X.; Wang, J.; Yue, T.; Li, Z. Surface molecularly imprinted polymer capped Mn-doped ZnS quantum dots as a phosphorescent nanosensor for detecting patulin in apple juice. *Food Chem.* **2017**, *232*, 145–154. [CrossRef]
166. Deng, P.; Lu, L.-Q.; Cao, W.-C.; Tian, X.-K. Phosphorescence detection of manganese(VII) based on Mn-doped ZnS quantum dots. *Spectrochim. Acta Part A Mol. Biomol. Spectrosc.* **2017**, *173*, 578–583. [CrossRef]
167. Lv, J.; Miao, Y.; Yang, J.; Qin, J.; Li, D.; Yan, G. A DNA probe based on phosphorescent resonance energy transfer for detection of transgenic 35S promoter DNA. *Biosens. Bioelectron.* **2017**, *91*, 560–565. [CrossRef]
168. Li, D.; Qin, J.; Yan, G. A phosphorescent sensor for detection of Micrococcal nuclease base on phosphorescent resonance energy transfer between quantum dots and DNA-ROX. *Sens. Actuators B Chem.* **2018**, *255*, 529–535. [CrossRef]
169. Wei, X.; Yu, M.; Li, C.; Gong, X.; Qin, F.; Wang, Z. Magnetic nanoparticles coated with a molecularly imprinted polymer doped with manganese-doped ZnS quantum dots for the determination of 2,4,6-trichlorophenol. *Microchim. Acta* **2018**, *185*, 208. [CrossRef] [PubMed]
170. Liu, C.L.; Hou, C.J.; Huo, D.Q. Enhanced Room-Temperature Phosphorescence of Mn-Doped ZnS Quantum Dots Composited with PDDA for Detection of Adriamycin. *J. Nanosci. Nanotechnol.* **2018**, *18*, 2993–2999. [CrossRef]
171. Zou, W.S.; Deng, M.Y.; Wang, Y.Q.; Zhao, X.; Li, W.H.; Huang, X.H. Alginate capped and manganese doped ZnS quantum dots as a phosphorescent probe for time-resolved detection of copper(II). *Mikrochim. Acta* **2018**, *186*, 41. [CrossRef]
172. Qin, J.; Zheng, J.; Fang, X.; Yan, G. Detection of resveratrol by phosphorescence quantum dots without conjunction and mutual impact exploration. *RSC Adv.* **2018**, *8*, 25997–26003. [CrossRef]

173. Luo, S.; Miao, Y.; Guo, J.; Sun, X.; Yan, G. Phosphorimetric determination of 4-nitrophenol using mesoporous molecular imprinting polymers containing manganese(II)-doped ZnS quantum dots. *Mikrochim. Acta* **2019**, *186*, 249. [CrossRef]
174. Chen, S.; Li, Y.; Wu, S.; Jiang, X.; Yang, H.; Su, X.; He, L.; Zou, L.; Ao, X.; Liu, S.; et al. A phosphorescent probe for cephalexin consisting of mesoporous thioglycolic acid-modified Mn:ZnS quantum dots coated with a molecularly imprinted polymer. *Mikrochim. Acta* **2019**, *187*, 40. [CrossRef]
175. Zhao, J.; Fan, Z. Aggregation-induced phosphorescence quenching method for the detection of picric acid based on melamine-passivated Mn-doped ZnS quantum dots. *Spectrochim. Acta Part A Mol. Biomol. Spectrosc.* **2019**, *223*, 117323. [CrossRef] [PubMed]
176. Liu, Z.; Hou, J.; He, Q.; Luo, X.; Huo, D.; Hou, C. New application of Mn-doped ZnS quantum dots: Phosphorescent sensor for the rapid screening of chloramphenicol and tetracycline residues. *Anal. Methods* **2020**, *12*, 3513–3522. [CrossRef] [PubMed]
177. Madurangika Jayasinghe, G.D.T.; Domínguez-González, R.; Bermejo-Barrera, P.; Moreda-Piñeiro, A. Room temperature phosphorescent determination of aflatoxins in fish feed based on molecularly imprinted polymer—Mn-doped ZnS quantum dots. *Anal. Chim. Acta* **2020**, *1103*, 183–191. [CrossRef] [PubMed]
178. Jinadasa, K.K.; Peña-Vázquez, E.; Bermejo-Barrera, P.; Moreda-Piñeiro, A. Synthesis and application of a surface ionic imprinting polymer on silica-coated Mn-doped ZnS quantum dots as a chemosensor for the selective quantification of inorganic arsenic in fish. *Anal. Bioanal. Chem.* **2020**, *412*, 1663–1673. [CrossRef]
179. Lv, X.; Gao, P. An optical sensor for selective detection of phenol via double cross-linker precipitation polymerization. *RSC Adv.* **2020**, *10*, 25402–25407. [CrossRef]
180. Liu, S.; Li, D.; Shi, D.; Zhang, G.; Luo, X.; Xu, Q.; Zhao, L.; Guo, J.; Yan, G. Construction of a room-temperature phosphorescent quantum dot probe and quantitative detection of thyroxine and carbamazepine. *J. Mol. Struct.* **2021**, *1229*, 129582. [CrossRef]
181. Qin, G.; Zuo, L.; Wei, Y.; Wang, L.; Bodwell, G. Highly sensitive detection for alkaline phosphatase using doped ZnS quantum dots with room temperature phosphorescence and its logic gate function. *Colloids Surf. B Biointerfaces* **2021**, *206*, 111968. [CrossRef]
182. Chen, S.; Su, X.; Yuan, C.; Jia, C.Q.; Qiao, Y.; Li, Y.; He, L.; Zou, L.; Ao, X.; Liu, A.; et al. A magnetic phosphorescence molecularly imprinted polymers probe based on manganese-doped ZnS quantum dots for rapid detection of trace norfloxacin residual in food. *Spectrochim. Acta Part A Mol. Biomol. Spectrosc.* **2021**, *253*, 119577. [CrossRef] [PubMed]
183. Kong, W.; Liu, M.; Zhang, J.; Wu, H.; Wang, Y.; Su, Q.; Li, Q.; Zhang, J.; Wu, C.; Zou, W.-S. Room-temperature phosphorescence and fluorescence nanocomposites as a ratiometric chemosensor for high-contrast and selective detection of 2,4,6-trinitrotoluene. *Anal. Chim. Acta* **2023**, *1282*, 341930. [CrossRef] [PubMed]
184. Yang, J.; Zhe, L.; Bo, Y.; Xiangdong, G.; Feiyong, Q.; Yanping, M.; Dehui, L.; Hongmei, G.; Li, W. Determination of Lead (II) by Phosphorescence Using Manganese-Doped Zinc Sulfide Quantum Dots. *Anal. Lett.* **2025**, *58*, 1495–1505. [CrossRef]
185. Forcada, S.; Sánchez-Visedo, A.; Melendreras, C.; Menéndez-Miranda, M.; Costa-Fernández, J.M.; Royo, L.J.; Soldado, A. Design and Evaluation of a Competitive Phosphorescent Immunosensor for Aflatoxin M1 Quantification in Milk Samples Using Mn:ZnS Quantum Dots as Antibody Tags. *Chemosensors* **2022**, *10*, 41. [CrossRef]
186. Hua, L.; Han, H.; Zhang, X. Size-dependent electrochemiluminescence behavior of water-soluble CdTe quantum dots and selective sensing of l-cysteine. *Talanta* **2009**, *77*, 1654–1659. [CrossRef] [PubMed]
187. Nakayama, M.; Kitano, T.; Ye, J.; Jin, J. Anodic Electrochemiluminescence of CdTe Quantum Dots Using Tripropylamine as Coreactant: Size-dependent Effect. *Anal. Sci.* **2020**, *36*, 859–863. [CrossRef]
188. Zhang, Y.; Clapp, A. Overview of Stabilizing Ligands for Biocompatible Quantum Dot Nanocrystals. *Sensors* **2011**, *11*, 11036–11055. [CrossRef]
189. Wang, S.; Du, L.; Jin, Z.; Xin, Y.; Mattoussi, H. Enhanced Stabilization and Easy Phase Transfer of $CsPbBr_3$ Perovskite Quantum Dots Promoted by High-Affinity Polyzwitterionic Ligands. *J. Am. Chem. Soc.* **2020**, *142*, 12669–12680. [CrossRef]
190. Kovalenko, M.V.; Protesescu, L.; Bodnarchuk, M.I. Properties and potential optoelectronic applications of lead halide perovskite nanocrystals. *Science* **2017**, *358*, 745–750. [CrossRef]
191. Sanjayan, C.G.; Jyothiab, M.S.; Balakrishna, R.G. Stabilization of CsPbBr3 quantum dots for photocatalysis, imaging and optical sensing in water and biological medium: A review. *J. Mater. Chem. C* **2022**, *10*, 6935–6956. [CrossRef]
192. Yu, M.; Yang, J.; Zhang, X.; Yuan, M.; Zhang, J.; Gao, L.; Tang, J.; Lan, X. In-Synthesis Se-Stabilization Enables Defect and Doping Engineering of HgTe Colloidal Quantum Dots. *Adv. Mater.* **2024**, *36*, 2311830. [CrossRef] [PubMed]
193. Parani, S.; Pandian, K.; Oluwafemi, O.S. Gelatin stabilization of quantum dots for improved stability and biocompatibility. *Int. J. Biol. Macromol.* **2018**, *107*, 635–641. [CrossRef] [PubMed]
194. Wang, M.; Zhang, M.; Qian, J.; Zhao, F.; Shen, L.; Scholes, G.D.; Winnik, M.A. Enhancing the Photoluminescence of Polymer-Stabilized CdSe/CdS/ZnS Core/Shell/Shell and CdSe/ZnS Core/Shell Quantum Dots in Water through a Chemical-Activation Approach. *Langmuir* **2009**, *25*, 11732–11740. [CrossRef] [PubMed]
195. Nie, Q.; Tan, W.B.; Zhang, Y. Synthesis and characterization of monodisperse chitosan nanoparticles with embedded quantum dots. *Nanotechnology* **2006**, *17*, 140. [CrossRef]

196. Ali, M.; Zayed, D.; Ramadan, W.; Kamel, O.A.; Shehab, M.; Ebrahim, S. Synthesis, characterization and cytotoxicity of polyethylene glycol-encapsulated CdTe quantum dots. *Int. Nano Lett.* **2019**, *9*, 61–71. [CrossRef]
197. Hens, B.; Smothers, J.; Rizvanovic, H.; Patel, R.; Wu, Q.; Kim, K. The Future of Anticancer Drugs: A Cytotoxicity Assessment Study of CdSe/ZnS Quantum Dots. *J. Nanotheranostics* **2020**, *1*, 19–38. [CrossRef]
198. Cheng, Q.; Duan, Y.; Fan, W.; Li, D.; Zhu, C.; Ma, T.; Liu, J.; Yu, M. Cellular uptake, intracellular behavior, and acute/sub-acute cytotoxicity of a PEG-modified quantum dot with promising in-vivo biomedical applications. *Heliyon* **2023**, *9*, e20028. [CrossRef]
199. Zhang, J.; Zhou, W.; Qi, H.; He, X. Deep-Learning-Assisted Digital Fluorescence Immunoassay on Magnetic Beads for Ultrasensitive Determination of Protein Biomarkers. *Anal. Chem.* **2025**, *97*, 2393–2401. [CrossRef]
200. Saren, G.; Zhu, L.; Han, Y. Quantitative Detection of Gastrointestinal Tumor Markers Using a Machine Learning Algorithm and Multicolor Quantum Dot Biosensor. *Comput. Intell. Neurosci.* **2022**, *2022*, 9022821. [CrossRef]
201. Zhang, S.; Zhu, W.; Zhang, X.; Mei, L.; Liu, J.; Wang, F. Machine learning-driven fluorescent sensor array using aqueous $CsPbBr_3$ perovskite quantum dots for rapid detection and sterilization of foodborne pathogens. *J. Hazard. Mater.* **2025**, *483*, 136655. [CrossRef]

International Journal of
Molecular Sciences

MDPI

Review

Cardiovascular Toxicity of Metal-Based Nanoparticles

Eun-Hye Kim [1], Sehyeon Park [2] and Ok-Nam Bae [2,*]

[1] College of Pharmacy, Kyungsung University, Busan 48434, Republic of Korea; dldh615@ks.ac.kr
[2] College of Pharmacy, Institute of Pharmaceutical Science and Technology, Hanyang University, Ansan 15588, Republic of Korea; pshsh6656@naver.com
* Correspondence: onbae@hanyang.ac.kr; Tel.: +82-31-400-5805

Abstract: The rapid development of nanotechnology has led to increased human exposure to metal-based nanoparticles (MNPs) through inhalation, ingestion, and dermal contact, raising growing concerns on their potential health effects. Due to their nanoscale size and unique physicochemical properties, the MNPs can translocate from the initial exposure sites to the circulatory system and accumulate in the body. This review focuses on MNP-induced cardiovascular toxicity, highlighting its biodistribution, cytotoxic mechanisms, and pathological impact associated with various cardiovascular diseases. MNPs disrupt endothelial function, promote oxidative stress, and induce apoptosis and ferroptosis in cardiovascular cells. Furthermore, MNPs increase endothelial permeability, impair blood–brain barrier integrity, and enhance procoagulant activity, thereby contributing to vascular and cardiac dysfunction. The particles and their released metal ions play a synergistic role in mediating these toxic effects. Here, we focused on the effects of nano-sized particles while incorporating recent in vitro and in vivo studies that address the cardiovascular impacts and mechanisms of MNP-induced toxicity. This comprehensive review will help understand and explain the potentially toxic effects of MNPs on the cardiovascular system.

Keywords: nanoparticle; metal-based nanoparticles (MNPs); cardiovascular system; cardiovascular toxicity

1. Introduction

Recent developments in nanotechnology have led to the emergence of nanomaterials that are utilized in various fields, including research, industry, and medicine [1–3]. Therefore, people are easily exposed to various nanomaterials through nanomedicine and consumer goods, including cosmetics, sunscreen, pharmaceuticals, food additives, and paints [4]. The use of consumer goods containing nanomaterials is closely linked to human exposure to nanoparticles. In the United States, the amounts of discharged nanoparticles from consumer goods were at least 2.67–3.1 $\times$ 10^3 metric tons/year into landfills and the environment in 2014, equivalent to about 9.42 g/person per year. Among them, 36–43%, 0.7–0.8%, 28–32%, and 24–36% were identified in landfills, air, water bodies, and soil systems, respectively [5]. Exposure to nanomaterials occurs through ingestion, inhalation, and direct contact [6]. Due to their nanosize, they can translocate to the circulatory system from primary target organs, including the gastrointestinal tract, lungs, and skin [7,8]. Circulating nanoparticles either move to secondary target organs, such as the kidney, heart, and brain, or continuously circulate in the bloodstream. Magnetite nanoparticles were detected in brain tissue, indicating that they can be transported through the olfactory bulb or the blood–brain barrier (BBB) and in the bloodstream of healthy humans [9]. These findings suggest that circulating nanoparticles can affect cardiovascular components, including

https://doi.org/10.3390/ijms26125816

endothelial and blood cells. The impaired cardiovascular system is closely linked with the development of cardiovascular diseases, which are significant causes of morbidity and mortality worldwide [10,11]. Previous reviews have focused on the contribution of nanomaterials to the toxicity and pathological alterations of cells and tissues in exposed organs. Our review suggests that nanomaterial-induced toxicity is associated with cardiovascular damage and disease.

Metal-based nanoparticles (MNPs) are produced from metals by destructive or constructive methods [12]. MNPs are majorly derived from aluminum, lead, silver, gold, iron, cobalt, zinc, titanium, silica, cadmium, and copper, which are widely used in various applications (Table 1) [13]. As the application of MNPs grows, regulatory perspectives and safety guidelines are also evolving. In the European Union (EU), MNPs are subject to specific provisions under the REACH and CLP regulations for cosmetics and food, which require separate registration and safety data for nanoforms [14]. Also, the European Food Safety Authority (EFSA) provides technical guidelines for evaluating nanoparticle exposure and toxicity [15]. In the United States, the Environmental Protection Agency (EPA) regulates nanomaterials under the Toxic Substances Control Act (TSCA), while the Food and Drug Administration (FDA) and the Occupational Safety and Health Administration (OSHA) oversee their use in food, drugs, and workplace safety [16,17].

MNPs have unique characteristics due to their size, which ranges from 10 nm to 100 nm. MNPs can be classified into two categories: metal and metal oxide nanoparticles [18]. Investigations of miners and refinery workers exposed to MNPs and having pulmonary disease have demonstrated a correlation between exposure to MNPs and disease incidence [19–21]. MNPs induce various toxicity via various mechanisms, including oxidative stress, inflammation, and cell death (Table 1) [22–25]. MNPs can release metal ions from particles, and it is essential to understand the mechanisms of MNP-induced toxicity due to the released metal ions. In this review, we focused on investigating whether MNP-induced toxic mechanisms involve particles themselves, ions released from the particles, or their synergistic activity.

Table 1. Types of metal-based nanoparticles, their applications, and toxicity.

MNPs	Usage	Toxicity
Aluminum oxide nanoparticles (AlO-NPs)	Cosmetics [26], Solid rocket propellants, Lubrication, and Drug delivery [27]	Neurotoxicity [28,29]
Cadmium oxide nanoparticles (CdO-NPs)	Paint pigments [30], Solar cells, and Phototransistors [31]	Genotoxicity [30], Developmental toxicity [32]
Copper nanoparticles (Cu-NPs)	Wastewater treatment [33]	Reproductive toxicity [34]
Copper oxide nanoparticles (CuO-NPs)	Nanofertilizers [35], Antifungal and antibacterial agent [36], Food packaging [37]	Respiratory toxicity [38], Neurotoxicity [39]
Gold nanoparticles (Au-NPs)	Photothermal therapy [40], Gene delivery and Targeted drug delivery [41], Biolabels [42]	Hepatotoxicity [43,44]
Iron oxide nanoparticles [1] (IO-NPs)	Cancer immunotherapy [45], Drug delivery [46]	Neurotoxicity [47,48]
Nickel oxide nanoparticles (NiO-NPs)	Lithium-ion batteries [49], Fuel cells, Drug delivery, and Antibiotics [50]	Hepatotoxicity [51], Respiratory toxicity [52]
Palladium nanoparticles (Pd-NPs)	Organic catalysis, Fuel cells, Biosensors and Electrocatalysis [53]	Immunotoxicity [54]
Rhenium nanoparticles (Re-NPs)	Tumor treatment therapy and Coatings [55]	-
Silver nanoparticles (Ag-NPs)	Anticancer therapy [56], Antiinflammatory drugs and Antibiotics [57]	Developmental toxicity [58], Genotoxicity [59], Hepatotoxicity [60]
Titanium dioxide nanoparticles (TiO_2-NPs)	Photodynamic therapy [61], Toothpaste [62], Food additives [63], Sunscreen [64]	Respiratory toxicity [65], Neurotoxicity [66], Developmental toxicity [67]
Zinc oxide nanoparticles (ZnO-NPs)	Cosmetics, Sunscreen and Textile finishes [68], Drug carriers [69], Food packaging [26]	Developmental toxicity [70], Respiratory toxicity [71], Immunotoxicity [72]

[1] IO-NPs represent Fe_2O_3 and Fe_3O_4 nanoparticles.

2. Exposure and Biodistribution of MNPs

Fine MNPs in the air are inhaled into the respiratory tract [73]. Welders who perform metal joining in the industrial manufacturing sector can easily inhale ultrafine metal particles, which can lead to several health problems. The inhalation of zinc oxide nanoparticles (ZnO-NPs) during welding contributes to metal fume fever, an occupational disease [74,75]. Due to their unique characteristics and benefits, various types of MNPs are used in the food industry. Among them, metal oxide nanoforms, including copper oxide nanoparticles, triiron tetraoxide nanoparticles, magnesium oxide nanoparticles, titanium dioxide nanoparticles (TiO_2-NPs), and ZnO-NPs, are used in food nanopackaging due to their advantages over general metal nanoparticles [76–78]. Therefore, humans intake MNPs through food contaminated with nanoparticles. In the medical field, MNPs are utilized for biomedical applications, including diagnosis, assessment, and treatment, as well as the development of new diagnostic, therapeutic, and prognostic methods [79,80]. Gold nanoparticles (Au-NPs) were approved for use in genetic technology by the Food and Drug Administration [81]. Superparamagnetic iron oxide nanoparticles have been introduced as safe and efficient magnetic resonance imaging (MRI) contrast agents for therapeutic evaluation and targeted molecule targeting [82]. Despite disagreements regarding long-term safety and risks, the use of MNPs in cancer and hereditary diseases is growing. Accordingly, MNPs are present everywhere and enter the body via various routes.

2.1. TiO_2-NPs

In the UK, the ingestion of TiO_2-NPs was estimated at 37.5 mg (median), indicating approximately 0.04 mg/kg body weight per day for a 70 kg adult. Similar to the UK, people over 7 years of age intake 0.06–0.17 mg of TiO_2-NPs/kg body weight per day [83]. When mice were orally injected with TiO_2-NPs, accumulated TiO_2-NPs were identified in the lymphoid tissue [84]. After oral ingestion of 6.25, 62.5, or 625 mg TiO_2-NPs/kg body weight for 18 weeks, the particles were observed in the basal cells of murine Peyer's patches. Additionally, TiO_2-NPs can translocate into the circulatory system and move to various new targets, including brain microvessels [85]. In rats, an intravenous (IV) injection was administered to evaluate the biodistribution of TiO_2-NPs. The liver, spleen, and lungs significantly accumulated TiO_2-NPs after IV administration, showing significantly increased titanium concentrations in organs compared with non-injected rats from 30 min to 1 year. In addition, TiO_2-NPs have been detected in the circulatory system after IV administration. In the brain microvasculature endothelial cells, the titanium burden was significantly detected from 5 min, which was approximately six-fold higher than that in the non-injected groups. Moreover, 24 h after TiO_2-NPs injection, a significant accumulation of TiO_2-NPs in the brain was observed. Mabondzo et al. suggested that titanium concentration in the injected group (261.40 ± 28.86 ng/g) was higher than that of the control group (68.25 ± 6.56 ng/g) 5 min after IV administration of TiO_2-NPs [85]. While TiO_2-NPs showed a significant toxic effect using rodents, there is a limit to completely extrapolating these results to humans. They are considered different conditions because there are differences in physiology, metabolism, genetics, biochemistry, diet, and environment between animals and humans. In addition, since animal experiments have limitations in reflecting the actual routes and amounts of exposure, the results of animal experiments are carefully interpreted.

2.2. Ag-NPs

Although the occupational exposure limit of Ag-NPs is 0.19 $\mu g/m^3$, the concentration of Ag-NPs reached up to 1.35 $\mu g/m^3$ during manufacturing and integration [86]. Since Ag-NPs have applications in various fields, including biomedical and commercial applications, they can enter the body via inhalation, ingestion, and contact [87]. Long-term exposure of

Ag-NPs to medaka fish, from the embryo stage to adulthood, resulted in the accumulation of silver in various tissues, including the liver, gills, intestine, ovary, and brain [88]. In mice, the concentrations of silver in the lungs, heart, brain, kidneys, spleen, and liver were measured after one and seven days following intratracheal (IT) instillation [89]. Although the concentrations of silver in the heart, kidney, spleen, and liver decreased, the brain and lungs showed either the same or increased concentrations of silver on day 7 compared with those on day 1. Lee et al. demonstrated altered gene expression in the brains of mice administered 1.91×10^7 particles/cm^3 of 20 nm Ag-NPs via nose only [90].

2.3. *Au-NPs*

The blood and urine from the healthy male who inhaled 116 ± 12 μg/m^3 of Au-NPs for 2 h contained a detectable concentration of gold [91]. Moreover, 24 h after inhalation, most subjects presented with gold in their bloodstream. In addition, approximately 1 ng/g (blood) and 100 ng/L (urine) of gold were detected for up to three months. Depending on their size, small Au-NPs were accumulated in the blood, urine, and liver when injected into mice through pulmonary instillation. Administration routes via injection, including IT instillation, gavage, and IV, are crucial for Au-NP accumulation in secondary target organs, such as the liver, spleen, kidney, brain, and others, as well as in the blood [92,93]. In cases of IT instillation and IV injection, none of the secondary target organs showed an increased Au-NP uptake at 24 h compared to 1 h after application. In contrast, gavage injection of Au-NPs resulted in longer detection in secondary target organs, excluding the liver, at 24 h compared to 1 h [92]. Since Au-NPs can enter the body through inhalation, ingestion, and direct injection into the bloodstream via nanomedicine applications, these data reflect the primary exposure routes of Au-NPs in humans. Nevertheless, since biological and interspecies differences exist between rodents and humans, animal experimental results have limitations in fully extrapolating to human toxicity.

2.4. *ZnO-NPs*

Inhalation of ZnO-NPs causes airway inflammation and metal fume fever [94]. Mice incubated in the chamber to inhale 1.93×10^6/cm^3 of ZnO-NPs for 3 days showed significantly increased Zn content [95]. Zn concentrations deposited in the group exposed to ZnO-NPs were measured at approximately 20.6 μg/g, which was 1.5-fold higher than that of the control group. Similar to these results, the Zn concentration was immediately increased in bronchoalveolar lavage fluid isolated from mice incubated in a whole-body chamber to inhale 3.6 ± 0.5 mg/m^3 of ZnO-NPs [96]. The exposed and sham groups showed approximately 73 μg/L and 16 μg/L Zn, respectively. A significant (1.2-fold) increase in Zn concentration in the lungs of exposed mice compared with sham mice was observed. In addition to the respiratory tract, a higher Zn concentration was observed in the heart of mice exposed to ZnO-NP inhalation (approximately 160 μg/g vs. 115 μg/g). While single-gavage of ZnO-NPs did not significantly alter the Zn concentration in rats, multiple-gavage of ZnO-NPs for 90 days resulted in an increased Zn concentration in the bones compared with that in control rats [97].

2.5. *Iron Oxide Nanoparticles (IO-NPs)*

Iron oxide-based nanoparticles are mainly divided into maghemite (Fe_2O_3) and magnetite (Fe_3O_4) nanoparticles. In this review, IO-NPs represent both the Fe_2O_3 and Fe_3O_4 nanoparticles. IO-NPs are widely used in the medical field, including magnetic drug delivery, replacement therapy, and hyperthermia targeting specific organs [98]. Shan et al. showed that IV-injected IO-NPs mainly accumulated in the mouse liver (23.7% of injected dose (ID)/g) and spleen (12.6% ID/g) at an early point (0–6 h) based on the nanoparticle biodistribution coefficient [99]. This is consistent with biodistribution, hepatotoxicity, and pulmonary toxicity in rats after IO-NP inhalation [100]. IO-NPs were detected in the lungs, heart, kidneys, muscles, and brain, showing slower absorption and lower uptake in mice. Additionally, particle size plays a role in the accumulation and clearance of IO-NPs in the bloodstream [101]. The clearance of particles smaller than 100 nm from the bloodstream was more rapid than that of particles larger than 100 nm. However, IO-NPs from 100 to 200 nm in size circulated longer in the bloodstream and were retained in vascular fenestrations.

3. Cellular Effects of MNPs in the Cardiovascular System

3.1. Cell Death

3.1.1. Apoptosis

Cellular death is broadly categorized into programmed cell death, which includes apoptosis, and non-programmed cell death [102]. Apoptosis is the process of programmed cell death to eliminate unwanted cells. Apoptosis requires energy and plays a crucial role in the pathogenesis of cardiovascular diseases [103,104]. Palladium nanoparticles induced apoptosis in human cardiac microvascular endothelial cells (HCMECs) in endothelial cell models [105]. However, TiO_2-NP exposure caused severe apoptosis in HCMECs when treated to the same extent as palladium nanoparticles. In addition, human umbilical vein endothelial cells (HUVECs) underwent apoptosis induced by TiO_2-NPs as determined by apoptotic markers, including Bax, Bcl-2, and the caspase family [106]. The apoptotic pathways are composed of the extrinsic pathway, initiated by signals originating from outside the cell, and the intrinsic pathway, which occurs within the injured cell [107]. Mitochondria are essential initiators of the intrinsic apoptosis pathway, activating intracellular caspase proteases that control cell death [108]. Human brain microvascular endothelial cells (HBMECs) and HUVECs were identified to exhibit mitochondrial damage induced by aluminum nanoparticles and TiO_2-NPs, respectively [109,110]. Both studies suggested that nanoparticles have toxic effects on mitochondria, as evidenced by decreased mitochondrial membrane potential, altered mitochondrial morphology, and reduced intracellular ATP levels. Consistent with mitochondrial dysfunction, aluminum nanoparticles and TiO_2-NPs significantly decreased cell viability as determined using the MTT assay for the detection of mitochondrial activity.

3.1.2. Ferroptosis

Ferroptosis is a type of programmed cell death dependent on iron ion levels and increased intracellular lipid peroxidation [111]. Ferroptosis affects various cells of the cardiovascular system, including cardiac and vascular cells, and it has garnered attention as a novel risk factor and therapeutic target for cardiovascular diseases [112]. In HUVECs, IN-OPs and ZnO-NPs induced ferroptosis through upregulating phospholipid peroxides [113,114]. Along with cellular ferroptosis, ZnO-NPs promoted ferritinophagy and vascular inflammation mediated by ferroptosis in mouse models [114].

3.2. Oxidative Stress

Oxidative stress is caused by the imbalance between free radicals and antioxidants [115]. Several intracellular oxidative stress markers exist, including reactive oxygen species (ROS), serum superoxide dismutase, catalase, malondialdehyde, and glutathione [116,117]. Oxidative stress plays an essential role in the development and progression of cardiovascular diseases such as myocardial infarction, ischemia/reperfusion, and heart failure [118–121]. Brain endothelial (bEnd.3) cells, a component of the BBB, promoted oxidative stress induced by Ag-NPs as determined by measuring intracellular ROS levels [122]. In addition to bEnd.3 cells, small-sized Ag-NPs (5 nm) led to increased levels of heat shock protein 70 kDa (HSP-70) and heme oxygenase-1 protein, which is related to intracellular ROS levels in EA.hy926 cells (human umbilical vein cells) compared with 100 nm-sized particles [123]. Intensive intracellular oxidative stress contributes to cell death via necrotic, apoptotic, or autophagic pathways. The promotion of apoptosis was mediated by the increased intracellular oxidative stress in HUVECs exposed to IO-NPs, Ag-NPs, or TiO_2-NPs [124–126].

3.3. Hyperpermeability

The endothelium acts as a physical barrier that separates and protects the surrounding tissues from the bloodstream [127]. The endothelial barrier function is facilitated by the selective vascular permeability, which is due to tight and adherent junction proteins [128]. Dysfunctional endothelium is the first step in vascular disease, leading to the injury of other cardiovascular cells [129]. Increased endothelial permeability is associated with the development and aggravation of cardiovascular diseases such as atherosclerosis, hypertension, stroke, and heart diseases. Brain endothelial cells are significant components of the BBB, acting as physical barriers and contributing to a highly selective, semipermeable membrane [130]. BBB plays a critical role in maintaining brain homeostasis and neurovascular health. The reduction in tight junction proteins and increased permeability result in a disrupted BBB, which contributes to neurovascular and cardiovascular damage. As brain endothelial cells prevent injury and toxicity to the central nervous system, BBB disruption leads to cardiovascular diseases and neurological dysfunction, including neurodegenerative diseases [131]. Aluminum nanoparticles and Ag-NPs reduced the expression of tight junction proteins, including ZO-1, occludin, and claudin-5, in HBMECs and bEnd.3, respectively [109,122]. Consistent with the decrease in tight junction protein expression, endothelial permeability increased in an in vitro BBB model exposed to Ag-NPs and TiO_2-NPs [132,133].

3.4. Procoagulant Activity

Red blood cells (RBCs) are representative blood cells that transport oxygen from the lungs to body tissues [134]. Injured RBCs with altered functions have implications for cardiovascular damage and diseases, such as venous thrombosis, cardiac injury, and ischemia/reperfusion in diabetes [135–137]. Increased procoagulant activity of RBCs, characterized by the externalization of phosphatidylserine (PS) on the RBC membrane, is a key factor in triggering various pathological circulatory conditions, including thrombus formation, venous thrombosis, and erythrophagocytosis by immune and endothelial cells [136,138]. The significance of PS-exposed blood cells in cardiovascular diseases has also been demonstrated in clinical patients [139,140]. Exposure of human RBCs to Ag-NPs, ZnO-NPs, and TiO_2-NPs increased the procoagulant and thrombotic activities associated with venous thrombosis [141–143]. ZnO-NP-exposed RBCs exhibited endothelial hyperpermeability and cytotoxicity attributed to excessive erythrophagocytosis by brain endothelial cells [141].

4. Pathological Conditions Induced by MNPs in the Cardiovascular System

4.1. Inflammation

Systemic and local inflammation plays an essential role in the development and progression of cardiovascular diseases [144]. IO-NP treatment promotes inflammation in human whole blood by increasing inflammatory cytokines, such as tumor necrosis factor (TNF)-α, interleukin (IL)-6, IL-1β, and IL-8 [145]. Along with whole blood, endothelial cells incubated with whole blood in combination with IO-NPs showed significantly increased levels of TNF-α, IL-6, IL-1β, and IL-8, indicating cardiovascular inflammation. Non-human primates and healthy persons exhibited systemic inflammation and metal fume fever after exposure to ZnO-NPs via IT injection or inhalation routes, respectively [146,147]. Inflammation plays a critical role in the pathophysiology of atherosclerosis and its transition into chronic inflammation [148]. Human aortic endothelial cells (HAECs) showed significantly increased nitric oxides and inflammatory factors, including intercellular adhesion molecule 1, IL-8, and monocyte chemoattractant protein-1 after treatment of IO-NPs, yttrium oxide nanoparticles, ZnO-NPs, and cerium oxide nanoparticles [149–151]. IO-NP-treated HAECs suggested that endothelial inflammation and dysfunction were promoted through increased cytokine expression, adhesion molecule expression, monocyte recruitment, and nitric oxide production, indicating risks for atherosclerosis [149].

4.2. Cardiotoxicity

Cardiovascular diseases encompass a range of cardiac and blood vessel disorders. The heart is the main secondary target organ of nanoparticles translocated from the circulatory system [8]. The oral uptake of MNPs, including IO-NPs, Ag-NPs, ZnO-NPs, and aluminum nanoparticles, has been shown to cause cardiotoxicity in animal models [26,152]. Long-term exposure to IO-NPs, Ag-NPs, or their mixture induced oxidative stress and increased the levels of pro-inflammatory cytokines such as TNF-α and IL-6 in cardiac tissues [152]. Additionally, myocardial degeneration occurred in isolated hearts from rats that led to changes in the myocardial fibers. IO-NPs promoted ROS levels in the cardiac tissues and had more fatal effects on mouse survival after a single IV injection than that of Au-NPs [153]. The heart weight is crucial in several age-related pathological conditions and cardiovascular diseases. After long-term exposure to ZnO-NPs, aluminum nanoparticles, or their combination, the heart weight decreased, and levels of TNF-α and IL-6 increased in cardiac tissues isolated from rats [26].

4.3. Vascular Toxicity

The circulatory system consists of vessels that carry blood and lymphatic fluid. The vascular system is continuously in contact with the absorbed NPs during the translocation of NPs from primary to secondary target organs. Additionally, MNPs, such as ZnO-NPs and TiO_2-NPs, are attached to and located within circulating RBCs [154]. In rats, IV injections of nickel nanoparticles into the dorsal penile vein resulted in an altered lipid profile, as indicated by increased cholesterol and apolipoprotein E levels, which are associated with atherosclerosis [155]. Blood cells in the bloodstream contribute to vasculopathy, such as venous thrombosis. IV injection of Ag-NPs, TiO_2-NPs, or ZnO-NPs led to thrombus formation in the veins in an in vivo rat venous model [141–143]. When rats were exposed to TiO2-NPs for 30 days via the oral route, the counts of white blood cells and granulocytes in the blood increased [156]. Moreover, TiO_2-NP exposure in rats initiated an inflammatory response, as shown by the increased concentrations of TNF-α and IL-6 in the serum.

5. Comparative Toxicity Between Metal Ions and MNPs

Because MNPs are derived from metals, they have unique characteristics due to the release of nano-sized metal ions. In experimental models, MNPs are dissolved in aqueous media, releasing metal ions into the surrounding media [157]. To identify the potential risks of MNPs, it is necessary to understand whether the toxic effects are caused by the particles themselves, the released metal ions, or their combination. Different perspectives on MNP-induced toxicity exist; some studies suggest that released ions are the primary or sole cause of toxic effects [158,159], whereas other studies suggest that particles, rather than released ions from MNPs, are the major contributors to toxic effects [160] (Table 2).

Table 2. Representative examples of different biological effects between metal ions and particles per se.

Category	Metal-Ions	Particles
Distribution	Rapid distribution into various tissues via circulation	Accumulate in tissues and cells over extended periods
Toxicity onset	Rapid onset (e.g., $CuCl_2$ toxicity within 24 h)	Delayed toxicity (e.g., Cu-NPs effects after 48 h)
Toxic mechanisms	Mainly due to free metal ions	Primarily caused by the particles per se (e.g., glial cells)
Pathological effects	Strong acute toxicity (e.g., AgOAc)	Some show no pathological symptoms (e.g., Ag-NPs)

Metal salts, such as $ZnCl_2$, are distributed into various tissues through circulation rather than ZnO-NPs [161]. In addition, while oral uptake of $CuCl_2$ induced toxicity within 24 h, it was observed that the toxic effects induced by copper nanoparticles were delayed until 48 h [162]. Rats orally injected with Ag-NPs showed no pathological symptoms. However, silver acetate (AgOAc) injection led to toxic effects, including weight loss, linitis plastica, tissue pigmentation, and death in animal models [163]. $ZnCl_2$, which has a lower zinc concentration than ZnO-NPs, caused severe toxicity and led to pathological conditions, such as anemia and tissue injuries in the liver, kidneys, lungs, and intestine [161].

Although metal ions released from MNPs primarily contribute to acute toxicity in animal models, particles accumulate in tissues and cells for more extended periods than metal salts [122,162]. ZnO-NPs and TiO2-NPs were identified as the primary sources that induced hemolysis of RBCs and oxidative stress rather than their released ions [164]. In addition, the cytotoxicity of ZnO-NPs resulted from the particles rather than the released zinc ions in glial cells [165]. Lower concentrations of ZnO-NPs induced changes in cellular morphology and showed increased cell death compared with $ZnSO_4$. The increased concentrations of intracellular zinc ions and silver ions derived from accumulated particles were identified in ZnO-NPs-treated HUVECs and Ag-NPs-treated bEnd.3 cells compared with treatments with $ZnCl_2$ and Ag ions alone, respectively [122,166]. In a study on the procoagulant activity of ZnO-NPs on RBCs, TPEN, a zinc chelator, promoted PS exposure to a certain extent by binding to zinc ions released from ZnO-NPs [141]. However, this study had a limitation in that the concentration of Zn ions released from $ZnCl_2$ treatment differed from that released from ZnO-NPs due to the pH of the surrounding media. These reports suggest that the accumulation of particles in cells and tissues, as well as the concentration of released ions, are closely related, and MNP-induced toxicity can be attributed to either factor or their synergistic activation (Figure 1).

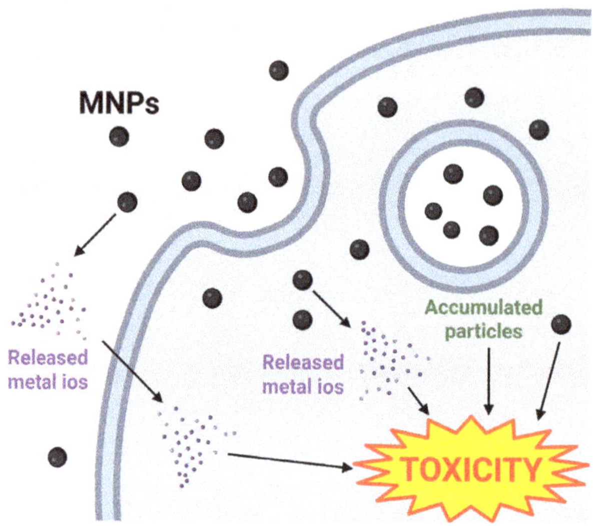

Figure 1. Comparison of MNP pathways contributing to toxicity. MNPs lead to toxic effects through metal ions, particles per se, or their combination. MNPs release their metal ions in aqueous media or the cytoplasm. Entered particles are accumulated in the organelle, as indicated by the released ions. These released ions and intracellular accumulated particles participate in the toxicity induced by MNPs.

6. Future Directions

These studies support that MNPs contribute to the development of cardiovascular diseases (Figure 2). Evidence indicates that the cardiovascular toxicity of MNPs is essential to establish an experimental model to explain this potential toxicity. Future studies should delineate the distinct toxic mechanisms of nanoparticles per se, released metal ions, and their combinations through carefully designed comparative studies. To provide different perspectives on cumulative risks, it is necessary to investigate the long-term biodistribution and persistence of MNPs in cardiovascular tissues.

To align nanotoxicology research more closely with regulatory sciences and enhance its utility for both academic and policy purposes, we suggest the following points. The incorporation of advanced methods for nanotoxicity assessment, including high-throughput screening (HTS), omics technologies (e.g., transcriptomics, proteomics, and metabolomics), and 3D organ-on-a-chip models, would offer valuable insights into the mechanistic understanding of nanoparticle-induced toxicity and facilitate predictive modeling. Additionally, the influence of surface modifications (e.g., PEGylation, ligand conjugation, or charge alteration) on the biological interactions and toxicity profiles of metal-based nanoparticles should be discussed. Surface chemistry plays a pivotal role in nanoparticle uptake, biodistribution, and immune response, thereby having direct implications for both safety and therapeutic potential.

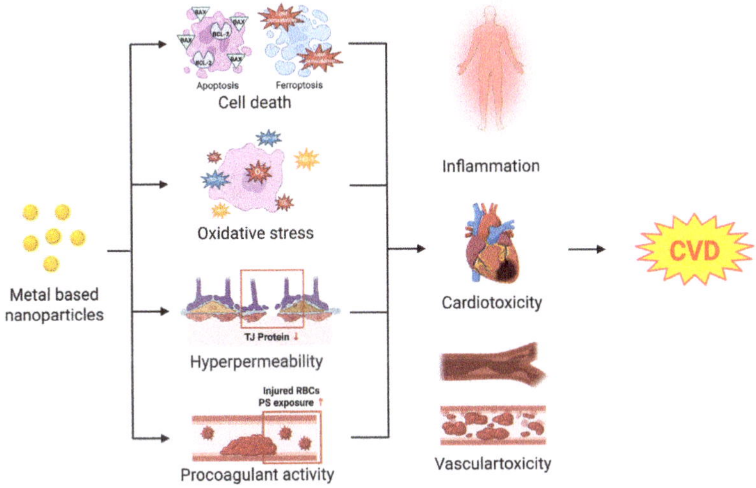

Figure 2. Cardiovascular diseases caused by MNPs. Circulating MNPs induce cellular death, oxidative stress, endothelial dysfunction, hyperpermeability, as well as procoagulant activity in the cardiovascular system. Cytotoxicity and altered cellular functions are closely associated with inflammation, cardiotoxicity, and vascular toxicity. These pathological conditions result from MNPs contributing to the development of cardiovascular diseases.

Standard in vivo models and exposure conditions are required to simulate real-world human exposure more accurately, particularly in chronic low-dose exposure scenarios. To simulate the proper experimental models, we need to establish risk assessments of MNPs to acquire clear exposure quantifications. In addition, most current studies have several limitations in applying the actual exposure routes of MNPs. Therefore, it is essential to consider both the exposure routes and the exposure amount to implement the actual exposure scenario accurately. In addition, including real-world environmental and occupational exposure scenarios, such as inhalation risks in industrial settings or chronic low-dose exposure through consumer products, would provide a more grounded understanding of public health implications. There are currently regulatory measures and attempts being made specifically to consider nanoparticles worldwide. Despite these efforts, challenges persist in consistently defining nanoparticles, assessing real-world exposure, and establishing legally binding limits. As such, regulatory bodies worldwide continue to refine risk assessment methodologies, develop nano-specific safety guidelines, and promote safe-by-design principles to ensure the responsible development and use of metal-based nanomaterials. Through these further studies, we aim to highlight the "safe-by-design" strategies for nanoparticles that integrate safety considerations early in the material design phase, thereby minimizing adverse outcomes such as cardiovascular risks, which have been increasingly associated with nanoparticle exposure.

Author Contributions: Conceptualization, E.-H.K. and O.-N.B.; writing—original draft preparation, E.-H.K. and S.P.; writing—review and editing, E.-H.K., S.P. and O.-N.B.; funding acquisition, O.-N.B. All authors have read and agreed to the published version of the manuscript.

Funding: This study was funded by the National Research Foundation of Korea (RS-2022-NR070655) and the Manufacturing Human Cell-based Artificial Blood and Platform Technology Development for Transfusion funded by the Multi-Ministerial Research Project of Korea (RS-2023-KH140699).

Conflicts of Interest: The authors declare no conflicts of interest.

Abbreviations

AgOAc	Silver Acetate
Ag-NP	Silver Nanoparticle
Au-NP	Gold Nanoparticle
BBB	Blood–brain Barrier
HAEC	Human Aortic Endothelial Cell
HBMEC	Human Brain Microvascular Endothelial Cell
HCMEC	Human Cardiac Microvascular Endothelial Cell
HUVEC	Human Umbilical Vein Endothelial Cell
HSP	Heat Shock Protein
IL	Interleukin
IO-NP	Iron Oxide Nanoparticle
IT	Intratracheal
IV	Intravenous
MNP	Metal-based Nanoparticle
MRI	Magnetic Resonance Imaging
PS	Phosphatidylserine
RBC	Red Blood Cell
TiO_2-NP	Titanium Dioxide Nanoparticle
TNF	Tumor Necrosis Factor
ZnO-NP	Zinc Oxide Nanoparticle

References

1. Malik, S.; Muhammad, K.; Waheed, Y. Emerging Applications of Nanotechnology in Healthcare and Medicine. *Molecules* **2023**, *28*, 6624. [CrossRef] [PubMed]
2. Malik, S.; Muhammad, K.; Waheed, Y. Nanotechnology: A Revolution in Modern Industry. *Molecules* **2023**, *28*, 661. [CrossRef] [PubMed]
3. Saeed, S.; Ud Din, S.R.; Khan, S.U.; Gul, R.; Kiani, F.A.; Wahab, A.; Zhong, M. Nanoparticle: A Promising Player in Nanomedicine and its Theranostic Applications for the Treatment of Cardiovascular Diseases. *Curr. Probl. Cardiol.* **2023**, *48*, 101599. [CrossRef] [PubMed]
4. Gupta, V.; Mohapatra, S.; Mishra, H.; Farooq, U.; Kumar, K.; Ansari, M.J.; Aldawsari, M.F.; Alalaiwe, A.S.; Mirza, M.A.; Iqbal, Z. Nanotechnology in Cosmetics and Cosmeceuticals-A Review of Latest Advancements. *Gels* **2022**, *8*, 173. [CrossRef]
5. Ameen, F.; Alsamhary, K.; Alabdullatif, J.A.; Alnadhari, S. A review on metal-based nanoparticles and their toxicity to beneficial soil bacteria and fungi. *Ecotoxicol. Environ. Saf.* **2021**, *213*, 112027. [CrossRef]
6. Xuan, L.; Ju, Z.; Skonieczna, M.; Zhou, P.K.; Huang, R. Nanoparticles-induced potential toxicity on human health: Applications, toxicity mechanisms, and evaluation models. *MedComm* **2023**, *4*, e327. [CrossRef]
7. Yee, M.S.-L.; Hii, L.-W.; Looi, C.K.; Lim, W.-M.; Wong, S.-F.; Kok, Y.-Y.; Tan, B.-K.; Wong, C.-Y.; Leong, C.-O. Impact of Microplastics and Nanoplastics on Human Health. *Nanomaterials* **2021**, *11*, 496. [CrossRef]
8. Oberdorster, G.; Oberdorster, E.; Oberdorster, J. Nanotoxicology: An emerging discipline evolving from studies of ultrafine particles. *Environ. Health Perspect* **2005**, *113*, 823–839. [CrossRef]
9. Maher, B.A.; Ahmed, I.A.; Karloukovski, V.; MacLaren, D.A.; Foulds, P.G.; Allsop, D.; Mann, D.M.; Torres-Jardon, R.; Calderon-Garciduenas, L. Magnetite pollution nanoparticles in the human brain. *Proc. Natl. Acad. Sci. USA* **2016**, *113*, 10797–10801. [CrossRef]
10. Khan, S.U.; Suleman, M.; Khan, M.U.; Khan, M.S.; Arbi, F.M.; Hussain, T.; Alsuhaibani, A.M.; Refat, M.S. Natural Allies for Heart Health: Nrf2 Activation and Cardiovascular Disease Management. *Curr. Probl. Cardiol.* **2024**, *49 Pt B*, 102084. [CrossRef]
11. Ullah, M.; Bibi, A.; Wahab, A.; Hamayun, S.; Rehman, M.U.; Khan, S.U.; Awan, U.A.; Riaz, N.-U.; Naeem, M.; Saeed, S.; et al. Shaping the Future of Cardiovascular Disease by 3D Printing Applications in Stent Technology and its Clinical Outcomes. *Curr. Probl. Cardiol.* **2024**, *49 Pt A*, 102039. [CrossRef] [PubMed]
12. Li, H.; Chen, X.; Shen, D.; Wu, F.; Pleixats, R.; Pan, J. Functionalized silica nanoparticles: Classification, synthetic approaches and recent advances in adsorption applications. *Nanoscale* **2021**, *13*, 15998–16016. [CrossRef] [PubMed]
13. Burlec, A.F.; Corciova, A.; Boev, M.; Batir-Marin, D.; Mircea, C.; Cioanca, O.; Danila, G.; Danila, M.; Bucur, A.F.; Hancianu, M. Current Overview of Metal Nanoparticles' Synthesis, Characterization, and Biomedical Applications, with a Focus on Silver and Gold Nanoparticles. *Pharmaceuticals* **2023**, *16*, 1410. [CrossRef] [PubMed]

14. Zabeo, A.; Basei, G.; Tsiliki, G.; Peijnenburg, W.; Hristozov, D. Ordered weighted average based grouping of nanomaterials with Arsinh and dose response similarity models. *NanoImpact* **2022**, *25*, 100370. [CrossRef]
15. Hardy, A.; Benford, D.; Halldorsson, T.; Jeger, M.J.; Knutsen, H.K.; More, S.; Naegeli, H.; Noteborn, H.; Ockleford, C.; Ricci, A.; et al. Guidance on risk assessment of the application of nanoscience and nanotechnologies in the food and feed chain: Part 1, human and animal health. *EFSA J* **2018**, *16*, e05327.
16. Petersen, E.J.; Ceger, P.; Allen, D.G.; Coyle, J.; Derk, R.; Garcia-Reyero, N.; Gordon, J.; Kleinstreuer, N.C.; Matheson, J.; McShan, D.; et al. U.S. Federal Agency interests and key considerations for new approach methodologies for nanomaterials. *ALTEX* **2022**, *39*, 183–206. [CrossRef]
17. Erdely, A.; Dahm, M.M.; Schubauer-Berigan, M.K.; Chen, B.T.; Antonini, J.M.; Hoover, M.D. Bridging the gap between exposure assessment and inhalation toxicology: Some insights from the carbon nanotube experience. *J. Aerosol. Sci.* **2016**, *99*, 157–162. [CrossRef]
18. Yadavalli, T.; Shukla, D. Role of metal and metal oxide nanoparticles as diagnostic and therapeutic tools for highly prevalent viral infections. *Nanomedicine* **2017**, *13*, 219–230. [CrossRef]
19. Kornberg, T.G.; Stueckle, T.A.; Antonini, J.A.; Rojanasakul, Y.; Castranova, V.; Yang, Y.; Wang, L. Potential Toxicity and Underlying Mechanisms Associated with Pulmonary Exposure to Iron Oxide Nanoparticles: Conflicting Literature and Unclear Risk. *Nanomaterials* **2017**, *7*, 307. [CrossRef]
20. Morimoto, Y.; Izumi, H.; Tomonaga, T.; Nishida, C.; Higashi, H. Adverse effects of nanoparticles on humans. *J. Occup Health* **2025**, *67*, uiaf002. [CrossRef]
21. Zhou, X.; Jin, W.; Ma, J. Lung inflammation perturbation by engineered nanoparticles. *Front. Bioeng Biotechnol.* **2023**, *11*, 1199230. [CrossRef] [PubMed]
22. Wang, Y.-L.; Lee, Y.-H.; Chou, C.-L.; Chang, Y.-S.; Liu, W.-C.; Chiu, H.-W. Oxidative stress and potential effects of metal nanoparticles: A review of biocompatibility and toxicity concerns. *Environ. Pollut.* **2024**, *346*, 123617. [CrossRef]
23. Zhang, N.; Xiong, G.; Liu, Z. Toxicity of metal-based nanoparticles: Challenges in the nano era. *Front. Bioeng Biotechnol.* **2022**, *10*, 1001572. [CrossRef] [PubMed]
24. Mortezaee, K.; Najafi, M.; Samadian, H.; Barabadi, H.; Azarnezhad, A.; Ahmadi, A. Redox interactions and genotoxicity of metal-based nanoparticles: A comprehensive review. *Chem. Biol. Interact* **2019**, *312*, 108814. [CrossRef] [PubMed]
25. Xiong, P.; Huang, X.; Ye, N.; Lu, Q.; Zhang, G.; Peng, S.; Wang, H.; Liu, Y. Cytotoxicity of Metal-Based Nanoparticles: From Mechanisms and Methods of Evaluation to Pathological Manifestations. *Adv. Sci.* **2022**, *9*, e2106049. [CrossRef]
26. Yousef, M.; Roychoudhury, S.; Jafaar, K.; Slama, P.; Kesari, K.; Kamel, M. Aluminum oxide and zinc oxide induced nanotoxicity in rat brain, heart, and lung. *Physiol. Res.* **2022**, *71*, 677–694. [CrossRef]
27. Park, S.-H.; Lim, J.-O.; Kim, W.-I.; Park, S.-W.; Lee, S.-J.; Shin, I.-S.; Moon, C.; Kim, J.-H.; Heo, J.-D.; Kim, J.-C. Subchronic Toxicity Evaluation of Aluminum Oxide Nanoparticles in Rats Following 28-Day Repeated Oral Administration. *Biol. Trace Element Res.* **2022**, *200*, 3215–3226. [CrossRef]
28. Huang, T.; Guo, W.; Wang, Y.; Chang, L.; Shang, N.; Chen, J.; Fan, R.; Zhang, L.; Gao, X.; Niu, Q.; et al. Involvement of Mitophagy in Aluminum Oxide Nanoparticle-Induced Impairment of Learning and Memory in Mice. *Neurotox Res.* **2021**, *39*, 378–391. [CrossRef]
29. Zhang, Y.; Jia, L.; Wang, Z.; Guo, W.; Qin, X.; Ge, C.; Niu, Q.; Zhang, Q. Alumina nanoparticles induce learning and memory impairment in a particle size-dependent and time-dependent manner. *Ecotoxicol. Environ. Saf.* **2025**, *295*, 118177. [CrossRef]
30. Demir, E.; Qin, T.; Li, Y.; Zhang, Y.; Guo, X.; Ingle, T.; Yan, J.; Orza, A.I.; Biris, A.S.; Ghorai, S.; et al. Cytotoxicity and genotoxicity of cadmium oxide nanoparticles evaluated using in vitro assays. *Mutat. Res. Toxicol. Environ. Mutagen.* **2020**, *850–851*, 503149. [CrossRef]
31. Dabour, K.; Al Naggar, Y.; Masry, S.; Naiem, E.; Giesy, J.P. Cellular alterations in midgut cells of honey bee workers (*Apis millefera* L.) exposed to sublethal concentrations of CdO or PbO nanoparticles or their binary mixture. *Sci. Total. Environ.* **2019**, *651 Pt 1*, 1356–1367. [CrossRef] [PubMed]
32. Blum, J.L.; Xiong, J.Q.; Hoffman, C.; Zelikoff, J.T. Cadmium associated with inhaled cadmium oxide nanoparticles impacts fetal and neonatal development and growth. *Toxicol. Sci.* **2012**, *126*, 478–486. [CrossRef] [PubMed]
33. Ameh, T.; Sayes, C.M. The potential exposure and hazards of copper nanoparticles: A review. *Environ. Toxicol. Pharmacol.* **2019**, *71*, 103220. [CrossRef] [PubMed]
34. Anima, B.; Mondal, P.; Gurusubramanian, G.; Roy, V.K. Mechanistic study of copper nanoparticle (CuNP) toxicity on the mouse uterus via apelin signaling. *Environ. Sci. Pollut Res. Int.* **2023**, *30*, 88824–88841. [CrossRef]
35. Devaraji, M.; Thanikachalam, P.V.; Elumalai, K. The potential of copper oxide nanoparticles in nanomedicine: A comprehensive review. *Biotechnol. Notes* **2024**, *5*, 80–99. [CrossRef]
36. Letchumanan, D.; Sok, S.P.M.; Ibrahim, S.; Nagoor, N.H.; Arshad, N.M. Plant-Based Biosynthesis of Copper/Copper Oxide Nanoparticles: An Update on Their Applications in Biomedicine, Mechanisms, and Toxicity. *Biomolecules* **2021**, *11*, 564. [CrossRef]

37. Dash, K.K.; Deka, P.; Bangar, S.P.; Chaudhary, V.; Trif, M.; Rusu, A. Applications of Inorganic Nanoparticles in Food Packaging: A Comprehensive Review. *Polymers* **2022**, *14*, 521. [CrossRef]
38. Zhang, Y.; Zhang, Z.; Mo, Y.; Zhang, Y.; Yuan, J.; Zhang, Q. MMP-3 mediates copper oxide nanoparticle-induced pulmonary inflammation and fibrosis. *J. Nanobiotechnol.* **2024**, *22*, 428. [CrossRef]
39. Goma, A.A.; El Okle, O.S.; Tohamy, H.G. Protective effect of methylene blue against copper oxide nanoparticle-induced neurobehavioral toxicity. *Behav. Brain. Res.* **2021**, *398*, 112942. [CrossRef]
40. Singh, P.; Pandit, S.; Mokkapati, V.; Garg, A.; Ravikumar, V.; Mijakovic, I. Gold Nanoparticles in Diagnostics and Therapeutics for Human Cancer. *Int. J. Mol. Sci.* **2018**, *19*, 1979. [CrossRef]
41. Golchin, K.; Golchin, J.; Ghaderi, S.; Alidadiani, N.; Eslamkhah, S.; Eslamkhah, M.; Davaran, S.; Akbarzadeh, A. Gold nanoparticles applications: From artificial enzyme till drug delivery. *Artif. Cells Nanomed. Biotechnol.* **2018**, *46*, 250–254. [CrossRef] [PubMed]
42. Daraee, H.; Eatemadi, A.; Abbasi, E.; Fekri Aval, S.; Kouhi, M.; Akbarzadeh, A. Application of gold nanoparticles in biomedical and drug delivery. *Artif Cells Nanomed. Biotechnol.* **2016**, *44*, 410–422. [CrossRef] [PubMed]
43. Enea, M.; Pereira, E.; Costa, J.; Soares, M.E.; Dias da Silva, D.; Bastos, M.L.; Carmo, H.F. Cellular uptake and toxicity of gold nanoparticles on two distinct hepatic cell models. *Toxicol In Vitro* **2021**, *70*, 105046. [CrossRef] [PubMed]
44. Yang, Y.; Fan, S.; Chen, Q.; Lu, Y.; Zhu, Y.; Chen, X.; Xia, L.; Huang, Q.; Zheng, J.; Liu, X. Correction: Acute exposure to gold nanoparticles aggravates lipopolysaccharide-induced liver injury by amplifying apoptosis via ROS-mediated macrophage-hepatocyte crosstalk. *J. Nanobiotechnol.* **2023**, *21*, 239. [CrossRef]
45. Meng, Y.Q.; Shi, Y.N.; Zhu, Y.P.; Liu, Y.Q.; Gu, L.W.; Liu, D.D.; Ma, A.; Xia, F.; Guo, Q.Y.; Xu, C.C.; et al. Recent trends in preparation and biomedical applications of iron oxide nanoparticles. *J. Nanobiotechnol.* **2024**, *22*, 24. [CrossRef]
46. Chamorro, S.; Gutierrez, L.; Vaquero, M.P.; Verdoy, D.; Salas, G.; Luengo, Y.; Brenes, A.; Jose Teran, F. Safety assessment of chronic oral exposure to iron oxide nanoparticles. *Nanotechnology* **2015**, *26*, 205101. [CrossRef]
47. Dora, M.F.; Taha, N.M.; Lebda, M.A.; Hashem, A.E.; Elfeky, M.S.; El-Sayed, Y.S.; Jaouni, S.A.; El-Far, A.H. Quercetin Attenuates Brain Oxidative Alterations Induced by Iron Oxide Nanoparticles in Rats. *Int. J. Mol. Sci.* **2021**, *22*, 3829. [CrossRef]
48. Patel, S.; Jana, S.; Chetty, R.; Thakore, S.; Singh, M.; Devkar, R. Toxicity evaluation of magnetic iron oxide nanoparticles reveals neuronal loss in chicken embryo. *Drug Chem. Toxicol.* **2019**, *42*, 1–8. [CrossRef]
49. More, S.L.; Kovochich, M.; Lyons-Darden, T.; Taylor, M.; Schulte, A.M.; Madl, A.K. Review and Evaluation of the Potential Health Effects of Oxidic Nickel Nanoparticles. *Nanomaterials* **2021**, *11*, 642. [CrossRef]
50. Berhe, M.G.; Gebreslassie, Y.T. Biomedical Applications of Biosynthesized Nickel Oxide Nanoparticles. *Int. J. Nanomed.* **2023**, *18*, 4229–4251. [CrossRef]
51. Adiguzel, C.; Karaboduk, H.; Apaydin, F.G.; Kalender, S.; Kalender, Y. Comparison of nickel oxide nano and microparticles toxicity in rat liver: Molecular, biochemical, and histopathological study. *Toxicol. Res.* **2023**, *12*, 741–750. [CrossRef] [PubMed]
52. Morimoto, Y.; Hirohashi, M.; Ogami, A.; Oyabu, T.; Myojo, T.; Hashiba, M.; Mizuguchi, Y.; Kambara, T.; Lee, B.W.; Kuroda, E.; et al. Pulmonary toxicity following an intratracheal instillation of nickel oxide nanoparticle agglomerates. *J. Occup. Health* **2011**, *53*, 293–295. [CrossRef] [PubMed]
53. Leso, V.; Iavicoli, I. Palladium Nanoparticles: Toxicological Effects and Potential Implications for Occupational Risk Assessment. *Int. J. Mol. Sci.* **2018**, *19*, 503. [CrossRef] [PubMed]
54. Gurunathan, S.; Kang, M.H.; Jeyaraj, M.; Kim, J.H. Palladium Nanoparticle-Induced Oxidative Stress, Endoplasmic Reticulum Stress, Apoptosis, and Immunomodulation Enhance the Biogenesis and Release of Exosome in Human Leukemia Monocytic Cells (THP-1). *Int. J. Nanomed.* **2021**, *16*, 2849–2877. [CrossRef]
55. Kepceoglu, A.; Gundogdu, Y.; Sarilmaz, A.; Ersoz, M.; Ozel, F.; Kilic, H.S. Rhenium/rhenium oxide nanoparticles production using femtosecond pulsed laser ablation in liquid. *Turk. J. Chem.* **2021**, *45*, 485–492. [CrossRef]
56. Xu, L.; Wang, Y.Y.; Huang, J.; Chen, C.Y.; Wang, Z.X.; Xie, H. Silver nanoparticles: Synthesis, medical applications and biosafety. *Theranostics* **2020**, *10*, 8996–9031. [CrossRef]
57. Almatroudi, A. Unlocking the Potential of Silver Nanoparticles: From Synthesis to Versatile Bio-Applications. *Pharmaceutics* **2024**, *16*, 1232. [CrossRef]
58. Quevedo, A.C.; Lynch, I.; Valsami-Jones, E. Silver nanoparticle induced toxicity and cell death mechanisms in embryonic zebrafish cells. *Nanoscale* **2021**, *13*, 6142–6161. [CrossRef]
59. AshaRani, P.V.; Low Kah Mun, G.; Hande, M.P.; Valiyaveettil, S. Cytotoxicity and genotoxicity of silver nanoparticles in human cells. *ACS Nano* **2009**, *3*, 279–290. [CrossRef]
60. Assar, D.H.; Mokhbatly, A.-A.A.; Ghazy, E.W.; Elbialy, Z.I.; Gaber, A.A.; Hassan, A.A.; Nabil, A.; Asa, S.A. Silver nanoparticles induced hepatoxicity via the apoptotic/antiapoptotic pathway with activation of TGFbeta-1 and alpha-SMA triggered liver fibrosis in Sprague Dawley rats. *Environ. Sci. Pollut Res. Int.* **2022**, *29*, 80448–80465. [CrossRef]
61. Cesmeli, S.; Biray Avci, C. Application of titanium dioxide (TiO_2) nanoparticles in cancer therapies. *J. Drug Target.* **2019**, *27*, 762–766. [CrossRef] [PubMed]

62. Shakeel, M.; Jabeen, F.; Shabbir, S.; Asghar, M.S.; Khan, M.S.; Chaudhry, A.S. Toxicity of Nano-Titanium Dioxide (TiO_2-NP) Through Various Routes of Exposure: A Review. *Biol. Trace Element Res.* **2016**, *172*, 1–36. [CrossRef] [PubMed]
63. Baranowska-Wojcik, E.; Szwajgier, D.; Oleszczuk, P.; Winiarska-Mieczan, A. Effects of Titanium Dioxide Nanoparticles Exposure on Human Health-a Review. *Biol. Trace Element Res.* **2020**, *193*, 118–129. [CrossRef] [PubMed]
64. Schneider, S.L.; Lim, H.W. A review of inorganic UV filters zinc oxide and titanium dioxide. *Photodermatol. Photoimmunol. Photomed.* **2019**, *35*, 442–446. [CrossRef]
65. Jeong, J.S.; Jegal, H.; Ko, J.W.; Kim, J.W.; Kim, J.H.; Chung, E.H.; Boo, S.Y.; Lee, S.H.; Lee, G.W.; Park, S.M.; et al. Titanium Dioxide Nanoparticle Exposure Provokes Greater Lung Inflammation in Females Than Males in the Context of Obesity. *Int. J. Nanomed.* **2025**, *20*, 5321–5336. [CrossRef]
66. Eid, A.; Ghaleb, S.S.; Zaki, A.; Ibrahim, M.; Farghali, A.A.; Ali, L.E.; Abdelgawad, M.A.; Ghoneim, M.M.; Al-Serwi, R.H.; Hassan, R.M.; et al. Hesperidin Attenuates Titanium Dioxide Nanoparticle-Induced Neurotoxicity in Rats by Regulating Nrf-2/TNF-alpha Signaling Pathway, the Suppression of Oxidative Stress, and Inflammation. *ACS Omega* **2023**, *8*, 37584–37591. [CrossRef]
67. Mohamadzadeh, N.; Zirak Javanmard, M.; Karimipour, M.; Farjah, G. Developmental Toxicity of the Neural Tube Induced by Titanium Dioxide Nanoparticles in Mouse Embryos. *Avicenna J. Med. Biotechnol.* **2021**, *13*, 74–80. [CrossRef]
68. Jiang, J.; Pi, J.; Cai, J. The Advancing of Zinc Oxide Nanoparticles for Biomedical Applications. *Bioinorg. Chem. Appl.* **2018**, *2018*, 1062562. [CrossRef]
69. Mishra, P.K.; Mishra, H.; Ekielski, A.; Talegaonkar, S.; Vaidya, B. Zinc oxide nanoparticles: A promising nanomaterial for biomedical applications. *Drug Discov. Today* **2017**, *22*, 1825–1834. [CrossRef]
70. Kteeba, S.M.; El-Adawi, H.I.; El-Rayis, O.A.; El-Ghobashy, A.E.; Schuld, J.L.; Svoboda, K.R.; Guo, L. Zinc oxide nanoparticle toxicity in embryonic zebrafish: Mitigation with different natural organic matter. *Environ. Pollut* **2017**, *230*, 1125–1140. [CrossRef]
71. Morimoto, Y.; Izumi, H.; Yoshiura, Y.; Tomonaga, T.; Oyabu, T.; Myojo, T.; Kawai, K.; Yatera, K.; Shimada, M.; Kubo, M.; et al. Evaluation of Pulmonary Toxicity of Zinc Oxide Nanoparticles Following Inhalation and Intratracheal Instillation. *Int. J. Mol. Sci.* **2016**, *17*, 1241. [CrossRef] [PubMed]
72. Senapati, V.A.; Gupta, G.S.; Pandey, A.K.; Shanker, R.; Dhawan, A.; Kumar, A. Zinc oxide nanoparticle induced age dependent immunotoxicity in BALB/c mice. *Toxicol. Res.* **2017**, *6*, 342–352. [CrossRef] [PubMed]
73. Andujar, P.; Lanone, S.; Brochard, P.; Boczkowski, J. Respiratory effects of manufactured nanoparticles. *Rev. Mal. Respir.* **2011**, *28*, e66–e75. [CrossRef] [PubMed]
74. Szucs-Somlyo, E.; Lehel, J.; Majlinger, K.; Lorincz, M.; Kovago, C. Metal-oxide inhalation induced fever-Immuntoxicological aspects of welding fumes. *Food Chem. Toxicol.* **2023**, *175*, 113722. [CrossRef]
75. Monse, C.; Raulf, M.; Jettkant, B.; van Kampen, V.; Kendzia, B.; Schurmeyer, L.; Seifert, C.E.; Marek, E.M.; Westphal, G.; Rosenkranz, N.; et al. Health effects after inhalation of micro- and nano-sized zinc oxide particles in human volunteers. *Arch. Toxicol.* **2021**, *95*, 53–65. [CrossRef]
76. Adeyemi, J.O.; Fawole, O.A. Metal-Based Nanoparticles in Food Packaging and Coating Technologies: A Review. *Biomolecules* **2023**, *13*, 1092. [CrossRef]
77. Zhou, X.; Zhou, X.; Zhou, L.; Jia, M.; Xiong, Y. Nanofillers in Novel Food Packaging Systems and Their Toxicity Issues. *Foods* **2024**, *13*, 2014. [CrossRef]
78. Hashemifard Dehkordi, P.; Moshtaghi, H.; Abbasvali, M. Effects of magnesium oxide and copper oxide nanoparticles on biofilm formation ofEscherichia coliandListeria monocytogenes. *Nanotechnology* **2023**, *34*, 155102. [CrossRef]
79. Roshani, M.; Rezaian-Isfahni, A.; Lotfalizadeh, M.H.; Khassafi, N.; Abadi, M.; Nejati, M. Metal nanoparticles as a potential technique for the diagnosis and treatment of gastrointestinal cancer: A comprehensive review. *Cancer Cell Int.* **2023**, *23*, 280. [CrossRef]
80. Scafa Udriste, A.; Burdusel, A.C.; Niculescu, A.G.; Radulescu, M.; Grumezescu, A.M. Metal-Based Nanoparticles for Cardiovascular Diseases. *Int. J. Mol. Sci.* **2024**, *25*, 1001. [CrossRef]
81. Sibuyi, N.R.S.; Moabelo, K.L.; Fadaka, A.O.; Meyer, S.; Onani, M.O.; Madiehe, A.M.; Meyer, M. Multifunctional Gold Nanoparticles for Improved Diagnostic and Therapeutic Applications: A Review. *Nanoscale Res. Lett.* **2021**, *16*, 174. [CrossRef] [PubMed]
82. Wahajuddin; Arora, S. Superparamagnetic iron oxide nanoparticles: Magnetic nanoplatforms as drug carriers. *Int. J. Nanomed.* **2012**, *7*, 3445–3471. [CrossRef] [PubMed]
83. Lomer, M.C.; Hutchinson, C.; Volkert, S.; Greenfield, S.M.; Catterall, A.; Thompson, R.P.; Powell, J.J. Dietary sources of inorganic microparticles and their intake in healthy subjects and patients with Crohn's disease. *Br. J. Nutr.* **2004**, *92*, 947–955. [CrossRef] [PubMed]
84. Riedle, S.; Wills, J.W.; Miniter, M.; Otter, D.E.; Singh, H.; Brown, A.P.; Micklethwaite, S.; Rees, P.; Jugdaohsingh, R.; Roy, N.C.; et al. A Murine Oral-Exposure Model for Nano- and Micro-Particulates: Demonstrating Human Relevance with Food-Grade Titanium Dioxide. *Small* **2020**, *16*, e2000486. [CrossRef]

85. Disdier, C.; Devoy, J.; Cosnefroy, A.; Chalansonnet, M.; Herlin-Boime, N.; Brun, E.; Lund, A.; Mabondzo, A. Tissue biodistribution of intravenously administrated titanium dioxide nanoparticles revealed blood-brain barrier clearance and brain inflammation in rat. *Part. Fibre Toxicol.* **2015**, *12*, 27. [CrossRef]
86. Nie, P.; Zhao, Y.; Xu, H. Synthesis, applications, toxicity and toxicity mechanisms of silver nanoparticles: A review. *Ecotoxicol. Environ. Saf.* **2023**, *253*, 114636. [CrossRef]
87. Ferdous, Z.; Nemmar, A. Health Impact of Silver Nanoparticles: A Review of the Biodistribution and Toxicity Following Various Routes of Exposure. *Int. J. Mol. Sci.* **2020**, *21*, 2375. [CrossRef]
88. Yan, N.; Wang, W.X. Maternal transfer and biodistribution of citrate and luminogens coated silver nanoparticles in medaka fish. *J. Hazard. Mater.* **2022**, *433*, 128862. [CrossRef]
89. Ferdous, Z.; Al-Salam, S.; Yuvaraju, P.; Ali, B.H.; Nemmar, A. Remote effects and biodistribution of pulmonary instilled silver nanoparticles in mice. *NanoImpact* **2021**, *22*, 100310. [CrossRef]
90. Lee, H.-Y.; Choi, Y.-J.; Jung, E.-J.; Yin, H.-Q.; Kwon, J.-T.; Kim, J.-E.; Im, H.-T.; Cho, M.-H.; Kim, J.-H.; Kim, H.-Y.; et al. Genomics-based screening of differentially expressed genes in the brains of mice exposed to silver nanoparticles via inhalation. *J. Nanopart. Res.* **2010**, *12*, 1567–1578. [CrossRef]
91. Miller, M.R.; Raftis, J.B.; Langrish, J.P.; McLean, S.G.; Samutrtai, P.; Connell, S.P.; Wilson, S.; Vesey, A.T.; Fokkens, P.H.; Boere, A.J.F. Inhaled Nanoparticles Accumulate at Sites of Vascular Disease. *ACS Nano* **2017**, *11*, 4542–4552. [CrossRef] [PubMed]
92. Schmid, G.; Kreyling, W.G.; Simon, U. Toxic effects and biodistribution of ultrasmall gold nanoparticles. *Arch. Toxicol.* **2017**, *91*, 3011–3037. [CrossRef] [PubMed]
93. Irvin-Choy, N.S.; Nelson, K.M.; Dang, M.N.; Gleghorn, J.P.; Day, E.S. Gold nanoparticle biodistribution in pregnant mice following intravenous administration varies with gestational age. *Nanomedicine* **2021**, *36*, 102412. [CrossRef] [PubMed]
94. Huang, K.L.; Chang, H.L.; Tsai, F.M.; Lee, Y.H.; Wang, C.H.; Cheng, T.J. The effect of the inhalation of and topical exposure to zinc oxide nanoparticles on airway inflammation in mice. *Toxicol. Appl. Pharmacol.* **2019**, *384*, 114787. [CrossRef]
95. Rossner, P.; Vrbova, K.; Strapacova, S.; Rossnerova, A.; Ambroz, A.; Brzicova, T.; Libalova, H.; Javorkova, E.; Kulich, P.; Vecera, Z.; et al. Inhalation of ZnO Nanoparticles: Splice Junction Expression and Alternative Splicing in Mice. *Toxicol. Sci.* **2019**, *168*, 190–200. [CrossRef]
96. Adamcakova-Dodd, A.; Stebounova, L.V.; Kim, J.S.; Vorrink, S.U.; Ault, A.P.; O'sHaughnessy, P.T.; Grassian, V.H.; Thorne, P.S. Toxicity assessment of zinc oxide nanoparticles using sub-acute and sub-chronic murine inhalation models. *Part. Fibre Toxicol.* **2014**, *11*, 15. [CrossRef]
97. Liang, C.; Fang, J.; Hu, J.; Geng, X.; Liu, H.; Feng, Y.; Wang, W.; Cui, W.; Yu, Z.; Jia, X. Toxicokinetics of zinc oxide nanoparticles and food grade bulk-sized zinc oxide in rats after oral dosages. *NanoImpact* **2022**, *25*, 100368. [CrossRef]
98. Dadfar, S.M.; Roemhild, K.; Drude, N.I.; von Stillfried, S.; Knuchel, R.; Kiessling, F.; Lammers, T. Iron oxide nanoparticles: Diagnostic, therapeutic and theranostic applications. *Adv. Drug Deliv. Rev.* **2019**, *138*, 302–325. [CrossRef]
99. Kumar, M.; Kulkarni, P.; Liu, S.; Chemuturi, N.; Shah, D.K. Nanoparticle biodistribution coefficients: A quantitative approach for understanding the tissue distribution of nanoparticles. *Adv. Drug Deliv. Rev.* **2023**, *194*, 114708. [CrossRef]
100. Sadeghi, L.; Yousefi Babadi, V.; Espanani, H.R. Toxic effects of the Fe_2O_3 nanoparticles on the liver and lung tissue. *Bratisl. Lek. Listy* **2015**, *116*, 373–378. [CrossRef]
101. Alphandery, E. Biodistribution and targeting properties of iron oxide nanoparticles for treatments of cancer and iron anemia disease. *Nanotoxicology* **2019**, *13*, 573–596. [CrossRef] [PubMed]
102. Shen, S.; Shao, Y.; Li, C. Different types of cell death and their shift in shaping disease. *Cell Death Discov.* **2023**, *9*, 284. [CrossRef] [PubMed]
103. Elmore, S. Apoptosis: A review of programmed cell death. *Toxicol. Pathol.* **2007**, *35*, 495–516. [CrossRef] [PubMed]
104. Lee, Y.; Gustafsson, A.B. Role of apoptosis in cardiovascular disease. *Apoptosis* **2009**, *14*, 536–548. [CrossRef]
105. Fromell, K.; Johansson, U.; Abadgar, S.; Bourzeix, P.; Lundholm, L.; Elihn, K. The effect of airborne Palladium nanoparticles on human lung cells, endothelium and blood-A combinatory approach using three in vitro models. *Toxicol. Vitr.* **2023**, *89*, 105586. [CrossRef]
106. Zeng, C.; Feng, Y.; Wang, W.; Zhou, F.; Liao, F.; Liu, Y.; Feng, S. The size-dependent apoptotic effect of titanium dioxide nanoparticles on endothelial cells by the intracellular pathway. *Environ. Toxicol.* **2018**, *33*, 1221–1228. [CrossRef]
107. Jan, R.; Chaudhry, G.E. Understanding Apoptosis and Apoptotic Pathways Targeted Cancer Therapeutics. *Adv. Pharm. Bull.* **2019**, *9*, 205–218. [CrossRef]
108. Tait, S.W.; Green, D.R. Mitochondrial regulation of cell death. *Cold Spring Harb. Perspect. Biol.* **2013**, *5*, a008706. [CrossRef]
109. Chen, L.; Yokel, R.A.; Hennig, B.; Toborek, M. Manufactured aluminum oxide nanoparticles decrease expression of tight junction proteins in brain vasculature. *J. Neuroimmune Pharmacol.* **2008**, *3*, 286–295. [CrossRef]
110. Huang, F.; Feng, Y.; Wang, Z.A.; Cao, Y.; Yan, Q.; Wang, W.; Feng, S. Environmentally Relevant Concentrations of Commercial Titanium Dioxide Nanoparticles Induce Ferroptosis in HUVECs. *Environ. Toxicol.* **2025**. [CrossRef]

111. Stockwell, B.R.; Friedmann Angeli, J.P.; Bayir, H.; Bush, A.I.; Conrad, M.; Dixon, S.J.; Fulda, S.; Gascon, S.; Hatzios, S.K.; Kagan, V.E.; et al. Ferroptosis: A Regulated Cell Death Nexus Linking Metabolism, Redox Biology, and Disease. *Cell* **2017**, *171*, 273–285. [CrossRef] [PubMed]
112. Wu, X.; Li, Y.; Zhang, S.; Zhou, X. Ferroptosis as a novel therapeutic target for cardiovascular disease. *Theranostics* **2021**, *11*, 3052–3059. [CrossRef] [PubMed]
113. Zhang, X.; Kong, F.; Wang, T.; Huang, X.; Li, W.; Zhang, M.; Wen, T.; Liu, J.; Zhang, Y.; Meng, J.; et al. Iron oxide nanoparticles cause surface coating- and core chemistry-dependent endothelial cell ferroptosis. *Nanotoxicology* **2022**, *16*, 829–843. [CrossRef] [PubMed]
114. Qin, X.; Zhang, J.; Wang, B.; Xu, G.; Yang, X.; Zou, Z.; Yu, C. Ferritinophagy is involved in the zinc oxide nanoparticles-induced ferroptosis of vascular endothelial cells. *Autophagy* **2021**, *17*, 4266–4285. [CrossRef]
115. Pizzino, G.; Irrera, N.; Cucinotta, M.; Pallio, G.; Mannino, F.; Arcoraci, V.; Squadrito, F.; Altavilla, D.; Bitto, A. Oxidative Stress: Harms and Benefits for Human Health. *Oxid. Med. Cell Longev.* **2017**, *2017*, 8416763. [CrossRef]
116. Tekin, S.; Seven, E. Assessment of serum catalase, reduced glutathione, and superoxide dismutase activities and malondialdehyde levels in keratoconus patients. *Eye* **2022**, *36*, 2062–2066. [CrossRef]
117. Jena, A.B.; Samal, R.R.; Bhol, N.K.; Duttaroy, A.K. Cellular Red-Ox system in health and disease: The latest update. *Biomed. Pharmacother.* **2023**, *162*, 114606. [CrossRef]
118. Panda, P.; Verma, H.K.; Lakkakula, S.; Merchant, N.; Kadir, F.; Rahman, S.; Jeffree, M.S.; Lakkakula, B.; Rao, P.V. Biomarkers of Oxidative Stress Tethered to Cardiovascular Diseases. *Oxid. Med. Cell Longev.* **2022**, *2022*, 9154295. [CrossRef]
119. Duan, D.; Li, H.; Chai, S.; Zhang, L.; Fan, T.; Hu, Z.; Feng, Y. The relationship between cardiac oxidative stress, inflammatory cytokine response, cardiac pump function, and prognosis post-myocardial infarction. *Sci. Rep.* **2024**, *14*, 8985. [CrossRef]
120. Aimo, A.; Castiglione, V.; Borrelli, C.; Saccaro, L.F.; Franzini, M.; Masi, S.; Emdin, M.; Giannoni, A. Oxidative stress and inflammation in the evolution of heart failure: From pathophysiology to therapeutic strategies. *Eur. J. Prev. Cardiol.* **2020**, *27*, 494–510. [CrossRef]
121. de Vries, D.K.; Kortekaas, K.A.; Tsikas, D.; Wijermars, L.G.; van Noorden, C.J.; Suchy, M.T.; Cobbaert, C.M.; Klautz, R.J.; Schaapherder, A.F.; Lindeman, J.H. Oxidative damage in clinical ischemia/reperfusion injury: A reappraisal. *Antioxid. Redox. Signal.* **2013**, *19*, 535–545. [CrossRef] [PubMed]
122. Chen, I.C.; Hsiao, I.L.; Lin, H.C.; Wu, C.H.; Chuang, C.Y.; Huang, Y.J. Influence of silver and titanium dioxide nanoparticles on in vitro blood-brain barrier permeability. *Environ. Toxicol. Pharmacol.* **2016**, *47*, 108–118. [CrossRef] [PubMed]
123. Jang, J.; Park, S.; Choi, I.H. Increased Interleukin-11 and Stress-Related Gene Expression in Human Endothelial and Bronchial Epithelial Cells Exposed to Silver Nanoparticles. *Biomolecules* **2021**, *11*, 234. [CrossRef] [PubMed]
124. Siddiqui, M.A.; Wahab, R.; Saquib, Q.; Ahmad, J.; Farshori, N.N.; Al-Sheddi, E.S.; Al-Oqail, M.M.; Al-Massarani, S.M.; Al-Khedhairy, A.A. Iron oxide nanoparticles induced cytotoxicity, oxidative stress, cell cycle arrest, and DNA damage in human umbilical vein endothelial cells. *J. Trace Elem. Med. Biol.* **2023**, *80*, 127302. [CrossRef]
125. Shi, J.; Sun, X.; Lin, Y.; Zou, X.; Li, Z.; Liao, Y.; Du, M.; Zhang, H. Endothelial cell injury and dysfunction induced by silver nanoparticles through oxidative stress via IKK/NF-kappaB pathways. *Biomaterials* **2014**, *35*, 6657–6666. [CrossRef]
126. Gholinejad, Z.; Khadem Ansari, M.H.; Rasmi, Y. Titanium dioxide nanoparticles induce endothelial cell apoptosis via cell membrane oxidative damage and p38, PI3K/Akt, NF-kappaB signaling pathways modulation. *J. Trace. Elem. Med. Biol.* **2019**, *54*, 27–35. [CrossRef]
127. Stevens, T.; Garcia, J.G.; Shasby, D.M.; Bhattacharya, J.; Malik, A.B. Mechanisms regulating endothelial cell barrier function. *Am. J. Physiol. Lung. Cell Mol. Physiol.* **2000**, *279*, L419–L422. [CrossRef]
128. Claesson-Welsh, L.; Dejana, E.; McDonald, D.M. Permeability of the Endothelial Barrier: Identifying and Reconciling Controversies. *Trends Mol. Med.* **2021**, *27*, 314–331. [CrossRef]
129. Medina-Leyte, D.J.; Zepeda-Garcia, O.; Dominguez-Perez, M.; Gonzalez-Garrido, A.; Villarreal-Molina, T.; Jacobo-Albavera, L. Endothelial Dysfunction, Inflammation and Coronary Artery Disease: Potential Biomarkers and Promising Therapeutical Approaches. *Int. J. Mol. Sci.* **2021**, *22*, 3850. [CrossRef]
130. Wu, D.; Chen, Q.; Chen, X.; Han, F.; Chen, Z.; Wang, Y. The blood-brain barrier: Structure, regulation, and drug delivery. *Signal. Transduct. Target Ther.* **2023**, *8*, 217. [CrossRef]
131. Chen, T.; Dai, Y.; Hu, C.; Lin, Z.; Wang, S.; Yang, J.; Zeng, L.; Li, S.; Li, W. Cellular and molecular mechanisms of the blood-brain barrier dysfunction in neurodegenerative diseases. *Fluids Barriers CNS* **2024**, *21*, 60. [CrossRef] [PubMed]
132. Xu, L.; Dan, M.; Shao, A.; Cheng, X.; Zhang, C.; Yokel, R.A.; Takemura, T.; Hanagata, N.; Niwa, M.; Watanabe, D. Silver nanoparticles induce tight junction disruption and astrocyte neurotoxicity in a rat blood-brain barrier primary triple coculture model. *Int. J. Nanomed.* **2015**, *10*, 6105–6118.
133. Chan, Y.J.; Liao, P.L.; Tsai, C.H.; Cheng, Y.W.; Lin, F.L.; Ho, J.D.; Chen, C.Y.; Li, C.H. Titanium dioxide nanoparticles impair the inner blood-retinal barrier and retinal electrophysiology through rapid ADAM17 activation and claudin-5 degradation. *Part. Fibre Toxicol.* **2021**, *18*, 4. [CrossRef] [PubMed]

134. Zhang, X.; Lin, Y.; Xin, J.; Zhang, Y.; Yang, K.; Luo, Y.; Wang, B. Red blood cells in biology and translational medicine: Natural vehicle inspires new biomedical applications. *Theranostics* **2024**, *14*, 220–248. [CrossRef]
135. Zhou, Z. Purinergic interplay between erythrocytes and platelets in diabetes-associated vascular dysfunction. *Purinergic. Signal.* **2021**, *17*, 705–712. [CrossRef]
136. Byrnes, J.R.; Wolberg, A.S. Red blood cells in thrombosis. *Blood* **2017**, *130*, 1795–1799. [CrossRef]
137. Pernow, J.; Mahdi, A.; Yang, J.; Zhou, Z. Red blood cell dysfunction: A new player in cardiovascular disease. *Cardiovasc. Res.* **2019**, *115*, 1596–1605. [CrossRef]
138. Sun, J.; Vyas, P.; Mann, S.; Paganini-Hill, A.; Nunes, A.C.F.; Lau, W.L.; Cribbs, D.H.; Fisher, M.J.; Sumbria, R.K. Insights Into the Mechanisms of Brain Endothelial Erythrophagocytosis. *Front. Cell Dev. Biol.* **2021**, *9*, 672009. [CrossRef]
139. Zlamal, J.; Singh, A.; Weich, K.; Jaffal, H.; Uzun, G.; Pelzl, L.; Althaus, K.; Bakchoul, T. Platelet phosphatidylserine is the critical mediator of thrombosis in heparin-induced thrombocytopenia. *Haematologica* **2023**, *108*, 2690–2702. [CrossRef]
140. Wang, L.; Bi, Y.; Yu, M.; Li, T.; Tong, D.; Yang, X.; Zhang, C.; Guo, L.; Wang, C.; Kou, Y.; et al. Phosphatidylserine-exposing blood cells and microparticles induce procoagulant activity in non-valvular atrial fibrillation. *Int. J. Cardiol.* **2018**, *258*, 138–143. [CrossRef]
141. Kim, E.H.; Baek, S.M.; Choi, S.; Cho, J.; Tahmasebi, S.; Bae, O.N. Promoted coagulant activity and disrupted blood-brain barrier depending on phosphatidylserine externalization of red blood cells exposed to ZnO nanoparticles. *Environ. Pollut* **2024**, *362*, 124921. [CrossRef] [PubMed]
142. Bian, Y.; Kim, K.; Ngo, T.; Kim, I.; Bae, O.N.; Lim, K.M.; Chung, J.H. Silver nanoparticles promote procoagulant activity of red blood cells: A potential risk of thrombosis in susceptible population. *Part. Fibre Toxicol.* **2019**, *16*, 9. [CrossRef] [PubMed]
143. Bian, Y.; Chung, H.Y.; Bae, O.N.; Lim, K.M.; Chung, J.H.; Pi, J. Titanium dioxide nanoparticles enhance thrombosis through triggering the phosphatidylserine exposure and procoagulant activation of red blood cells. *Part. Fibre Toxicol.* **2021**, *18*, 28. [CrossRef] [PubMed]
144. Henein, M.Y.; Vancheri, S.; Longo, G.; Vancheri, F. The Role of Inflammation in Cardiovascular Disease. *Int. J. Mol. Sci.* **2022**, *23*, 12906. [CrossRef]
145. Gerogianni, A.; Bal, M.; Mohlin, C.; Woodruff, T.M.; Lambris, J.D.; Mollnes, T.E.; Sjostrom, D.J.; Nilsson, P.H. In vitro evaluation of iron oxide nanoparticle-induced thromboinflammatory response using a combined human whole blood and endothelial cell model. *Front. Immunol.* **2023**, *14*, 1101387. [CrossRef]
146. Park, E.J.; Kim, S.N.; Yoon, C.; Cho, J.W.; Lee, G.H.; Kim, D.W.; Park, J.; Choi, I.; Lee, S.H.; Song, J.; et al. Repeated intratracheal instillation of zinc oxide nanoparticles induced pulmonary damage and a systemic inflammatory response in cynomolgus monkeys. *Nanotoxicology* **2021**, *15*, 621–635. [CrossRef]
147. Monse, C.; Hagemeyer, O.; Raulf, M.; Jettkant, B.; van Kampen, V.; Kendzia, B.; Gering, V.; Kappert, G.; Weiss, T.; Ulrich, N.; et al. Concentration-dependent systemic response after inhalation of nano-sized zinc oxide particles in human volunteers. *Part. Fibre Toxicol.* **2018**, *15*, 8. [CrossRef]
148. Ajoolabady, A.; Pratico, D.; Lin, L.; Mantzoros, C.S.; Bahijri, S.; Tuomilehto, J.; Ren, J. Inflammation in atherosclerosis: Pathophysiology and mechanisms. *Cell Death Dis.* **2024**, *15*, 817. [CrossRef]
149. Zhu, M.T.; Wang, B.; Wang, Y.; Yuan, L.; Wang, H.J.; Wang, M.; Ouyang, H.; Chai, Z.F.; Feng, W.Y.; Zhao, Y.L. Endothelial dysfunction and inflammation induced by iron oxide nanoparticle exposure: Risk factors for early atherosclerosis. *Toxicol. Lett.* **2011**, *203*, 162–171. [CrossRef]
150. Gojova, A.; Guo, B.; Kota, R.S.; Rutledge, J.C.; Kennedy, I.M.; Barakat, A.I. Induction of inflammation in vascular endothelial cells by metal oxide nanoparticles: Effect of particle composition. *Environ. Health Perspect* **2007**, *115*, 403–409. [CrossRef]
151. Gojova, A.; Lee, J.T.; Jung, H.S.; Guo, B.; Barakat, A.I.; Kennedy, I.M. Effect of cerium oxide nanoparticles on inflammation in vascular endothelial cells. *Inhal. Toxicol.* **2009**, *21* (Suppl. S1), 123–130. [CrossRef] [PubMed]
152. Yousef, M.I.; Abuzreda, A.A.; Kamel, M.A. Cardiotoxicity and lung toxicity in male rats induced by long-term exposure to iron oxide and silver nanoparticles. *Exp. Ther. Med.* **2019**, *18*, 4329–4339. [CrossRef] [PubMed]
153. Wu, L.; Wen, W.; Wang, X.; Huang, D.; Cao, J.; Qi, X.; Shen, S. Ultrasmall iron oxide nanoparticles cause significant toxicity by specifically inducing acute oxidative stress to multiple organs. *Part. Fibre. Toxicol.* **2022**, *19*, 24. [CrossRef] [PubMed]
154. Avsievich, T.; Popov, A.; Bykov, A.; Meglinski, I. Mutual interaction of red blood cells influenced by nanoparticles. *Sci. Rep.* **2019**, *9*, 5147. [CrossRef]
155. Magaye, R.R.; Yue, X.; Zou, B.; Shi, H.; Yu, H.; Liu, K.; Lin, X.; Xu, J.; Yang, C.; Wu, A.; et al. Acute toxicity of nickel nanoparticles in rats after intravenous injection. *Int. J. Nanomed.* **2014**, *9*, 1393–1402.
156. Chen, Z.; Wang, Y.; Zhuo, L.; Chen, S.; Zhao, L.; Luan, X.; Wang, H.; Jia, G. Effect of titanium dioxide nanoparticles on the cardiovascular system after oral administration. *Toxicol. Lett.* **2015**, *239*, 123–130. [CrossRef]
157. Odzak, N.; Kistler, D.; Behra, R.; Sigg, L. Dissolution of metal and metal oxide nanoparticles in aqueous media. *Environ. Pollut* **2014**, *191*, 132–138. [CrossRef]

158. Zhang, H.; Ji, Z.; Xia, T.; Meng, H.; Low-Kam, C.; Liu, R.; Pokhrel, S.; Lin, S.; Wang, X.; Liao, Y.P.; et al. Use of metal oxide nanoparticle band gap to develop a predictive paradigm for oxidative stress and acute pulmonary inflammation. *ACS Nano* **2012**, *6*, 4349–4368. [CrossRef]
159. Wehmas, L.C.; Anders, C.; Chess, J.; Punnoose, A.; Pereira, C.B.; Greenwood, J.A.; Tanguay, R.L. Comparative Metal Oxide Nanoparticle Toxicity Using Embryonic Zebrafish. *Toxicol. Rep.* **2015**, *2*, 702–715. [CrossRef]
160. Xiao, Y.; Vijver, M.G.; Chen, G.; Peijnenburg, W.J. Toxicity and accumulation of Cu and ZnO nanoparticles in Daphnia magna. *Environ. Sci. Technol.* **2015**, *49*, 4657–4664. [CrossRef]
161. Yazdanshenas, M.R.; Rezaei, M.R.; Kharkan, J. Comparative toxicity of zinc oxide nanoparticles and zinc salts in male mice: Hematological, biochemical, and histopathological impacts. *Toxicol. Rep.* **2025**, *14*, 102003. [CrossRef] [PubMed]
162. Lee, I.C.; Ko, J.W.; Park, S.H.; Lim, J.O.; Shin, I.S.; Moon, C.; Kim, S.H.; Heo, J.D.; Kim, J.C. Comparative toxicity and biodistribution of copper nanoparticles and cupric ions in rats. *Int. J. Nanomed.* **2016**, *11*, 2883–2900.
163. Boudreau, M.D.; Imam, M.S.; Paredes, A.M.; Bryant, M.S.; Cunningham, C.K.; Felton, R.P.; Jones, M.Y.; Davis, K.J.; Olson, G.R. Differential Effects of Silver Nanoparticles and Silver Ions on Tissue Accumulation, Distribution, and Toxicity in the Sprague Dawley Rat Following Daily Oral Gavage Administration for 13 Weeks. *Toxicol. Sci.* **2016**, *150*, 131–160. [CrossRef] [PubMed]
164. Khan, M.; Naqvi, A.H.; Ahmad, M. Comparative study of the cytotoxic and genotoxic potentials of zinc oxide and titanium dioxide nanoparticles. *Toxicol. Rep.* **2015**, *2*, 765–774. [CrossRef]
165. Valdiglesias, V.; Alba-Gonzalez, A.; Fernandez-Bertolez, N.; Touzani, A.; Ramos-Pan, L.; Reis, A.T.; Moreda-Pineiro, J.; Yanez, J.; Laffon, B.; Folgueira, M. Effects of Zinc Oxide Nanoparticle Exposure on Human Glial Cells and Zebrafish Embryos. *Int. J. Mol. Sci.* **2023**, *24*, 12297. [CrossRef]
166. Zhang, C.; Liu, Z.; Zhang, Y.; Ma, L.; Song, E.; Song, Y. "Iron free" zinc oxide nanoparticles with ion-leaking properties disrupt intracellular ROS and iron homeostasis to induce ferroptosis. *Cell Death Dis.* **2020**, *11*, 183. [CrossRef]

International Journal of
Molecular Sciences

MDPI

Review

Advancements in Lithography Techniques and Emerging Molecular Strategies for Nanostructure Fabrication

Prithvi Basu [1], Jyoti Verma [1], Vishnuram Abhinav [2], Ratneshwar Kumar Ratnesh [3], Yogesh Kumar Singla [4] and Vibhor Kumar [1,*]

1 Department of Electrical Engineering, Texas A&M University, College Station, TX 77843, USA; prithvi.basu@exchange.tamu.edu (P.B.); jyotiverma607@tamu.edu (J.V.)
2 Department of Electrical Engineering, Indian Institute of Technology Bombay, Mumbai 400076, India; vishnuram.abhinav@gmail.com
3 Department of Electronics and Communication Engineering, Meerut Institute of Engineering and Technology, Meerut 250005, India; ratnes123@gmail.com
4 School of Engineering, Math and Technology, Navajo Technical University, Crownpoint, NM 87313, USA; ysingla@navajotech.edu
* Correspondence: vibhorsoni123@gmail.com

Abstract: Lithography is crucial to semiconductor manufacturing, enabling the production of smaller, more powerful electronic devices. This review explores the evolution, principles, and advancements of key lithography techniques, including extreme ultraviolet (EUV) lithography, electron beam lithography (EBL), X-ray lithography (XRL), ion beam lithography (IBL), and nanoimprint lithography (NIL). Each method is analyzed based on its working principles, resolution, resist materials, and applications. EUV lithography, with sub-10 nm resolution, is vital for extending Moore's Law, leveraging high-NA optics and chemically amplified resists. EBL and IBL enable high-precision maskless patterning for prototyping but suffer from low throughput. XRL, using synchrotron radiation, achieves deep, high-resolution features, while NIL provides a cost-effective, high-throughput method for replicating nanostructures. Alignment marks play a key role in precise layer-to-layer registration, with innovations enhancing accuracy in advanced systems. The mask fabrication process is also examined, highlighting materials like molybdenum silicide for EUV and defect mitigation strategies such as automated inspection and repair. Despite challenges in resolution, defect control, and material innovation, lithography remains indispensable in semiconductor scaling, supporting applications in integrated circuits, photonics, and MEMS/NEMS devices. Various molecular strategies, mechanisms, and molecular dynamic simulations to overcome the fundamental lithographic limits are also highlighted in detail. This review offers insights into lithography's present and future, aiding researchers in nanoscale manufacturing advancements.

Keywords: lithography; nanostructure; extreme UV lithography; electron beam lithography (EBL); X-ray lithography (XRL); ion beam lithography (IBL); nanoimprint lithography (NIL); nanofabrication

1. Introduction

Semiconductor chip manufacturing involves various techniques and processes performed on a substrate. Lithography is a crucial step accounting for nearly one-third of the total manufacturing cost. As chip density increases, device sizes must shrink to keep pace with Moore's Law. Achieving this miniaturization requires an optimal lithography method with the necessary resolution [1]. A lithography system comprises an optical system, a mask, resists, and an exposure setup. The optical system includes a lens array, a light source, and sensors [2]. Masks, typically made from glass or quartz, are coated with chromium or iron oxide. Patterns are generated on the mask using maskless writing technologies and then transferred to the resist on the wafer [3,4].

Resists are complex organic compounds that undergo chemical changes when exposed to light or photons. Photoresists are classified as positive or negative tone. In positive tone resists, the exposed regions are washed away during development [5]. This produces patterns identical to those on the mask. In contrast, negative tone resists undergo crosslinking, reducing solubility in the exposed regions and creating patterns that are the inverse of the mask [6,7]. Due to their superior resolution, positive tone resists are widely used in IC fabrication. Optical lithography historically employs UV lamps with wavelengths of 365 nm and 436 nm for exposure [8].

To construct nanometer-sized devices for cutting-edge semiconductor integrated circuits (ICs) and almost any other fundamental application across multiple scientific disciplines, the production or imprint of nano-sized patterns on a substrate, known as nanolithography, is essential [9,10]. This technique is not only a fundamental driving technology in semiconductor and integrated circuit manufacturing, but it also plays an important role in the fabrication of commercially available microelectromechanical system (MEMS) and nanoelectromechanical system (NEMS) devices, which are used in automobiles, optical switches, microphones, mobile phone chips, and biosensors. Nanolithography has been used to successfully create semiconductor devices such as MOSFET, CMOS, and NMOS, as well as memory devices such as DRAM and MRAM [11,12]. For the reasons stated above, excimer lasers were introduced, which used DUV wavelengths of 248 and 193 nm [13]. However, the resolution was limited by diffraction effects; therefore, more advanced techniques such as short-wavelength EUV lithography, electron beam lithography, X-ray lithography, ion beam lithography, and nanoimprint lithography gained popularity.

One of the methods to fabricate complex features with variable dimensions ranging from a few microns to a sub-100 nm scale is the Mix and Match (M&M) approach [14]. This technique involves combining two or more patterning techniques to improve throughput. The Mix and Match (M&M) approach is a potential technique to manufacture devices in advanced research areas like quantum technologies and photonic integrated circuits (PICs) [15,16]. Other direct-writing technologies like two-photon polymerization (2PP) and laser direct writing are capable of fabricating high-resolution 3D structures [17]. 2PP is a non-linear technique used in 3D printing and microfabrication, where polymer materials are selectively crosslinked using laser light. 2PP has been implemented to fabricate structured biocompatible 3D nanogrid structures to enhance neuronal directional growth [18,19]. The versatility of the 2PP technique makes it an excellent choice for advanced applications in fields such as microelectronics, biotechnology, and optics [20].

Laser direct write lithography using near-i-line wavelength is a high-throughput and cost-effective maskless technique for microscale fabrication, directed towards low-volume production. Laser direct writing applications including processors, energy harvesting and storage, sensing, and bioelectronics have been explored [21,22]. Soft lithography, a cost-effective, flexible process, consists of a wide variety of techniques based on printing, molding, and embossing with an elastomeric stamp [23]. Wet lithography is another technique that is capable of patterning large areas of microstructures of organic and inorganic soluble materials and biological compounds in buffer solutions. This technique does not require any specialized facilities or tools and can be achieved using compact discs and microscopy grids [24]. The discovery of Scanning Probe Microscopy (SPM) also boosted the advancements of nanotechnology in mapping the surface topography and studying the physical and chemical properties of sub-nanometer molecular structures [25].

The first manufacturable multi-beam system for scanning electron microscopy was invented in 2005, featuring 100 beams of sub-10 nm size. Around the same time, IMS Nanofabrication (Wien, Austria) introduced its own multi-beam concept for projection mask-less lithography and later for photomask patterning [26]. IMS's electron multi-beam system was an evolution of its earlier ion multi-beam technology. By 2015, IMS's electron multi-beam program achieved its goals, demonstrating the ability to pattern a full 6-inch mask in high-volume manufacturing, marking a significant shift in photomask patterning technology [26,27]. This innovation arrived just in time to complement the deployment of EUV lithography, overcoming limitations of resolution and write time seen with Variable Shaped Beam (VSB) technology [28,29].

In 2016, IMS launched the MBMW-101, the first high-volume manufacturing (HVM) version, marking a breakthrough in mask writing technology. This was followed by the introduction of the MBMW-201 in 2019, a second-generation tool that further enhanced throughput and resolution [30]. The MBMW-201 has been used for EUV mask production at 7 nm, 5 nm, and 3 nm nodes, as well as research for the 2 nm node [31]. NuFlare Technology (NFT) has developed several mask writers, including the MBM-2000 and MBM-2000PLUS, which are used for producing advanced optical and EUV masks. In 2023, NFT launched the MBM-3000, designed for the N2 technology node [32]. More recently, researchers have developed a mask process correction (MPC) system integrated into the multi-beam writer. This advancement, coupled with the introduction of inverse lithography technology (ILT), has ushered in a new era of photomask design, transitioning mask patterns from rectilinear to curvilinear shapes [33].

In recent decades, much research and development has been conducted to advance nano-lithographic processes. Even though optical lithography has accounted for more than 40% of lithographic operations in the last five years, alternative nano-lithographic techniques have achieved substantial advances in terms of resolution capabilities, new and superior resists, and developing materials [34]. In the current review, a variety of modern lithography techniques like extreme ultraviolet lithography, electron beam lithography, X-ray lithography, ion beam lithography, and nanoimprint lithography are discussed along with the role of alignment marks in ensuring precise layer-to-layer registration, emphasizing the innovations and various techniques suitable for lithography systems. The mask fabrication process is also touched upon, detailing the materials, types, innovations in various mask materials, and challenges in producing defect-free masks with emphasis on advancements in their achievable resolution in recent years. Finally, the emerging molecular strategies including the molecular structure and properties of resist materials, the advanced photochemical reactions involved in the photoresist chemistry, and predicting the material behavior at nanoscale via molecular simulations are also described in detail.

2. Extreme Ultraviolet Lithography

Extreme ultraviolet lithography (EUVL), with a wavelength of approximately 13.5 nm, was introduced in 1988 and has been widely employed in the semiconductor industry since 2018. Because of the Rayleigh diffraction effect, smaller wavelengths of light have been used to achieve a reduced feature size with excellent resolution via a lens system, resulting in a smaller semiconductor device. Multi-patterning technologies enable IC production to extend beyond 3 nm [35]. Nguyen et al. [36] fabricated NMOS transistors with minimal gate length variations of 0.075, 0.1, 0.11, 0.12, and up to 0.18 μm to quantitatively assess the lithographic efficacy of the EUVL laboratory apparatus. This experiment utilized three distinct gate oxide thicknesses: 25 Å, 40 Å, and 55 Å. Each device field comprised 24 devices, 23 transistors, and a $50 \times 50\ \mu m^2$ capacitor, with a maximum transistor gate length of 20 μm.

An EUV system comprises a source, a patterning mask, an exposure apparatus, and multilayer mirrors. The EUV source primarily comprises discharge-produced plasma (DPP) sources or laser-produced plasma (LPP) sources capable of generating electron temperatures over 100,000 K. The EUV source, utilizing laser-produced plasma (LPP) technology, comprises multiple components, including a carbon dioxide (CO_2) laser. Research demonstrated that tin plasma EUV can be stimulated by a CO_2 laser pulse at a wavelength of 10.6 µm due to its superior conversion efficiency, reduced debris generation, and elevated average power levels, all while mitigating significant beam distortion issues typically associated with solid-state lasers at high intensities. Elevated conversion efficiency is achieved through the use of brief laser pulses and increased repetition rates. The conversion efficiency of the EUV source is dependent on the effectiveness of plasma formation and heating [37]. The schematics illustrating the laser-droplet interaction in an industrial EUV light source module are depicted in Figure 1 [38].

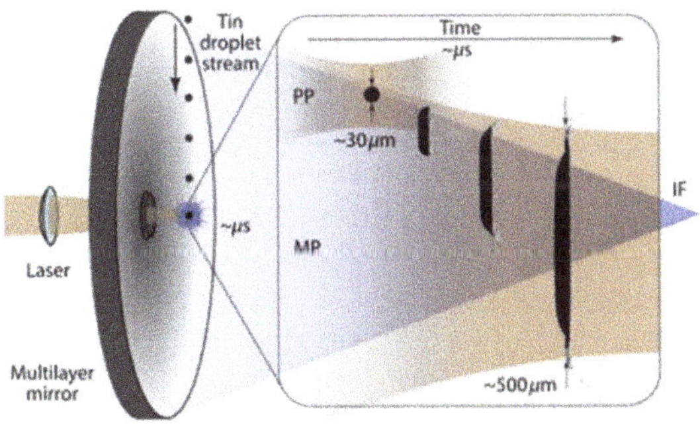

Figure 1. Schematic diagram of a laser–droplet interaction in an industrial EUV light source module.

Reflective mirrors, rather than refractive lenses, are utilized for the optical components that facilitate the imaging functions of the scanner. These mirrors feature surface coatings comprising up to 100 alternating layers of silicon and molybdenum, which reflect light through interlayer interference. Notably, these coatings demonstrate exceptional performance with light at a wavelength of 13.5 nm. However, each mirror is theoretically limited to reflecting only 72% of the incident EUV light, with the remainder being absorbed by the mirror [39].

To selectively filter 13.5 nm EUV from the plasma and enhance reflective rates, the collector surface is coated with several Mo/Si layers. The reflection wavelength λ can be accurately determined by adjusting layer thickness and reflection angle. For a wavelength of 13.5 nm, approximately 40 pairs of multilayer mirrors are employed, featuring a thickness of about 270 nm, an angle of incidence of 6°, and a bilayer spacing close to 7 nm. A standard multilayer mirror for the EYVL system consists of around 40 pairs of Mo/Si layers.

Various multilayer mirrors, including Ru/B_4C, Ru/C, Pd/B_4C, V/Sc, Cr/V, and Pd/Y multilayer mirrors, have been produced by a research group at Tongji University using direct current magnetron sputtering. The theoretical reflectance for a 40-pair Mo/Si multilayer at a wavelength of 13.36 nm is calculated to be 74.47% [40]. A capping layer positioned above the Mo/Si multilayer is implemented to enhance the endurance and reliability of the EUV collector. Ruthenium has been identified as an excellent material choice for this capping layer due to its stability in EUV and hydrogen environments. Additionally, a research group at ASML has explored other advanced materials that exhibit strong EUV transmission and thermal stability, concluding that polysilicon and metal silicide films perform well in high-temperature conditions [41].

ASML has recently devised an innovative technique for merging many small droplets into larger ones at increased separation distances. ASML has launched its third-generation droplet generator, including a droplet diameter of 27 μm and an average lifespan of around 2700 h [42]. ASML is presently examining the next phase of laser power enhancement. Recent advancements have led to the achievement of a dose-controlled EUV power output of 420 W at a high duty cycle. In addition, ASML's next-generation NXE:3600D has reached an open-loop EUV power level of 530 W, while the degradation of the EUV collector has been observed to remain below 0.05% [43].

The latest high-power CO_2 laser, with master oscillator power amplifier (MOPA) pre-pulse technology, is considered the optimal solution, capable of producing a laser power output exceeding 40 kW. MOPA pre-pulse technology can deliver a stabilized EUV power of 250 W, achieving a conversion efficiency of 6% and maintaining a stable dose with an error margin of less than 0.1%. Academic and industrial research and development teams are actively exploring innovative approaches for the advancement of EUV sources. One promising strategy involves the integration of a multifiber laser coherent synthesis system with a nano-Sn generator. This combination is expected to provide significant benefits in achieving high peak power, high average power, and enhanced conversion efficiency simultaneously [44].

2.1. EUV Lithography Resolution

The advent of EUV scanners, now reinforcing their role as benchmark high-volume manufacturing (HVM) system, enables the industry to surpass the 10 nm resolution barrier in a single exposure. To achieve the theoretical minimum limit of 0.25 for the imaging factor k_1, the Abbe–Lin equation can be utilized to calculate the critical dimension (CD) value as follows:

$$CD = k_1 \times (\lambda/NA) = 0.25 \times (13.5/0.33) = 10.3 \text{ nm}$$

The requirement of an increased numerical aperture (NA) of the optics is essential to exceed the 10 nm resolution limit while adhering to the parameters of the EUV lithographic framework (wavelength λ = 13.5 nm). The implementation of high-NA optics, functioning at NA = 0.55, as opposed to the prior NA = 0.33, enables the production of 8 nm features while utilizing a k_1 imaging factor of 0.325 [45]. The diminishment of the minimal feature size that may be printed, hence facilitating the ongoing contraction as prescribed by Moore's Law, represents but one perspective on the advantages of high-NA EUV lithography [46]. The enhancement of the numerical aperture in the imaging optics has enabled the production of features that exhibit greater lithographic contrast, as measured by the normalized image log-slope (NILS). Enhancing the NILS is as crucial as the reduction, as indicated by the relationship $LCDU = k_4 \times (1/NILS) \times (hv/dose)^{1/2}$ [45], where lithographic contrast is associated with local CD uniformity (LCDU), a metric of CD variability. To achieve a sufficiently high yield with minimal defectivity, it is imperative that LCDU remains low [45,47].

2.2. EUV Lithography Resists

Two primary categories of resists are utilized in EUV lithography: chemically amplified resists (CARs) and non-chemically amplified resists. The main challenges facing next-generation lithography resists include resolution, sensitivity, and line edge roughness (LER). Additionally, factors such as EUV absorption, defect density, and outgassing significantly influence the efficiency of the system. Chemically amplified resists are commonly employed in 248 nm and 193 nm optical lithography, typically comprising a polymeric or molecular matrix, photoacid generators (PAGs), and base quenchers. Upon exposure to irradiation, PAGs produce acids that alter the dissolution rate of the matrix during the post-exposure bake (PEB) stage. During this process, the photogenerated acids catalyze reactions within specific groups of the matrix, leading to changes in hydrophilicity, crosslinking, or backbone scission, which ultimately affects solubility, particularly in aqueous base developers [48].

The energy of a photon at 248 nm is 5 eV, while at 193 nm it is 6 eV; in contrast, a photon at 13 nm has an energy of 93 eV. This increased energy per photon contributes to shot noise and LER. A single-layer ultrathin resist applied over a hard mask can help control line edge roughness (LER). Research indicates that reducing PAG size and increasing PAG concentration can effectively lower LER [49].

On the other hand, metal oxide nanoparticles (NPs) have been proposed as next-generation photoresist materials for several key reasons related to their unique properties, which make them ideal candidates for advanced lithographic processes [50]. The main factors driving their potential are high photochemical stability, increased resolution and patterning ability, improved sensitivity and process control, reduced toxicity and environmental impact, and their compatibility with advanced lithographic techniques (e.g., 3 nm, 2 nm) [51]. The production of these nanoparticles involves carefully controlled hydrolysis of zirconium or hafnium alkoxides in an excess of carboxylic acid, followed by precipitation processes to yield ZrO_2-NP or HfO_2-NP with organic ligands. The NP size is kept below 3 nm, making it suitable for sub-20 nm lithography. These nanoparticles can generate both positive and negative tone patterns using either a photo radical initiator or PAG. They demonstrate significant etch resistance along with thermal and chemical stability [52]. A schematic diagram illustrating the ZrO_2 and HfO_2 nanoparticles, including their core metal oxides and surrounding organic ligands, is shown in Figure 2.

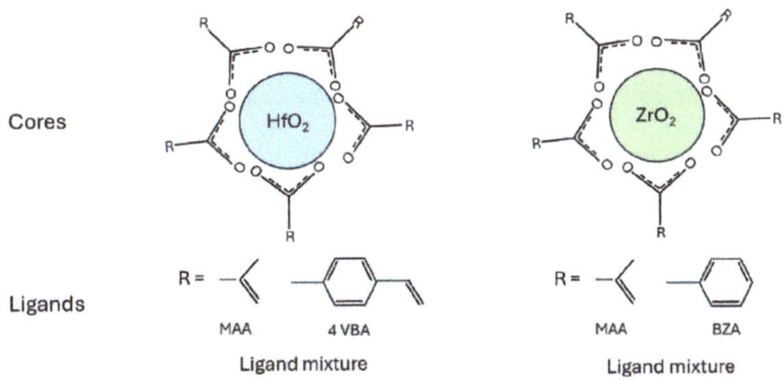

Figure 2. Schematic diagrams of ZrO_2 and HfO_2 nanoparticle with their core metal oxides and the organic ligands surrounding the cores.

The nanoparticle films utilizing 2,2-dimethoxy-2-phenyl acetophenone (DPAP) as a photo-initiator have been identified as capable of functioning as both positive and negative tone photoresists. Exposure to deep-UV light (254 nm or 193 nm) followed by development in isopropyl alcohol or t-amyl alcohol produces a negative tone image. In contrast, exposure followed by post-exposure baking (PEB) at 130 °C for 3 min, followed by development in tetramethylammonium hydroxide (TMAH), results in positive tone patterns [53].

In 2010, researchers at Cornell University introduced one of the earliest metal-oxide-based hybrid resists. This photoresist features an inorganic HfO_2 core surrounded by a shell of electro-reactive organic ligands. These resists exhibit dual-tone behavior, functioning as either positive or negative photoresists depending on the development conditions. The resist's performance in both positive and negative tones was evaluated under identical experimental conditions with exposure to a 254 nm wavelength, achieving resolutions of 0.9 μm for positive tones and 0.8 μm for negative tones.

As previously noted, HfO_2 has superior chemical and thermal stability and is resistant to oxidation, in contrast to organic polymers. Furthermore, due to their diminutive size (diameter under 1 nm), they demonstrate little light scattering and are applicable in both light-based lithography processes, namely DUV and EUV [54]. Figure 3 [55] shows the SEM images of 25 and 22 nm half-pitch patterns obtained by EUV lithography on a novel non-chemically amplified resist (nCAR) based on biphenyl iodonium perfluoro-1-butanesulfonate-modified polystyrene with a naphthalimide scaffold.

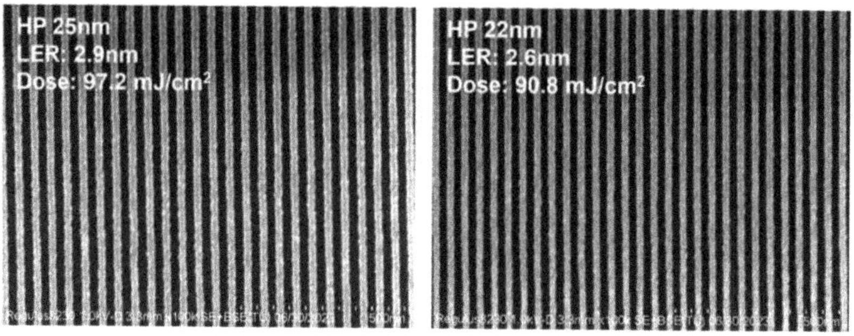

Figure 3. SEM images of 25 and 22 nm HP patterns by EUV lithography on n-CAR resist based on biphenyl iodonium perfluoro-1-butanesulfonate-modified polystyrene with a naphthalimide scaffold. Copyright 2024 Royal Society of Chemistry. Reproduced with permission from [55].

Enhancement of sensitivity by the incorporation of a metal sensitizer was noted with both EB and EUV methods. The enhancement in sensitivity arises from increased acid yield and electron efficiency, rather than elevated EUV photon absorption. Metal sensitizer salts significantly influence acid yield and dissolution properties. By carefully selecting a suitable sensitizer and optimizing its concentration, resist performance can be effectively enhanced. The incorporation of the sensitizer resulted in a 43% enhancement in sensitivity alongside a decrease in roughness. The enhancement in sensitivity is mostly attributable to the increased acid yield and electron efficiency [56].

A study investigating the stability of these clusters using spectroscopic techniques found that Zn-based oxoclusters, such as Zn(MA)(TFA), which are characterized by methacrylate and trifluoroacetate ligands, have labile ligands that undergo structural changes over a two-month period when kept as crystalline powder [57]. In contrast, when deposited as a thin film, these clusters undergo polymerization and/or hydrolysis within a few hours under ambient conditions. Nonetheless, the thin films exhibit stability in both air and vacuum for a sufficient duration for intact lithography applications and processing. The results elucidate the stability of resist systems, a crucial factor in developing new hybrid photoresists, and underscore that inorganic resists are prone to structural alterations that must be regulated to achieve reproducibility in lithographic performance [58].

2.3. Advantages and Limitations

To prolong Moore's Law while managing stochastic effects to acceptable levels, three critical parameters are aerial picture contrast, photoresist, and dosage. The innovative EUV system incorporates 0.55 NA projection optics, as opposed to the 0.33 NA of the last generation. One primary effect of the elevated numerical aperture (NA) is that the aerial image contrast on the wafer is enhanced when fabricating identical structures. The 0.55 high-NA system enhances contrast, and with modifications to the mask absorber, the contrast can be preserved while reducing resolutions. The extensive transmission and accelerated phases guarantee that the productivity of the high-NA scanner is sustained even at elevated dosages. The primary advantage of the increased numerical aperture is the resultant enhanced contrast [59].

Two critical lithography challenges, line width roughness (LWR) and pattern collapse, have been recognized as necessitating enhancement as extreme ultraviolet (EUV) technology advances toward high-volume production. Pattern collapse constitutes a critical challenge in EUV lithography, constraining the process window and resolution. Reducing the resist thickness and aspect ratio of the resultant images might alleviate collapse and enhance resolution; nevertheless, it produces images that frequently lack sufficient durability for contemporary etch and integration processes [60].

The efficacy of EUV must be enhanced at the wafer level to achieve cost-effectiveness in technology. The efficiency of the current EUV source power throughput is inadequate. To incorporate working integrated circuits, the defect density on reflecting masks must be reduced. The utilization of non-pellicle masks presents significant challenges, leading to their replacement with movable pellicle and thermophoretic protection [52].

3. Electron Beam Lithography

The enhancement of resolution is recognized as the primary challenge in the advancement of optical lithography. Specific issues, such as resist outgassing and line edge roughness (LER), are encountered by Extreme Ultraviolet Lithography (EUVL), limiting its effectiveness. The absorption of extreme ultraviolet light by resist materials is also seen as a constraint on the application of EUV lithography.

In contrast, electron beam lithography (EBL) is employed as a sophisticated maskless technique for creating intricate patterns at the nanoscale, which is difficult to achieve with traditional methods [61]. Since its introduction in the 1970s, significant advancements have been made in EBL. Direct patterning is enabled by this technique without the need for a mask, unlike conventional mask writing procedures [62].

The highest practical resolution in EBL is achieved due to the short wavelengths associated with electrons, which can be less than 0.1 nm for energies between 10 and 50 keV. The resolution in EBL is primarily influenced by the quality of the resist used, requiring that the electron beam's diameter be smaller than the dimensions of the final structure to attain enhanced resolution [63].

A concentrated electron beam is employed by EBL to delineate patterns on a substrate. The short wavelength of the electron beam enables a resolution of approximately 10 nm, free from diffraction limitations. A film or resist that is sensitive to electrons is patterned by applying it to the sample using a spin coating process, typically at speeds ranging from 1000 to 6000 rpm, to achieve a uniform layer [64]. When exposed to the electron beam, the resist's solubility is modified, enabling it to become selectively more soluble in either the exposed or unexposed areas, depending on the type of resist utilized. The removal of the resist is accomplished by immersing it in a solution known as a developer [65].

An electron gun is a device designed to generate, accelerate, focus, and direct a beam of electrons toward a substrate. The process begins with the emission of electrons from cathodes or electron sources. These emitted electrons are subsequently accelerated by electrostatic fields, which enhance their kinetic energy and form them into a high-energy beam. Beam optics, comprising electric and magnetic focusing lenses along with a deflection system, are crucial for directing the beam toward a specific target on the substrate [66]. Traditionally, the beam is regarded as possessing a Gaussian profile, with the beam size defined as the full-width half maximum (FWHM) of the Gaussian distribution. The beam form is elliptical rather than circular; hence, it is more precisely characterized by the FWHM of a Gaussian distribution along the X and Y axes. In typical practice, the beam size and shape are improved by either focusing to create a sharp image of a resolution target or by performing a focusing process [67,68]. The effective generation and transmission of an electron beam to the substrate necessitate a high-vacuum environment.

Two primary techniques are employed for pattern generation in EBL: raster scanning and vector scanning. In a raster-scanning system, the exposing beam is directed in one direction at a consistent rate while the substrate is moved beneath it by a controlled stage. To form a patterned design, the electron beam is turned on and off millions of times during each scan, similar to the raster scanning method used in televisions [69]. Throughput is enhanced by the vector-scanning technique, which directs the exposure beam exclusively to the areas of the substrate that need exposure. A comparison between raster and vector scanning methods is shown in Figure 4 [70].

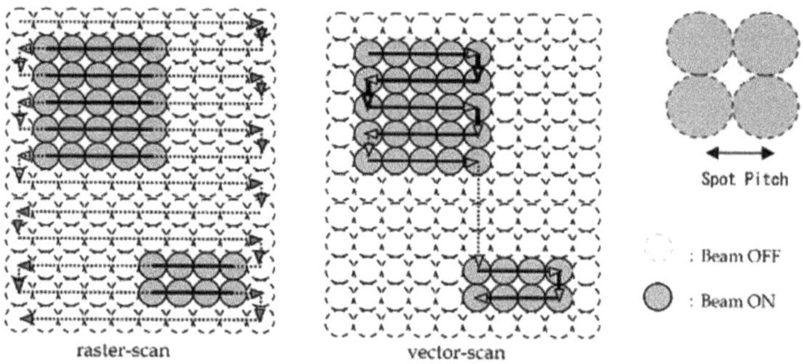

Figure 4. Comparison between raster and vector scanning methods in EBL. Copyright 2010, Cen Shawn Wu, Yoshiyuki Makiuchi and ChiiDong Chen. Originally published in [70] under https://creativecommons.org/licenses/by-nc-sa/3.0/ license (accessed on 20 January 2025). Available from DOI: 10.5772/8179.

The interaction between electrons and solid materials, such as electron beam resist, involves various scattering events. During this interaction, two types of scattering occur: small angle scattering, also known as forward-scattering, which causes an increase in the diameter of the initial beam, and large angle scattering, or back-scattering, which results in the proximity effect. This proximity effect arises when the dose received by a specific pattern feature is affected by electrons that scatter from nearby features [71]. Backscattering occurs as electrons penetrate through the resist and into the substrate. During these interactions, primary electrons lose energy, leading to the generation of secondary electrons with energies ranging from 2 to 50 eV [72].

The exposure process of the resist is primarily attributed to the interaction of electrons with the material. Since the range of these interactions in the resist is limited to a few nanometers, their contribution to the proximity effect is minimal. A shape is exposed by an electron beam writer through the discrete scanning of a focused electron beam across predefined primitives. Designers create CAD patterns that represent their devices.

A critical process in electron beam lithography (EBL) involves layout fracturing, which entails breaking down the layout pattern into multiple non-overlapping rectangles known as primitives. This decomposition is necessary before the pattern can be processed by an electron beam writer [73]. Understanding how a CAD pattern is converted into primitives allows for better control over the writing process. The exposure of a shape is achieved by scanning the focused electron beam across the pattern, resulting in a shot pattern that fills the shape. Typically, the spacing between each shot is kept smaller than the beam size, ensuring solid exposure of the shape. Each shape is exposed individually until the entire pattern is completed [74].

3.1. Electron Beam Lithography Resolution

The ultimate resolution of electron beam lithography (EBL) is influenced not by the capabilities of electron optical systems, which can achieve resolutions down to 0.1 nm, but rather by the characteristics of the resist and the subsequent fabrication process. It was previously thought that this limit was dictated by electron scattering within the resist and possibly by the properties of the resist itself. There are two main types of scattering that can be analyzed separately and represented using Gaussian distributions. The first phenomenon, known as forward-scattering, involves minor angle inelastic scattering of electrons as they penetrate the resist, characterized by a narrower distribution. For thin resists measuring less than 0.1 μm and under high accelerating voltages exceeding 100 kV, this width becomes negligible, around 1 nm.

The second phenomenon, referred to as back-scattering, involves significant angle elastic scattering primarily occurring in the substrate beneath the resist. In this case, electrons are reflected from the substrate back into the resist layer. The total exposure contributed by back-scattered electrons is roughly equivalent to that from incoming electrons but is spread over a diameter that corresponds to the electron range within the substrate [75]. Figure 5 [70] shows the Monte Carlo simulation of electron forward-scattering for different beam energies of 30 keV and 100 keV. It can be seen that forward-scattering is much higher at 30 keV than 100 keV. The resist sensitivity is observed to decrease with higher accelerating voltages, while lower accelerating voltages are advantageous from a throughput perspective. However, this comes at the cost of reduced resolution. An effective strategy to mitigate this limitation is the use of a thin imaging layer [76].

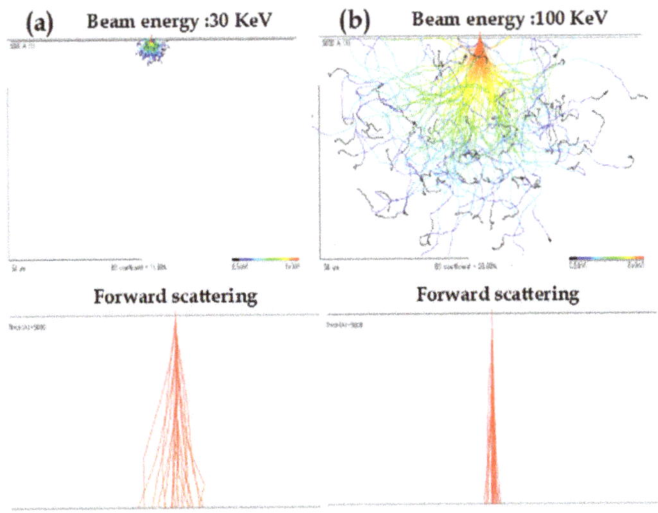

Figure 5. Monte Carlo simulation of electron forward-scattering for (**a**) 30 keV and (**b**) 100 keV beam energies. Forward scattering is higher at beam energy of 30 keV than 100 keV. Copyright 2010, Cen Shawn Wu, Yoshiyuki Makiuchi and ChiiDong Chen. Originally published in [70] under https://creativecommons.org/licenses/by-nc-sa/3.0/ license (accessed on 20 January 2025). Available from DOI: 10.5772/8179.

Using electron beam lithography (EBL), high-performance CMOS devices with a gate length of 36 nm have been successfully fabricated. EBL has demonstrated effectiveness in fabricating fully scaled 0.5 μm CMOS devices, as well as 0.25 μm NMOS and PMOS devices. Additionally, it has been utilized for the production of 100 nm T-gate GaAs PHEMTs and single electron tunneling (SET) devices, all known for their capabilities in high-density, low-power memory applications [77].

3.2. Electron Beam Lithography Resists

PMMA (polymethyl methacrylate) is the primary resist employed in electron beam lithography due to its excellent contrast and resolution. It can function as both a positive and negative resist throughout the fabrication process. When exposed to an electron beam, PMMA undergoes chain scission, leading to the formation of smaller fragments that are subsequently removed by developers such as diluted methyl isobutyl ketone (MIBK) and a mixture of isopropyl alcohol (IPA) and deionized water, thus acting as a positive tone resist.

If PMMA is subjected to high doses exceeding 10 mC/cm^2 at 30 keV, crosslinking occurs, resulting in a dense molecular network that exhibits resistance to developers and functions as a negative tone resist. Despite its advantages, PMMA has limited dry etch resistance, which can impede high-throughput applications. This resist has been extensively utilized in various applications, including electronic devices, nanoimprinting, photon sieves, masks, metal nanowires, and quantum wire production. Studies indicate that while PMMA remains the preferred choice for electron beam lithography due to its favorable properties, ZEP has emerged as a strong alternative [78].

Figure 6 [79] shows the contrast curves for PMMA for different resist thickness, different developer solutions, and different doses to clear the resist. Curve 1 shows an initial PMMA thickness of 300 nm, with a developer of H_2O:IPA 3:7 solution, with a dose to clear = 122 μC/cm^2; Curve 2 shows an initial PMMA thickness of 1134 nm, with a developer of H_2O:IPA 3:7 solution, with a dose to clear = 147 μC/cm^2; Curve 3 shows an initial PMMA thickness of 300 nm, with a developer of MIBK:IPA 1:3 solution, with a dose to clear = 157 μC/cm^2; Curve 4 shows an initial PMMA thickness of 1165 nm, with a developer of MIBK:IPA 1:3 solution, with a dose to clear = 182 μC/cm^2. Varius optical microscopy and SEM images of metal nanopatterns fabricated using EBL with different dose ranges are shown in Figure 7 [80].

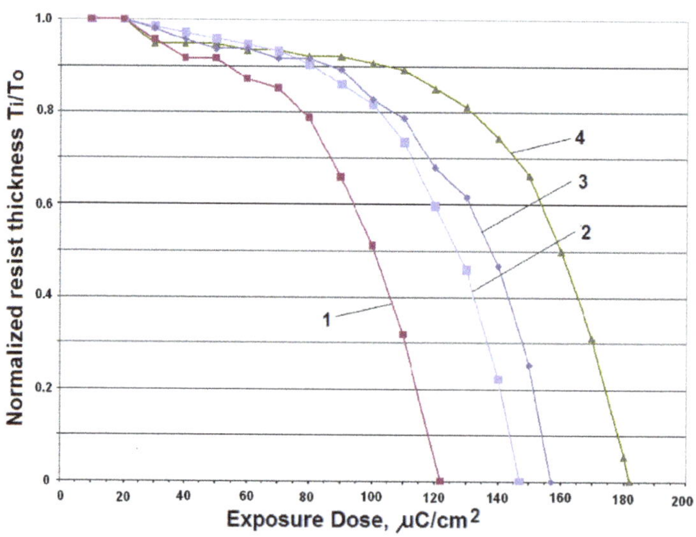

Figure 6. Different contrast curves for PMMA on silicon using EBL. Copyright 2018 IOP Publishing. Reproduced with permission from [79].

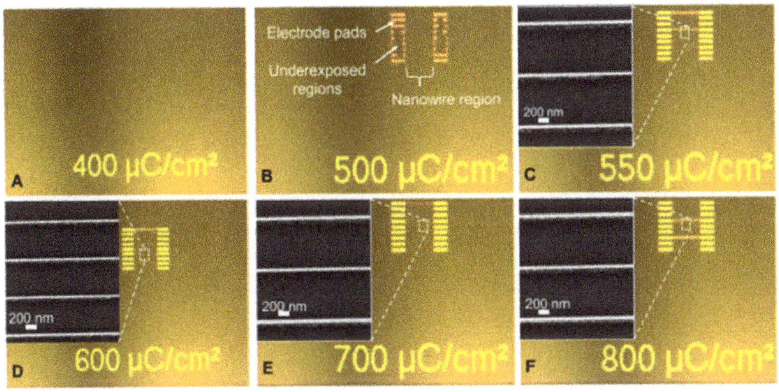

Figure 7. Optical microscopy and SEM images of metal nanopatterns fabricated with EBL with a dose range varying from (**A**) 400 μC/cm^2, (**B**) 500 μC/cm^2, (**C**) 550 μC/cm^2, (**D**) 600 μC/cm^2, (**E**) 700 μC/cm^2, (**F**) 800 μC/cm^2.

Chemically amplified resists (CARs) have become prominent as effective materials for lithography due to their high sensitivity and resolution. Typically, CARs are composed of an acid-reactive polymer, often an epoxy-based polymer, in combination with a photoacid generator (PAG). A newly developed PAG-bound polymer that incorporates glycidyl methacrylate (GMA), methyl methacrylate (MMA), and triphenyl sulfonium salt methacrylate (TPSMA) has been utilized as a negative tone resist in electron beam lithography (EBL) [81]. However, CARs encounter challenges related to the release of photoacid when exposed to an electron beam, which serves as a catalyst and influences sensitivity.

To overcome this limitation, a novel MAPDST–MMA copolymer that includes the sulfonium group has been developed as a non-chemically amplified negative resist (n-CAR) for advanced lithography applications. This innovative material exhibits exceptional sensitivity to electron beam radiation, achieving high-resolution features down to 20 nm. The MAPDST–MMA copolymer demonstrates impressive performance, achieving a resolution of 20 nm for a 1:2 line/space ratio and maintaining low line edge roughness (LER) between 1.8 and 2.3 nm. Furthermore, this resist material shows strong etch resistance against plasma etch chemistry [82].

Several new resist materials have been developed using radiation-sensitive sulfonium groups, with MAPDST serving as the primary ingredient. These materials were polymerized with methyl methacrylate (MMA), 4-carboxy styrene (STYCOOH), or N-vinylcarbazole (NVK) as co-units through a free-radical polymerization process [83]. Sodium PSS, or water-soluble poly (sodium 4-styrenesulfonate), can function as a negative electron beam blocker and is produced using water. Due to the presence of metal sodium, PSS exhibits significantly greater resistance to dry etching compared to PMMA. However, it is important to note that sodium PSS lacks the sensitivity and contrast of organic resists like PMMA, limiting its use in patterning thick or high-resolution structures. Despite these limitations, feature sizes for sparse patterns down to 40 nm have been achieved with this material [84].

Hydrogen silsesquioxane (HSQ) inorganic electron beam resists are recognized for their capability to achieve both high sensitivity and high resolution in lithographic applications. They demonstrate sensitivity levels comparable to chemically amplified resists (CARs) while providing excellent contrast. Features as small as 15 nm have been successfully written with a line width roughness (LWR) of 1.9 nm. Additionally, HSQ has shown to etch slowly compared to thermally grown SiO_2 during reactive ion etching [85]. When utilized in electron beam lithography (EBL), HSQ can reach resolutions of approximately

5 nm. Despite its effectiveness, the low sensitivity of HSQ often limits its application in various scenarios. While HSQ is considered one of the best EBL resists due to its performance, it is not as sensitive as some organic resists like PMMA.

Another category of resist includes water-based conducting polyanilines, which also fit within this group. Furthermore, eco-friendly options such as biomaterials derived from silk fibers have been explored as potential EBL resists [86]. Using 3D EBL techniques, functional and arbitrary 3D nanostructures have been fabricated with precision below 15 nm using only water as a medium. The innovative approach of genetically engineering recombinant spider silk proteins as a resist allows for the creation of any 3D object at the nanoscale with high resolution and strength. This method enables the precise shaping of polymorphic spider silk proteins at nearly the molecular level by quantitatively describing structural transitions induced by energetic electrons at varying depths within the 3D protein matrix [87]. Table 1 shows advanced in resist materials and improvement in resolution with EBL.

Table 1. Advanced resists for electron beam lithography.

Resist	Resist Type	Developer Solution	Sensitivity ($\mu C/cm^2$)	Contrast	Resolution (nm)	Ref.
40XT	Positive	PEDOT: PSS (0% dilution for 5 s)	8	8 ± 2	95 ± 10	[88]
40XT	Positive	PEDOT: PSS (33% dilution for 40 s)	9	10 ± 0.5	85 ± 5	[88]
$ZircSO_x$	Negative	TMAH for 4 min	7.6	2.6	<10	[85]
$HafSO_x$	Negative	TMAH for 1.5 min	21	2.5	<10	[85]
ma-N 2400	Negative	ma-D 532	60–160	1.7–2.8	<30	[89]
AR-N 7700	Negative	AR 300-46	8	>5	40–100	[89]
P(HEMA-co-MAAEMA)	Negative	methanol	0.5	1.2	100–200	[89]
Poly (GMA-co-MMA-co-TPSMA)	Negative	7:3 IPA/DI water solution	120		15	[81]

3.3. *Advantages and Limitations*

A focused electron beam serves as the most precise writing tool available, capable of creating pattern features as small as a few nanometers. Unlike traditional photolithography, electron beam lithography (EBL) does not require a pre-existing pattern mask; it can directly write patterns based on stored data. EBL remains the preferred method for producing nanometer-scale objects in small quantities due to its high resolution and flexibility in pattern modification [90]. This technique surpasses traditional photolithography, which is limited by diffraction effects that restrict spatial precision to a few hundred nanometers. Although EBL does have a diffraction limit, the high-energy electrons utilized in various studies enable a diffraction limit in the nanometer or even picometer range. Research has demonstrated that EBL can achieve isolated feature sizes down to 2 nm and half-pitch dimensions of 5 nm using advanced materials like hydrogen silsesquioxane (HSQ) resist. The resolution capabilities of EBL are primarily constrained by factors such as electron scattering within the resist and imperfections in the electron optics, rather than by the size of the electron beam itself [91].

The biggest task for electron beam lithography is to increase throughput. A single fine beam has struggled to achieve both higher sharpness and higher throughput simultaneously. While optical lithography can design a wafer in minutes, several tens of hours are required for EBL to accomplish the same task due to its serial process. To address this issue, variable shaped beam systems and character/cell projection methods have been developed. However, the throughput of the most advanced systems has not yet reached the necessary levels [76]. As discussed in earlier sections, electron scattering (both forward- and back-scattering) limits the resolution of EBL. Primary electrons are more likely to move forward when the acceleration voltage of the electron beam is low, as lower voltages enhance forward-scattering. The proximity effect, caused by back-scattered electrons that lead to increased exposure, further deteriorates clarity. Additionally, secondary electrons with energies ranging from 2 to 50 eV contribute to reduced resolution. These challenges make EBL less practical compared to other methods. To circumvent these issues, new next-generation lithographic techniques such as X-rays and ion-beam lithography are being developed.

4. X-Ray Lithography

X-ray lithography (XRL) was developed in the early 1970s to enhance the resolution of existing lithography techniques used in the semiconductor industry. At that time, UV projection lithography was the leading technology, capable of resolving features down to one micron. However, as the microelectronics industry aimed for smaller sizes, specifically around 250 nm, UV lithography was perceived as having limited future prospects due to its inability to overcome the physical limits imposed by diffraction. To align with Moore's Law and the International Technology Roadmap for Semiconductors (ITRS), there was a clear need for radiation with shorter wavelengths to facilitate further miniaturization of devices [92].

H.I. Smith and Spears from MIT Lincoln Labs were among the first to propose X-ray lithography (XRL) in the 1970s [93]. Following this, Bell Laboratories and various U.S. industrial organizations developed X-ray systems utilizing palladium (Pd) targets [94]. The wavelengths of radiation employed in XRL range from 0.4 nm to 5 nm. XRL is also the initial step in the LIGA process, which involves placing metal electrodes on the created resist structure to form a mold or electrode for subsequent replication processes, including electro discharge machining or molding.

There are two categories of X-rays: soft and hard. Soft X-rays have energy levels between 150 eV and approximately 2 keV, while hard X-rays possess energies exceeding 5 keV. Soft X-rays are effective for high-resolution designs requiring less resist thickness. Deep X-ray lithography (DXRL) is frequently utilized in the LIGA process and for heating thick resists [95]. A schematic diagram of a LIGA X-ray beamline is presented in Figure 8 [96].

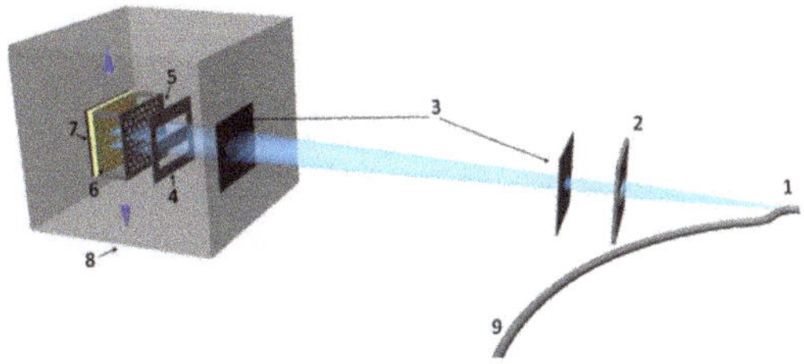

Figure 8. Schematic diagram of a LIGA X-ray beamline. 1: radiation area; 2: front-end absorber. 3: beryllium windows; 4: beam stop and spectral filter; 5: X-ray mask; 6: substrate; 7: motorized translation stage; 8: exposure chamber; 9: electron beam orbit. Copyright 2021 IOP Publishing. Reproduced with permission from [96]. All rights reserved.

Various applications of X-ray lithography (XRL) include IBM's fabrication of 64 MB DRAMs utilizing XRL for the gate level, Mitsubishi's development of a 1 GB DRAM test site featuring 0.14 μm gates and a stacked capacitor with a capacitance exceeding 25 fF/cell, and a collaboration between Toshiba and NTT to create a 4 GB DRAM test site with functional cells at a pitch of 0.24 μm. Additionally, IBM has produced CMOS logic devices, including a fully functional 64 KB SRAM with 0.2 μm features, while NTT has fabricated CMOS logic devices featuring a 12 KB SRAM and a 48 × 48 bit multiplier, also with 0.2 μm features. Motorola has developed CMOS logic devices, including a fully functional 1 MB SRAM, using three levels of XRL with feature sizes of 0.375 μm. Furthermore, IBM has fabricated CMOS test circuits, including ring oscillators with feature sizes of 0.1 μm [97]. Silverman et al. [97] successfully created a polysilicon gate measuring 50 nm by etching a resist line that was 70 nm wide.

X-ray lithography (XRL) involves transferring patterns to a wafer substrate using a mask and a highly collimated X-ray beam. An X-ray radiation system comprises several components: a high-power X-ray source, a beamline, a mask, a substrate coated with resist, and additional exposure elements. Key factors that contribute to the reliability of an XRL system include resolution, critical dimension (CD) control, and overlay accuracy. Historically, X-ray tubes served as sources of X-ray radiation; however, their large spot size and lower beam intensity rendered them inadequate for advanced technological requirements. To address these limitations, plasma sources were introduced, offering higher radiation intensity with smaller spot sizes and beam diameters [98,99]. In the late 1970s, the Naval Research Laboratory conducted experiments demonstrating the effectiveness of plasma-generated X-ray sources for lithographic applications [100].

There are two primary types of plasma sources used in this field: laser-induced plasma sources and discharge plasma sources [101,102]. While discharge plasma sources provide significantly higher radiation intensity, laser plasma sources have lower intensity levels. At that time, the absence of affordable, high-power lasers led to the widespread use of discharge plasma sources. However, the adoption of plasma sources for X-ray radiation has diminished due to challenges such as electrode erosion and damage to masks and wafers caused by these sources [52].

Recent advancements in X-ray sources have led to the development of synchrotron radiation sources, which generate high-intensity X-rays using magnetic storage rings. In these rings, electrons emit a highly collimated beam of synchrotron radiation, achieving a maximum depth of focus of approximately 0.3 mm [103]. The initial lithography experiments utilizing synchrotron radiation were conducted in 1974 by Spiller and Feder [104]. By 1980, the X-ray resonator at IBM Yorktown was exploring the potential of using storage rings as X-ray sources.

Over several years, IBM established an X-ray lithography (XRL)-dedicated beamline at Brookhaven National Laboratory. In 1991, IBM constructed an Advanced Lithography Facility in East Fishkill, New York, to accommodate the HELIOS superconducting storage ring, which was provided by Oxford Instruments. The ring operated with a small injector at approximately 200 MeV and reached 686 MeV at 165 mA. With standard resist materials, the X-ray flux from HELIOS, which had an effective wavelength of around 0.8 nm, could undergo a second exposure. Throughout the 1990s, IBM expanded its X-ray initiatives, including the production of high-resolution DRAM chips that showcased the capabilities of XRL [105].

The radiation produced by a synchrotron radiation source travels through an ultrahigh vacuum beamline [106]. Given that the length of a typical beamline can extend to several tens of meters, electromagnetic waves can be approximated as planar at the surface of the X-ray mask membrane. Throughout the beamline, the radiation interacts with various optical elements arranged in distinct configurations. The beam passes through a series of foils made from different materials and strikes one or more mirrors, which are crucial for selecting the appropriate wavelength spectrum before reaching the mask [107]. This process occurs while accounting for the unavoidable scattering of X-rays caused by residual gas molecules present in the beamline [92]. To enhance X-ray diffraction, a stepped attenuator is employed, which increases in thickness as the distance from the optical axis of the X-ray lens grows. This attenuator, made from a 100 μm thick polyimide sheet, is placed in front of the X-ray mask. By varying the dose through attenuators of different thicknesses and splitting the X-ray beam used for structuring, new properties can be imparted to the resulting microstructures [108].

4.1. X-Ray Lithography Resolution

The creation of a photoelectron, which possesses energy equal to the X-ray energy minus the binding energy of the electron to its atom, represents the fundamental interaction that occurs when a low-energy X-ray is absorbed by a solid. During the resist process, over ninety percent of the emitted photoelectrons originate from an inner shell, with their binding energy typically in the range of several hundred volts. After the photoelectron is ejected, the atom remains in an excited state. This excited state can decay either by emitting another photon, known as fluorescence, or by releasing an Auger electron [109]. Historically, it was believed that the resolution of X-ray lithography was constrained by the maximum range of the photoelectrons and Auger electrons released upon X-ray absorption [110].

When employing X-ray lithography (XRL), the typical resolution achieved is around 15 nm, with minimal false scattering occurring in the process. Figure 9 [96] illustrates SEM images of channels fabricated using XRL in polyethylene terephthalate (PET), a material with low sensitivity. In the images, Figure 9a,b exhibit aspect ratios of approximately 7 and 20, respectively. In lithography, particles such as ultraviolet photons and electrons often experience significant spurious scattering, resulting in unintended exposure in areas that should remain unaffected [111]. Soft X-rays, particularly those in the energy range of 140 to 500 eV, are favored because its mask is easier to fabricate compared to hard X-rays. Hard X-rays need thick metal to block it, and it is difficult to make such a mask with very narrow line width through a very thick metal film. The application of soft XRL is suitable for achieving high-resolution structures. Soft X-rays, particularly those in the energy range of 140 to 500 eV, are favored as they provide a satisfactory balance between diffraction-limited resolution and resolution limited by photoelectron blur and penumbral effects [95].

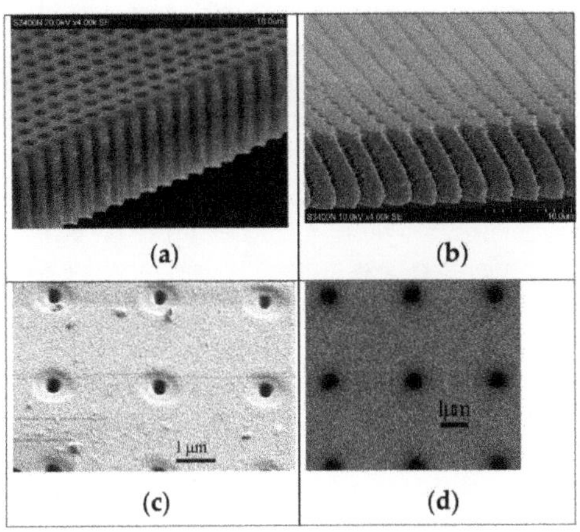

Figure 9. (**a**–**c**) SEM images of channels fabricated in PET film using X-ray lithography and (**d**) X-ray mask. Copyright 2021 IOP Publishing. Reproduced with permission from [96]. All rights reserved.

4.2. X-Ray Lithography Resists

To successfully transfer patterns using X-ray lithography, a radiation-sensitive organic substance known as resist is required. To achieve optimal performance in the X-ray lithography (XRL) process, the resist material must possess several key characteristics, including submicron resolution of less than 0.1 μm, a sufficient aspect ratio, excellent thickness uniformity, thermal stability, minimal defects, strong contrast, and tolerance to dosage variations. The sensitivity of positive tone/chain scission-type resists to X-rays is relatively low, typically falling between 100 and 2000 mJ/cm^2 at a wavelength of 8 Å [112]. In contrast, negative resists have been developed that exhibit sensitivities around 10 mJ/cm^2 at a wavelength of 4.37 Å [112,113]. A comparatively thin layer of resist is required to obtain a consistent X-ray exposure through the resist thickness. This is because the X-ray absorption is essentially linear with thickness when the resist layer is relatively thin. Certain operations necessitate the presence of resist layers that are as thick as 3 μm. The need for uniform absorption in this scenario is a relatively short wavelength, which is less than 10 Å. To enable operation at longer wavelengths, it is essential to utilize a thin resist layer along with one or more layers of etch-resistant thin films [114]. Resist sensitivity is proportional to absorption coefficient α, higher absorption results in higher sensitivity, and absorption coefficient α depends greatly on wavelength.

X-ray lithography and electron beam lithography are two methods that are extremely similar in terms of how they expose polymer films. In both instances, the energy of the particles when they collide with one another is significantly greater than the energy that is necessary to either make or break a chemical bond. The interaction between X-rays and the resist is significantly less powerful than that of electrons. Following the absorption of an X-ray photon, a shower of secondary electrons is produced. These electrons carry the majority of the energy from the Incident photon, leading to significantly stronger interactions with the resist material. It can be inferred that secondary electrons are primarily responsible for the chemical transformations occurring within the resist. Essentially, an X-ray exposure of a resist can be viewed as an electron exposure. The key distinction between an X-ray exposure system and an electron beam exposure system lies in the energy levels: incident electrons in an electron beam system typically range from 10 to 50 keV, while secondary electrons generated by soft X-ray exposure possess much lower energies. Consequently, any electron beam resist can also function as an X-ray resist [115].

Polymethyl methacrylate (PMMA) is widely recognized as one of the most effective resists for X-ray lithography due to its ability to achieve the highest resolution among all resists, reaching down to 100 Å. However, PMMA's sensitivity is relatively low, around 1000 mJ/cm^2, and it exhibits poor etch stability. Figure 10 illustrates cross-sectional SEM images of cone- and pyramid-shaped silicon absorbers that were transferred to a PMMA structure using X-ray lithography [116]. The IBM Research XRL program successfully utilized novolak positive resist, achieving linewidth control of 7 nm on device wafers at a dosage of 700 mJ/cm^2. Most device processing has been confirmed to be compatible with this resist material [117].

It is highly advantageous for future lithography technologies to utilize standard novolak-based resists (dissolution inhibitor/matrix resin) due to their demonstrated stability across various applications. Unfortunately, when synchrotron radiation sources are processed similarly to optical lithography, their sensitivity to X-rays falls short by at least a factor of ten. However, if the processing of the resist, especially during the development phase, is optimized for X-ray lithography, certain optical novolak/diazotype resists can yield favorable results regarding resolution and stability, even with increased sensitivity. The development step plays a crucial role in this optimization. It was determined that HPR 204 and HUNT WX 214 produced the most favorable outcomes in terms of resolution and stability. Additional chemical amplification was implemented later in order to further enhance the sensitivity of novolak resists [118].

Figure 10. Cross-sectional SEM images of cone (**a**,**c**)- and pyramid (**b**,**d**)-shaped Si absorbers transferred to a PMMA structure using XRL. Copyright 2015 MDPI. Reproduced with permission from [116].

The formulation of a dual tone resist that was devised based on acid-catalyzed deprotection was accomplished by combining poly (t-butoxycarbonyloxystyrene) (PBOCST) with triphenylsulfonium hexafiuoroantimonate (15 weight percent of the total solid) in Arcosolv PM Acetate. Irradiation of the onium salt results in the production of strong Bronsted acids, which are responsible for catalyzing the transformation of the lipophilic polyhydroxystyrene (PBOCST) into the hydrophilic (even acidic) polyhydroxystyrene (PHOST).

At the lithography beamline of the Stanford Synchrotron Radiation Laboratory (SSRL), where the incident dosage necessary for achieving a final thickness of one hundred percent was thirteen millijoules per square centimeter, the sensitivity characterization was carried out. In the range of 15 to 20 percent, the resist contrast that was obtained was extremely high [119]. Further down the line, excellent results were obtained by various research groups in a number of commercially available resists, such as Apex-E, UVII-HS, UV-4, TDUR-N908, and SAL601 [97]. Different types of X-ray lithography resists and their resolution are shown in Table 2.

Table 2. Types of resists for X-Ray Lithography.

Resist	Resist Type	Incident Dose (mJ/cm^2)	Resolution (nm)	Wavelength (nm)	Ref.
PMMA	Positive	500	5	0.1–10	[115]
PMMA	Positive	330	35	0.1–10	[115]
CoP (MMA MAA)	Positive	50	<50	1–5	[115]
TIP(MMA-MAA)	Positive	24	<50	1–5	[120]
DQN	Positive	1000	50	0.4–4	[120]
PBOCST	Positive	5	50	0.4–4	[121]
DCIPA	Negative	7.8	500	0.5–5	[121]
Epoxidized polybutadiene	Negative	1.5	1000	0.5–5	[121]

4.3. Advantages and Limitations

When applied to thick, single-layer resists with a high aspect ratio and depth of focus, X-ray lithography (XRL) can pattern features with resolutions of 15 nm or less while maintaining a high level of fidelity. The wavelength of X-rays is under 10 nm, resulting in a relatively modest diffraction limit. This characteristic allows XRL to circumvent the resolution limitations posed by diffraction that are encountered in EUV lithography. Additionally, XRL benefits from a substantial number of experienced vendors who have developed XRL tools and possess a deep understanding of the technology, its challenges, and the necessary tooling requirements, along with a considerable base of knowledgeable end users [122].

However, mask stability presents a significant challenge for XRL systems, particularly concerning radiation damage that accumulates over the mask's operational life.

Recent studies indicate that under extreme conditions, distortions of around 20 nm can be observed. The masks used in XRL are particularly vulnerable to damage or bending due to their thinness [123]. Concerns regarding device radiation damage during X-ray lithography have also been raised by researchers in this area. While X-ray synchrotron sources can generate high-power X-rays, they come with substantial maintenance costs and physical space requirements [124]. Consequently, other advanced lithographic techniques are being explored.

5. Ion Beam Lithography

Ion beam lithography (IBL) is an advanced development in the field of nanofabrication due to its potential to achieve high-resolution patterning beyond the capabilities of traditional optical lithography. One of the key advantages of IBL is its ability to bypass the diffraction limit, which has constrained optical methods. The ability to produce patterns smaller than 100 nm marks a significant leap forward.

However, IBL faces challenges similar to those encountered in electron beam lithography, primarily the proximity effect but only for very light ions [125]. The lateral scattering of ions and the secondary electrons they generate can cause the beam to spread, reducing the precision of the patterning. This limits the ability to create very fine features with high resolution. To overcome these challenges, ongoing research focuses on improving ion beam control, resist materials, and techniques to minimize scattering and improve pattern fidelity. This in turn limits the spreading of exposure features in a resist to less than 10 nm [126].

In most cases, the range of light ions in resist in the energy domain spanning from 50 to 150 keV is comparable to resist thickness. Ions are therefore effective at exposing resist, and the minimum dose required for exposure is between 10^{12} and 10^{13} ions per square centimeter. The range of light ions like H^+ or He^+ is typically very predictable and straight because of their relatively low mass and charge, and hence there is minimal lateral scattering. A schematic diagram of a helium in beam system utilizing a gas field ion source is shown in Figure 11 [127]. It is because of this that the dose at a particular location does not depend on the exposure of areas that are adjacent to it; in other words, there is not much of a proximity effect [128,129]. As a result of the fact that ion/electron collisions are the primary interaction for rapid light ions, it is necessary for an ion to undergo many collisions before it finally arrives to a state of permanent rest.

The major method of contact for ions is the transfer of momentum between the slow ions and the surface atoms. This results in the physical sputtering of the material, which causes the atoms to reorganize themselves and brings about structural changes. There are now three unique and independent ion beam techniques that can create structures. These methods are as follows: (1) Focused ion beam (FIB), which uses a finely focused ion beam (typically heavy ions like gallium at 30 keV) to locally sputter or deposit material on a surface and finds its use in microfabrication, milling, and modification of surfaces; (2) Proton beam writing (PBW), in which fast protons (generally MeV) are used to write precise, deep, three-dimensional patterns into a resist material; (3) Ion projection lithography (IPL), which involves projecting medium-energy ions (around 100 keV) through a patterned mask onto a surface, typically used for rapid production of microstructures [130].

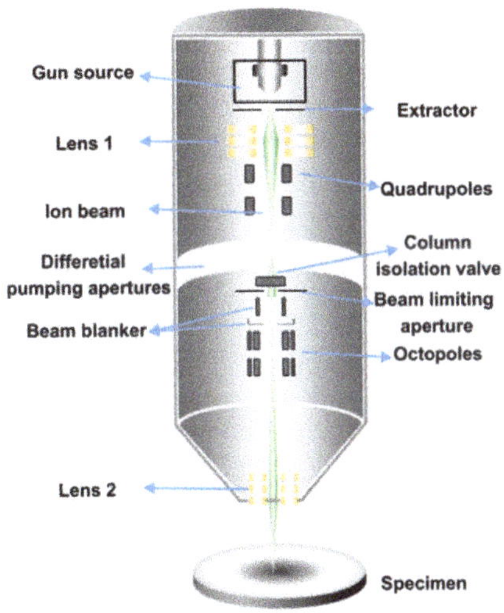

Figure 11. Schematic diagram of a helium ion beam system utilizing gas field ion source. Copyright 2021 IOP Publishing. Reproduced with permission from [127]. All rights reserved.

5.1. Focused Ion Beam (FIB)

The introduction of focused ion beam (FIB) technology dates to the 1970s, when the source of ions was a standard implanter, and the beam diameter that was attained was in the range of 3 μm. In the time since then, it has become the most advanced and commonly utilized technique for ion beam lithography. There are three primary motivating factors that contribute to the effectiveness of FIB techniques in nanofabrication. The direct write approach, which combines milling, depositing, etching, and other processes into a single piece of equipment, is the first method that simplifies the sample preparation process and reduces total throughput. The ability to create 3D structures with a high aspect ratio is another significant benefit, especially in fields like microelectronics, materials science, and nanotechnology.

The high precision and control offered by FIB allow for detailed and intricate designs at the nanoscale, making it invaluable for things like device prototyping, failure analysis, and even sample preparation for tunneling electron microscopy (TEM). Finally, the nanoscale interaction volume is crucial in understanding how the ion beam interacts with the material. This volume typically ranges to few tens of nanometers for 5–50 keV gallium ions [131]. The ions induce various processes such as sputtering, deposition, and even changes in the material's atomic structure within that interaction volume. This makes FIB not just a tool for material removal or structuring, but also one that can provide insight into the material properties at the microscopic and nanoscopic levels.

Patterns may be created in nearly any material using FIB, which is a unique feature in comparison to other lithography processes, which are restricted to patterning resist materials due to their limitations. On the other hand, the process is relatively sluggish, and a gallium ion source with a power of 30 keV can remove between one and ten atoms for every incident ion. Both the electrons and the atoms absorb energy from the incident ion, which causes the ion to lose energy. Due to Coulomb contact, the electrons in the material are excited to transition into bound states or continuum states when the ion passes through. On the other hand, the loss of atoms is partially attributable to a limited number of random collisions, which results in a large loss of energy during the event. Due to this, the ion is diverted away from its intended path, and the atom in the solid is moved away from its position in the lattice [132].

There has been a significant amount of progress made in the creation of ion sources for the operation of FIB. Earlier FIB instruments had a low current density and a low brightness of the ion sources, which resulted in poor efficiency and a limited application range. The liquid metal ion source has been the most popular option of FIB ion source ever since the introduction of the liquid gallium metal ion source. This source generates a brightness of around 10^6 $Acm^{-2}Sr$, making it the most preferred choice.

In addition, new ion sources are being developed, including gold, silicon, germanium, and other liquid ion sources. These new ion sources have the potential to improve the long-term stability of ion beams thanks to their characteristics. In addition, various ion sources, such as liquid metal alloy ion sources (LMAISs), have been utilized for specialized applications, such as high-resolution FIB implantation. These ion sources include AuSi, AuSiB, PdAs, PdAsB, NiB, and NiAs. Investigations are also being conducted on additional promising LMAIS candidates such as $Ga_{35}Bi_{60}Li_5$ and $Co_{31}Nd_{64}B_5$ [133].

The earlier FIBs that were integrated with an inductively coupled plasma (ICP) ion source could perform large-volume milling operations at a removal rate that was approximately one hundred times quicker than that of a Ga^+ source. This was achieved by delivering up to two microamperes of Xe^+ ions that were concentrated into a sub-five micrometer spot. Nevertheless, the bigger spot size that is produced by a plasma FIB source results in a reduction in the quality of the cross-sectional area, and the beam currents are not sufficiently high to satisfy the requirements of the advanced level [134].

Recent developments in the field of nanofabrication have been made possible by gaseous field ion sources (GFISs), which are capable of producing minuscule beam currents of less than 2 pA and producing a spot size of approximately 0.35 nm. For instance, with the advanced He^+/Ne^+ gas field ion source microscope resolutions can reach 0.25 nm, and feature sizes can be less than 5 nm, pushing the limits of nanofabrication. Although Ga^+ ions are still used, their maximum penetration depth is limited to about 80 nm. Furthermore, their trajectory is no longer as directed downward, and they experience more back-scattering, which may affect the quality of the process.

When compared to gallium metal ions, the He^+ beam has a significantly smaller spot diameter than a Ga^+ beam, sometimes as small as 0.3 nm, which allows for higher-resolution imaging. This is important in nanoscale characterization because it enables finer details to be resolved. The larger penetration depth of He^+ ions in PMMA (polymethyl methacrylate) can be advantageous for imaging or material modification, as it can interact deeper within the sample compared to Ga^+ ions, which may have shallower penetration. Also, He^+ ions do not contribute to any contamination [133].

Recently, FIB was used for the bit milling of 2D and 3D nanostructures [135]. FIB-milled structures consisting of 40 nm thick segmented gold islands for surface plasmon polaritons and for resonant plasmons was demonstrated [136]. Through the use of FIB milling, regular and periodic lines measuring 200 and 300 nm in wavelength have been milled at 70 pA with a high degree of fidelity in PDMS structures. Another application of FIB was the fabrication of a field emitter tip in a diamond crystal, which had a radius of less than 100 nm [137]. Mosberg et al. [138] demonstrated the growth of self-catalyzed GaAsSb nanowires using different FIB patterning conditions as shown in Figure 12 [138]. In addition to ion implantation of doping species in semiconductors and targeted ion etching on a submicron scale, the other significant applications of FIB technology are scanning ion microscopy (SIM) through secondary ion mass spectroscopy (SIMS). There are now other applications of FIB technology that are being developed.

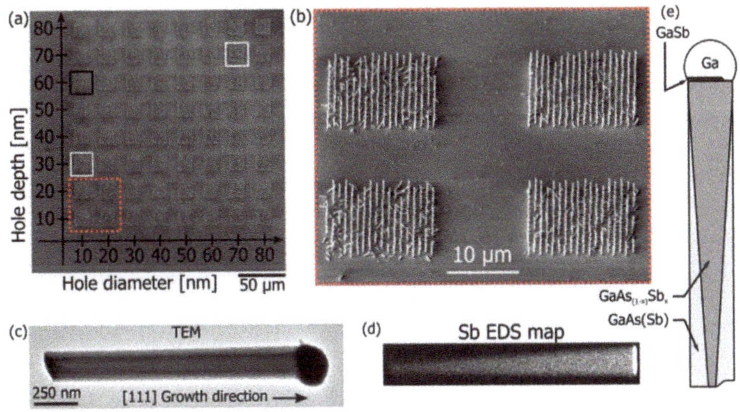

Figure 12. GaAsSb nanowires grown by varying different FIB conditions. (**a**) 8 × 8 hole depth -diameter matrix; (**b**) arrays of 18 × 15 nanowires; (**c**) TEM of broken defect-free nanowire with (**d**) corresponding Sb EDS map; (**e**) schematic of nanowire features including GaSb crystal in Ga droplet. Copyright 2023 IOP Publishing. Reproduced with permission from [138]. All rights reserved.

5.2. Proton Beam Writing

The Centre for Ion Beam Applications (CIBA) at the National University of Singapore is responsible for the development of a cutting-edge nano-lithographic technique known as proton beam writing (PBW), which is a technology that allows for direct writing in three dimensions. The use of accelerated protons (in the energy range of MeV), rather than electrons (in the energy range of keV), is a key aspect of PBW. Protons can penetrate deeper into materials like PMMA or SU-8 with a more defined path, making them a great choice for creating high-resolution, 3D microstructures as opposed to electrons in this approach [139].

In a single process, PBW can create well-defined structures with smooth and vertical walls, which is crucial for maintaining dimensional accuracy in high-resolution fabrication. PBW can produce features with a high aspect ratio and with dimensions smaller than 100 nm. This makes it a promising method for fabricating ultra-fine, complex structures. The penetration depth of the proton beam in the resist material (like PMMA) is dependent on the energy of the protons. A 1 MeV proton penetrates 20 μm deep, while a 3.5 MeV proton can go much deeper, up to 160 μm. This gives flexibility in controlling the depth of the exposure, which is important when creating structures with different depths or for writing multilayer devices [140].

The large mass difference between protons and electrons plays a significant role in why protons travel relatively undisturbed compared to lighter particles like electrons. Since the proton is much more massive, it retains most of its momentum during collisions with electrons, which results in minimal deflection or change in the proton's trajectory. The concept of electronic stopping is key here—it refers to the energy loss due to collisions between the proton and the electrons in the material. However, because the proton has so much more mass and momentum, these interactions do not result in major changes to its path. Despite radiation damage causing secondary electrons to be produced, these electrons stay near the proton beam's axis, leading to a less pronounced proximity effect than in electron-based lithography [141].

This property of protons is one reason why proton beams are commonly used in proton therapy for cancer treatment. Their ability to deliver high doses of energy with minimal scattering allows for precise targeting of tumors while minimizing damage to surrounding healthy tissue. Using 2 MeV proton beam nickel pillars of 400 nm and 5 μm have been successfully fabricated using PMMA as a resist [142]. It has been established that 3D nickel stamps may be created by utilizing PBW writing in conjunction with nickel electroplating. These stamps feature a nickel wall that is 100 nm broad and 2 μm high, which corresponds to an aspect ratio of 20. Manufacturing of high-aspect-ratio silicon needles with nanometer-sized tips has been accomplished with the use of PBW [143]. High-aspect-ratio test structures have been fabricated using PBW in SU-8 negative resist, demonstrating features of 60 nm with a depth of 10 μm [130].

Metallic stamps of aspect ratio 20:1 (100 nm wide and 2 μm deep) were successfully fabricated in CIBA using PBW [139]. A high-brightness nano-aperture electron impact gas ion source (NAIS) which can generate ions with high spatial resolution from a small virtual source of 100 nm in an ionization chamber was experimentally examined by Liu et.al. [144]. Figure 13 [145] shows arrays of silicon pillars with a periodicity of 2 μm and 1.2 μm, respectively. It was possible to generate an axial Ar ion beam current of approximately 50 pA by employing an angular beam current density of 6.6×10^5 A/Sr on the beam. According to the report, the greatest brightness was measured to be 750 Am^{-2} Sr, while the ion source energy spread was around 1 eV. Sub-10 nm H^{2+}/H^{+} ion lithography is becoming increasingly possible with the present PBW; however, it is restricted by the low RF ion source brightness [144].

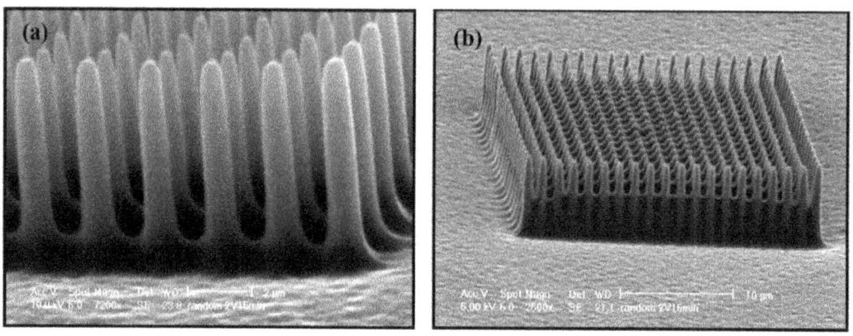

Figure 13. Arrays of silicon pillars with (**a**) 2 μm and (**b**) 1.2 μm periodicity fabricated by PBW. Copyright 2006 IAEA. Reproduced with permission from [145].

Nanoimprinting, silicon machining, medicinal applications, and photonics are some of the areas in which PBW is utilized, highlighting its potential for high-precision patterning in advanced manufacturing processes. PBW stands out due to its ability to create high-aspect-ratio structures with minimal proximity effects, which is a common challenge with techniques like electron beam lithography (EBL). Its use as an etch stop for creating porous silicon and its potential in biosensor and photonic device fabrication (like waveguides and microlens arrays) are also exciting, as these applications demand extreme precision at very small scales. The fact that PBW is still relatively new, with limited commercial instruments available, suggests there is a lot of room for future development. The possibility of reducing proton spot diameters to under 30 nm opens up exciting opportunities for 3D direct writing, potentially revolutionizing the fabrication of intricate micro- and nanoscale device [140].

5.3. Ion Projection Lithography

Ion projection lithography (IPL) offers a promising alternative or complement to current lithography techniques used in semiconductor manufacturing. By using ions such as protons, H_2^+, He^+, and Ar^+, IPL allows for finer patterning and the potential to continue scaling down feature sizes on chips, especially as traditional photolithography approaches its resolution limits. One of the main advantages of using ions instead of light photons in IPL is the precision with which the ions interact with the resist material. The light ions, due to their lower forward-scattering, can deliver a more defined dose to the resist without significant diffusion [146]. This means the printed features are more precise, and there is less risk of distortion, which is crucial when working at extremely small scales.

Additionally, IPL can operate in the 50 to 150 keV energy range, which is suitable for achieving high resolution. The ability to project these ions through a mask pattern and onto a substrate offers advantages in terms of depth of focus and stability in producing fine details, especially when compared to traditional optical lithography, which suffers from diffraction limits. The technique's ability to work with medium-energy ions also means there is less energy transferred to secondary electrons, which could otherwise cause unwanted effects like proximity bias or pattern distortions. These properties make IPL particularly suited for producing advanced semiconductor devices with small geometries while maintaining high fidelity and low defect rates. Figure 14 [147] shows TEM images of cross-sections of silicon substrates exposed with He ion beam energy of 25 keV and various doses. A study by Melngailis et al. [128] demonstrates patterning down to 70 nm line-space pairs, along with innovations like the pattern lock servo system to address drift. Additionally, field distortion of less than 0.15 μm over 8 × 8 mm is measured, which agrees with the calculations.

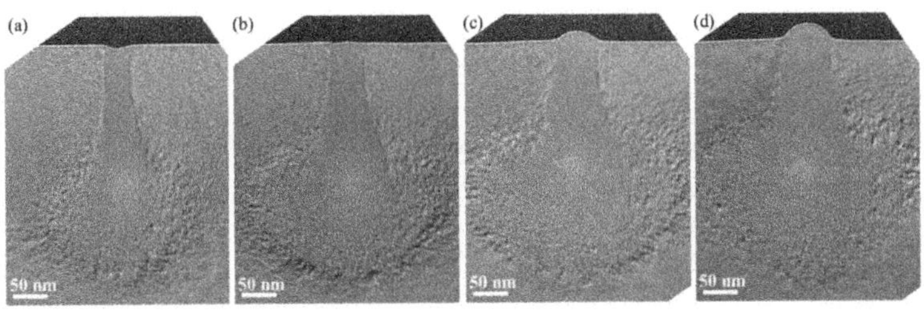

Figure 14. TEM images of silicon substrate cross-sections treated with He ion beam energy of 25 keV and ion doses of (**a**) 0.02 nC/μm; (**b**) 0.03 nC/μm; (**c**) 0.04 nC/μm; (**d**) 0.05 nC/μm. Copyright 2020 MDPI. Reproduced with permission from [147].

On the basis of these accomplishments, a new generation of ion lithography machine has been built. This machine utilizes three times the demagnification and exposes a field measuring 20 mm by 20 mm with minimum dimensions of 0.12 μm. Furthermore, the ion optics introduce less than 10 nm of distortion [148]. These systems have demonstrated a significant number of the characteristics that are necessary for high-throughput lithography. With a single beam exposure, the ability to achieve a resolution of 50 nm has been demonstrated in certain areas of the exposure field and 75 nm across the entire field, which is 12.5 × 12.5 mm^2 [149].

To perform ion projection lithography (IPL), it is necessary to have an ion source that possesses beam properties that are adequate. A small virtual source size (<10 μm) with a high emission current density (>0.5 A/cm^2) and a finite energy spread (~1–2 eV) is the ideal ion source for intense pulsed light (IPL) [150,151]. This characteristic is referred to as the figure of merit. IPL systems use volume-plasma sources, and this process can be carried out with or without demagnification. In ion optical systems, chromatic aberration arises when ions with different energies do not focus on the same point in space. This means that ions with slightly different velocities have different focal points, leading to a blurred or distorted pattern on the substrate. This energy spread can distort the focused image, making it less sharp and precise, which is a significant problem when aiming for high-resolution lithography such as in IPL.

The multicusp volume-plasma source has largely replaced the duoplasmatron volume source in ion-projection printing because it produces a more stable and uniform ion beam, which helps reduce chromatic aberration and improve pattern resolution [152]. There are currently two IPL systems that are utilizing coaxial multicusp ion sources. One of these systems is known as the ALG-1000, a key IPL system developed by the Advanced Lithography Group (ALG), a consortium that includes industrial and research institutes from both the United States and Europe. Process Development Tool (PDT) is another IPL system that incorporates the coaxial multicusp ion source and is part of the MEDEA program, an international IPL development initiative [153]. These systems are part of the broader push to explore IPL as a potential solution for high-resolution patterning in semiconductor manufacturing, especially as feature sizes continue to shrink below the limits of traditional optical lithography.

5.4. Ion Beam Lithography Resists

The best formed features in ion exposed resist are typically found in PMMA, which is a positive tone organic resist. This is like the situation that occurs in X-ray or EBL applications. It is particularly effective for patterning in helium ion beam lithography (HIBL) because of its affordability, adhesion to silicon substrates, and good resolution capabilities. At the end of the 1970s, polymethyl methacrylate (PMMA) with an average molecular weight of 1.85 × 105 g per mole was patterned to a feature size of 2.7 μm using a helium ion dose ~17 $\mu C/cm^2$ [154]. Shi et al. [155] explored the performance of higher molecular variants of PMMA for finer features utilizing both EBL and HIBL.

High contrast and higher etch resistance are two of the improved features that inorganic resists possess, in contrast to organic resists. HSQ (hydrogen silsesquioxane) is an inorganic negative-tone resist that has become an important material in high-resolution patterning, particularly for HIBL. HSQ is often diluted with a solvent like methyl isobutyl ketone (MIBK) to achieve the desired consistency and film thickness. Dilutions of 1:10 are used for thinner films (~5 nm), while a 1:1 dilution is used for thicker films (55–70 nm). Studies show that dot patterns were successfully created in HSQ films that were either 5 nm or 55 nm thick and the pitch of the patterns was 98 nm. For the 5 nm thick films, the dot width was around 6 nm, and for the 55 nm thick films, the dot width was around 14 nm. As a result of these findings, it was revealed that the HSQ resist exhibits a sensitivity that is approximately 4.4 times greater for helium ions (Ds~31 $\mu C/cm^2$) compared to electrons (Ds ~137 $\mu C/cm^2$), but the contrast values are nearly identical, approximately 2.1, for both forms of exposure. Helium ions have a higher yield of secondary electrons inside the thin resist compared to electrons of the same energy, which could enhance the exposure process and explain the greater sensitivity of HSQ in HIBL compared to EBL [156].

Additionally, Liu et al. [119] carried out lithographic patterning on hafnium-based inorganic resists. $HafSO_x$ (a hafnium-based resist), like HSQ, is another advanced material used for high-resolution patterning, particularly in HIBL. $HafSO_x$ shows a significant difference in sensitivity between helium ion exposure and electron beam exposure [157]. For helium ions, the required exposure dose is only about 4 $\mu C/cm^2$ whereas for electrons, a much higher dose of 420 $\mu C/cm^2$ is needed. This difference is likely due to the higher yield of secondary electrons generated by helium ions, which enhances the material's sensitivity to ion beams.

The alumina-based resist is yet another type of inorganic resist that, when subjected to a helium ion beam, produces isolated lines of 5 nm in width and 20 nm in pitch. By using a 30 keV Focused Helium Ion Beam (FHIB), researchers have been able to control different pitch sizes, with pitches of 20 nm, 40 nm, and 64 nm being established. The dosages ranged from 200 to 700 μC/cm^2, and the pitches were established. The lifespan of the resist solution, on the other hand, causes the resist to have some limitations. Cattoni et al. [158] demonstrated the fabrication of fine features (<10 nm) with alumina-based resist using HIBL.

For nanofabrication, a wide variety of chemically amplified resist compounds are being utilized in conjunction with various lithography techniques. Both chemically amplified resists (CARs) and nonchemically amplified resists (n-CAR) have been taken into consideration for use in HIBL operations. Numerous unique hybrids of organic and inorganic n-carbohydrates are now being produced to achieve this goal. According to this point of view, a new n-CAR MAPDSA-co-MAPDST hybrid resist for HIBL is a promising step forward in nanofabrication, particularly for achieving sub-20 nm features. Its hybrid nature and high sensitivity to helium ions offer significant advantages in terms of resolution and efficiency compared to traditional resists. The successful patterning of features as small as 20 nm at 60 μC/cm^2 marks a key milestone, and this resist could potentially become an important tool for advanced semiconductor fabrication, nanotechnology, and emerging technologies. This resist exhibited an extremely low sensitivity of 7.2 μC/cm^2 and a line edge roughness (LER) of 1.27 ± 0.31 nm [159].

One of the principal resists that were examined during the previous studies of proton beam writing (PBW) at CIBA was polymethyl methacrylate (PMMA). As a result of these investigations, the better patterning capabilities of PBW on PMMA, which are below 100 nm, have become widely distributed. According to the findings, trench widths of 65 nm and walls as narrow as 50 nm can be patterned in PMMA. CIBA has also been successful in achieving superior outcomes for the nickel electroplating process by using main PMMA molds that have been patterned with proton beams. Three-dimensional stamps with 100 nm features were fabricated for pattern transfer into PMMA films. These stamps consist of two raised platforms connected by several high-aspect-ratio ridges. The ridges are 100 nm wide, 2 μm deep, and 30 μm long [160]. The stamps are used to create test patterns by transferring them into PMMA, and this process involves physical imprinting of the features, which could potentially be combined with beam writing methods like PBW.

Sakai et al. [161] worked with a 75 μm thick acrylic film and optimized the fluence (the ion dose) for PBW. The optimal fluence for a 3 MeV proton beam was found to be approximately 9.0×10^{13} ions/cm^2. The fluence is an important parameter in PBW because it dictates how many protons interact with the resist material, which in turn influences the resolution and quality of the patterned features. The fluence also needs to be carefully controlled to achieve the desired pattern resolution without overexposing or damaging the resist. Bolhius et al. [162] investigated the use of GG-developer, a commonly used developer in the LIGA process. They found that the GG-developer was capable of developing proton beam written structures in PMMA, with feature sizes down to 133 nm in a 2.4 μm thick PMMA layer. The GG-developer is specifically optimized to develop high-resolution patterns in thick PMMA films, making it well suited for PBW processes where deep patterns with fine features need to be produced.

Cutroneo et al. [163] worked on producing micro-channels in PMMA resist using PBW with ion fluences ranging from 3.1×10^{13} ions/cm^2 to 3.7×10^{14} ions/cm^2. The variation in fluence helps optimize the resolution and depth of the channels created in PMMA, showcasing the versatility of PBW in terms of exposure control and patterning capability. Erps et al. [164] reported the fabrication of SU-8 micro-pillars with a diameter of 20 μm and a height of 96 μm, which results in an aspect ratio of almost 1:5. Van Kan et al. [165] fabricated structures of 1.5 μm width with an aspect ratio of 24 using negative tone SU-8 resist and fabricated various other 3D structures with different proton energies.

HSQ is highlighted as another high-resolution negative resist for PBW. HSQ is known for its excellent resolution and high sensitivity to ion beams, making it ideal for sub-20 nm feature sizes. In one study, 20 nm wide structures were fabricated in 20 nm thick HSQ using 10 keV H_3^+ ions at a fluence of 7.5×10^{13} ions/cm^2 [160,166]. For HSQ resist in thicker layers, 22 nm and 19 nm wide structures were achieved in 850 nm and 100 nm thick HSQ, respectively, using 1 MeV proton ions [160]. These results further show that PBW with HSQ can achieve sub-20 nm resolution even in relatively thick resist layers, which is important for multilayer patterning or when the resist needs to withstand multiple processing steps. Various other negative tone resists like ma-N, ma-P, AR-P 3200 series, TADEP, and KMPR have been studied with different proton beam energies and proton fluences to fabricate high-aspect-ratio 3D structures. Table 3 shows different types of resists used in ion beam lithography.

Table 3. Types of resists for Ion Beam Lithography.

Resist	Resist Type	Sensitivity (μC/cm^2)	Contrast	Ref.
PMMA	Positive	2	2.7	[52]
AZ-5206	Positive	1.9	3.3	[52]
OiR-897	Negative	1.4	5	[159]
ARCH	Negative	6.9	30	[159]
AZ-114-PN	Negative	0.12	3.5	[167]
SAL-601	Negative	0.11	3.2	[167]
HSQ	Negative	>10	2–4	[168]
HiPR-6512	Negative	1.0	5.0	[168]
HPR-506	Negative	1.8	3.5	[168]

5.5. Advantages and Limitations

The fact that ions experience very little or no lateral scattering is one of the most significant advantages of ion beam lithography in comparison to electron beam lithography. Hence, the proximity effect in IBL is quite little. In the field of IBL, where the de-Broglie wavelength of a 150 keV H^+ ion is approximately 10^{-4} nm, there are diffraction effects that are essentially non-existing. For diffraction, the depth of focus is approximately one millimeter. The IBL system is equipped with electrostatic lenses that can be tuned and have linear focal characteristics. These lenses also can electronically modify magnification. As a result of the low expected total system cost and the high throughput of 50–120 wafers per hour on 200 mm wafers, the tool operating cost is particularly appealing [127].

However, the presence of charged particles presents a hurdle for IBL systems, since it often results in pattern blur and inaccurate pattern placement. Because some ions can transmit through the resist and impinge on the substrate, ion beams can damage the substrates that lie beneath them if the energy is not effectively controlled. Additional research and development are required for the silicon membrane stencil mask technology that is utilized for IPL. The employment of stencil mask technology allows for the avoidance of issues involving ion scattering in transmission masks; however, it necessitates the creation of intricate patterns, which in turn necessitates the use of a secondary complimentary mask [169]. IBL is a promising technology with its own set of pros and cons, and other nano-lithographic techniques like nanoimprint lithography are developed.

6. Nanoimprint Lithography

A high-resolution parallel patterning method with cheap costs and great throughput, nanoimprinting lithography (NIL) relies on the naive idea of imprinting a pattern of nanoscale characteristics onto a substrate from a mold. This method was suggested by Stephen Chou et al. [170]. Initially, by replicating sub-25 nm vias and trenches to a depth of

100 nm, NIL demonstrated its great potential and fulfilled its promise of high-throughput patterning of nanostructures. Nanoimprint lithography (NIL) is a precision manufacturing technique used to create nanoscale patterns on substrates. At its core, the NIL process requires four essential components: a substrate, an imprint polymer (also known as resist), an imprint mold (or stamp), and an impression mask. The process begins by pressing the imprint mold, which contains the desired pattern, into the imprint polymer. This direct mechanical contact between the mold and polymer results in the transfer of patterns from the mold to the polymer, maintaining a one-to-one ratio between the mold features and the substrate [171,172].

During this imprinting process, a thin residual fluid layer is intentionally left behind between the mold and the substrate. This residual layer plays a critical role by preventing direct contact between the mold's relief features and the substrate surface, acting as a buffer to reduce the risk of damage or collisions. The presence of this fluid layer ensures that the mold's intricate patterns are transferred without distortion or impact. After the imprinting step, the imprint polymer undergoes a curing or hardening process either by photocuring or thermal curing depending on the type of polymer used [173]. Once the polymer is solidified, the mold can be removed, leaving behind the desired pattern imprinted on the polymer surface. To complete the pattern transfer, the residual layer is removed using an etching process, often an anisotropic plasma etch. This etching technique effectively strips away the remaining polymer layer, leaving the final, imprinted pattern transferred to the underlying substrate. The result is a highly detailed and accurate nanoscale pattern that can be used for a wide range of applications, such as semiconductor fabrication, photonics, and biotechnology [174]. Thermal NIL, ultraviolet NIL, and roll-to-roll NIL are three common techniques of NIL.

Thermal NIL is a high-resolution, cost-effective technique used to transfer nanoscale patterns onto a thermoplastic polymer (resist) on a substrate using heat and mechanical pressure. The process begins by applying a thin layer of resist, typically made from materials like Polymethyl Methacrylate (PMMA) or Polycarbonate, onto a substrate such as silicon or glass [175]. The resist is heated above its glass transition temperature (T_g) to soften it, allowing it to conform to the mold's pattern under pressure. After imprinting, the system is cooled, and the resist solidifies, locking the pattern in place. Any residual layer of polymer is removed through etching techniques like plasma or chemical etching to clean the final imprinted pattern. Thermal NIL is ideal for creating high-precision patterns, particularly for applications in semiconductor manufacturing, optics, and biotechnology [176]. The use of a planar mold in a thermal NIL process is illustrated in Figure 15 [175]. Thermal NIL has evolved to achieve increasingly smaller patterns, from 25 nm dot patterns to 6 nm, and ultimately 10 nm dots with a 40 nm pitch on PMMMA [177]. Despite challenges such as mold wear and the need for careful thermal and alignment control, thermal NIL offers significant advantages in terms of resolution, throughput, and material flexibility. As the technology continues to advance, it holds immense potential for the future of nanoscale fabrication and high-performance device production.

UV nanoimprint lithography (UV NIL) is a variant of NIL that uses ultraviolet (UV) light to cure the resist material instead of relying on heat, as in thermal NIL. It is a high-resolution, cost-effective lithographic technique used for fabricating nanoscale patterns on substrates, particularly for semiconductor, photonics, and biosensing applications [178]. The process starts with preparing the substrate, which can be made of materials like silicon, glass, or flexible substrates depending on the application. The substrate is coated with a thin layer of UV-sensitive resist material. The resist used in UV NIL is typically a UV-curable polymer that undergoes crosslinking when exposed to UV light. Common resists include UV-sensitive acrylates, epoxy-based resists, or PMMA doped with photosensitive components [179]. The resist is applied using spin coating or other techniques to ensure a uniform and thin layer. The mold used in UV NIL contains the nanoscale pattern to be transferred onto the substrate.

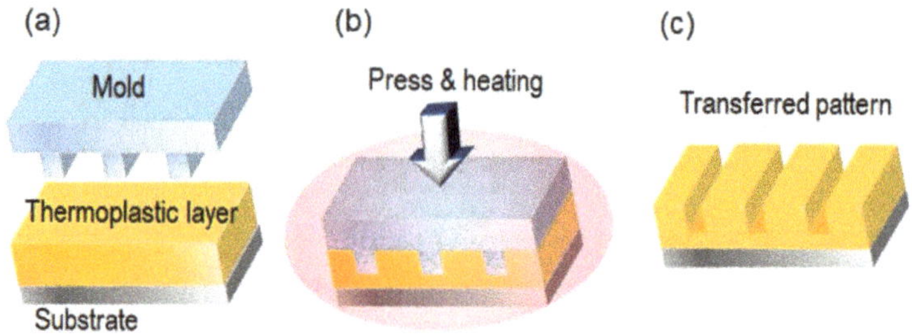

Figure 15. Thermal nanoimprint lithography process with a planar mold depicting (**a**) alignment; (**b**) pressing and heating; (**c**) pattern transfer and mold release. Copyright 2023 MDPI. Reproduced with permission from [175].

The mold is made from materials like silicon, quartz, or nickel, and it has finely patterned features that can range from several nanometers to microns in size. Once the mold is in contact with the resist, UV light is passed through the mold to expose the resist. The UV light causes the resist to undergo photopolymerization, curing and crosslinking the exposed regions [180]. This process solidifies the resist, locking the pattern into place. UV light is typically applied for a few seconds to a minute, depending on the resist material and light intensity. After the resist is cured, the mold is carefully removed [181]. Due to the solidified nature of the resist, the patterned structure is retained on the substrate. The mold removal is crucial to avoid damaging the newly imprinted features. The cured resist layer now serves as a mask for subsequent etching steps. Techniques like plasma etching or wet chemical etching can be used to transfer the pattern from the resist onto the substrate material, such as silicon or metal [182]. A 3D UV-curable NIL was demonstrated by Mohamed et al. [183], and perfusable channel array in a sub-micrometer to micrometer scale was fabricated by integrating NIL and UV lithography by Isobe et al. [184].

Roll-to-roll nanoimprint lithography (R2R NIL) is an advanced nanofabrication technique that combines the principles of NIL with continuous manufacturing processes. R2R NIL employs a continuous process that involves a flexible substrate (such as plastic film or foil) that is unwound from a roll, imprinted with nanoscale patterns using a mold, and then rewound after processing [185]. This system allows for high-volume, cost-effective production, which is a significant advantage over traditional photolithography techniques. R2R NIL is particularly attractive for applications that require large-area patterning on flexible or lightweight substrates [186]. The substrate, typically a flexible film (such as polyethylene terephthalate (PET), polyimide, or metal foils), is unwound from a roll. This substrate is then cleaned and coated with a thin layer of resist material, which is a polymer sensitive to the imprinting process [187]. This can be a thermoplastic polymer (used in thermal NIL) or a UV-curable polymer (used in UV NIL). A uniform layer of resist is applied to the flexible substrate, often using spin coating, doctor blade coating, or slot-die coating techniques. The application process is crucial to ensuring uniformity of the resist layer across the entire substrate.

Once the resist is applied, the continuous roll of substrate is passed through an imprinting station. The substrate is brought into contact with a mold that contains nanoscale features. The mold, typically made of silicon, quartz, or metal alloys, carries the pattern that needs to be transferred onto the substrate. In the imprinting process, pressure is applied to ensure that the softened resist fills the mold's nanoscale features [188]. The imprint method can either be thermal NIL (where the resist is heated above its glass transition temperature) or UV NIL (where the resist is cured using UV light). After the imprinting process, the resist is either cooled (for thermal NIL) or exposed to UV light (for UV NIL) to solidify the resist, locking in the nanoscale pattern [189].

Once the resist is hardened, the mold is carefully removed, leaving behind the imprinted pattern on the substrate. The substrate moves continuously through the system, where each roll of material is imprinted with the nanoscale pattern and then rewound after processing. This allows for high-throughput production of large areas with a uniform pattern [190]. After the imprinting and curing steps, additional processing (such as etching, pattern transfer, or film deposition) may be performed to finalize the pattern and transfer it into the underlying substrate. Groh et al. developed an in-line metrology system for the patterning of 3D holographic structures using R2R NIL [191].

6.1. Nanoimprint Lithography Resolution

Nanoimprint lithography (NIL) has emerged as a powerful technique for high-resolution nanofabrication, demonstrating significant advancements in recent decades [192]. Chou et al. [193] achieved sub-10 nm resolution, imprinting arrays of 10 nm diameter holes with a 40 nm period in polymethylmethacrylate (PMMA), with the smallest feature reaching 6 nm. They successfully transferred these patterns to metal using liftoff and fabricated silicon quantum devices and nano-compact disks with a data density of 400 Gbits/in^2, showcasing NIL's potential for nano-device fabrication. Austin et al. [194] further advanced NIL by achieving 5 nm linewidth and 14 nm line pitch at room temperature and low pressure, demonstrating high uniformity over a 4-inch wafer and highlighting NIL as a cost-effective, high-throughput alternative to electron beam lithography for sub-35 nm applications. Recent studies have extended NIL's resolution limits to sub-10 nm features using molecular dynamics simulations of polymer crosslinking to address shape retention challenges. Strategies such as optimizing resist formulations and introducing sacrificial bridge structures have shown promise in achieving sub-15 nm half-pitch structures with improved shape retention [195]. These developments underscore NIL's potential for scalable, precise, and cost-efficient nanofabrication, making it a key technology for next-generation semiconductor devices and memory applications.

6.2. Advancement of New Materials in Nanoimprint Lithography

Nanoimprint lithography (NIL) is a cost-effective and high-throughput technique for creating polymer nanostructures. This method relies on the mechanical deformation of resist materials, which allows for resolutions beyond the limitations of traditional lithography [196]. The advancements in NIL depend heavily on the materials used for both the imprinting mold and the resist. Molds typically require high strength and durability, while resists need to be easily deformable, possess good mold-releasing properties, and exhibit sufficient mechanical strength [197]. Recent material developments include siloxane copolymers, which offer excellent mold-releasing properties, fast thermally curable liquid resists based on PDMS for rapid crosslinking, and UV-curable liquid resists that enable room-temperature nanoimprinting, expanding the applicability and efficiency of NIL [197,198].

Advancements in NIL have been propelled by crosslinked polymers, offering advantages over traditional thermoplastics like PMMA. These polymers, including thermally crosslinked mr-I 9000 and photochemically crosslinked mr-L 6000, can be processed from organic solvents, similar to thermoplastics, but achieve high thermal and mechanical stability through polymerization. Thermally crosslinked polymers undergo crosslinking during imprinting, dependent on the polymerization rate, while photochemically crosslinked polymers allow for lower imprint temperatures and shorter times. The use of these polymers has enabled the successful imprinting of structures with resolutions down to 50 nm, demonstrating their potential for creating patterns for etch masks and permanent applications [199]. Cedeno et al. [200] demonstrated the use of NIL for patterning organic semiconductors directly. Addressing mold release and pattern transfer challenges, siloxane copolymers, such as poly (dimethyl siloxane)-block-polystyrene (PDMS-b-PS) and related compounds, leverage PDMS properties like low surface energy and high silicon content, enhancing etching resistance in oxygen plasmas for high-resolution, high-aspect-ratio nanostructures [201].

Further progress relies on materials science, developing novel polymers and resins with optimized viscosity, crosslinking density, and surface energy to improve pattern transfer and reduce defects, alongside surface modification techniques and new template materials with enhanced release [202]. Recent efforts focus on liquid precursors curable at room temperature via UV light or heating, with UV-curable epoxysilicone materials undergoing cationic crosslinking, offering oxygen insensitivity, low shrinkage, and good dry-etching resistance, demonstrating successful patterning of features down to 20 nm at low pressures [198].

Superplastic micro-forming of metallic glasses, such as Palladium-based amorphous alloy $Pd_{40}Cu_{30}Ni_{10}P_{20}$, has been explored, replicating nanometer-sized die shapes with high geometrical transferability, suggesting potential for mass production of nanodevices [203]. Metallic glasses, in general, offer superior mechanical properties for robust molds, capable of direct nanopatterning through hot embossing and crystallization for pattern replication [204]. NIL's scope has expanded to encompass diverse materials like shape memory materials, covalent adaptive polymers, and porous materials, enabling applications in photovoltaics, water filtration, and smart materials, highlighting its growing impact [205]. Figure 16 [192] shows a C-shaped SERS nanopattern fabricated using UV-NIL and Figure 17 [196] demonstrates SEM images of a patterned surface after UV-NIL with different imprint speeds where the amount of air defects was found to increase on increasing the imprinting speed.

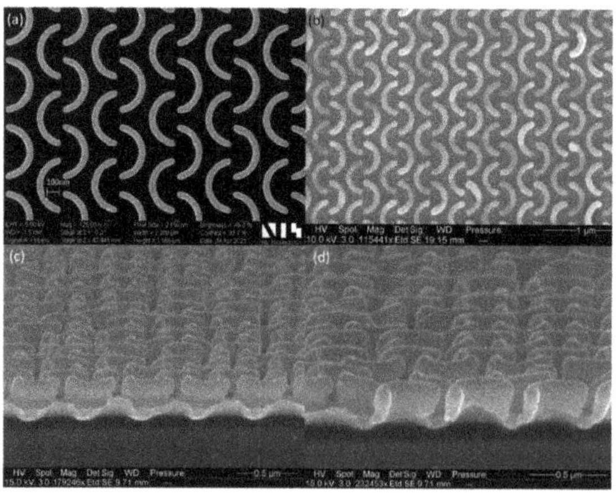

Figure 16. SEM images of (**a**) top view of the Si master; (**b**) top view of the C-shaped SERS pattern fabricated with UV-NIL; (**c**,**d**) cross-sections of UV-NIL SERS substrate. Copyright 2023 MDPI. Reproduced with permission from [192].

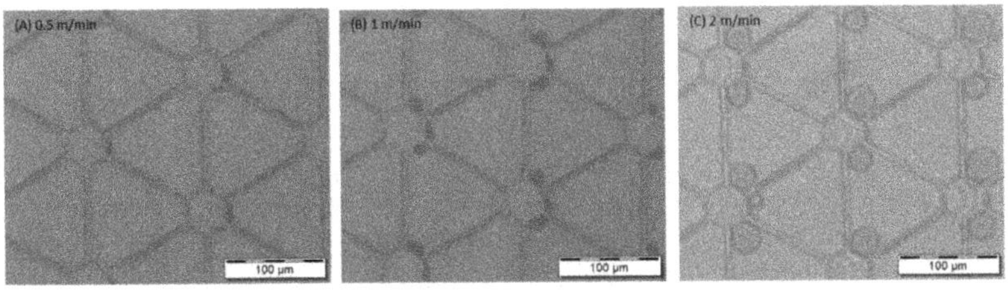

Figure 17. SEM images of a patterned surface after UV-NIL with different imprint speeds: (**A**) 0.5 m/min, (**B**) 1 m/min, (**C**) 2 m/min. Copyright 2021 MDPI. Reproduced with permission from [196].

NIL is also being explored for organic electronics, utilizing materials like α-sexithiophene, poly(4-diphenyl-aminostyrene) (PDAS), and poly(phenyl-bis-4-aminostyrene) (PBAS) to modify thin film morphology, inducing smoother surfaces, and enabling patterned substrates with controlled optical and electrical properties, holding potential for advancing organic electronics, especially in FET fabrication. Metal-assisted chemical imprinting (Mac-Imprint) is a contact-based wet etching process that combines metal-assisted chemical etching (MACE) and NIL [206].

The Mac-Imprint method utilizes mesoporous metal catalysts to replicate silicon from polymeric molds, enhancing diffusion pathways and pattern transfer fidelity [207,208]. Gray-scale direct imprinting of porous substrates (DIPSs) enables three-dimensional patterning of porous nanomaterials, directly tailoring nanomaterial properties without intermediate resists, applicable in diffractive and plasmonic sensing [209]. These innovations aim to enhance NIL resolution, throughput, and reliability, expanding its applications across electronics, photonics, and biotechnology [206]. Table 4 displays the process conditions for various NIL materials.

Table 4. Types of materials for Nanoimprint Lithography and their process conditions.

Material	Processing Temperature (°C)	Pressure (MPa)	Process Timing	Ref
PMMA	200	10	3 min	[175]
Polystyrene	160	10	3 min	[175]
Polypropylene	165–225	5	5 min for melting, 30 min for press	[210]
Polyethylene naphthalate	290	2.5	10 min	[211]
Cyclic–olefin copolymer	150	5	5 min	[212]
Polyethylene terephthalate	75–150	2	5 min	[213]
Polycarbonate	160	5	5 min	[214]
Polycarbonate	180	10	10 min	[214]
Polyimide	200	3	2 min	[215]
Epoxy	95	1.2	10 min	[216]
Polyetherimide	285	1	3 min	[217]
Fluorinated Ethylene Propylene	270	0.18	5 min	[218]
Ethylene tetrafluoroethylene	250	1.38	10 s	[219]
Cellulose Acetate	115	13.6	0.2 m/min	[220]

6.3. *Advantages and Limitations*

Nanoimprint lithography (NIL) is a cutting-edge lithographic technique that overcomes the constraints of electron beam lithography and ultraviolet (EUV) lithography. It does not require any designed instruments and therefore is a lucrative technique that can manufacture high-resolution features. NIL is used to generate nanoscale patterns with high throughput, with a vast variety of use, along with simplicity and a high degree of adaptability, and it finds application in the production of LEDs and optoelectronic devices, solar cells, memory devices, and biosensors [221]. NIL, which is capable of replicating features with a size of less than 10 nm on wide area substrates, is a formidable candidate for enabling nanoscale manufacturing for the advancement of integrated circuits. The adhesion at the mold resist interface, on the other hand, raises the possibility of introducing damage to the nanopatterns; hence, the process of releasing the mold requires a considerable deal of care. In addition, the mold is prone to quick degradation due to substantial pressure and temperature that is applied between mold and substrate, which ultimately results in the mold needing to be replaced more frequently [222].

7. Alignment Marks in Lithography

Alignment marks are critical to precision and accuracy in nanoscale lithography. This ensures that the device navigates the correct layer layout and hence delivers appropriate device performance [223]. By aligning these marks, lithography tools compensate for distortions or changes in position caused by imperfections in the equipment, physical stress, or environmental factors [224]. Due to the size shrinking to the nano-level, slight deviations can cause significant deviations. This has led to the development of increasingly sophisticated alignment mark designs and detection methods tailored to various lithographic technologies, such as EUV lithography, EBL, and NIL [52].

Alignment markers in modern lithography have evolved from simple macroscale structures to complex fabrications that combine nanoscale features and high-contrast materials. For example, optical alignment systems use diffraction-based methods to detect the position of marks, while electron-beam systems directly image the mark structures with nanoscale precision. These marks also provide information about the quality of alignment. The system can be used to make real-time adjustments. Also, alignment marks not only serve as positional references but also offer data on alignment quality to quantify overlay errors [122]. Advanced systems use this feedback for real-time adjustments, improving lithography precision. Additionally, these marks are designed for compatibility with various substrates and conditions, enhancing their versatility across applications [225].

Additionally, improvements in materials used for alignment markers have greatly improved their durability and effectiveness [226]. High-contrast materials such as chromium and titanium nitride are now widely used to enhance marker visibility in various detection systems. These materials are resistant to chemical etching and heat treatment. This ensures that the markers last even in harsh production conditions. Emerging materials, including advanced polymers and nanocomposites, are also being investigated for their potential to improve alignment efficiency further [227].

7.1. Alignment Mark Designs

The development of unique alignment marks suited to applications and technologies—such as basic cross-section points adequate for aligning base layers in lithography—has resulted from the necessity to manufacture a broad variety of instruments. Advanced technologies like extreme ultraviolet (EUV) lithography require more complex design for nanoscale precision. The type of alignment mark used also depends on the processing environment—marks must withstand conditions such as high temperatures or corrosive chemicals. Additionally, complex multilayer fabrication may use intricate marks like box-in-box designs to correct for translational and rotational misalignments. As lithographic techniques evolve, custom and application-specific marks tailored to unique process requirements and substrate characteristics become crucial, enabling precise alignment critical for modern, complex nanostructures. Figure 18 shows a schematic diagram of different types of alignment mark designs as in (a) crosses; (b) crosses and boxes; (c) box in box; and (d) vernier marks.

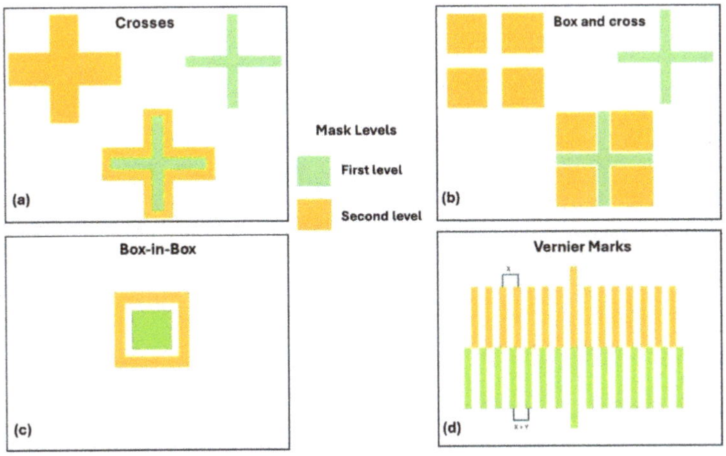

Figure 18. Schematic diagram of different alignment mark designs: (**a**) crosses; (**b**) crosses and boxes; (**c**) box in box; and (**d**) vernier marks.

- **Crosses and Boxes**: The simplest and most used forms of alignment marks are crosses and dots. These marks are favored for their straightforward design and ease of use. Cross marks, consisting of two perpendicular lines, provide a clear intersection point that can be easily detected by alignment systems. Dots, on the other hand, are small points used as reference markers. Both types are used extensively in both lithography and electron beam lithography due to their simplicity and effectiveness in providing a clear, detectable reference for alignment purposes [228].
- **Box-in-Box Marks**: These marks are more complex and provide multiple reference points within a single mark. Typically, they consist of two concentric squares or rectangles, where the alignment is performed based on the position and orientation of the inner shape relative to the outer shape. This design allows for more detailed error analysis and correction, as it can provide information on both translational and rotational misalignments. These marks are widely used in multilayer processes where each layer's exact positioning is crucial for the device's functionality [229,230].
- **Segmented or Vernier Marks**: Named after the vernier scale used for precise measurements in mechanical engineering, these alignment marks are designed to measure misalignment more finely. They consist of two sets of lines where one set is slightly offset from the other. The degree of misalignment can be determined by observing the point at which the lines from the two sets appear to be continuous. This type of mark is extremely useful in fine-tuning the alignment process and is often used in conjunction with other types of marks to achieve high-precision overlays [231].
- **Grating Patterns**: For applications requiring higher precision, grating patterns are employed. These marks consist of a series of parallel lines or a grid, and they are particularly useful in advanced lithography techniques such as extreme ultraviolet (EUV) lithography. Grating patterns allow for the utilization of diffraction-based methods to precisely measure and correct alignment errors. The interference patterns generated by these gratings provide a more detailed feedback mechanism, enabling sub-nanometer precision in layer alignment. This type of mark is essential in applications where the overlay accuracy needs to be within a few nanometers [226,232].

7.2. Lithography Alignment Techniques

Lithography alignment techniques are pivotal in the semiconductor manufacturing process, ensuring the precise transfer of patterns from masks to wafers. These techniques are crucial for maintaining the integrity and functionality of the final semiconductor devices, and they vary significantly based on their operational principles and the technologies employed. Below is an in-depth look at the primary types of lithography alignment techniques utilized in the industry.

1. **Geometric Image Alignment**: This technique leverages the geometric properties of images projected during the lithography process. It involves aligning the patterns based on their shapes and relative positions as viewed through the lithographic optical system. The effectiveness of geometric image alignment is dependent on several factors, including the design of the alignment marks and potential distortions introduced by the optical lenses. These distortions can affect the precision of the alignment, making the quality of the optical components and the design of the alignment marks critical to the success of this technique [233].
2. **Light Intensity-Based Alignment**: This method utilizes variations in light intensity to facilitate precise alignment. It typically involves the detection of light reflected from specially designed alignment marks on the wafer, with the intensity of this reflected light compared against expected values to determine alignment accuracy. While straightforward, this technique can be susceptible to fluctuations in lighting conditions, which may impact its reliability. However, its simplicity makes it attractive for many standard lithographic applications where extreme precision is not the primary concern [234].
3. **Phase Shift Alignment**: This sophisticated technique uses phase-shifting masks to manipulate the phase of light waves passing through the mask. By creating points of destructive interference, phase shift alignment enhances the resolution of the patterning process, allowing to produce smaller features on the wafer. This method is particularly valuable in advanced lithography processes where reducing feature sizes is crucial for increasing the density and functionality of semiconductor devices [235].
4. **Moiré Fringe Alignment**: Figure 19 [232] shows grating pattern alignment marks for marginally distinct periods $P2 \sim 1.1P1$. Moiré fringe techniques employ the interference patterns that arise when two sets of gratings on the mask and the wafer overlap. These patterns are highly sensitive to misalignment, making them useful for achieving precise alignment with accuracies up to 10 nm and alignment ranges up to 500 μm. The technique involves a combination of coarse and fine alignment steps integrated into the same imaging process, allowing for both broad and highly precise adjustments. Moiré fringes can be generated using either physical or digital gratings, making this method versatile across different lithographic processes [232].
5. **Front-to-Back Alignment (FTBA)**: FTBA is essential for applications requiring the alignment of multiple layers, such as in the fabrication of microelectromechanical systems (MEMSs). This technique projects alignment marks from the back side of the wafer to its front side, ensuring precise layer-to-layer registration throughout the lithography process. Its ability to maintain high throughput makes it ideal for applications where speed and accuracy are both critical [236].
6. **Maskless Lithography Techniques**: These innovative methods, which include electron beam lithography and focused ion beam lithography, do not rely on traditional photomasks. Instead, they directly write patterns onto the wafer surface using beams of electrons or ions. While these techniques offer high resolution and are ideal for prototyping or manufacturing with low volume requirements, their throughput is generally lower than that of traditional masked lithography methods. This makes them

less suitable for high-volume production but invaluable for custom or experimental applications where flexibility and precision are paramount [237].

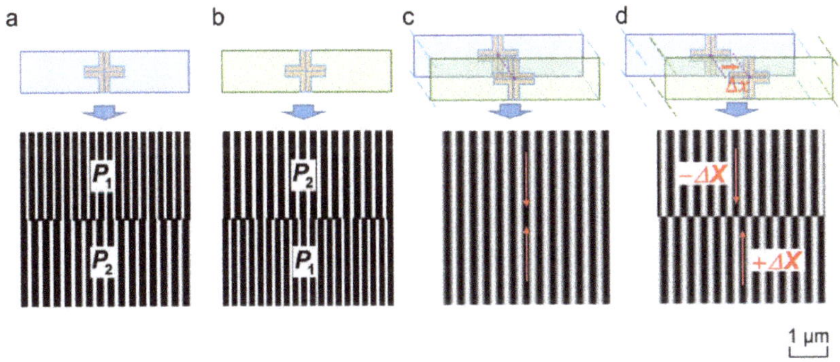

Figure 19. Grating pattern alignment marks for (**a**) wafer; (**b**) mask marks; (**c**) perfectly aligned fringe; (**d**) misaligned fringe. Copyright 2024 MDPI. Reproduced with permission from [232].

7.3. Techniques for Implementing Alignment Marks

7.3.1. Design Considerations

The design of alignment marks is critical in determining both the yield and performance of semiconductor devices. These marks must be carefully crafted to ensure they are detectable under various imaging conditions and resistant to processing steps that could degrade their visibility or geometric integrity. The size, shape, and materials used for alignment marks are selected based on their visibility in the chosen lithography system and their compatibility with the processing environment. For instance, smaller marks are less obtrusive and conserve valuable wafer space but may be harder to detect, especially under less-than-ideal imaging conditions. Conversely, larger or more complex marks can be easier to detect but consume more wafer real estate. The contrast between the marks and the substrate is also vital; high-contrast materials such as chromium or titanium nitride might be used to ensure marks are easily discernible against the silicon background. Moreover, the design must account for potential distortions due to lens aberrations, focus variations, and process variations such as etching or deposition inconsistencies. Marks must be robust enough to maintain their integrity through multiple layers of processing, as any degradation can lead to misalignment and thus lower the yield and performance of the final device [238,239].

7.3.2. Placement Strategies

The placement of alignment marks on masks is another strategic consideration that can significantly influence the accuracy and effectiveness of the lithography process. Optimal placement strategies ensure that marks are accessible and readable by alignment systems throughout the fabrication process, even after various processing steps that might obscure or damage earlier marks. One common strategy is to place alignment marks at the corners of the exposure field as well as near the center. This arrangement allows for comprehensive coverage and ensures that any distortions or shifts in the wafer can be detected and corrected. It is crucial that the marks are placed away from areas of critical device structures to avoid interference but still close enough to areas of interest to provide effective alignment cues.

In multilayer processes, alignment marks may need to be replicated on each layer, or new marks may be added in subsequent layers to account for any interlayer misalignment. The alignment marks on each layer must be precisely aligned relative to those on previous layers to ensure accurate stacking of features. This can involve the use of interlayer alignment strategies, where marks are designed to interlock or align with marks on adjacent layers, enhancing the overall alignment accuracy. Advanced techniques, such as the use of "global" and "local" alignment marks, are also employed. Global marks are used for the initial rough alignment of the wafer, while local marks, placed closer to the areas of active device fabrication, are used for fine alignment within a specific region of the wafer. This tiered approach allows for high-precision alignment that is crucial for advanced device structures, particularly at nodes below 10 nm where even minor misalignments can be catastrophic. Furthermore, the alignment mark placement must consider the entire process flow, including any potential for mark degradation due to subsequent processing steps such as etching, chemical mechanical polishing, or high temperature anneals. The strategic placement and robust design of alignment marks are essential not only for maintaining alignment accuracy but also for maximizing the yield and performance of the fabricated devices [239].

8. Overview of Mask Fabrication

The mask making process is fundamental to lithography in semiconductor manufacturing, where masks serve as critical stencils for transferring circuit patterns, defining features at micro- and nanoscale levels. This highly precise process utilizes advanced materials and technologies, beginning with the design and layout of circuit patterns via computer-aided design (CAD) software. These patterns are then meticulously transferred onto mask substrates through lithography and etching. The accuracy of these masks is crucial, as any imperfections can significantly impact the yield and functionality of the final semiconductor products. Understanding the various types of photomasks and the materials used for their fabrication is vital due to their central role in lithography. This knowledge is essential for choosing suitable mask types for specific lithographic requirements and for driving the development of innovative mask technologies to address the evolving challenges of modern semiconductor design and production.

8.1. Photomask Types

Photomasks are essential tools in the lithography process for semiconductor manufacturing, functioning as precise templates that transfer circuit patterns onto wafers. The type of photomask chosen is critical as it directly influences the resolution and quality of the final semiconductor devices [240]. There are several types of photomasks, each designed to meet specific requirements of modern lithography systems:

- **Binary Masks**: These are the most straightforward type of masks, consisting of clear and opaque regions. The opaque areas block the light, while the clear areas allow light to pass through, directly transferring the pattern onto the wafer. Binary masks are widely used for their simplicity and effectiveness in less complex lithography techniques.
- **Phase-Shifting Masks (PSMs)**: Phase-shifting masks enhance the resolution and image quality by shifting the phase of light passing through the mask. This shift creates constructive and destructive interference patterns, sharply defining the edges of the features on the wafer. PSMs are particularly effective in advanced lithography, where feature sizes are close to or below the wavelength of the exposing light.
- **Attenuated Phase-Shifting Masks**: These are a variant of the traditional phase-shifting masks, incorporating a partially transparent film that reduces the intensity of the transmitted light. This attenuation helps in improving the process latitude and contrast of the image on the wafer, allowing for even finer resolution. These masks are beneficial in applications requiring extremely small feature sizes and high-density patterns [241].

Each type of photomask plays a specific role in semiconductor manufacturing, depending on the complexity of the design and the resolution requirements of the lithography process. The selection of the appropriate mask type is driven by factors such as the desired feature size, the optical properties of the lithography system, and the specific constraints of the semiconductor device being fabricated. As semiconductor technology continues to advance, the development and refinement of photomask types remain central to achieving higher resolutions and better device performance.

8.2. Materials Used in Photomask Fabrication

The choice of materials used in photomask fabrication is crucial, as it directly impacts the performance, durability, and quality of the masks used in semiconductor lithography. The materials must possess specific properties such as optical transparency, chemical resistance, and mechanical stability. Here are some of the primary materials employed in the construction of photomasks:

- **Quartz**: Quartz is the most commonly used substrate material for photomasks due to its excellent optical transparency in the ultraviolet (UV) range, which is crucial for lithography processes involving deep UV and extreme ultraviolet (EUV) light. Quartz also offers superior dimensional stability and resistance to thermal expansion, which are essential for maintaining pattern accuracy during the intense UV exposure in lithography processes. Its high chemical resistance ensures that it can withstand the rigorous cleaning and processing chemicals used during mask fabrication and wafer processing [242].
- **Chrome**: Chromium is used to form the opaque regions on photomasks. It is deposited on the quartz substrate to create the patterned features that define the mask's circuit designs. Chrome's high opacity to UV light makes it an ideal choice for creating sharp, well-defined lines that block light effectively. Additionally, chrome adheres strongly to quartz, providing durability and resistance to peeling or flaking during the mask usage. Its ability to be finely patterned with high precision through processes like electron beam lithography ensures that even the most intricate designs can be accurately rendered [243].
- **Titanium Nitride**: Used in advanced photomasks, titanium nitride (TiN) serves as an alternative to chromium for certain applications, particularly in phase-shifting masks. TiN can be used to form attenuated phase-shifting areas due to its semi-transparent properties at specific lithography wavelengths. This material is also known for its mechanical hardness and chemical stability, which are beneficial for the longevity and reuse of masks in high-volume production environments [244].
- **Molybdenum and Silicon thin films**: Another material used in advanced mask technologies, particularly for EUV lithography, is molybdenum and silicon thin films. It is chosen for its ability to withstand the high-energy photons used in EUV lithography without degrading. This material provides a low degree of absorption and thermal stability, which are crucial for maintaining mask integrity under the intense conditions of EUV exposure [245].

The selection of materials for photomask fabrication is guided by the specific requirements of the lithography process, including the exposure wavelength, feature size precision, and environmental durability. Advances in material science continue to play a pivotal role in the development of new photomask technologies that push the boundaries of what can be achieved in semiconductor manufacturing [246].

8.3. Challenges in Mask Fabrication

8.3.1. Resolution Limits

In the realm of semiconductor manufacturing, the quest for miniaturization continues to push the limits of lithography and mask-making. As feature sizes decrease, achieving high resolution in mask-making becomes significantly more challenging. This drive towards smaller features confronts a fundamental barrier in optical lithography, where the resolution is constrained by the wavelength of light used for exposure. The resolution limit, often described by the Rayleigh criterion, delineates the smallest feature size that can be reliably resolved. In practical terms, as the industry moves towards node sizes below 10 nm, traditional optical lithography struggles to maintain fidelity, prompting a shift towards advanced techniques such as extreme ultraviolet (EUV) lithography. EUV uses significantly shorter wavelengths, allowing for finer patterning, but it also introduces complexities in mask fabrication. Mask features must now incorporate intricate, smaller-scale designs that are extremely challenging to produce and replicate accurately [247]. To push the boundaries of resolution, ongoing innovations in both materials and mask-making technologies are critical. This includes the development of new resist materials that can capture finer details and the use of sophisticated etching techniques that can accurately transfer these minute patterns onto the mask substrate. Additionally, advances in electron beam lithography, which can bypass some of the limitations of light-based lithography, offer another avenue for enhancing resolution in mask production [248,249].

8.3.2. Defect Mitigation

Defects in mask-making, such as particulates, pinholes, or inaccuracies in pattern generation, pose significant challenges to the lithographic process. These defects can propagate through to the wafer, leading to malfunctioning circuits and reduced yield. The impact of such defects becomes exponentially more significant as feature sizes shrink, where even the smallest flaw can render a device non-functional [250]. To ensure the high quality and reliability of masks, several strategies and technologies are employed. Advanced inspection systems play a crucial role in this aspect. Techniques such as optical and electron microscopy are used to scrutinize masks at various stages of the fabrication process, detecting anomalies that could affect performance [251]. Once defects are identified, precise repair technologies are deployed. These may involve material deposition to fill pinholes or focused ion beams to correct pattern inaccuracies. Moreover, cleanroom environments are meticulously maintained to minimize the introduction of particulates during mask production [252]. Automation in mask handling and storage also reduces the risk of human-induced defects. In the broader scope, statistical process control and continuous improvement practices are integral to refining mask-making processes, thereby reducing the incidence of defects over time [253]. Overall, addressing these challenges involves a combination of cutting-edge technology, stringent process control, and ongoing innovation. These efforts are vital for sustaining the advancement of semiconductor technology, enabling the production of more powerful and reliable electronic devices.

9. Molecular Strategies in Nanoscale Lithography

As the quest for smaller, more powerful devices continues, advanced lithography techniques face increasingly stringent demands for resolution, precision, and throughput. Traditional lithographic materials and methods are reaching their fundamental limits, necessitating a paradigm shift towards molecular-level control and innovative material design. The incorporation of molecular strategies has become essential to overcoming these challenges and realizing the full potential of nanoscale fabrication [254].

This section focuses on the critical role of molecular materials in pushing the boundaries of lithography. We explore how the design and properties of these materials, engineered at the molecular level, directly influence the performance and capabilities of advanced lithographic processes. By harnessing the principles of molecular chemistry and materials science, researchers are developing novel resists, templates, and techniques that enable unprecedented precision in pattern transfer and device manufacturing.

Our discussion begins with an examination of resist materials, the cornerstone of lithography. We delve into the molecular composition, properties, and reaction mechanisms that govern their behavior during exposure and development. Furthermore, we explore nanostructured films, with an emphasis on self-assembling materials, and discuss how the molecular-level organization of these films can be leveraged to create intricate patterns and functional devices.

9.1. Molecular Materials for Lithography

9.1.1. Resist Materials: Molecular Design and Properties

Molecular resists represent an alternative to traditional polymer resists in lithography, particularly for extreme ultraviolet (EUV) and electron beam lithography (EBL) [255]. These materials are designed with the goal of improving resolution and reducing line edge roughness (LER), which are critical challenges in next-generation lithographic techniques [256]. Molecular resists offer the potential for reduced pixel size and greater synthetic control, as they are composed of single molecules with precisely defined structures. The design of molecular resists often involves incorporating all the functionalities needed for a chemically amplified resist (CAR) into a single molecule. This can include a photoacid generator (PAG) moiety bonded to a protected polyphenol derivative. Having a single-component resist can ensure uniform PAG density and better control over acid diffusion, which influences both resolution and LER [257]. Wang et al. [258] synthesized and characterized a novel molecular glass (TPSiS) with a photoacid generator for use as a single-component molecular resist in electron beam lithography. The TPSiS resist demonstrates high resolution, resolving 25 nm dense line/space patterns, and exhibits potential for application due to its etching selectivity.

Y. Wang et al. [259] designed and synthesized molecular resists based on a bis(4-butoxyphenyl) sulfone core attached to a varying number of radiation-sensitive triphenylsulfonium units (BPSSn, where n = 2, 3, and 4). These materials were evaluated for their physical properties, including solubility, film-forming ability, and thermal stability, to assess their viability as photoresist materials. The BPSS4 resist demonstrated high resolution (16/13 nm) and low line edge roughness (2.5/2.5 nm) in e-beam and EUV dense line patterning, respectively. The BPSS4 resist also exhibited high etch resistance and accurate pattern transfer capabilities, making it suitable for high-resolution lithography applications. Kim et al. [260] synthesized a novel molecular resist material based on polyhedral oligomeric silsesquioxane with diazoketo groups for deep UV lithography. Initial lithographic evaluations suggest the new material platform's potential for next-generation resists.

Molecular resists can be designed for both positive and negative tone imaging. In negative tone imaging, the BPSS materials allowed for negative patterning through organic development in both e-beam and EUV lithography [261]. In positive-tone resists, the material becomes soluble upon exposure and subsequent development, while negative-tone resists become insoluble. The development of molecular resists often involves optimizing the dissolution behavior to ensure high chemical contrast and reduced swelling during the development process. A new family of fullerene-based negative tone chemically amplified e-beam resists, using industry compatible solvents, has been developed. The molecular resist core is often synthesized through the acid-catalyzed condensation of phenol with a ketone or aldehyde and then functionalized [262]. The molecular structures of a new family of phenol-based fullerene derivatives are shown in Figure 20 [262].

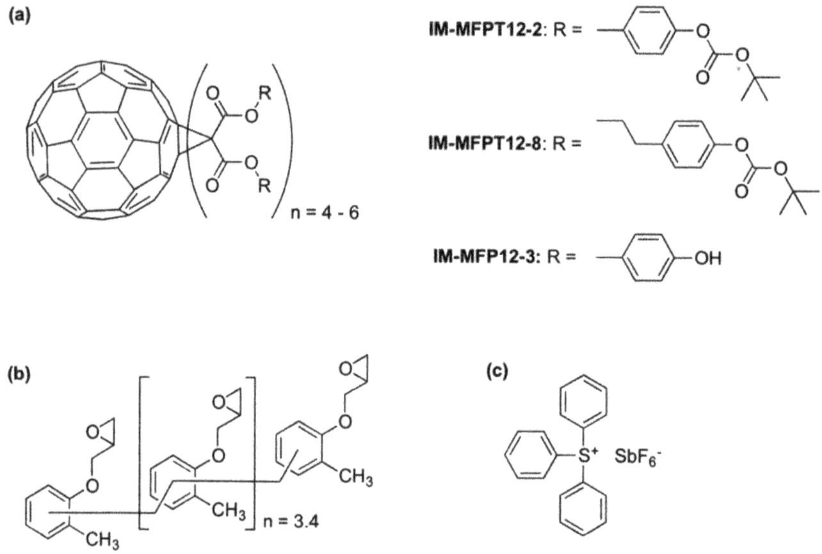

Figure 20. Molecular structure of phenol-based fullerene chemically amplified resists. (**a**) IM –MFP12 –2, IM –MFP12 –8, IM –MFP12 –3; (**b**) IM –MFP12 –2 and IM –MFP12 –8 combined with an epoxy crosslinker; (**c**) IM –MFP12 –2 and IM –MFP12 –8 combined with triphenyl sulfonium hexafluoro antimonate. Copyright 2014 Royal Society of Chemistry. Reproduced with permission from [262].

The PhD thesis of Dr. Jedsada Manyam explores Next-Generation Lithography (NGL) technologies, particularly focusing on the development and optimization of chemically amplified fullerene-based resists for electron beam lithography [263]. It addresses the limitations of conventional polymeric resists and aims to achieve high resolution, sensitivity, and etch resistance for advanced microfabrication. A sensitivity of ~40 $\mu C/cm^2$ was achieved at 20 keV. Isolated features with a line width of 13.6 nm as well as ~20 nm lines on a 36 nm pitch were patterned, whilst one variant demonstrated resolution to 15 nm half-pitch at slightly higher dose.

While early molecular resists have demonstrated promising results, including high resolution and low LER, challenges remain [264]. These include achieving sufficient sensitivity, preventing pattern collapse, and optimizing the development process [265]. Further research and development are focused on addressing these challenges and exploring the full potential of molecular resists for advanced lithography. Some approaches involve blending molecular resist components to adjust sensitivity. Molecular resists, including oligomers, molecular glasses, discotic liquid crystals, and inorganic materials, have garnered attention as promising candidates [266].

9.1.2. Nanostructured Resists: Self-Assembling Materials for Advanced Patterning

Nanostructured resists leverage the self-assembly properties of materials to create patterns at the nanoscale, offering a route to high-resolution and high-throughput lithography. This approach moves beyond traditional resist materials and processes by utilizing the spontaneous organization of molecules or nanostructures into ordered arrays or templates [86]. One approach involves using structures generated by dip-pen nanolithography (DPN) as positive resists for fabricating nanohole arrays [267]. In this method, the DPN-generated nanostructures serve as a mask, protecting the underlying material during etching. Polymer pen lithography (PPL) is a cantilever-free scanning probe-based technique used for molecular printing that can generate sub-100 nm molecular features in a massively parallel fashion [268]. PPL enables the creation of combinatorial arrays of nanostructures composed of proteins and metals, making it suitable for high-throughput screening and biologically relevant applications.

Another avenue in nanostructured resists is the use of block copolymers (BCPs), which self-assemble into periodic nanostructures with domain sizes that can be controlled by the molecular weight and composition of the polymer [269]. These self-assembled nanostructures can be used directly as a resist or as a template for further pattern transfer. Beyond block copolymers, other self-assembling materials, such as liquid crystals and supramolecular assemblies, are being explored as potential nanostructured resists. These materials offer a diverse range of morphologies and functionalities, enabling the creation of complex patterns with nanoscale precision. Chang et al. [270] review self-assembled nanocomposites and nanostructures for environmental and energy applications, highlighting their use in wastewater purification, hydrogen production, energy storage, and energy harvesting. The review discusses the classification, synthesis, characterization, and properties of these materials, aiming to demonstrate the advantages of self-assembled nanostructures and provide perspectives for future research. Self-assembled nanomaterials offer improved ionic transport and electronic conductivities and can tolerate high currents, offering a solution for high-power energy storage.

E-beam lithography (EBL) is a powerful tool used alongside nanostructured resists for generating nanostructures and fabricating nanodevices with fine features approaching a few nanometers in size. EBL does not require a template, can easily write quickly, and can expose a thick resist without ion contamination [86]. While nanostructured resists hold significant promise, challenges remain in controlling the self-assembly process, achieving high pattern fidelity, and integrating these materials into existing lithographic workflows. However, the potential benefits in terms of resolution, throughput, and cost-effectiveness make nanostructured resists an active area of research and development for advanced lithography [53].

9.2. Molecular Mechanisms in Lithography

9.2.1. Photoresist Chemistry: Photochemical Reactions and Processes

Photoresist chemistry is a critical factor in extreme ultraviolet (EUV) lithography, requiring the development of materials with high sensitivity to achieve efficient patterning at reduced feature sizes [271]. Out-of-Band (OoB) radiation, approximately 4% of the total radiation from EUV sources, can negatively impact resist performance [272,273]. Photoacid generators (PAGs) with selectivity to EUV radiation are designed and synthesized to minimize the effect of OoB by decreasing deep UV (DUV) absorption through incorporation of insensitive cations [224]. Both blended PAG and polymer-bound PAG resist formulations utilizing these PAGs have demonstrated improved EUV sensitivity [274].

Significant effort is focused on chemically amplified resists (CARs) for EUV lithography [48]. Studies show that metal salt sensitizers improve both EUV photon absorption and electron yield, resulting in higher sensitivity. For instance, the incorporation of a specific metal sensitizer with high EUV absorbance could improve sensitivity by 17% without significantly impacting LWR or local critical dimension uniformity (LCDU) [266]. Yamamoto et al. [56] improved sensitivity by 43% by adding a metal sensitizer. Metal sensitizers improve both EUV photon absorption and electron yield resulting in higher sensitivity [275]. Higher absorption of anions increases electron yield by merely higher absorption but has limited impacts on electron generation efficiency; however, cations of metal sensitizer dictate electron generation efficiency. Halogenated sensitizers, such as fluorine and iodine, also help with electron generation due to their higher absorption, but sensitivity depends heavily on the chemical environment of the halogen bonds [273]. However, higher electron yield does not always translate to higher sensitivity for all halogen sensitizers [276]. The chemical environment where these halogens are attached has a significant impact on the sensitivity.

Molecular CAR systems are also actively researched. Cyclic low-molecular-weight (CLM) resists have achieved sub-30 nm half-pitch resolution with high sensitivity [266]. Noria derivatives with pendant adalantyl ester groups have produced 25 nm resolution patterns using EUV lithography at doses less than 10 mJ/cm^2 [277]. Modified Noria molecules with oxetane crosslinking moieties have been used in negative-tone CARs, achieving 20 nm resolution with 3.2 nm LER [278]. Molecular glass (MG) resists based on calixarene cores are also being explored as alternatives to polymeric resists, achieving 28 nm 1:1 space line using EUV exposure [279]. The multi-trigger resist concept, which involves regenerating photoacids, is being explored to achieve high sensitivity solubility changes above a certain dose threshold and sharper lines with lower LER [280].

Inorganic resists, including metal oxides, are also under investigation, with some hybrid materials showing 26 nm lines with a 4.2 mJ/cm^2 EUV dose [281] and directly patternable metal oxide hardmasks achieving 13 nm half-pitch at 35 mJ/cm^2 and 11 nm hp with 1.7 nm LWR [282]. Tin-oxo cages have been found to have dual tone property which showed positive tone behavior when irradiated with a low dose of EUV or E-beam and a negative tone behavior when it is irradiated at higher dosages [283]. Organohydrogensilsesquioxane-based EUV resists have shown excellent etch selectivity and ability to form patterns by using industry standard TMAH development process, as well as low LWR (<2 nm) with sufficient sensitivity (40–60 mJ/cm^2) [284]. Figure 21 [48] shows a summary of the evolution of the EUV resists optimized by electron-induced chemistry.

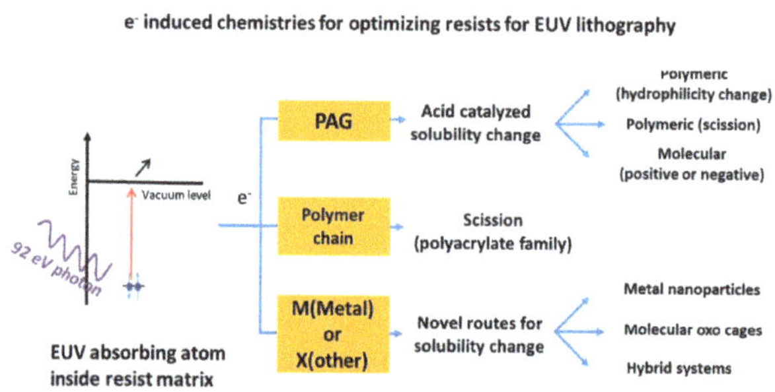

Figure 21. Schematic diagram of a summary of the evolution of the EUV resists optimized by electron-induced chemistry. Copyright 2020 MDPI. Reproduced with permission from [48].

9.2.2. Molecular Dynamics Simulations: Predictive Modeling of Nanoscale Behavior

Molecular dynamics (MD) simulations are indispensable for understanding and predicting material behavior at the nanoscale, providing insights into lithographic processes that are challenging to observe experimentally [285]. MD enables the modeling of atom and molecule dynamics, crucial for investigating phenomena such as resist material behavior, EUV-induced processes, etching, and self-assembly. Regarding EUV resists and lithography, simulations enable study of the following:

- **Resist Materials**: MD can illuminate how resist materials behave during lithographic steps such as exposure, development, and etching. For example, MD can provide insights into the dynamics of polymer chains during pattern formation and the interactions between resist components and solvents. Such insight may lead to improved resist formulations. A study by Kim et al. [286] used coarse-grained MD simulations to investigate the effect of chain conformation on line-edge roughness (LER) formation in EUV photoresists, revealing the importance of balancing chain length and polymer–developer interactions to minimize roughness [286,287]. This simulation of EUV photoresists is useful in designing a material and process for sub-1x nm patterning, as the critical dimension is almost the molecular size [288].

- **Nanoelectrode Lithography**: MD simulations can provide important insights into the underlying mechanisms of novel lithography techniques. Reactive Force Field Molecular Dynamics (ReaxFF MD) simulations have been applied to study nanoelectrode lithography (NEL) for the oxidation of silicon surfaces, including the effects of pulse duration and electric field strength on oxide formation [289].
- **Etch Processes**: MD simulations are employed to model plasma etching, simulating the interactions of ions and radicals with resist materials and substrates. This aids in optimizing etching conditions and designing etch-resistant materials, as demonstrated in studies using MD to analyze the effects of plasma composition and process parameters on etch selectivity and pattern transfer [290–293]. Figure 22 [291] shows a plasma etch simulation of SiO_2 with HF with different ion energies of (a) 20 eV, (b) 30 eV, (c) 40 eV, and (d) 80 eV. Simulation results show that the amount of SiO_2 etched increases with the increase in incident ion energy of each HF molecule.

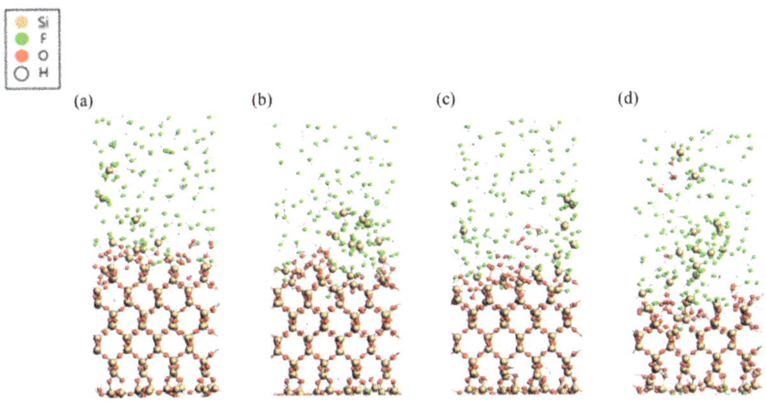

Figure 22. Simulation results of plasma etching of SiO_2 with HF with different ion energies of (**a**) 20 eV, (**b**) 30 eV, (**c**) 40 eV, and (**d**) 80 eV. Copyright 2021 American Chemical Society. Reproduced with permission from [291].

- **Self-Assembly**: MD simulations are valuable for studying the self-assembly of nanostructures like block copolymers used in nanostructured resists. These simulations provide insights into the factors governing the morphology and ordering of self-assembled structures and molecular self-assembly of photoresist materials, aiding in the design of materials with tailored properties [294,295].
- **Photolithography Development Process**: MD can predict the behavior of surfactants in photoresist during development process [296]. Results found that the concentration of surfactant affected the surface structure, which could potentially affect the yield of manufactured devices [297,298].

MD simulations provide atomic-level details, making them a powerful tool for understanding lithography's complex molecular mechanisms. The predictive capabilities of MD simulations can also accelerate the discovery and optimization of lithographic materials and processes, paving the way for advanced nanofabrication techniques. The increasing computational power and improved accuracy of force fields make

MD simulations an essential component of modern lithography research, but they also present challenges. Overcoming these limitations will further enhance the role of MD in advancing nanofabrication [299,300].

9.3. Molecular Scale Patterning Techniques

9.3.1. DNA Origami: Bottom–Up Fabrication of Complex Nanostructures

DNA origami is a bottom–up nanofabrication method that allows for the creation of arbitrary two- and three-dimensional shapes at the nanoscale. This technique utilizes the specificity of base pairing between complementary DNA strands to construct complex structures by folding a long single-stranded DNA molecule (often from a viral source) with the help of numerous shorter "staple" strands [301]. These staple strands bind to specific sequences on the scaffold strand, causing it to fold into a desired shape [302]. DNA origami serves as a molecular breadboard, integrating bottom–up (bio)chemistry with macroscopic devices created by top–down lithography [303]. Martynenko et al. [304] explored how DNA origami serves as a molecular breadboard, integrating bottom–up (bio)chemistry with top–down lithography at nanometer resolution. By using self-assembled colloidal nanoparticles instead of top–down patterning, the complexity of manufacturing is reduced, making the positioning of molecules and nanoscale objects on macroscopic arrays more accessible.

The DNA origami technique was developed by Paul Rothemund at the California Institute of Technology [305]. The process involves designing the base sequences of DNA to enable self-assembly into a variety of shapes. Computer-aided design (CAD) software, such as caDNAno, assists in designing the placement of staple strands, ensuring accurate folding and reducing errors. The precisely designed staple strand sequences are synthesized in a lab and then mixed with the scaffold strand in a buffer solution. Subjecting this mixture to a specific temperature cycle allows the staple strands to find and bind to their complementary sequences on the scaffold strand, causing it to fold into the desired shape through hydrogen bonding. Created designs can be observed through methods including electron microscopy, atomic force microscopy, or fluorescence microscopy [306]. Tørring et al. [307] highlighted DNA origami as a promising method for the spatially controlled positioning of functional materials via self-assembly, a central goal in nanotechnology. Their tutorial review discusses the basic design principles, organization of functional materials, and recent implementations in DNA robotics, along with future challenges and opportunities.

DNA origami structures can act as templates for nanoscale patterning, with applications in biotechnology [308]. They can be used to create plasmonic, entirely metallic nanostructures in a parallel manner on different substrates. DNA origami can also organize individual molecules and nanostructures with nanoscale spatial resolution and single-molecule control. Functionalized DNA origami can be selectively placed into nanoarrays using thermal scanning probe lithography (t-SPL), enabling the construction of hetero-functionalized biomolecular nanoarrays with applications in bionanotechnology and materials science [303]. Figure 23 [306] shows a schematic diagram of DNA origami engineered nanomaterials and various applications.

DNA origami leverages programmable self-assembly of DNA strands to create precise 2D/3D nanostructures, enabling bottom–up fabrication of sub-10 nm features [309]. This technique complements top–down lithography methods like nanoimprint lithography (NIL) by providing molecularly precise templates for patterning metallic nanowires or quantum dots [310,311]. For example, DNA origami scaffolds guide the deposition of lithographic resists, enhancing resolution in EUV lithography by reducing line-edge roughness through controlled nanoparticle placement [312]. Its ability to generate complex, customizable geometries make it ideal for prototyping nanoscale devices such as biochips while avoiding the high costs of traditional mask-based lithography.

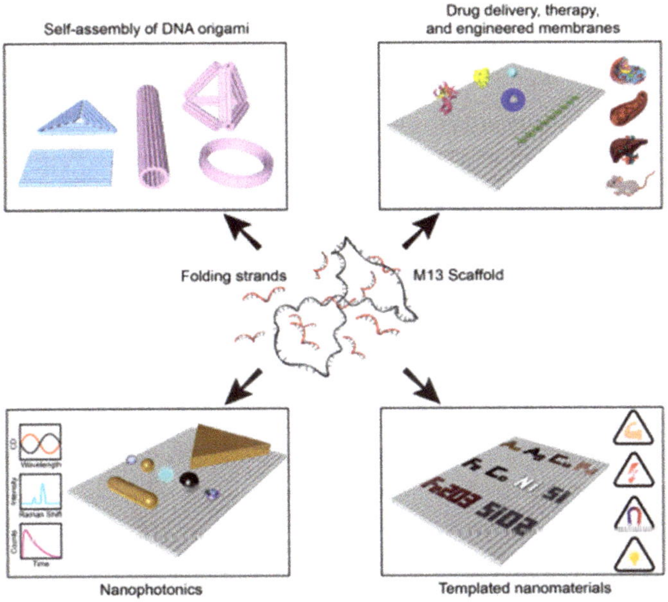

Figure 23. Schematic diagram of DNA origami applications and its engineered nanomaterials. Copyright 2023 American Chemical Society. Reproduced with permission from [306].

9.3.2. Biomolecular Templates: Guiding Assembly at the Molecular Level

Biomolecular templates harness the inherent self-assembly properties of biological molecules to direct the organization of materials at the nanoscale [313]. This bottom–up approach leverages the precise interactions between biomolecules like DNA, proteins, and viral particles to create defined arrangements, offering an alternative to top–down lithographic techniques. Molecular self-assembly is a process where molecules adopt a defined arrangement without external guidance, driven by non-covalent interactions such as hydrogen bonding, hydrophobic forces, and electrostatic interactions. Biomolecular assembly is a key process that drives structure, information storage, and communication in living organisms. Scientists emulate these natural assemblies to create modified or synthetic architectures for technological applications [314,315].

Researchers utilize both rational design and laboratory evolution strategies in biomolecular engineering. Rational design involves intuitive, modular, and computer-aided design to modify specific interactions within the assembly [316]. The use of biomolecular templates offers several advantages, including the ability to create complex architectures and to organize materials with nanoscale precision. For example, viral coat proteins, such as those from the tobacco mosaic virus (TMV), can be used as templates for the self-assembly of plasmonic nanoparticles. Petrescu et al. [317] demonstrated the use of tobacco mosaic virus (TMV) coat protein as a template to self-assemble plasmonic nanoparticles into highly symmetrical plasmonic nanorings. This approach exploits the self-assembling properties and chemical addressability of TMV coat protein, using site-directed mutagenesis and bioconjugation strategies, to arrange nanoparticles spatially and control material properties through collective interactions. TMV coat protein is known for its ability to self-assemble into supramolecular nanoparticles, as either protein discs or rods. TMV can also be modified with a small molecular fluorous ponytail at specific sites and self-assemble into spherical nanoparticles through fluorous interaction-induced self-assembly [318]. These resulting macromolecular assemblies can have interesting structural and mechanical properties. Molecular dynamics simulations are valuable for visualizing the stabilization of nuclei at an atomic level during the initial stages of condensation.

Biomolecular templates, including proteins, peptides, and viral particles, direct the assembly of nanomaterials with atomic precision, bridging gaps in conventional lithography. In nanoimprint lithography, flexible biomolecular molds (e.g., elastomer–protein composites) improve conformality on rough substrates, reducing defects [205]. For EUV lithography, peptide-based additives enhance resist sensitivity and etch resistance by modulating acid diffusion and secondary electron yield [319]. Techniques like polymer pen lithography (PPL) exploit biomolecular inks to achieve sub-100 nm resolution with combinatorial patterning capabilities, enabling high-throughput screening of nanostructures for applications in biosensing and nanoelectronics [320]. These templates also support hybrid approaches, integrating self-assembly with optical or electron beam lithography for sub-5 nm node fabrication.

10. Conclusions and Future Scope

Lithography technology faces significant challenges in applications like LSI circuit manufacturing and nanophotonics, primarily due to the limits of optical resolution and the desire for miniaturization. As device features shrink below the wavelength of light, traditional photolithography struggles with diffraction limits, necessitating techniques like EUV lithography and multiple patterning to achieve smaller features. Additionally, increasing alignment and overlay precision is crucial as feature sizes shrink, with misalignments leading to yield loss. Advances in metrology, multi-patterning, and machine learning-based alignment systems help address these issues. However, limitations in photoresist materials and mask defects remain a challenge, with the potential for maskless lithography methods to help. Furthermore, the environmental impact of lithography processes, particularly with EUV and CMP, raises concerns, prompting research into greener, more sustainable materials and processes. Overall, overcoming these challenges requires continuous innovation in materials, techniques, and sustainability.

The techniques explored in this review—extreme ultraviolet lithography (EUV), electron beam lithography (EBL), X-ray lithography (XRL), ion beam lithography (IBL), and nanoimprint lithography (NIL)—demonstrate distinct strengths and limitations, each addressing specific challenges in achieving nanoscale precision. EUV lithography emerges as a dominant technique for high-volume manufacturing at advanced nodes, leveraging shorter wavelengths, high-NA optics, and chemically amplified resists to achieve sub-10 nm resolutions. Achieving higher resolution and lower defect rate in EUV and other specialized lithography systems requires precise control of photons and energy transfer. This involves advancing optical systems such as multilayer optics and phase-shifting masks, improving resist materials, optimizing photon source stability and brightness (laser plasma sources), and implementing better process control (quantum control of photons, optical proximity correction, in situ feedback loops). By combining these approaches, it is possible to overcome the limitations of current technologies and continue scaling down to the next generation of semiconductor manufacturing. However, it requires a significant investment in both equipment and operational costs, and while its resolution is ideal, the cost of masks remains a challenge.

Meanwhile, EBL and IBL excel in flexibility and prototyping but face significant throughput constraints. XRL and NIL, with their unique capabilities in high-aspect-ratio patterning and cost-effective replication, offer promising alternatives for niche applications. NIL offers ultra-high resolution and low equipment costs but suffers from process complexity, making it better suited for niche applications or small-scale production. Critical to the success of lithography are innovations in resist materials, mask-making processes, and alignment techniques. Advanced materials such as molybdenum and silicon thin films for EUV masks, hybrid organic–inorganic resists, and metal-assisted chemical etching (MACE) techniques for NIL have pushed the boundaries of what is achievable. Alignment marks and systems now enable real-time corrections, improving overlay precision and throughput. Despite these advancements, challenges such as defect mitigation, mask durability, and stochastic effects persist, particularly as feature sizes approach atomic dimensions. Table 5 shows a comparison of different lithography techniques.

Table 5. Comparison of different lithography techniques.

	EUV Lithography	Electron Beam Lithography	X-Ray Lithography	Ion Beam Lithography	Nanoimprint Lithography
Wavelength	~13.5 nm	~1–10 nm	~1–10 nm (soft X-rays)	~10–100 nm	N/A
Throughput	High (production level)	Slow	Moderate	Slow	Very fast
Resolution	~5–7 nm	sub 10 nm	~10 nm	~10 nm	~10–20 nm
Cost	High	High	High	High	Moderate
Masks	Required	No masks	Required	FIB: No masks PBW: No masks IPL: Required	No masks
Advantages	High resolution, well-suited for mass production	High resolution, not for mass production	High resolution	High precision	High throughput
Disadvantages	Complex and expensive infrastructure, limited material availability	Slow speed, expensive	Complex, limited material interactions	Slow, expensive, resolution limited by beam interactions	Limited to specific materials, requires molds

Looking forward, the integration of computational techniques like AI-driven lithographic optimization, the development of alternative lithographic methods such as maskless direct-write systems, and the exploration of emerging materials will be pivotal in overcoming current limitations. Additionally, hybrid approaches combining different lithographic techniques could provide tailored solutions for complex manufacturing needs. In conclusion, while lithography faces significant technical and economic challenges, it remains indispensable in pushing the boundaries of semiconductor technology. As research and development continue to address its limitations, lithography will undoubtedly play a central role in shaping the next generation of electronic devices, from advanced integrated circuits to nanoscale sensors and beyond. Some potential research hotspots and breakthrough points include hybrid lithography systems, advanced resists and materials, fabricating quantum circuits and photonic devices, and developing more sustainable and energy-efficient lithography techniques, such as reducing the power consumption of EUV machines or finding eco-friendly resist materials. This review highlights the ongoing efforts and innovations that ensure lithography remains at the forefront of nanoscale manufacturing, inspiring further exploration and breakthroughs in this critical field.

Author Contributions: Conceptualization, data curation, writing—review and editing, writing—original draft, P.B.; data visualization and graphics optimization, writing—review and editing, writing—original draft, J.V.; data visualization, writing—review and editing, writing—original draft, V.A.; writing—review and editing, R.K.R.; review and editing, Y.K.S.; review and editing, supervision, conceptualization, data curation, V.K.; All authors have read and agreed to the published version of the manuscript.

Funding: This research received no external funding.

Institutional Review Board Statement: Not applicable.

Informed Consent Statement: Not applicable.

Data Availability Statement: The authors will provide available data or assist in finding them in the databases.

Conflicts of Interest: The authors declare no conflicts of interest.

References

1. Pease, R.F.; Chou, S.Y. Lithography and Other Patterning Techniques for Future Electronics. *Proc. IEEE* **2008**, *96*, 248–270. [CrossRef]
2. Thompson, L.F. An Introduction to Lithography. In *Introduction to Microlithography*; ACS Symposium Series; American Chemical Society: Washington, DC, USA, 1983; Volume 219, pp. 1–13.
3. Chan, K.F.; Feng, Z.; Yang, R.; Ishikawa, A.; Mei, W. High-resolution maskless lithography. *J. Micro/Nanolithography MEMS MOEMS* **2003**, *2*, 331–339.
4. Pease, R.F. Maskless lithography. *Microelectron. Eng.* **2005**, *78–79*, 381–392. [CrossRef]
5. Chiu, G.L.T.; Shaw, J.M. Optical lithography: Introduction. *IBM J. Res. Dev.* **1997**, *41*, 3–6. [CrossRef]
6. Bratton, D.; Yang, D.; Dai, J.; Ober, C.K. Recent progress in high resolution lithography. *Polym. Adv. Technol.* **2006**, *17*, 94–103. [CrossRef]
7. Seisyan, R.P. Nanolithography in microelectronics: A review. *Tech. Phys.* **2011**, *56*, 1061–1073. [CrossRef]
8. Harriott, L.R. Limits of lithography. *Proc. IEEE* **2001**, *89*, 366–374. [CrossRef]
9. Ito, T.; Okazaki, S. Pushing the limits of lithography. *Nature* **2000**, *406*, 1027–1031. [CrossRef]
10. Sanders, D.P. Advances in Patterning Materials for 193 nm Immersion Lithography. *Chem. Rev.* **2010**, *110*, 321–360. [CrossRef]
11. Hasan, R.M.M.; Luo, X. Promising Lithography Techniques for Next-Generation Logic Devices. *Nanomanufacturing Metrol.* **2018**, *1*, 67–81. [CrossRef]
12. Kim, S.S.; Yong, S.K.; Kim, W.; Kang, S.; Park, H.W.; Yoon, K.J.; Sheen, D.S.; Lee, S.; Hwang, C.S. Review of Semiconductor Flash Memory Devices for Material and Process Issues. *Adv. Mater.* **2023**, *35*, 2200659. [CrossRef]
13. Rothschild, M.; Forte, A.R.; Horn, M.W.; Kunz, R.R.; Palmateer, S.C.; Sedlacek, J.H.C. 193-nm lithography. *IEEE J. Sel. Top. Quantum Electron.* **1995**, *1*, 916–923. [CrossRef]
14. Canpolat-Schmidt, C.; Heldt, G.; Helke, C.; Voigt, A.; Reuter, D. *Lithographic Performance of Resist ma-N 1402 in an e-Beam/i-Line Stepper Intra-Level Mix and Match Approach*; SPIE: Bellingham, WA, USA, 2022; Volume 12472.
15. Helke, C.; Canpolat-Schmidt, C.H.; Heldt, G.; Schermer, S.; Hartmann, S.; Voigt, A.; Reuter, D. Intra-level mix and match lithography with electron beam lithography and i-line stepper combined with resolution enhancement for structures below the CD-limit. *Micro Nano Eng.* **2023**, *19*, 100189. [CrossRef]
16. Schermer, S.; Helke, C.; Reinhardt, M.; Hartmann, S.; Tank, F.; Wecker, J.; Heldt, G.; Voigt, A.; Reuter, D. Characterization of negative tone photoresist mr-EBL 6000.5 for i-line stepper and electron beam lithography for the Intra-Level Mix & Match Approach. *Micro Nano Eng.* **2024**, *23*, 100264. [CrossRef]
17. Lewis, J.A.; Gratson, G.M. Direct writing in three dimensions. *Mater. Today* **2004**, *7*, 32–39. [CrossRef]
18. Wang, H.; Pan, C.-F.; Li, C.; Menghrajani, K.S.; Schmidt, M.A.; Li, A.; Fan, F.; Zhou, Y.; Zhang, W.; Wang, H.; et al. Two-photon polymerization lithography for imaging optics. *Int. J. Extrem. Manuf.* **2024**, *6*, 042002. [CrossRef]
19. Harinarayana, V.; Shin, Y.C. Two-photon lithography for three-dimensional fabrication in micro/nanoscale regime: A comprehensive review. *Opt. Laser Technol.* **2021**, *142*, 107180. [CrossRef]
20. Faraji Rad, Z.; Prewett, P.D.; Davies, G.J. High-resolution two-photon polymerization: The most versatile technique for the fabrication of microneedle arrays. *Microsyst. Nanoeng.* **2021**, *7*, 71. [CrossRef]
21. Pinheiro, T.; Morais, M.; Silvestre, S.; Carlos, E.; Coelho, J.; Almeida, H.V.; Barquinha, P.; Fortunato, E.; Martins, R. Direct Laser Writing: From Materials Synthesis and Conversion to Electronic Device Processing. *Adv. Mater.* **2024**, *36*, 2402014. [CrossRef]
22. Wahl, M.; Michelmann, J.; Sailer, H.; Schneider, A.; Dionisio, N. *Advanced Patterning Strategies for Maskless Laser Direct Write Lithography*; SPIE: Bellingham, WA, USA, 2024; Volume 12956.
23. Qin, D.; Xia, Y.; Whitesides, G.M. Soft lithography for micro-and nanoscale patterning. *Nat. Protoc.* **2010**, *5*, 491.
24. Cavallini, M.; Gentili, D.; Greco, P.; Valle, F.; Biscarini, F. Micro- and nanopatterning by lithographically controlled wetting. *Nat. Protoc.* **2012**, *7*, 1668–1676. [CrossRef]
25. Samorì, P. *Scanning Probe Microscopies Beyond Imaging: Manipulation of Molecules and Nanostructures*; John Wiley & Sons: Hoboken, NJ, USA, 2006.
26. Shamoun, B.; Chandramouli, M.; Liu, B.; Juday, R.; Bucay, I.; Sowers, A.; Abboud, F. *Multi-Beam Mask Writer in EUV Era: Challenges and Opportunities*; SPIE: Bellingham, WA, USA, 2021; Volume 11610.

27. Tomandl, M.; Spengler, C.; Hudek, P.; Klein, C.; Loeschner, H.; Platzgummer, E. Multi-beam mask writing opens up new fields of application, including curvilinear mask pattern for high numerical aperture extreme ultraviolet lithography. *J. Micro/Nanopatterning Mater. Metrol.* **2024**, *23*, 011205. [CrossRef]
28. Matsumoto, H.; Yasuda, J.; Motosugi, T.; Kimura, H.; Kojima, Y.; Yamashita, H.; Saito, M.; Nakayamada, N. *Multi-Beam Mask Writer MBM-3000 for Next Generation EUV Mask Production*; SPIE: Bellingham, WA, USA, 2024; Volume 12956.
29. Ceballos, A.; MacWilliams, K.; Prescop, T.; Loewen, R. *Full-Wafer, Maskless Patterning with Sub-50nm Resolution and Large Depth-of-Focus Enabled by Multicolumn Electron Beam Lithography*; SPIE: Bellingham, WA, USA, 2024; Volume 12956.
30. Lee, P.P.; Chang, P.J.; Tsai, F.F.; Ku, H.C.; Cheng, M.-D.; Wang, J.J.; Lii, M.J.; Atanelov, J.; Dang, A.-D.; Kvasnica, S. *Make a Core Engine Ready to Empower High Quality Multi-Beam Mask Writer for N2 Node and Beyond*; SPIE: Bellingham, WA, USA, 2024; Volume 13177.
31. Ou, T.W.; Ho, W.K.; Lai, T.S.; Lu, J.L.; Chen, A.C.; Egl, A.; Kühmayer, M.; Brenner, F. *New Applications on Multi-Beam Mask Writers to Enable Mask-Making in 3nm and Beyond*; SPIE: Bellingham, WA, USA, 2024; Volume 13216.
32. Aibara, I.; Matsumoto, H.; Yasuda, J.; Yasui, K.-i.; Motosugi, T.; Kimura, H.; Kawaguchi, M.; Kojima, Y.; Saito, M.; Nakayamada, N. *Recent Progress of Multi-Beam Mask Writer MBM-3000*; SPIE: Bellingham, WA, USA, 2024; Volume 13273.
33. Noriaki, N.; Haruyuki, N.; Yasuo, K.; Kenichi, Y.; Abhishek, S.; Nagesh, S.; Yukihiro, M.; Aki, F. Curvilinear mask process correction embedded on multi-beam mask writer. *J. Micro/Nanopatterning Mater. Metrol.* **2024**, *23*, 011206. [CrossRef]
34. Rothschild, M.; Forte, A.R.; Kunz, R.R.; Palmateer, S.C.; Sedlacek, J.H.C. Lithography at a wavelength of 193 nm. *IBM J. Res. Dev.* **1997**, *41*, 49–55. [CrossRef]
35. Fu, N.; Liu, Y.; Ma, X.; Chen, Z. EUV lithography: State-of-the-art review. *J. Microelectron. Manuf.* **2019**, *2*, 1–6. [CrossRef]
36. Nguyen, K.B.; Cardinale, G.F.; Tichenor, D.A.; Kubiak, G.D.; Berger, K.; Ray-Chaudhuri, A.K.; Perras, Y.; Haney, S.J.; Nissen, R.; Krenz, K.; et al. Fabrication of metal–oxide–semiconductor devices with extreme ultraviolet lithography. *J. Vac. Sci. Technol. B Microelectron. Nanometer Struct. Process. Meas. Phenom.* **1996**, *14*, 4188–4192. [CrossRef]
37. Hansson, B.; Fomenkov, I.; Böwering, N.; Ershov, A.; Partlo, W.; Myers, D.; Khodykin, O.; Bykanov, A.; Rettig, C.; Hoffman, J.; et al. *LPP EUV Source Development for HVM*; SPIE: Bellingham, WA, USA, 2006; Volume 6151.
38. Versolato, O.O. Physics of laser-driven tin plasma sources of EUV radiation for nanolithography. *Plasma Sources Sci. Technol.* **2019**, *28*, 083001. [CrossRef]
39. Wagner, C.; Harned, N. Lithography gets extreme. *Nat. Photonics* **2010**, *4*, 24–26. [CrossRef]
40. Yang, D.-K.; Wang, D.; Huang, Q.-S.; Song, Y.; Wu, J.; Li, W.-X.; Wang, Z.-S.; Tang, X.-H.; Xu, H.-X.; Liu, S. The development of laser-produced plasma EUV light source. *Chip* **2022**, *1*, 100019. [CrossRef]
41. van Zwol, P.J.; Nasalevich, M.; Voorthuijzen, W.P.; Kurganova, E.; Notenboom, A.; Vles, D.; Peter, M.; Symens, W.; Giesbers, A.J.M.; Klootwijk, J.H.; et al. *Pellicle Films Supporting the Ramp to HVM with EUV*; SPIE: Bellingham, WA, USA, 2017; Volume 10451.
42. Brandt, D.; Fomenkov, I.; Stewart, J. *Progress in Availably of NXE:3400B EUVL Sources in the Field and Power Scaling Towards 500W (Conference Presentation)*; SPIE: Bellingham, WA, USA, 2020; Volume 11323.
43. Brandt, D.; Purvis, M.; Fomenkov, I.; Brown, D.; Schafgans, A.; Mayer, P.; Rafac, R. *Advances Toward High Power EUV Sources for EUVL Scanners for HVM in the Next Decade and Beyond*; SPIE: Bellingham, WA, USA, 2021; Volume 11609.
44. Mourou, G.A.; Hulin, D.; Galvanauskas, A. The Road to High Peak Power and High Average Power Lasers: Coherent-Amplification-Network (CAN). In Proceedings of the AIP Conference Proceedings, Varenna, Italy, 19–24 September 2006; pp. 152–163.
45. Bilski, B.; Zimmermann, J.; Roesch, M.; Liddle, J.; van Setten, E.; Bottiglieri, G.; van Schoot, J. *High-NA EUV Imaging: Challenges and Outlook*; SPIE: Bellingham, WA, USA, 2019; Volume 11177.
46. van Eelco, S.; Gerardo, B.; John, M.; van Jan, S.; Kars, T.; Joseph, Z.; Timon, F.; Stephen, H.; Joerg, Z.; Matthias, R.; et al. High NA EUV lithography: Next step in EUV imaging. In *Extreme Ultraviolet (EUV) Lithography X*; SPIE: Bellingham, WA, USA, 2019; p. 1095709.
47. Borman, S.; Gustas, D.; van Loo, H.; Gang, T.; Oorschot, D.; Brouwer, A.; Colina, A.; Horsten, F.; van Rhee, T. *Measurement of Image Fading Impact on 20P40 Contact Hole LCDU*; SPIE: Bellingham, WA, USA, 2023; Volume PC12750.
48. Manouras, T.; Argitis, P. High Sensitivity Resists for EUV Lithography: A Review of Material Design Strategies and Performance Results. *Nanomaterials* **2020**, *10*, 1593. [CrossRef]
49. Cao, H.; Yueh, W.; Rice, B.; Roberts, J.; Bacuita, T.; Chandhok, M. *Sources of Line-Width Roughness for EUV Resists*; SPIE: Bellingham, WA, USA, 2004; Volume 5376.
50. Hasan, M.W.; Deeb, L.; Kumaniaev, S.; Wei, C.; Wang, K. Recent Advances in Metal-Oxide-Based Photoresists for EUV Lithography. *Micromachines* **2024**, *15*, 1122. [CrossRef] [PubMed]
51. Dong, X.; Shao, Y.a.; Ping, H.; Tong, X.; Wu, Y.; Zhang, Y.; Wang, M.; Zheng, Z.; Zhao, J.; Wang, J.; et al. Effect of Metal Oxide Deposition on the Sensitivity and Resolution of E-Beam Photoresist. *ACS Appl. Mater. Interfaces* **2024**, *16*, 56019–56030. [CrossRef]

52. Sharma, E.; Rathi, R.; Misharwal, J.; Sinhmar, B.; Kumari, S.; Dalal, J.; Kumar, A. Evolution in Lithography Techniques: Microlithography to Nanolithography. *Nanomaterials* **2022**, *12*, 2754. [CrossRef]
53. Ober, C.; Xu, H.; Kosma, V.; Sakai, K.; Giannelis, E. *EUV Photolithography: Resist Progress and Challenges*; SPIE: Bellingham, WA, USA, 2018; Volume 10583.
54. Hassaan, M.; Saleem, U.; Singh, A.; Haque, A.J.; Wang, K. Recent Advances in Positive Photoresists: Mechanisms and Fabrication. *Materials* **2024**, *17*, 2552. [CrossRef] [PubMed]
55. Yao, X.; Lian, P.; Chen, J.; Zeng, Y.; Yu, T.; Wang, S.; Guo, X.; Hu, R.; Tian, P.; Vockenhuber, M.; et al. Iodonium functionalized polystyrene as non-chemically amplified resists for electron beam and extreme ultraviolet lithography. *RSC Appl. Polym.* **2024**, *2*, 870–879. [CrossRef]
56. Yamamoto, H.; Vesters, Y.; Jiang, J.; Simone, D.D.; Vandenberghe, G.; Kozawa, T. Role of Metal Sensitizers for Sensitivity Improvement in EUV Chemically Amplified Resist. *J. Photopolym. Sci. Technol.* **2018**, *31*, 747–751. [CrossRef]
57. Sanjay; Kumar, V.; Vohra, A. Triple and quadruple metal gate work function engineering to improve the performance of junctionless double surrounding gate In0.53Ga0.47As nanotube MOSFET for the upcoming Sub 3 nm technology node. *Phys. Lett. A* **2024**, *527*, 130011. [CrossRef]
58. Thakur, N.; Tseng, L.-T.; Vockenhuber, M.; Ekinci, Y.; Castellanos, S. Stability studies on a sensitive EUV photoresist based on zinc metal oxoclusters. *J. Micro/Nanolithography MEMS MOEMS* **2019**, *18*, 043504.
59. Van Schoot, J.; Lok, S.; van Setten, E.; Maas, R.; Troost, K.; Peeters, R.; Finders, J.; Stoeldraijer, J.; Benschop, J.; Graeupner, P.; et al. *High-NA EUV Lithography Exposure Tool: Advantages and Program Progress*; SPIE: Bellingham, WA, USA, 2021; Volume 11517.
60. Petrillo, K.; Huang, G.; Ashworth, D.; Georger, J.; Ren, L.; Cho, K.Y.; Montgomery, W.; Wurm, S.; Kawakami, S.; Dunn, S.; et al. *Line Width Roughness Control and Pattern Collapse Solutions for EUV Patterning*; SPIE: Bellingham, WA, USA, 2011; Volume 7969.
61. Groves, T.R.; Pickard, D.; Rafferty, B.; Crosland, N.; Adam, D.; Schubert, G. Maskless electron beam lithography: Prospects, progress, and challenges. *Microelectron. Eng.* **2002**, *61–62*, 285–293. [CrossRef]
62. Vieu, C.; Carcenac, F.; Pépin, A.; Chen, Y.; Mejias, M.; Lebib, A.; Manin-Ferlazzo, L.; Couraud, L.; Launois, H. Electron beam lithography: Resolution limits and applications. *Appl. Surf. Sci.* **2000**, *164*, 111–117. [CrossRef]
63. Tseng, A.A.; Kuan, C.; Chen, C.D.; Ma, K.J. Electron beam lithography in nanoscale fabrication: Recent development. *IEEE Trans. Electron. Packag. Manuf.* **2003**, *26*, 141–149. [CrossRef]
64. Cui, Z. (Ed.) Electron Beam Lithography. In *Nanofabrication: Principles, Capabilities and Limits*; Springer International Publishing: Berlin/Heidelberg, Germany, 2024; pp. 83–139.
65. Sebastian, E.M.; Jain, S.K.; Purohit, R.; Dhakad, S.K.; Rana, R.S. Nanolithography and its current advancements. *Mater. Today: Proc.* **2020**, *26*, 2351–2356. [CrossRef]
66. Chen, Y. Nanofabrication by electron beam lithography and its applications: A review. *Microelectron. Eng.* **2015**, *135*, 57–72. [CrossRef]
67. Pfeiffer, H. *Direct Write Electron Beam Lithography: A Historical Overview*; SPIE: Bellingham, WA, USA, 2010; Volume 7823.
68. Kumar, V.; Maan, A.S. Improvement in reverse bias leakage current of Ni/4H-nSiC Schottky barrier diodes via MeV selective ion irradiation. *IOP Conf. Ser. Mater. Sci. Eng.* **2018**, *331*, 012016. [CrossRef]
69. Hohn, F.J. Electron beam lithography: Its applications. *J. Vac. Sci. Technol. B Microelectron. Process. Phenom.* **1989**, *7*, 1405–1411. [CrossRef]
70. Cen Shawn, W.; Yoshiyuki, M.; ChiiDong, C. High-energy Electron Beam Lithography for Nanoscale Fabrication. In *Lithography, Michael, W., Ed.*; IntechOpen: Rijeka, Hrvatska, 2010; p. 241.
71. Regina, F.; Jeff, S.; Alan, B.; Paul, P.; Mark, M.; Kurt, R.; Luc, H.; Bart, V. Demonstration of lithography patterns using reflective e-beam direct write. In *Alternative Lithographic Technologies III*; SPIE: Bellingham, WA, USA, 2011; p. 79701T.
72. Ahmed, H. Physical principles of electron beam lithography. *Sci. Prog.* **1986**, *70*, 473–487.
73. Nabity, J.C.; Wybourne, M.N. A versatile pattern generator for high-resolution electron-beam lithography. *Rev. Sci. Instrum.* **1989**, *60*, 27–32. [CrossRef]
74. Yu, B.; Jhih-Rong, G.; Pan, D.Z. L-shape based layout fracturing for e-beam lithography. In Proceedings of the 2013 18th Asia and South Pacific Design Automation Conference (ASP-DAC), Yokohama, Japan, 22–25 January 2013; pp. 249–254.
75. Broers, A.N.; Hoole, A.C.F.; Ryan, J.M. Electron beam lithography—Resolution limits. *Microelectron. Eng.* **1996**, *32*, 131–142. [CrossRef]
76. Okazaki, S. High resolution optical lithography or high throughput electron beam lithography: The technical struggle from the micro to the nano-fabrication evolution. *Microelectron. Eng.* **2015**, *133*, 23–35. [CrossRef]
77. Liu, M.; Qiu, Y.; Chen, b.; Xiu, Q.; Zheng, Y. Electron beam lithography and its application in fabricating nano-device. In Proceedings of the 7th International Conference on Solid-State and Integrated Circuits Technology, Beijing, China, 18–21 October 2004; Volume 561, pp. 563–566.
78. Pan, J.C.; Lang, F.C.; Zhao, X.P.; Hou, X.H.; Li, J.J. Nanomechanical properties of the ZEP-520 electron beam resist film. *Mater. Today Commun.* **2021**, *27*, 102442. [CrossRef]

79. Koleva, E.; Vutova, K.; Asparuhova, B.; Kostic, I.; Cvetkov, K.; Gerasimov, V. Modeling approaches for electron beam lithography. *J. Phys. Conf. Ser.* **2018**, *1089*, 012016. [CrossRef]
80. Sahin, O.; Albayrak, O.M.; Yapici, M.K. Optimization of e-beam lithography parameters for nanofabrication of sub-50 nm gold nanowires and nanogaps based on a bilayer lift-off process. *Nanotechnology* **2024**, *35*, 395301. [CrossRef]
81. Yoo, J.B.; Park, S.-W.; Kang, H.N.; Mondkar, H.S.; Sohn, K.; Kim, H.-M.; Kim, K.-B.; Lee, H. Triphenylsulfonium salt methacrylate bound polymer resist for electron beam lithography. *Polymer* **2014**, *55*, 3599–3604. [CrossRef]
82. Singh, V.; Satyanarayana, V.S.V.; Sharma, S.K.; Ghosh, S.; Gonsalves, K.E. Towards novel non-chemically amplified (n-CARS) negative resists for electron beam lithography applications. *J. Mater. Chem. C* **2014**, *2*, 2118–2122. [CrossRef]
83. Satyanarayana, V.S.V.; Kessler, F.; Singh, V.; Scheffer, F.R.; Weibel, D.E.; Ghosh, S.; Gonsalves, K.E. Radiation-Sensitive Novel Polymeric Resist Materials: Iterative Synthesis and Their EUV Fragmentation Studies. *ACS Appl. Mater. Interfaces* **2014**, *6*, 4223–4232. [CrossRef]
84. Abbas, A.S.; Alqarni, S.; Shokouhi, B.B.; Yavuz, M.; Cui, B. Water soluble and metal-containing electron beam resist poly(sodium 4-styrenesulfonate). *Mater. Res. Express* **2014**, *1*, 045102. [CrossRef]
85. Stowers, J.; Keszler, D.A. High resolution, high sensitivity inorganic resists. *Microelectron. Eng.* **2009**, *86*, 730–733. [CrossRef]
86. Gangnaik, A.S.; Georgiev, Y.M.; Holmes, J.D. New Generation Electron Beam Resists: A Review. *Chem. Mater.* **2017**, *29*, 1898–1917. [CrossRef]
87. Qin, N.; Qian, Z.-G.; Zhou, C.; Xia, X.-X.; Tao, T.H. 3D electron-beam writing at sub-15 nm resolution using spider silk as a resist. *Nat. Commun.* **2021**, *12*, 5133. [CrossRef]
88. Kofler, J.; Schmoltner, K.; Klug, A.; List-Kratochvil, E.J.W. Highly robust electron beam lithography lift-off process using chemically amplified positive tone resist and PEDOT:PSS as a protective coating. *J. Micromechanics Microengineering* **2014**, *24*, 095010. [CrossRef]
89. Canalejas-Tejero, V.; Carrasco, S.; Navarro-Villoslada, F.; García Fierro, J.L.; Capel-Sánchez, M.d.C.; Moreno-Bondi, M.C.; Barrios, C.A. Ultrasensitive non-chemically amplified low-contrast negative electron beam lithography resist with dual-tone behaviour. *J. Mater. Chem. C* **2013**, *1*, 1392–1398. [CrossRef]
90. Groves, T.R. 3—Electron beam lithography. In *Nanolithography*; Feldman, M., Ed.; Woodhead Publishing: Sawston, UK, 2014; pp. 80–115.
91. Smith, D.J. Ultimate resolution in the electron microscope? *Mater. Today* **2008**, *11*, 30–38. [CrossRef]
92. Tormen, M.; Grenci, G.; Marmiroli, B.; Romanato, F. X-ray Lithography. In *Nano-Lithography*; Wiley: Hoboken, NJ, USA, 2013; pp. 1–86.
93. Smith, H.I.; Spears, D.L.; Bernacki, S.E. X-Ray Lithography: A Complementary Technique to Electron Beam Lithography. *J. Vac. Sci. Technol.* **1973**, *10*, 913–917. [CrossRef]
94. Maldonado, J.R.; Poulsen, M.E.; Saunders, T.E.; Vratny, F.; Zacharias, A. X-ray lithography source using a stationary solid Pd target. *J. Vac. Sci. Technol.* **1979**, *16*, 1942–1945. [CrossRef]
95. Bharti, A.; Turchet, A.; Marmiroli, B. X-Ray Lithography for Nanofabrication: Is There a Future? *Front. Nanotechnol.* **2022**, *4*, 835701. [CrossRef]
96. Nazmov, V.; Goldenberg, B.; Vasiliev, A.; Asadchikov, V. Optimization of x-ray lithography conditions for fabrication of large arrays of high-aspect-ratio submicron pores. *J. Micromechanics Microengineering* **2021**, *31*, 055011. [CrossRef]
97. Silverman, J.P. X-ray lithography: Status, challenges, and outlook for 0.13 μm. *J. Vac. Sci. Technol. B Microelectron. Nanometer Struct. Process. Meas. Phenom.* **1997**, *15*, 2117–2124. [CrossRef]
98. Nagel, D.J. Chapter 6—Plasma Sources for X-Ray Lithography. In *VLSI Electronics Microstructure Science*; Einspruch, N.G., Brown, D.M., Eds.; Elsevier: Amsterdam, The Netherlands, 1984; Volume 8, pp. 137–170.
99. Okada, I.; Saitoh, Y.; Itabashi, S.; Yoshihara, H. A plasma x-ray source for x-ray lithography. *J. Vac. Sci. Technol. B Microelectron. Process. Phenom.* **1986**, *4*, 243–247. [CrossRef]
100. Kalantar, D.H.; Hammer, D.A.; Mittal, K.C.; Qi, N.; Maldonado, J.R.; Vladimirsky, Y. Intense pulsed plasma x-ray source for lithography. *J. Vac. Sci. Technol. B Microelectron. Nanometer Struct. Process. Meas. Phenom.* **1991**, *9*, 3245–3249. [CrossRef]
101. Chaker, M.; Pépin, H.; Bareau, V.; Lafontaine, B.; Toubhans, I.; Fabbro, R.; Faral, B. Laser plasma x-ray sources for microlithography. *J. Appl. Phys.* **1988**, *63*, 892–899. [CrossRef]
102. Chaker, M.; La Fontaine, B.; Côté, C.Y.; Kieffer, J.C.; Pépin, H.; Talon, M.H.; Enright, G.D.; Villeneuve, D.M. Laser plasma sources for proximity printing or projection x-ray lithography. *J. Vac. Sci. Technol. B Microelectron. Nanometer Struct. Process. Meas. Phenom.* **1992**, *10*, 3239–3242. [CrossRef]
103. Kohn, V.G. On the theory of synchrotron radiation nanofocusing with planar compound refractive lenses. *J. Synchrotron Radiat.* **2022**, *29*, 615–621. [CrossRef]
104. Haelbich, R.P.; Silverman, J.P.; Warlaumont, J.M. Synchrotrom radiation X-ray lithography. *Nucl. Instrum. Methods Phys. Res.* **1984**, *222*, 291–301. [CrossRef]

105. Maldonado, J.R.; Peckerar, M. X-ray lithography: Some history, current status and future prospects. *Microelectron. Eng.* **2016**, *161*, 87–93. [CrossRef]
106. Pearlman, J.; Riordan, J. *Bright Discharge Plasma Sources For X-Ray Lithography*; SPIE: Bellingham, WA, USA, 1985; Volume 0537.
107. Kitayama, T.; Hayasaka, T.; Yoshihara, H.; Ishihara, S. *Synchrotron X-ray Lithography System Using A Compact Source*; SPIE: Bellingham, WA, USA, 1989; Volume 1089.
108. Nazmov, V.P.; Goldenberg, B.G.; Reznikova, E.F.; Boerner, M. Self-aligned single-exposure deep x-ray lithography. *AIP Conf. Proc.* **2020**, 2299. [CrossRef]
109. Feldman, M.; Sun, J. Resolution limits in x-ray lithography. *J. Vac. Sci. Technol. B Microelectron. Nanometer Struct. Process. Meas. Phenom.* **1992**, *10*, 3173–3176. [CrossRef]
110. Smith, H.I.; Schattenburg, M.L. X-ray lithography from 500 to 30 nm: X-ray nanolithography. *IBM J. Res. Dev.* **1993**, *37*, 319–329. [CrossRef]
111. Zalevsky, Z.; Abdulhalim, I. Chapter 4—Fabrication Aspects of Integrated Devices. In *Integrated Nanophotonic Devices*, 2nd ed.; Zalevsky, Z., Abdulhalim, I., Eds.; William Andrew Publishing: Oxford, UK, 2014; pp. 103–115.
112. Thompson, L.F.; Feit, E.D.; Bowden, M.J.; Lenzo, P.V.; Spencer, E.G. Polymeric Resists for X-Ray Lithography. *J. Electrochem. Soc.* **1974**, *121*, 1500. [CrossRef]
113. Schenk, R.; Halle, O.; Müllen, K.; Ehrfeld, W.; Schmidt, M. Highly sensitive resist material for deep X-ray lithography. *Microelectron. Eng.* **1997**, *35*, 105–108. [CrossRef]
114. Moldonado, J.R. X-ray lithography, where it is now and where it is going. *J. Electron. Mater.* **1990**, *19*, 699–709. [CrossRef]
115. Spille, E.; Feder, R. X-ray lithography. In *X-Ray Optics: Applications to Solids*; Queisser, H.-J., Ed.; Springer: Berlin, Heidelberg, 1977; pp. 35–92.
116. Mekaru, H. Performance of SU-8 Membrane Suitable for Deep X-Ray Grayscale Lithography. *Micromachines* **2015**, *6*, 252–265. [CrossRef]
117. Warlaumont, J. X-ray lithography: On the path to manufacturing. *J. Vac. Sci. Technol. B Microelectron. Process. Phenom.* **1989**, *7*, 1634–1641. [CrossRef]
118. Heuberger, A. X-ray lithography. *Microelectron. Eng.* **1986**, *5*, 3–38. [CrossRef]
119. Seligson, D.; Ito, H.; Willson, C.G. The impact of high-sensitivity resist materials on x-ray lithography. *J. Vac. Sci. Technol. B Microelectron. Process. Phenom.* **1988**, *6*, 2268–2273. [CrossRef]
120. Haller, I.; Feder, R.; Hatzakis, M.; Spiller, E. Copolymers of Methyl Methacrylate and Methacrylic Acid and Their Metal Salts as Radiation Sensitive Resists. *J. Electrochem. Soc.* **1979**, *126*, 154. [CrossRef]
121. Lenzo, P.V.; Spencer, E.G. High-speed low-power x-ray lithography. *Appl. Phys. Lett.* **1974**, *24*, 289–291. [CrossRef]
122. Stokes, K.; Clark, K.; Odetade, D.; Hardy, M.; Goldberg Oppenheimer, P. Advances in lithographic techniques for precision nanostructure fabrication in biomedical applications. *Discov Nano* **2023**, *18*, 153. [CrossRef]
123. Silverman, J.P. Challenges and progress in X-ray lithography. *J. Vac. Sci. Technol. B Microelectron. Nanometer Struct. Process. Meas. Phenom.* **1998**, *16*, 3137–3141. [CrossRef]
124. Murphy, J.B.; White, D.L.; MacDowell, A.A.; Wood, O.R. Synchrotron radiation sources and condensers for projection x-ray lithography. *Appl. Opt.* **1993**, *32*, 6920–6929. [CrossRef]
125. Mah, A.H.; Mei, H.; Basu, P.; Laws, T.S.; Ruchhoeft, P.; Verduzco, R.; Stein, G.E. Swelling responses of surface-attached bottlebrush polymer networks. *Soft Matter* **2018**, *14*, 6728–6736. [CrossRef]
126. Brown, W.L.; Venkatesan, T.; Wagner, A. Ion beam lithography. *Nucl. Instrum. Methods Phys. Res.* **1981**, *191*, 157–168. [CrossRef]
127. He, S.; Tian, R.; Wu, W.; Li, W.-D.; Wang, D. Helium-ion-beam nanofabrication: Extreme processes and applications. *Int. J. Extrem. Manuf.* **2021**, *3*, 012001. [CrossRef]
128. Melngailis, J.; Mondelli, A.A.; Berry, I.L., III.; Mohondro, R. A review of ion projection lithography. *J. Vac. Sci. Technol. B Microelectron. Nanometer Struct. Process. Meas. Phenom.* **1998**, *16*, 927–957. [CrossRef]
129. Kumar, V.; Verma, J.; Maan, A.S.; Akhtar, J. Epitaxial 4H–SiC based Schottky diode temperature sensors in ultra-low current range. *Vacuum* **2020**, *182*, 109590. [CrossRef]
130. Watt, F.; Bettiol, A.A.; van Kan, J.A.; Teo, E.J.; Breese, M.B.H. Ion Beam Lithography and Nanofabrication: A Review. *Int. J. Nanosci.* **2005**, *04*, 269–286. [CrossRef]
131. Joshi-Imre, A.; Bauerdick, S. Direct-Write Ion Beam Lithography. *J. Nanotechnol.* **2014**, *2014*, 170415. [CrossRef]
132. Melngailis, J. Focused ion beam technology and applications. *J. Vac. Sci. Technol. B Microelectron. Process. Phenom.* **1987**, *5*, 469–495. [CrossRef]
133. Li, P.; Chen, S.; Dai, H.; Yang, Z.; Chen, Z.; Wang, Y.; Chen, Y.; Peng, W.; Shan, W.; Duan, H. Recent advances in focused ion beam nanofabrication for nanostructures and devices: Fundamentals and applications. *Nanoscale* **2021**, *13*, 1529–1565. [CrossRef]
134. Smith, N.S.; Notte, J.A.; Steele, A.V. Advances in source technology for focused ion beam instruments. *MRS Bull.* **2014**, *39*, 329–335. [CrossRef]

135. Narro, J.; Diaz, R. A New Tool for Automation of Focused Ion Beam Bitmap Milling of Two-and Three-Dimensional Micro and Nanostructures. *Microsc. Microanal.* **2022**, *28*, 88–90. [CrossRef]
136. Rosa, L.; Sun, K.; Mizeikis, V.; Bauerdick, S.; Peto, L.; Juodkazis, S. 3D-Tailored Gold Nanoparticles for Light Field Enhancement and Harvesting over Visible-IR Spectral Range. *J. Phys. Chem. C* **2011**, *115*, 5251–5256. [CrossRef]
137. Stanishevsky, A. Focused ion beam patterning of diamondlike carbon films. *Diam. Relat. Mater.* **1999**, *8*, 1246–1250. [CrossRef]
138. Mosberg, A.B.; Ren, D.; Ahtapodov, L.; Weman, H.; Fimland, B.-O.; van Helvoort, A.T.J. Focused ion beam lithography for position-controlled nanowire growth. *Nanotechnology* **2023**, *34*, 335301. [CrossRef] [PubMed]
139. Ansari, K.; van Kan, J.A.; Bettiol, A.A.; Watt, F. Fabrication of high aspect ratio 100nm metallic stamps for nanoimprint lithography using proton beam writing. *Appl. Phys. Lett.* **2004**, *85*, 476–478. [CrossRef]
140. van Kan, J.A.; Bettiol, A.A.; Watt, F. Three-dimensional nanolithography using proton beam writing. *Appl. Phys. Lett.* **2003**, *83*, 1629–1631. [CrossRef]
141. Watt, F.; Breese, M.B.H.; Bettiol, A.A.; van Kan, J.A. Proton beam writing. *Mater. Today* **2007**, *10*, 20–29. [CrossRef]
142. van Kan, J.A.; Zhang, F.; Chiam, S.Y.; Osipowicz, T.; Bettiol, A.A.; Watt, F. Proton beam writing: A platform technology for nanowire production. *Microsyst. Technol.* **2008**, *14*, 1343–1348. [CrossRef]
143. Van Kan, J.A.; Bettiol, A.A.; Ansari, K.; Teo, E.J.; Sum, T.; Watt, F. Proton beam writing: A progress review. *Int. J. Nanotechnol.* **2004**, *1*, 464–479. [CrossRef]
144. Liu, N.; Xu, X.; Pang, R.; Santhana Raman, P.; Khursheed, A.; van Kan, J.A. Brightness measurement of an electron impact gas ion source for proton beam writing applications. *Rev. Sci. Instrum.* **2015**, 87. [CrossRef]
145. Teo, E.J.; Breese, M.B.H.; Mangaiyarkarasi, D.; Bettiol, A.A.; Champeaux, F.; Watt, F.; Blackwood, D.J. Proton beam writing in silicon. In Proceedings of the IAEA: International Atomic Energy Agency (IAEA), Vienna, Austria, 5–17 November 2006.
146. Tseng, A.A. Recent developments in nanofabrication using ion projection lithography. *Small* **2005**, *1*, 594–608. [CrossRef]
147. Chen, Q.; Shao, T.; Xing, Y. An Experiment-Based Profile Function for the Calculation of Damage Distribution in Bulk Silicon Induced by a Helium Focused Ion Beam Process. *Sensors* **2020**, *20*, 2306. [CrossRef] [PubMed]
148. Heuberger, A.; Bruenger, W. Ion Projection Lithography for Nano Patterning. *MRS Proc.* **1999**, *584*, 3. [CrossRef]
149. Bruenger, W.H.; Kaesmaier, R.; Loeschner, H.; Springer, R. Status of Ion Projection Lithography. *MRS Proc.* **2000**, *636*, D5.5.1. [CrossRef]
150. Kalbitzer, S.; Knoblauch, A. High-brightness source for ion and electron beams (invited). *Rev. Sci. Instrum.* **1998**, *69*, 1026–1031. [CrossRef]
151. Smith, N.S.; Skoczylas, W.P.; Kellogg, S.M.; Kinion, D.E.; Tesch, P.P.; Sutherland, O.; Aanesland, A.; Boswell, R.W. High brightness inductively coupled plasma source for high current focused ion beam applications. *J. Vac. Sci. Technol. B Microelectron. Nanometer Struct. Process. Meas. Phenom.* **2006**, *24*, 2902–2906. [CrossRef]
152. Lee, Y.; Gough, R.A.; Kunkel, W.B.; Leung, K.N.; Vujic, J.; Williams, M.D.; Wutte, D.; Zahir, N. Multicusp sources for ion beam projection lithography. *Rev. Sci. Instrum.* **1998**, *69*, 877–879. [CrossRef]
153. Guharay, S.K.; Wang, W.; Dudnikov, V.G.; Reiser, M.; Orloff, J.; Melngailis, J. High-brightness ion source for ion projection lithography. *J. Vac. Sci. Technol. B Microelectron. Nanometer Struct. Process. Meas. Phenom.* **1996**, *14*, 3907–3910. [CrossRef]
154. Komuro, M.; Atoda, N.; Kawakatsu, H. Ion Beam Exposure of Resist Materials. *J. Electrochem. Soc.* **1979**, *126*, 483. [CrossRef]
155. Shi, X.; Prewett, P.; Huq, E.; Bagnall, D.M.; Boden, S. A Quantitative Comparison Between Helium Ion and Electron Beam Lithography on PMMA Resist. 2016. Available online: https://eprints.soton.ac.uk/408123/ (accessed on 20 January 2025).
156. Sidorkin, V.; van Veldhoven, E.; van der Drift, E.; Alkemade, P.; Salemink, H.; Maas, D. Sub-10-nm nanolithography with a scanning helium beam. *J. Vac. Sci. Technol. B Microelectron. Nanometer Struct. Process. Meas. Phenom.* **2009**, *27*, L18–L20. [CrossRef]
157. Feixiang, L.; Viacheslav, M.; Mengjun, L.; Gavin, M.; Boris, Y.; Torgny, G.; David, J.; Eric, G. Helium ion beam lithography (HIBL) using HafSOx as the resist. In *Advances in Patterning Materials and Processes XXXIII*; SPIE: Bellingham, WA, USA, 2016; p. 977928.
158. Cattoni, A.; Mailly, D.; Dalstein, O.; Faustini, M.; Seniutinas, G.; Rösner, B.; David, C. Sub-10nm electron and helium ion beam lithography using a recently developed alumina resist. *Microelectron. Eng.* **2018**, *193*, 18–22. [CrossRef]
159. Ravi Kiran, N.; Chauhan, M.; Sharma, S.K.; Ghosh, S.; Gonsalves, K.E. Resists for Helium Ion Beam Lithography: Recent Advances. *ACS Appl. Electron. Mater.* **2020**, *2*, 3805–3817. [CrossRef]
160. van Kan, J.A.; Malar, P.; Wang, Y.H. Resist materials for proton beam writing: A review. *Appl. Surf. Sci.* **2014**, *310*, 100–111. [CrossRef]
161. Sakai, T.; Yasuda, R.; Iikura, H.; Nojima, T.; Koka, M.; Satoh, T.; Ishii, Y.; Oshima, A. Fabrication of fine imaging devices using an external proton microbeam. *Nucl. Instrum. Methods Phys. Res. Sect. B Beam Interact. Mater. At.* **2014**, *332*, 238–241. [CrossRef]
162. Bolhuis, S.; van Kan, J.A.; Watt, F. Enhancement of proton beam writing in PMMA through optimization of the development procedure. *Nucl. Instrum. Methods Phys. Res. Sect. B Beam Interact. Mater. At.* **2009**, *267*, 2302–2305. [CrossRef]
163. Cutroneo, M.; Havranek, V.; Mackova, A.; Semian, V.; Torrisi, L.; Calcagno, L. Micro-patterns fabrication using focused proton beam lithography. *Nucl. Instrum. Methods Phys. Res. Sect. B Beam Interact. Mater. At.* **2016**, *371*, 344–349. [CrossRef]

164. Van Erps, J.; Vervaeke, M.; Ottevaere, H.; Hermanne, A.; Thienpont, H. Deep Proton Writing for the rapid prototyping of polymer micro-components for optical interconnects and optofluidics. *Nucl. Instrum. Methods Phys. Res. Sect. B Beam Interact. Mater. At.* **2013**, *307*, 243–247. [CrossRef]
165. van Kan, J.A.; Sanchez, J.L.; Xu, B.; Osipowicz, T.; Watt, F. Resist materials for proton micromachining. *Nucl. Instrum. Methods Phys. Res. Sect. B Beam Interact. Mater. At.* **1999**, *158*, 179–184. [CrossRef]
166. Kumar, V.; Maan, A.S.; Akhtar, J. Electronic transport in epitaxial 4H–SiC based Schottky diodes modified selectively by swift heavy ions. *Mater. Sci. Semicond. Process.* **2020**, *115*, 105108. [CrossRef]
167. Wang, Y.H.; Malar, P.; van Kan, J.A. Resist evaluation for proton beam writing, Ni mold fabrication and nano-replication. *Microsyst. Technol.* **2014**, *20*, 2079–2088. [CrossRef]
168. Luo, C.; Xu, C.; Lv, L.; Li, H.; Huang, X.; Liu, W. Review of recent advances in inorganic photoresists. *RSC Adv.* **2020**, *10*, 8385–8395. [CrossRef]
169. Sossna, E.; Kassing, R.; Rangelow, I.W.; Herzinger, C.M.; Tiwald, T.E.; Woollam, J.A.; Wagner, T. Thickness analysis of silicon membranes for stencil masks. *J. Vac. Sci. Technol. B Microelectron. Nanometer Struct. Process. Meas. Phenom.* **2000**, *18*, 3259–3263. [CrossRef]
170. Lan, H.; Ding, Y.; Liu, H.; Lu, B. Review of the wafer stage for nanoimprint lithography. *Microelectron. Eng.* **2007**, *84*, 684–688. [CrossRef]
171. Chen, Y. Applications of nanoimprint lithography/hot embossing: A review. *Appl. Phys. A* **2015**, *121*, 451–465. [CrossRef]
172. Raza, A.; Saeed, Z.; Aslam, A.; Nizami, S.M.; Habib, K.; Malik, A.N. Advances, Application and Challenges of Lithography Techniques. In Proceedings of the 2024 5th International Conference on Advancements in Computational Sciences (ICACS), Lahore, Pakistan, 19–20 February 2024; pp. 1–6.
173. Schift, H. Nanoimprint lithography: 2D or not 2D? A review. *Appl. Phys. A* **2015**, *121*, 415–435. [CrossRef]
174. Traub, M.C.; Longsine, W.; Truskett, V.N. Advances in Nanoimprint Lithography. *Annu. Rev. Chem. Biomol. Eng.* **2016**, *7*, 583–604. [CrossRef] [PubMed]
175. Unno, N.; Mäkelä, T. Thermal Nanoimprint Lithography—A Review of the Process, Mold Fabrication, and Material. *Nanomaterials* **2023**, *13*, 2031. [CrossRef]
176. Kumar, V.; Pawar, S.; Maan, A.S.; Akhtar, J. Diameter dependent thermal sensitivity variation trend in Ni/4H-SiC Schottky diode temperature sensors. *J. Vac. Sci. Technol. B* **2015**, *33*, 052207. [CrossRef]
177. Chou, S.Y.; Krauss, P.R.; Renstrom, P.J. Imprint of sub-25 nm vias and trenches in polymers. *Appl. Phys. Lett.* **1995**, *67*, 3114–3116. [CrossRef]
178. Lan, H.; Liu, H. UV-Nanoimprint Lithography: Structure, Materials and Fabrication of Flexible Molds. *J. Nanosci. Nanotechnol.* **2013**, *13*, 3145–3172. [CrossRef]
179. Francone, A. *Materials and Anti-Adhesive Issues in UV-NIL*; Institut National Polytechnique de Grenoble—INPG: Saint-Martin-d'Hères, French, 2010.
180. He, X.; Zang, L.; Xin, Y.; Zou, Y. An overview of photopolymerization and its diverse applications. *Appl. Res.* **2023**, *2*, e202300030. [CrossRef]
181. Haslinger, M.J.; Mitteramskogler, T.; Kopp, S.; Leichtfried, H.; Messerschmidt, M.; Thesen, M.W.; Mühlberger, M. Development of a soft UV-NIL step & repeat and lift-off process chain for high speed metal nanomesh fabrication. *Nanotechnology* **2020**, *31*, 345301. [CrossRef] [PubMed]
182. Gates, B.D.; Xu, Q.; Stewart, M.; Ryan, D.; Willson, C.G.; Whitesides, G.M. New Approaches to Nanofabrication: Molding, Printing, and Other Techniques. *Chem. Rev.* **2005**, *105*, 1171–1196. [CrossRef]
183. Mohamed, K.; Alkaisi, M.M.; Blaikie, R.J. A Three-Dimensional Ultraviolet Curable Nanoimprint Lithography (3D UV-NIL). *AIP Conf. Proc.* **2009**, *1151*, 114–117. [CrossRef]
184. Isobe, G.; Kanno, I.; Kotera, H.; Yokokawa, R. Perfusable multi-scale channels fabricated by integration of nanoimprint lighography (NIL) and UV lithography (UVL). *Microelectron. Eng.* **2012**, *98*, 58–63. [CrossRef]
185. Kooy, N.; Mohamed, K.; Pin, L.T.; Guan, O.S. A review of roll-to-roll nanoimprint lithography. *Nanoscale Res. Lett.* **2014**, *9*, 320. [CrossRef] [PubMed]
186. Maury, P.; Turkenburg, D.; Stroeks, N.; Giesen, P.; Barbu, I.; Meinders, E.; van Bremen, A.; Iosad, N.; van der Werf, R.; Onvlee, H. Roll-to-roll UV imprint lithography for flexible electronics. *Microelectron. Eng.* **2011**, *88*, 2052–2055. [CrossRef]
187. Yi, P.; Wu, H.; Zhang, C.; Peng, L.; Lai, X. Roll-to-roll UV imprinting lithography for micro/nanostructures. *J. Vac. Sci. Technol. B* **2015**, 33. [CrossRef]
188. Kim, G.E.; Kim, H.; Woo, K.; Kang, Y.; Lee, S.-H.; Jeon, Y.; Lee, M.G.; Kwon, S. Uniform Pressing Mechanism in Large-Area Roll-to-Roll Nanoimprint Lithography Process. *Appl. Sci.* **2021**, *11*, 9571. [CrossRef]
189. Maize, K.; Mi, Y.; Cakmak, M.; Shakouri, A. Real-Time Metrology for Roll-To-Roll and Advanced Inline Manufacturing: A Review. *Adv. Mater. Technol.* **2023**, *8*, 2200173. [CrossRef]

190. Jambhapuram, M.; Good, J.K.; Azoug, A. Impact of winding on nanoimprinted surfaces in roll-to-roll nanoimprint lithography. *Polym. Eng. Sci.* **2022**, *62*, 2657–2667. [CrossRef]
191. Groh, B.; Lee, K.; Cullinan, M.; Chang, C. *Development of Joint Manufacturing and In-Line Metrology System for the Patterning of 3D Holographic Structures in Roll-to-Roll Processes*; TMS: Pittsburgh, PA, USA, 2024.
192. Milenko, K.; Dullo, F.T.; Thrane, P.C.V.; Skokic, Z.; Dirdal, C.A. UV-Nanoimprint Lithography for Predefined SERS Nanopatterns Which Are Reproducible at Low Cost and High Throughput. *Nanomaterials* **2023**, *13*, 1598. [CrossRef]
193. Chou, S.Y.; Krauss, P.R.; Zhang, W.; Guo, L.; Zhuang, L. Sub-10 nm imprint lithography and applications. *J. Vac. Sci. Technol. B Microelectron. Nanometer Struct. Process. Meas. Phenom.* **1997**, *15*, 2897–2904. [CrossRef]
194. Austin, M.D.; Ge, H.; Wu, W.; Li, M.; Yu, Z.; Wasserman, D.; Lyon, S.A.; Chou, S.Y. Fabrication of 5nm linewidth and 14nm pitch features by nanoimprint lithography. *Appl. Phys. Lett.* **2004**, *84*, 5299–5301. [CrossRef]
195. Cherala, A.; Pandya, P.N.; Liechti, K.M.; Sreenivasan, S.V. Extending the resolution limits of nanoshape imprint lithography using molecular dynamics of polymer crosslinking. *Microsyst. Nanoeng.* **2021**, *7*, 13. [CrossRef] [PubMed]
196. Lohse, M.; Thesen, M.W.; Haase, A.; Smolka, M.; Iceta, N.B.; Ayerdi Izquierdo, A.; Ramos, I.; Salado, C.; Schleunitz, A. Novel Concept of Micro Patterned Micro Titer Plates Fabricated via UV-NIL for Automated Neuronal Cell Assay Read-Out. *Nanomaterials* **2021**, *11*, 902. [CrossRef]
197. Guo, L.J. Nanoimprint Lithography: Methods and Material Requirements. *Adv. Mater.* **2007**, *19*, 495–513. [CrossRef]
198. Klement, W.; Willens, R.H.; Duwez, P.O.L. Non-crystalline Structure in Solidified Gold–Silicon Alloys. *Nature* **1960**, *187*, 869–870. [CrossRef]
199. Pfeiffer, K.; Reuther, F.; Fink, M.; Gruetzner, G.; Carlberg, P.; Maximov, I.; Montelius, L.; Seekamp, J.; Zankovych, S.; Sotomayor-Torres, C.M.; et al. A comparison of thermally and photochemically cross-linked polymers for nanoimprinting. *Microelectron. Eng.* **2003**, *67–68*, 266–273. [CrossRef]
200. Clavijo Cedeño, C.; Seekamp, J.; Kam, A.P.; Hoffmann, T.; Zankovych, S.; Sotomayor Torres, C.M.; Menozzi, C.; Cavallini, M.; Murgia, M.; Ruani, G.; et al. Nanoimprint lithography for organic electronics. *Microelectron. Eng.* **2002**, *61–62*, 25–31. [CrossRef]
201. Choi, P.; Fu, P.F.; Guo, L.J. Siloxane Copolymers for Nanoimprint Lithography. *Adv. Funct. Mater.* **2007**, *17*, 65–70. [CrossRef]
202. Cheng, X.; Guo, L.J.; Fu, P.F. Room-Temperature, Low-Pressure Nanoimprinting Based on Cationic Photopolymerization of Novel Epoxysilicone Monomers. *Adv. Mater.* **2005**, *17*, 1419–1424. [CrossRef] [PubMed]
203. Kumar, G.; Tang, H.X.; Schroers, J. Nanomoulding with amorphous metals. *Nature* **2009**, *457*, 868–872. [CrossRef]
204. Yoshimoto, M. Sub-nanoscale nanoimprint fabrication of atomically stepped glassy substrates of silicate glass and acryl polymer. *Appl. Phys. A* **2015**, *121*, 321–326. [CrossRef]
205. Cox, L.M.; Martinez, A.M.; Blevins, A.K.; Sowan, N.; Ding, Y.; Bowman, C.N. Nanoimprint lithography: Emergent materials and methods of actuation. *Nano Today* **2020**, *31*, 100838. [CrossRef]
206. Huang, Z.; Geyer, N.; Werner, P.; de Boor, J.; Gösele, U. Metal-Assisted Chemical Etching of Silicon: A Review. *Adv. Mater.* **2011**, *23*, 285–308. [CrossRef]
207. Hsu, K.H.; Schultz, P.L.; Ferreira, P.M.; Fang, N.X. Electrochemical Nanoimprinting with Solid-State Superionic Stamps. *Nano Lett.* **2007**, *7*, 446–451. [CrossRef]
208. Sharstniou, A.; Niauzorau, S.; Ferreira, P.M.; Azeredo, B.P. Electrochemical nanoimprinting of silicon. *Proc. Natl. Acad. Sci. USA* **2019**, *116*, 10264–10269. [CrossRef]
209. Ryckman, J.D.; Jiao, Y.; Weiss, S.M. Three-dimensional patterning and morphological control of porous nanomaterials by gray-scale direct imprinting. *Sci. Rep.* **2013**, *3*, 1502. [CrossRef] [PubMed]
210. Muanchan, P.; Suzuki, S.; Kyotani, T.; Ito, H. One-dimensional polymer nanofiber arrays with high aspect ratio obtained by thermal nanoimprint method. *Polym. Eng. Sci.* **2017**, *57*, 214–223. [CrossRef]
211. Gourgon, C.; Philippot, G.; Labau, S.; Tortai, J.H.; Benwadih, M.; Bablet, J. Influence of PEN thermal properties on flexible film patterned by nanoimprint lithography. *Microelectron. Eng.* **2011**, *88*, 1959–1963. [CrossRef]
212. Masciullo, C.; Sonato, A.; Romanato, F.; Cecchini, M. Perfluoropolyether (PFPE) Intermediate Molds for High-Resolution Thermal Nanoimprint Lithography. *Nanomaterials* **2018**, *8*, 609. [CrossRef] [PubMed]
213. Cecchini, M.; Signori, F.; Pingue, P.; Bronco, S.; Ciardelli, F.; Beltram, F. High-Resolution Poly(ethylene terephthalate) (PET) Hot Embossing at Low Temperature: Thermal, Mechanical, and Optical Analysis of Nanopatterned Films. *Langmuir* **2008**, *24*, 12581–12586. [CrossRef]
214. Zanut, A.; Cian, A.; Cefarin, N.; Pozzato, A.; Tormen, M. Nanoelectrode Arrays Fabricated by Thermal Nanoimprint Lithography for Biosensing Application. *Biosensors* **2020**, *10*, 90. [CrossRef] [PubMed]
215. Cui, B.; Cortot, Y.; Veres, T. Polyimide nanostructures fabricated by nanoimprint lithography and its applications. *Microelectron. Eng.* **2006**, *83*, 906–909. [CrossRef]
216. Wu, C.C.; Hsu, S.L.C. High density patterns fabricated in low-viscosity thermal-curable epoxy system for thermal-nanoimprint lithography. In Proceedings of the 2007 Digest of papers Microprocesses and Nanotechnology, Kyoto, Japan, 5–8 November 2007; pp. 312–313.

217. Kurose, T.; Shishido, H.; Ishigami, A.; Nemoto, A.; Ito, H. Fabrication of high-resolution conductive patterns on a thermally imprinted polyetherimide film by the capillary flow of conductive ink. *Microsyst. Technol.* **2020**, *26*, 1981–1986. [CrossRef]
218. Greer, A.I.M.; Vasiev, I.; Della-Rosa, B.; Gadegaard, N. Fluorinated ethylene–propylene: A complementary alternative to PDMS for nanoimprint stamps. *Nanotechnology* **2016**, *27*, 155301. [CrossRef] [PubMed]
219. Weiss, D.N.; Meyers, S.T.; Keszler, D.A. All-inorganic thermal nanoimprint process. *J. Vac. Sci. Technol. B* **2010**, *28*, 823–828. [CrossRef]
220. Vig, A.L.; Mäkelä, T.; Majander, P.; Lambertini, V.; Ahopelto, J.; Kristensen, A. Roll-to-roll fabricated lab-on-a-chip devices. *J. Micromechanics Microengineering* **2011**, *21*, 035006. [CrossRef]
221. Bhagoria, P.; Mathew Sebastian, E.; Kumar Jain, S.; Purohit, J.; Purohit, R. Nanolithography and its alternate techniques. *Mater. Today Proc.* **2020**, *26*, 3048–3053. [CrossRef]
222. Kwon, B.; Kim, J.H. Importance of Molds for Nanoimprint Lithography: Hard, Soft, and Hybrid Molds. *J. Nanosci.* **2016**, *2016*, 6571297. [CrossRef]
223. Zhao, M.; Xu, T.; Chen, B.; Niu, J. *Technology of Alignment Mark in Electron Beam Lithography*; SPIE: Bellingham, WA, USA, 2014; Volume 9285.
224. McPhillimy, J.; Jevtics, D.; Guilhabert, B.J.E.; Klitis, C.; Hurtado, A.; Sorel, M.; Dawson, M.D.; Strain, M.J. Automated Nanoscale Absolute Accuracy Alignment System for Transfer Printing. *ACS Appl. Nano Mater.* **2020**, *3*, 10326–10332. [CrossRef] [PubMed]
225. Jiang, L.; Feldman, M. *Accurate Alignment Technique for Nanoimprint Lithography*; SPIE: Bellingham, WA, USA, 2005; Volume 5752.
226. Jiang, W.; Wang, H.; Xie, W.; Qu, Z. Lithography Alignment Techniques Based on Moiré Fringe. *Photonics* **2023**, *10*, 351. [CrossRef]
227. Wu, Y.; Xiao, Z. The Recent Progress of Lithography Machine and the State-of-art Facilities. *Highlights Sci. Eng. Technol.* **2022**, *5*, 155–165. [CrossRef]
228. Park, J.; Jeong, J. Improved MSRN-Based Attention Block for Mask Alignment Mark Detection in Photolithography. *Appl. Sci.* **2022**, *12*, 2721. [CrossRef]
229. Nishi, Y.; Doering, R. *Handbook of Semiconductor Manufacturing Technology*; Taylor & Francis: Abingdon, UK, 2000.
230. Adel, M.; Ghinovker, M.; Poplawski, J.; Kassel, E.; Izikson, P.; Pollentier, I.; Leray, P.; Laidler, D. *Characterization of Overlay Mark Fidelity*; SPIE: Bellingham, WA, USA, 2003; Volume 5038.
231. Corno, F.; Prinetto, P.; Rebaudengo, M.; Reorda, M.S. Partial scan flip flop selection for simulation-based sequential ATPGs. In Proceedings of the International Test Conference 1996. Test and Design Validity, Washington, DC, USA, 20–25 October 1996; pp. 558–564.
232. Xu, F.; Ding, Y.; Chen, W.; Xia, H. An Improved Algorithm to Extract Moiré Fringe Phase for Wafer-Mask Alignment in Nanoimprint Lithography. *Micromachines* **2024**, *15*, 1408. [CrossRef]
233. Whang, A.J.W.; Gallagher, N.C. Synthetic approach to designing optical alignment systems. *Appl. Opt.* **1988**, *27*, 3534–3541. [CrossRef]
234. Tan, C.-B.; Yeo, S.-H.; Koh, H.P.; Koo, C.; Foong, Y.; Siew, Y. *Evaluation of Alignment Marks Using ASML ATHENA Alignment System in 90-nm BEOL Process*; SPIE: Bellingham, WA, USA, 2003; Volume 5038.
235. Nitayama, A.; Sato, T.; Hashimoto, K.; Shigemitsu, F.; Nakase, M. New phase shifting mask with self-aligned phase shifters for a quarter micron photolithography. In Proceedings of the International Technical Digest on Electron Devices Meeting, Washington, DC, USA, 3–6 December 1989; pp. 57–60.
236. Pal, P.; Kim, Y.-J.; Chandra, S. Front-to-Back Alignment Techniques in Microelectronics/MEMS Fabrication: A Review. *Sens. Lett.* **2006**, *4*, 1–10. [CrossRef]
237. Menon, R.; Patel, A.; Gil, D.; Smith, H.I. Maskless lithography. *Mater. Today* **2005**, *8*, 26–33. [CrossRef]
238. Suzuki, K.; Matsui, S.; Ochiai, Y. *Sub-Half-Micron Lithography for ULSIs*; Cambridge University Press: Cambridge, UK, 2000.
239. Franssila, S. *Introduction to Microfabrication*; Wiley: Hoboken, NJ, USA, 2005.
240. Smith, B.W.; Suzuki, K. *Microlithography: Science and Technology*, 2nd ed.; CRC Press: Boca Raton, FL, USA, 2018.
241. Reed, I. Lithography Essentials: Mastering Materials and Tools for Stunning Prints; Marcelo Marins Rodrigues. Available online: https://www.google.com/books/edition/Lithography_Essentials/jzM0EQAAQBAJ?hl=en (accessed on 20 January 2025).
242. Zinsmeister, G. *Glass and Quartz Substrates for Photomask Fabrication*; SPIE: Bellingham, WA, USA, 2023; Volume 12806.
243. Noll, H.; Johnson, P. *The Performance of Chrome Material with Use*; SPIE: Bellingham, WA, USA, 2023; Volume 12806.
244. Pierrat; Wong; Vaidya. Phase-shifting mask topography effects on lithographic image quality. In Proceedings of the 1992 International Technical Digest on Electron Devices Meeting, San Francisco, CA, USA, 13–16 December 1992; pp. 53–56.
245. Shigetomi, A.; Matsuda, S.; Watakabe, Y.; Kato, T. High-performance VLSI photomask with a molybdenum silicide film. *Microelectron. Eng.* **1991**, *14*, 73–86. [CrossRef]
246. Rizvi, S. *Handbook of Photomask Manufacturing Technology*; CRC Press: Boca Raton, FL, USA, 2018.
247. Liebmann, L. *Resolution Enhancement Techniques in Optical lithography: It's Not Just a Mask Problem*; SPIE: Bellingham, WA, USA, 2001; Volume 4409.
248. Fang, Y.; He, Y. Resolution technology of lithography machine. *J. Phys. Conf. Ser.* **2022**, *2221*, 012041. [CrossRef]

249. Cui, Z. (Ed.) Optical Lithography. In *Nanofabrication: Principles, Capabilities and Limits*; Springer International Publishing: Cham, Switzerland, 2024; pp. 9–81.
250. Lawrence, S.M.; Rik, J. Contribution of mask defectivity in stochastics of EUVL-based wafer printing. *J. Micro/Nanopatterning Mater. Metrol.* **2021**, *20*, 021003. [CrossRef]
251. Liebmann, L. *Mask Challenges in High-Resolution Lithography*; SPIE: Bellingham, WA, USA, 2017; Volume 10321.
252. Neisser, M.; Wurm, S. ITRS lithography roadmap: 2015 challenges. *Adv. Opt. Technol.* **2015**, *4*, 235–240. [CrossRef]
253. Ji, Y.; Wu, X.; Han, X.; Xie, W.; Ma, X. Phase defect detection algorithm for extreme ultraviolet mask blank based on watershed edge detection. In Proceedings of the 2021 International Workshop on Advanced Patterning Solutions (IWAPS), Foshan, China, 12–13 December 2021; pp. 1–4.
254. Mao, D.; Liu, L.; Zhang, C.; Liu, H.; Mao, C. Molecular Lithography on Silicon Wafers Guided by Porous, Extended Arrays of Small DNA Tiles. *Langmuir* **2023**, *39*, 11782–11787. [CrossRef] [PubMed]
255. Takemoto, I.; Ando, N.; Edamatsu, K.; Lee, Y.; Takashima, M.; Yokoyama, H. *Molecular Resists for EUV and EB Lithography*; SPIE: Bellingham, WA, USA, 2008; Volume 6923.
256. Frommhold, A.; Yang, D.; McClelland, A.; Roth, J.; Xue, X.; Rosamund, M.C.; Linfield, E.H.; Robinson, A.P.G. Novel Molecular Resist for EUV and Electron Beam Lithography. *J. Photopolym. Sci. Technol.* **2015**, *28*, 537–540. [CrossRef]
257. Brainard, R.L.; Barclay, G.G.; Anderson, E.H.; Ocola, L.E. Resists for next generation lithography. *Microelectron. Eng.* **2002**, *61–62*, 707–715. [CrossRef]
258. Wang, Y.; Yuan, J.; Chen, J.; Zeng, Y.; Yu, T.; Guo, X.; Wang, S.; Yang, G.; Li, Y. A Single-Component Molecular Glass Resist Based on Tetraphenylsilane Derivatives for Electron Beam Lithography. *ACS Omega* **2023**, *8*, 12173–12182. [CrossRef]
259. Wang, Y.; Chen, J.; Zeng, Y.; Yu, T.; Wang, S.; Guo, X.; Hu, R.; Tian, P.; Vockenhuber, M.; Kazazis, D.; et al. Nonchemically Amplified Molecular Resists Based on Sulfonium-Functionalized Sulfone Derivatives for Sub-13 nm Nanolithography. *ACS Appl. Nano Mater.* **2023**, *6*, 18480–18490. [CrossRef]
260. Kim, J.-B.; Ganesan, R.; Choi, J.-H.; Yun, H.-J.; Kwon, Y.-G.; Kim, K.-S.; Oh, T.-H. Photobleachable silicon-containing molecular resist for deep UV lithography. *J. Mater. Chem.* **2006**, *16*, 3448–3451. [CrossRef]
261. Sanjay; Kumar, V.; Vohra, A. DC and Analog/RF Performance Evaluation Using Dual Metal Gate Work Function Engineering of Junctionless Cylindrical Gate All Around Si Nanowire MOSFET Using NEGF Approach for Upcoming Sub 5 nm Technology Node. *Int. J. Precis. Eng. Manuf.* **2024**, *25*, 1885–1897. [CrossRef]
262. Yang, D.X.; Frommhold, A.; Xue, X.; Palmer, R.E.; Robinson, A.P.G. Chemically amplified phenolic fullerene electron beam resist. *J. Mater. Chem. C* **2014**, *2*, 1505–1512. [CrossRef]
263. Manyam, J. *Novel Resist Materials for next Generation Lithography*; University of Birmingham: Birmingham, UK, 2011.
264. Kumar, V.; Maan, A.S.; Akhtar, J. Barrier height inhomogeneities induced anomaly in thermal sensitivity of Ni/4H-SiC Schottky diode temperature sensor. *J. Vac. Sci. Technol. B* **2014**, *32*, 041203. [CrossRef]
265. Lawson, R.A. *Molecular Resists for Advanced Lithography-Design, Synthesis, Characterization, and Simulation*; Georgia Institute of Technology: Atlanta, GA, USA, 2011.
266. Cen, J.; Deng, Z.; Liu, S. Emerging trends in the chemistry of polymeric resists for extreme ultraviolet lithography. *Polym. Chem.* **2024**, *15*, 4599–4614. [CrossRef]
267. Salaita, K.S.; Lee, S.W.; Ginger, D.S.; Mirkin, C.A. DPN-Generated Nanostructures as Positive Resists for Preparing Lithographic Masters or Hole Arrays. *Nano Lett.* **2006**, *6*, 2493–2498. [CrossRef] [PubMed]
268. Eichelsdoerfer, D.J.; Liao, X.; Cabezas, M.D.; Morris, W.; Radha, B.; Brown, K.A.; Giam, L.R.; Braunschweig, A.B.; Mirkin, C.A. Large-area molecular patterning with polymer pen lithography. *Nat. Protoc.* **2013**, *8*, 2548–2560. [CrossRef] [PubMed]
269. Nealey, P.F.; Edwards, E.W.; Muller, M.; Stoykovich, M.P.; Solak, H.H.; Pablo, J.J.d. Self-assembling resists for nanolithography. In Proceedings of the IEEE InternationalElectron Devices Meeting, 2005. IEDM Technical Digest, Washington, DC, USA, 5 December 2005; pp. 4–359.
270. Chang, S.; Wang, X.; Hu, Q.; Sun, X.; Wang, A.; Dong, X.; Zhang, Y.; Shi, L.; Sun, Q. Self-Assembled Nanocomposites and Nanostructures for Environmental and Energy Applications. *Crystals* **2022**, *12*, 274. [CrossRef]
271. Brainard, R.L.; Bakshi, V. Photoresists for extreme ultraviolet lithography. *EUV Lithogr.* **2009**, *178*, 383.
272. Sakamoto, R.; Fujitani, N.; Onishi, R.; Nishita, T. The Novel Solution for Negative Impact of Out-of-band and Outgassing by Top Coat Materials(OBPL) in EUVL. *J. Photopolym. Sci. Technol.* **2013**, *26*, 685–689. [CrossRef]
273. Liu, J.; Wang, D.; Li, Y.; Wang, H.; Chen, H.; Wang, Q.; Kang, W. Exceptional Lithography Sensitivity Boosted by Hexafluoroisopropanols in Photoresists. *Polymers* **2024**, *16*, 825. [CrossRef]
274. Kang, Y.K.; Lee, S.J.; Eom, S.; Kim, B.G.; Hwang, C.-C.; Kim, M.-G. Recent progress of inorganic photoresists for next-generation EUV lithography. *J. Mater. Chem. C* **2024**, *12*, 15855–15887. [CrossRef]
275. Jiang, J.; Giordano, G.; Fallica, R.; DeSimone, D.; Vandenberghe, G. Sensitizer for EUV Chemically Amplified Resist: Metal versus Halogen. *J. Photopolym. Sci. Technol.* **2019**, *32*, 21–25. [CrossRef]

276. Yannick, V.; Jing, J.; Hiroki, Y.; Danilo De, S.; Takahiro, K.; Stefan De, G.; Geert, V. Sensitizers in extreme ultraviolet chemically amplified resists: Mechanism of sensitivity improvement. *J. Micro/Nanolithography MEMS MOEMS* **2018**, *17*, 043506. [CrossRef]
277. Kudo, H.; Nina, N.; Sato, T.; Oizumi, H.; Itani, T.; Miura, T.; Watanabe, T.; Kinoshita, H. Extreme Ultraviolet (EUV)-Resist Material Based on Noria (Water Wheel-like Macrocycle) Derivatives with Pendant Alkoxyl and Adamantyl Ester Groups. *J. Photopolym. Sci. Technol.* **2012**, *25*, 587–592. [CrossRef]
278. Prashant, K.K.; Ken, M.; Sara, K.; Scott, D.; Pradeep, P.; James, B.; Deirdre, O.; Paul, D.A. Sub-20nm lithography negative tone chemically amplified resists using cross-linker additives. In *Advances in Resist Materials and Processing Technology*; SPIE: Bellingham, WA, USA, 2013; p. 94250S.
279. Green, D.P.; Vipul, J.; Brad, B.; Mike, W.; Michael, C.; David, V.; Steve, L. Development of molecular resist derivatives for EUV lithography. In *Extreme Ultraviolet (EUV) Lithography IV*; SPIE: Bellingham, WA, USA, 2013; p. 867912.
280. Popescu, C.; Callaghan, G.O.; McClelland, A.; Roth, J.; Lada, T.; Robinson, A.P.G. Performance enhancements with high opacity multi-trigger resist. In *Advances in Patterning Materials and Processes XXXVII*; SPIE: Bellingham, WA, USA, 2020; p. 1132611.
281. Vasiliki, K.; Kazuki, K.; Hong, X.; Jérémy, O.; Christopher, K.O.; Emmanuel, P.G. Elucidating the patterning mechanism of zirconium-based hybrid photoresists. *J. Micro/Nanolithography MEMS MOEMS* **2017**, *16*, 041007. [CrossRef]
282. Andrew, G.; Jeremy, T.A.; Benjamin, L.C.; Peter De, S.; Joseph, E.; Michael, G.; Kai, J.; Michael, K.; Stephen, T.M.; Jason, K.S.; et al. Integrated fab process for metal oxide EUV photoresist. In *Advances in Patterning Materials and Processes XXXII*; SPIE: Bellingham, WA, USA, 2015; p. 94250S.
283. Zhang, Y.; Haitjema, J.; Baljozovic, M.; Vockenhuber, M.; Kazazis, D.; Jung, T.A.; Ekinci, Y.; Brouwer, A.M. Dual-tone Application of a Tin-Oxo Cage Photoresist Under E-beam and EUV Exposure. *J. Photopolym. Sci. Technol.* **2018**, *31*, 249–255. [CrossRef]
284. Thomas, G.; Nguyen Dang, L.; Markus, L.; Kimmo, K.; Oskari, K.; Emilia, K.; Dimitrios, K.; Yasin, E.; Juha, R. Advanced EUV negative tone resist and underlayer approaches exhibiting sub-20nm half-pitch resolution. In *Advances in Patterning Materials and Processes XXXVI*; SPIE: Bellingham, WA, USA, 2019; p. 109600B.
285. Li, Z.; Yan, Y.; Wang, J.; Geng, Y. Molecular Dynamics Study on Tip-Based Nanomachining: A Review. *Nanoscale Res. Lett.* **2020**, *15*, 201. [CrossRef]
286. Kim, T.-Y.; Kang, I.-H.; Park, J.; Kim, M.; Oh, H.-K.; Hur, S.-M. Coarse-Grained Modeling of EUV Patterning Process Reflecting Photochemical Reactions and Chain Conformations. *Polymers* **2023**, *15*, 1988. [CrossRef] [PubMed]
287. Kumar, V.; Kaminski, N.; Maan, A.S.; Akhtar, J. Capacitance roll-off and frequency-dispersion capacitance–conductance phenomena in field plate and guard ring edge-terminated Ni/SiO_2/4H-nSiC Schottky barrier diodes. *Phys. Status Solidi* **2016**, *213*, 193–202. [CrossRef]
288. Park, J.; Lee, S.-G.; Vesters, Y.; Severi, J.; Kim, M.; De Simone, D.; Oh, H.-K.; Hur, S.-M. Molecular Modeling of EUV Photoresist Revealing the Effect of Chain Conformation on Line-Edge Roughness Formation. *Polymers* **2019**, *11*, 1923. [CrossRef]
289. Hasan, R.M.M.; Politano, O.; Luo, X. ReaxFF molecular dynamics simulation study of nanoelectrode lithography oxidation process on silicon (100) surface. *Appl. Surf. Sci.* **2019**, *496*, 143679. [CrossRef]
290. Shim, S.; Vella, J.R.; Draney, J.S.; Na, D.; Graves, D.B. An examination of the performance of molecular dynamics force fields: Silicon and silicon dioxide reactive ion etching. *J. Vac. Sci. Technol. A* **2024**, *42*, 023207. [CrossRef]
291. Kim, D.H.; Kwak, S.J.; Jeong, J.H.; Yoo, S.; Nam, S.K.; Kim, Y.; Lee, W.B. Molecular Dynamics Simulation of Silicon Dioxide Etching by Hydrogen Fluoride Using the Reactive Force Field. *ACS Omega* **2021**, *6*, 16009–16015. [CrossRef] [PubMed]
292. Martirosyan, V.; Despiau-Pujo, E.; Dubois, J.; Cunge, G.; Joubert, O. Helium plasma modification of Si and Si3N4 thin films for advanced etch processes. *J. Vac. Sci. Technol. A* **2018**, *36*, 041301. [CrossRef]
293. Vella, J.R.; Hao, Q.; Donnelly, V.M.; Graves, D.B. Dynamics of plasma atomic layer etching: Molecular dynamics simulations and optical emission spectroscopy. *J. Vac. Sci. Technol. A* **2023**, *41*, 062602. [CrossRef]
294. Bystrov, V.; Likhachev, I.; Filippov, S.; Paramonova, E. Molecular Dynamics Simulation of Self-Assembly Processes of Diphenylalanine Peptide Nanotubes and Determination of Their Chirality. *Nanomaterials* **2023**, *13*, 1905. [CrossRef]
295. Zhou, Y.; Bore, S.L.; Tao, A.R.; Paesani, F.; Arya, G. Many-body potential for simulating the self-assembly of polymer-grafted nanoparticles in a polymer matrix. *npj Comput. Mater.* **2023**, *9*, 224. [CrossRef]
296. Xu, L. Molecular dynamics simulation of self-assembly in anionic and zwitterionic surfactant systems. *Anal. Chem. A J.* **2023**, *2*, 64–73.
297. Kong, D.; Meng, X.; Zhu, J.; Zhou, W. Molecular dynamics simulation of surfactant induced wettability alteration of shale reservoirs. *Front. Energy Res.* **2023**, *11*, 1272132. [CrossRef]
298. Hanzawa, M.; Ogura, T.; Tsuchiya, K.; Akamatsu, M.; Sakai, K.; Sakai, H. Anti-adsorption Mechanism of Photoresist by Pluronic Surfactants: An Insight into Their Adsorbed Structure. *Langmuir* **2023**, *39*, 7876–7883. [CrossRef] [PubMed]
299. Gerard, M.S.; Vivek, K.S.; Lewis, W.F.; Michael, D.S.; Sean, D.B.; Willson, C.G. Recent advances in a molecular level lithography simulation. In *Advances in Resist Technology and Processing XVII*; SPIE: Bellingham, WA, USA, 2000; pp. 675–685.
300. Uchida, H.; Imoto, R.; Ando, T.; Okabe, T.; Taniguchi, J. Molecular Dynamics Simulation of the Resist Filling Process in UV-nanoimprint Lithography. *J. Photopolym. Sci. Technol.* **2021**, *34*, 139–144. [CrossRef]

301. Zheng, T.; O'Neill, C.; Marshall, J.F.; Iskratsch, T.; Palma, M. Selective placement of functionalised DNA origami via thermal scanning probe lithography patterning. *Mater. Adv.* **2024**, *5*, 9376–9382. [CrossRef]
302. Haggenmueller, S.; Matthies, M.; Sample, M.; Šulc, P. How we simulate DNA origami. *arXiv* **2024**, arXiv:2409.13206.
303. Liu, W.; Duan, H.; Zhang, D.; Zhang, X.; Luo, Q.; Xie, T.; Yan, H.; Peng, L.; Hu, Y.; Liang, L.; et al. Concepts and Application of DNA Origami and DNA Self-Assembly: A Systematic Review. *Appl. Bionics Biomech.* **2021**, *2021*, 9112407. [CrossRef]
304. Martynenko, I.V.; Ruider, V.; Dass, M.; Liedl, T.; Nickels, P.C. DNA Origami Meets Bottom-Up Nanopatterning. *ACS Nano* **2021**, *15*, 10769–10774. [CrossRef]
305. Rothemund, P.W.K. Folding DNA to create nanoscale shapes and patterns. *Nature* **2006**, *440*, 297–302. [CrossRef]
306. Zhan, P.; Peil, A.; Jiang, Q.; Wang, D.; Mousavi, S.; Xiong, Q.; Shen, Q.; Shang, Y.; Ding, B.; Lin, C.; et al. Recent Advances in DNA Origami-Engineered Nanomaterials and Applications. *Chem. Rev.* **2023**, *123*, 3976–4050. [CrossRef] [PubMed]
307. Tørring, T.; Voigt, N.V.; Nangreave, J.; Yan, H.; Gothelf, K.V. DNA origami: A quantum leap for self-assembly of complex structures. *Chem. Soc. Rev.* **2011**, *40*, 5636–5646. [CrossRef]
308. Li, L.; Nie, S.; Du, T.; Zhao, J.; Chen, X. DNA origami technology for biomedical applications: Challenges and opportunities. *MedComm. Biomater. Appl.* **2023**, *2*, e37. [CrossRef]
309. Li, N.; Shang, Y.; Han, Z.; Wang, T.; Wang, Z.-G.; Ding, B. Fabrication of Metal Nanostructures on DNA Templates. *ACS Appl. Mater. Interfaces* **2019**, *11*, 13835–13852. [CrossRef]
310. Zhang, Y.; Chao, J.; Liu, H.; Wang, F.; Su, S.; Liu, B.; Zhang, L.; Shi, J.; Wang, L.; Huang, W.; et al. Transfer of Two-Dimensional Oligonucleotide Patterns onto Stereo controlled Plasmonic Nanostructures through DNA-Origami-Based Nanoimprinting Lithography. *Angew. Chem. Int. Ed.* **2016**, *55*, 8036–8040. [CrossRef]
311. Liu, Y.; Zeng, T.; Liu, C.; Fang, X.; Li, S.; Cao, X.; Lu, C.; Yang, H. DNA Origami-Based Letterpress Printing of Gold Nanostructures with Predesigned Morphologies. *Nano Lett.* **2023**, *23*, 11569–11577. [CrossRef]
312. Guo, M.; Qu, Z.; Min, F.; Li, Z.; Qiao, Y.; Song, Y. Advanced unconventional techniques for sub-100 nm nanopatterning. *InfoMat* **2022**, *4*, e12323. [CrossRef]
313. Lang, E.N.; Claridge, S.A. Biomolecular templates for interfacial nanomaterial assembly. In *Encyclopedia of Nanomaterials*, 1st ed.; Yin, Y., Lu, Y., Xia, Y., Eds.; Elsevier: Oxford, UK, 2023; pp. 214–229.
314. Koepnick, B.; Flatten, J.; Husain, T.; Ford, A.; Silva, D.-A.; Bick, M.J.; Bauer, A.; Liu, G.; Ishida, Y.; Boykov, A.; et al. De novo protein design by citizen scientists. *Nature* **2019**, *570*, 390–394. [CrossRef] [PubMed]
315. Ji, Z.; Huo, H.; Duan, L.; Wang, S. Design of robust malate dehydrogenases by assembly of motifs of halophilic and thermophilic enzyme based on interaction network. *Biochem. Eng. J.* **2023**, *190*, 108758. [CrossRef]
316. Wilson, C.J.; Bommarius, A.S.; Champion, J.A.; Chernoff, Y.O.; Lynn, D.G.; Paravastu, A.K.; Liang, C.; Hsieh, M.-C.; Heemstra, J.M. Biomolecular Assemblies: Moving from Observation to Predictive Design. *Chem. Rev.* **2018**, *118*, 11519–11574. [CrossRef]
317. Petrescu, D.S.; Zahr, O.K.; Abu-Baker, I.; Blum, A.S. Biomolecular Self-Assembly of Nanorings on a Viral Protein Template. *Biomacromolecules* **2022**, *23*, 3407–3416. [CrossRef]
318. Gao, S.; Liu, X.; Wang, Z.; Jiang, S.; Wu, M.; Tian, Y.; Niu, Z. Fluorous interaction induced self-assembly of tobacco mosaic virus coat protein for cisplatin delivery. *Nanoscale* **2018**, *10*, 11732–11736. [CrossRef]
319. Deng, J.; Kaefer, F.; Bailey, S.; Otsubo, Y.; Meng, Z.; Segalman, R.; Ober, C.K. New Approaches to EUV Photoresists: Studies of Polyacetals and Polypeptoids to Expand the Photopolymer Toolbox. *J. Photopolym. Sci. Technol.* **2021**, *34*, 71–74.
320. Huang, Z.; Li, S.; Zhang, J.; Pang, H.; Ivankin, A.; Wang, Y. Localized photoactuation of polymer pens for nanolithography. *Molecules* **2023**, *28*, 1171. [CrossRef]

International Journal of
Molecular Sciences

MDPI

Review

Applications of Green Carbon Dots in Personalized Diagnostics for Precision Medicine

Habtamu F. Etefa * and Francis B. Dejene

Department of Chemical and Physics Science, Walter Sisulu University, Private Bag X-1, Mthatha 5117, South Africa; fdejene@wsu.ac.za
* Correspondence: habtamufekadu24@gmail.com or hetefa@wsu.ac.za

Abstract: Green carbon dots (GCDs) have emerged as a revolutionary tool in precision medicine, offering transformative capabilities for personalized diagnostics and therapeutic strategies. Their unique optical and biocompatible properties make them ideal for non-invasive imaging, real-time monitoring, and integration with genomics, proteomics, and bioinformatics, enabling accurate diagnosis and tailored treatments based on patients' genetic and molecular profiles. This study explores the potential of GCDs in advancing individualized patient care by examining their applications in precision medicine. It evaluates their utility in non-invasive diagnostic imaging, targeted therapy delivery, and the formulation of personalized treatment plans, emphasizing their interaction with advanced genomic, proteomic, and bioinformatics platforms. GCDs demonstrated exceptional versatility in enabling precise diagnostics and delivering targeted therapies. Their integration with cutting-edge technologies showed significant promise in crafting personalized treatment strategies, enhancing their functionality and effectiveness in real-time monitoring and patient-specific applications. The findings underscore the pivotal role of GCDs in reshaping healthcare by advancing precision medicine and improving patient outcomes. The ongoing development and integration of GCDs with emerging technologies promise to further enhance their capabilities, paving the way for more effective, individualized medical care.

Keywords: GCDs; drug delivery systems; theranostics; genomics; proteomics

1. Introduction

Green carbon dots (GCDs), quantum dots (QDs), and graphene derivatives are all nanomaterials with unique properties and applications [1]. Below is a detailed comparison of these materials regarding efficacy, safety, and cost-effectiveness, followed by a table summarizing their competitive advantages and disadvantages. As seen in Table 1, green carbon dots (GCDs), quantum dots (QDs), and graphene derivatives each offer unique properties and applications. Still, they differ significantly in efficacy, safety, cost-effectiveness, and limitations. GCDs are highly biocompatible, tunable in photoluminescence, and eco-friendly, making them ideal for bioimaging, sensing, and catalysis, though they have lower brightness and limited electrical conductivity. In contrast, QDs exhibit superior brightness, narrow emission spectra, and high stability, but their high toxicity, biocompatibility issues, and costly production limit their applications, particularly in biomedical fields. Graphene derivatives excel in electrical conductivity, mechanical strength, and surface area, making them suitable for electronics and energy storage. Still, they face moderate toxicity, environmental concerns, and energy-intensive synthesis challenges. GCDs stand out for their safety, low cost, and scalability. At the same time, QDs and graphene derivatives offer superior performance in specific applications but are hindered by toxicity and higher production costs.

Table 1. Competitive Advantages and Disadvantages.

Property	Green Carbon Dots (GCDs)	Quantum Dots (QDs)	Graphene Derivatives
Efficacy	Tunable photoluminescence, High biocompatibility, Easy functionalization [2,3]	Superior brightness, Narrow emission spectra, High stability [4,5]	Excellent electrical conductivity, High mechanical strength, Large surface area [6]
Safety	Low toxicity, Biodegradable, Eco-friendly synthesis [7]	High toxicity (heavy metal-based), Biocompatibility issues [8]	Moderate toxicity, Environmental concerns [9]
Cost-Effectiveness	Low-cost synthesis and scalable production [10]	High production costs, Limited scalability [11]	Moderate cost, Energy-intensive synthesis [12]
Applications	- Bioimaging - Sensing - Catalysis [13,14]	- Displays - Solar cells - Biomedical imaging (limited) [15]	Electronics, Energy storage, and Sensors [16]
Limitations	- Lower brightness compared to QDs - Limited electrical conductivity [17]	- Toxicity concerns - High cost [18]	Toxicity in biological systems, High energy consumption [19]

As seen in Table 2, the green synthesis of carbon dots offers significant advantages over conventional methods in terms of energy consumption, environmental impact, and efficiency. Green synthesis operates at lower reaction temperatures (100–200 °C) compared to conventional high-temperature pyrolysis (300–800 °C), resulting in approximately 50–70% lower energy input and 60–75% reduced energy consumption per gram (0.5–1.5 kWh/g vs. 2–5 kWh/g). Additionally, the synthesis duration is significantly shorter (1–4 h vs. 6–12 h), enhancing efficiency. Environmentally, green synthesis utilizes renewable raw materials like biomass and waste, as opposed to non-renewable fossil fuels, and employs non-toxic solvents such as water, contrasting with the toxic organic solvents used in conventional methods. Waste generation is also drastically reduced, with minimal biodegradable by-products compared to significant hazardous waste, achieving an 80–90% reduction. Overall, green synthesis is a more sustainable, energy-efficient, and environmentally friendly approach to producing carbon dots.

Table 2. The eco-friendly synthesis of green carbon dots with conventional energy consumption and environmental impact methods.

Parameter	Green Synthesis of Carbon Dots	Conventional Methods	Comparison	Refs.
		Energy Consumption		
Reaction Temperature	100–200 °C (low-temperature processes)	300–800 °C (high-temperature pyrolysis)	~50–70% lower energy input	[20,21]
Energy Input (per gram)	0.5–1.5 kWh/g	2–5 kWh/g	~60–75% lower energy consumption	[22,23]
Duration of Synthesis	1–4 h	6–12 h	~50–70% shorter synthesis time	[24–26]
		Environmental Impact		
Raw Materials	Renewable (e.g., biomass, waste)	Non-renewable (e.g., fossil fuels)	100% renewable vs. non-renewable	[27,28]
Solvent Use	Water or mild solvents	Toxic organic solvents	Non-toxic vs. toxic solvents	[29,30]
Waste Generation	Minimal (biodegradable by-products)	Significant (hazardous waste)	~80–90% less waste	[31,32]

Carbon dots are a relatively new class of carbon nanomaterials, first reported in the early 2000s [33]. They are typically defined as quasi-spherical nanoparticles with sizes below 10 nanometers, composed primarily of carbon atoms, with traces of oxygen, hydrogen, and nitrogen. The photoluminescent properties of CDs have attracted considerable interest, as these materials emit light when excited, making them useful for bioimaging and diagnostic applications [34]. Specifically, GCDs have fluorescence emission in the green spectrum, which is particularly beneficial for in vivo imaging due to low tissue autofluorescence in this range, enabling clearer and more accurate imaging results [35]. The field of precision medicine is on the brink of transformative change due to advancements in nanotechnology, with carbon-based nanomaterials playing an increasingly pivotal role [36]. Among these, carbon dots (CDs), particularly green carbon dots (GCDs), have garnered attention for their unique properties and immense potential in medical applications. GCDs, a subset of carbon-based nanomaterials, are distinguished by their photoluminescent characteristics, eco-friendly synthesis, and biocompatibility, all of which make them suitable for diverse applications in medical science, especially in diagnostics and therapeutics [37,38]. The integration of GCDs into precision medicine holds the promise of more targeted, efficient, and safer healthcare solutions tailored to the individual characteristics of each patient.

Precision medicine, often termed personalized medicine, is a medical approach that emphasizes the customization of healthcare based on the genetic, environmental, and lifestyle differences among individuals [39]. Unlike conventional medicine, which often applies a one-size-fits-all strategy, precision medicine aims to provide more tailored interventions, optimizing therapeutic efficacy while minimizing side effects [40]. This approach requires precise and sensitive tools for disease detection, monitoring, and treatment. The versatility of GCDs, from their tunable optical properties to their robust biocompatibility, positions them as valuable assets in achieving these precision-driven objectives. An attractive feature of GCDs lies in their "green" synthesis routes. Unlike conventional CDs, which are often synthesized using toxic precursors and harsh conditions, GCDs can be produced from eco-friendly precursors through simple, cost-effective, and scalable methods [17].

Biomass-derived sources, including fruit peels, plant extracts, and other waste materials, have been widely explored for GCD synthesis [41]. The optical, chemical, and biological properties of GCDs contribute to their suitability in precision medicine. Their tunable fluorescence properties allow researchers to design GCDs with specific emission wavelengths, tailored to desired applications [42]. This customization is particularly relevant for bioimaging, where selecting the appropriate wavelength minimizes background noise, enhances image clarity, and provides higher resolution in cellular imaging. Additionally, GCDs exhibit high photostability compared to conventional organic dyes, which tend to bleach under prolonged exposure to light [43]. This attribute is crucial for real-time monitoring of biological processes in vivo, allowing researchers to track cellular interactions and therapeutic effects over extended periods. The surface functionalization of GCDs further extends their applicability in precision medicine [13,44]. For example, GCDs functionalized with tumor-targeting molecules can selectively accumulate in cancerous tissues, enabling precise delivery of therapeutics to diseased cells while minimizing off-target effects [45]. This characteristic aligns with the goals of precision medicine, which seeks to enhance therapeutic outcomes by delivering the right treatment to the right patient at the right time.

Moreover, GCDs demonstrate excellent biocompatibility, low toxicity, and high cellular uptake efficiency, making them ideal candidates for in vivo applications [46]. Numerous studies have reported minimal adverse effects of GCDs on living cells and tissues, with evidence suggesting that these nanoparticles are either metabolized or safely excreted from the body [47]. This biocompatibility is critical in precision medicine, where the safety profile of therapeutic and diagnostic agents is paramount. One of the most promising applications of GCDs in precision medicine is in disease diagnosis [48]. Fluorescent imaging, particularly in cancer diagnosis, can benefit greatly from the use of GCDs. Due to their strong luminescence and biocompatibility, GCDs can be employed to mark cancerous cells, providing highly specific and sensitive diagnostic results [49]. Additionally, in bacterial and viral infections, GCDs have been explored as fluorescent tags to visualize pathogens and understand the spread of infection within host cells.

Beyond diagnostics, GCDs offer therapeutic advantages as well. In photodynamic therapy (PDT), GCDs can generate reactive oxygen species (ROS) upon exposure to light, leading to the targeted destruction of cancer cells [50]. This therapy minimizes damage to healthy tissue, aligning with precision medicine's objective to reduce side effects associated with traditional therapies. The application of GCDs in drug delivery is another area of significant interest. By serving as drug carriers, GCDs can enhance the bioavailability and targeted delivery of therapeutic agents, improving treatment outcomes and reducing the risks associated with systemic drug distribution.

2. Fundamentals of Green Carbon Dots

An attractive aspect of GCDs lies in their eco-friendly synthesis methods [49]. Unlike other nanomaterials that may require toxic precursors or harsh reaction conditions, GCDs can be synthesized from natural, renewable sources, such as fruits, vegetables, and biomass waste [51]. For example, citric acid, glucose, and even plant extracts have been used as starting materials to produce GCDs in environmentally friendly processes [49]. Such eco-friendly synthesis approaches make GCDs safer for biological applications and align with the broader goals of sustainable development and green chemistry by minimizing waste and avoiding hazardous reagents. In contrast to many traditional fluorescent dyes and quantum dots, GCDs exhibit low toxicity and are often well tolerated by biological systems [52]. This is partly due to their carbon-based composition, which is more compatible with living tissues than metal-based nanoparticles. As a result, GCDs are considered safer for medical applications, particularly in diagnostics and therapy, where long-term exposure and accumulation of nanoparticles can raise toxicity concerns. In summary, these nanomaterials exhibit environmentally friendly synthesis methods, excellent biocompatibility, and chemical stability [53]. These properties and the capacity for surface functionalization make GCDs particularly well suited for applications in biomedical fields where safety, effectiveness, and sustainability are critical considerations. As research continues, GCDs are likely to play an increasingly important role in advancing diagnostic and therapeutic technologies, particularly within the emerging field of precision medicine. In summary, the fundamentals of GCDs include their unique green fluorescence, environmentally friendly synthesis methods, excellent biocompatibility, and chemical stability [53,54]. These properties and the capacity for surface functionalization make GCDs particularly well suited for applications in biomedical fields where safety, effectiveness, and sustainability are critical considerations. As research continues, GCDs are likely to play an increasingly important role in advancing diagnostic and therapeutic technologies, particularly within the emerging field of precision medicine.

Top-Down and Bottom-Up Approaches for Carbon Dots

GCDs can be synthesized using various methods, primarily divided into two main categories: top-down and bottom-up approaches (seen in Figure 1). Here are the most common methods used to produce GCDs. Carbon dots are produced by top-down synthesis, which breaks up or exfoliates larger carbonaceous materials [55]. The top-down method uses laser ablation, chemical and electrochemical oxidation, arc discharge, and ultrasonication techniques to break down a wide range of carbon nanomaterials (graphite, carbon soot, carbon nanotubes, activated carbon, and nanodiamonds) into smaller carbon nanoparticles. Carbon dots may be produced in a scalable and economical manner using top-down synthesis techniques [56]. They can produce carbon dots in significant amounts and use easily accessible carbon sources. Nevertheless, in contrast to bottom-up approaches, they might have less control over the surface chemistry and functionalization. Bottom-up approaches assemble molecular precursors from smaller carbon units, including carbohydrates, organic acids, polymers, natural products, and biomass, using hydrothermal, carbonization, microwave, solvothermal, pyrolysis, and thermal decomposition procedures, whereas the top-down strategy comprises the breaking down of a wide range of carbon nanomaterials (graphite, carbon soot, carbon nanotubes, activated carbon, and nanodiamonds) into smaller carbon nanoparticles via laser ablation, chemical and electrochemical oxidation, arc discharge, and ultrasonication procedures.

The hydrothermal method is among the most popular synthesis routes for GCDs from the bottom-up approaches due to its simplicity and ability to use eco-friendly precursors. In this method, a carbon source such as citric acid, glucose, or even biomass waste like fruit peels is placed in water or an organic solvent, then heated in a sealed reactor at high temperature and pressure (usually 150–250 °C). The solvothermal method is similar but uses an organic solvent instead of water, which can influence the size and luminescence properties of the resulting GCDs [13]. Microwave-Assisted Synthesis: Another eco-friendly approach is microwave-assisted synthesis, which involves heating carbon sources like citric acid or urea in a microwave [57]. This method is highly efficient, requiring only a few minutes to produce GCDs with consistent properties. The rapid heating causes carbonization, forming nanoparticles with desirable green emission and good dispersibility in aqueous solutions. Ultrasonic Synthesis: In ultrasonic synthesis, high-intensity ultrasound waves are applied to a carbon source, producing localized high temperatures and pressures that break down the precursor into nanoparticles. This method is considered low-energy and relatively straightforward, making it a popular choice for synthesizing GCDs with green emissions. Pyrolysis: Pyrolysis involves the thermal decomposition of carbon precursors in an oxygen-deficient environment. While this method can yield GCDs with excellent photoluminescent properties, it requires precise temperature and atmosphere control to avoid unwanted combustion and obtain the desired size and surface characteristics. Below is a schematic diagram showing various synthesis routes for GCDs, emphasizing the eco-friendly and low-toxicity methods that distinguish these materials in the realm of nanotechnology [58].

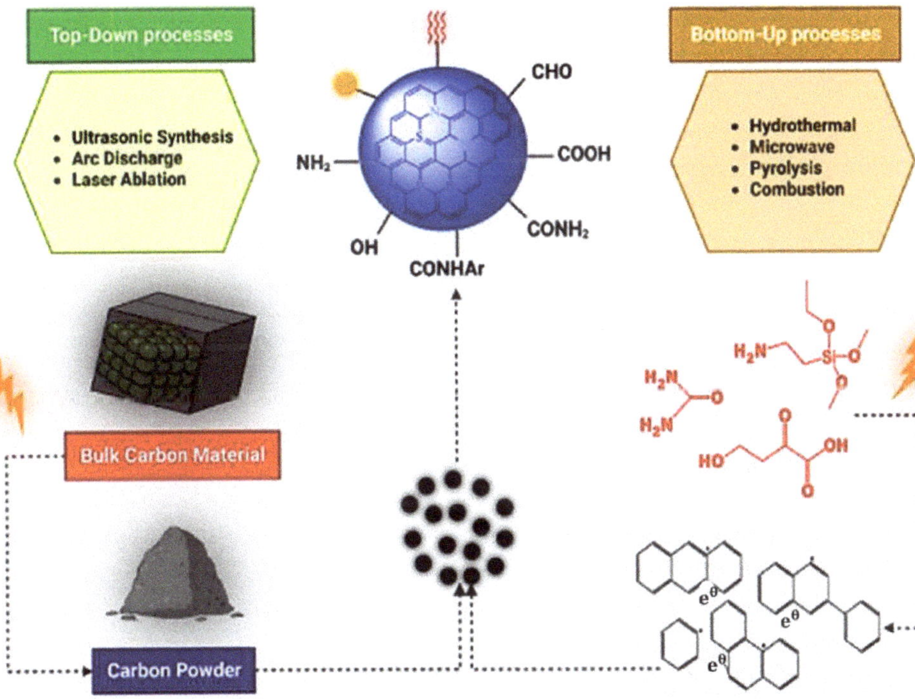

Figure 1. Top-down and bottom-up approaches for the fabrication of carbon dots [58]. Copyright 2023, Elsevier.

3. Role of Carbon Dots in Personalized Diagnostics

The field of personalized diagnostics seeks to deliver patient-specific medical insights, improving disease detection, monitoring, and treatment decision-making accuracy [36]. This approach tailors diagnostic tools and therapies based on individual genetic, molecular, and cellular profiles, enabling a more precise understanding of disease at the personal level (Figure 2). Among the novel materials driving advances in personalized diagnostics, GCDs have emerged as promising agents due to their unique properties such as fluorescence, biocompatibility, and eco-friendly synthesis. GCDs' ability to act as highly sensitive and specific diagnostic probes has positioned them at the forefront of innovations in bioimaging, biosensing, and targeted diagnostics, transforming how diseases are detected and monitored [47,59].

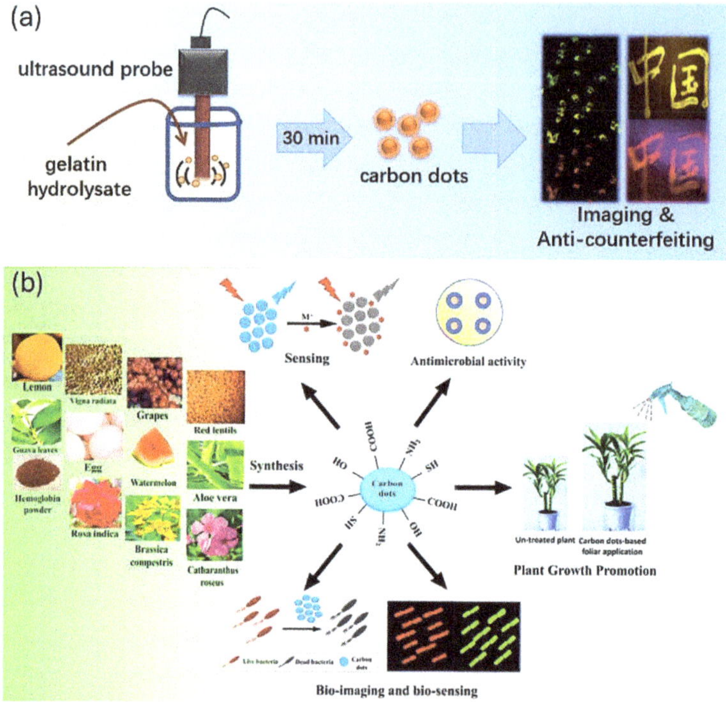

Figure 2. (**a**) Green carbon dots fluorescence imaging and anti-counterfeiting applications [60]. Copyright 2021, Published by American Chemical Society. (**b**) Environmentally friendly sources employed for the Green CDs and their applications for bio-imaging and bio-sensing [61]. Copyright Elsevier (2020).

Photoluminescent Properties in Imaging-Based Diagnostics

One of the defining features of GCDs is their stable green fluorescence, which plays a significant role in diagnostic imaging applications [62]. GCDs, however, exhibit high photostability, maintaining their fluorescence over extended periods. This stability makes them ideal for real-time, high-resolution imaging of cells and tissues, a critical requirement in personalized diagnostics where precise visualization of molecular and cellular details is essential [63]. In cancer diagnostics, for instance, GCDs can be functionalized with tumor-targeting ligands, such as antibodies or peptides that recognize cancer-specific receptors. Once functionalized, these GCDs can selectively bind to cancer cells, providing a high-contrast image of the tumor site while sparing healthy tissue [64]. This selective imaging aligns with the objectives of personalized diagnostics by allowing clinicians to detect even small, early-stage tumors with greater accuracy, leading to better-informed treatment decisions. Additionally, GCDs' tunable emission properties allow for multicolor imaging, facilitating the simultaneous tracking of multiple biomarkers or cellular processes, which is especially valuable for complex diseases like cancer and autoimmune disorders [65].

4. Applications of GCDs in Precision Medicine

The benefits and drawbacks of CDs' use in biomedicine include bioimaging, biosensing, chemical or fluorescence sensing, nanomedicine, electrocatalysis, and photocatalysis. CDs made using different methods have been employed in various applications. These characteristics have made GCDs increasingly valuable in precision medicine, especially in personalized diagnostics, imaging, biosensing, and molecular probing applications. In precision medicine, which seeks to tailor treatments and diagnostics to individual patients based on genetic, molecular, and environmental factors, GCDs are highly adaptable tools that can enhance accuracy, specificity, and sensitivity [66]. GCDs with strong biocompatibility and low cytotoxicity can readily enter cells and disperse throughout the cytoplasmic region, as shown in Figure 3a. CDs can efficiently track and deliver genes or drugs to specific targets, as shown in Figure 3b. Based on Figure 3c, there is a greater need than ever for innovative, improved, focused therapy because of the huge number of deaths brought on by infections and cancer. Photodynamic therapy (PDT) is one potential method to treat various clinical disorders in addition to traditional therapies and drugs. Because CDs can effectively track and deliver genes or medications, as shown in Figure 3d, they can also function as nanocarriers. Branching polyethyleneimine CQDs are one such gene delivery agent. Additionally, CDs are essential for regulating drug release; loading CDs with doxorubicin can restrict drug release in HeLa cells.

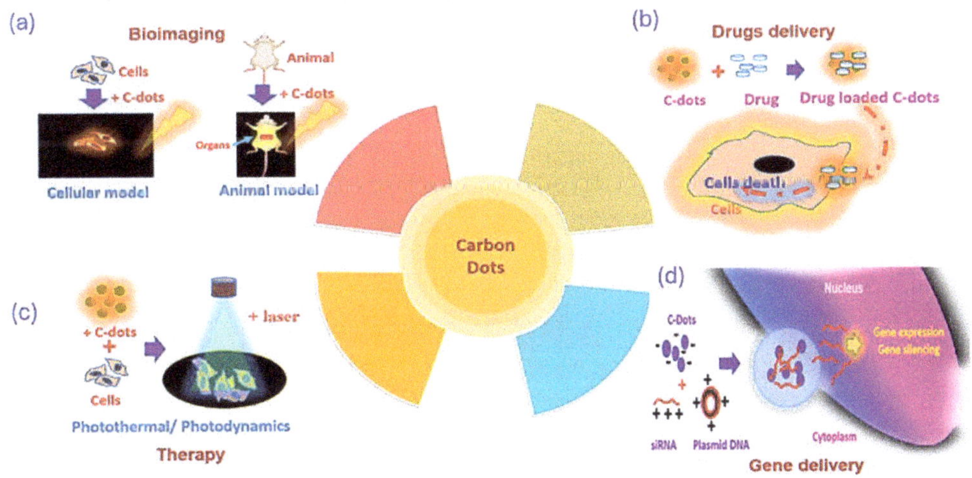

Figure 3. Various biomedical applications of carbon dots: (**a**) bioimaging, (**b**) drug delivery, (**c**) therapy, (**d**) gene delivery [67]. Copyright 2023, mdpi.com.

4.1. Personalized Diagnostics for Cancer

For the next phase of cancer research, diagnostic methods and chemotherapy administration based on nanotechnologies, including carbon dots, may be viable options [68,69]. GCDs have emerged as a promising tool in the field of personalized diagnostics due to their unique properties and eco-friendly synthesis methods [67]. They also exhibit excellent luminescence and biocompatibility, making them ideal for bioimaging applications. Their strong fluorescence allows for high-contrast imaging of biological tissues, aiding in early disease detection [67]. Due to their high stability and tunable fluorescence, GCDs are used in biosensors to detect various biomolecules [14]. This is crucial for personalized diagnostics as it enables the monitoring of specific biomarkers related to individual health conditions. This personalized approach ensures that medications are delivered directly to the affected area, minimizing side effects and improving treatment efficacy [14]. Biomedical applications have advanced significantly thanks to DNA nanostructures with various biological functions. Still missing, nevertheless, is a uniform approach to the effective synthesis of DNA nanostructures [70]. For example, in cancer diagnostics, GCDs functionalized with antibodies against tumor-specific antigens can identify cancer cells in their early stages, allowing clinicians to catch the disease before it progresses. In cardiovascular diseases, GCDs can be modified to detect early biomarkers like troponins or C-reactive proteins, offering a non-invasive means of diagnosis that can lead to faster interventions and improved patient outcomes [71,72]. In personalized diagnostics, the ability to monitor disease progression non-invasively is invaluable. GCDs, due to their stable fluorescence and surface functionalizability, can serve as biosensors that continuously detect disease biomarkers in bodily fluids like blood, urine, or saliva. This enables patients to undergo regular monitoring without invasive procedures, enhancing comfort and compliance with diagnostic protocols [73]. For example, GCD-based sensors can detect circulating tumor DNA (ctDNA) in blood samples [50,74–76], providing insights into cancer dynamics and allowing oncologists to adjust treatment plans based on the patient's real-time condition [77].

Imaging for Precision Medicine

Bioimaging is a critical application area for GCDs in precision medicine. Their stable green fluorescence, biocompatibility, and capacity for functionalization make them highly effective for various imaging modalities [78]. GCDs offer significant advantages over traditional imaging agents, which often suffer from photobleaching or exhibit high toxicity, making GCDs a superior choice for long-term imaging studies. GCDs are particularly effective in cancer imaging, where their high photostability allows for real-time, high-resolution visualization of tumors [35]. This targeted imaging is a marked improvement over conventional imaging agents, which cannot distinguish cancerous cells from healthy tissues accurately [79]. Furthermore, GCDs can be tuned to emit in the green emission range, which minimizes interference from the body's natural autofluorescence, thereby enhancing imaging clarity and contrast. In addition to emission imaging, GCDs can be engineered for multimodal imaging, integrating fluorescence with other imaging techniques like magnetic resonance imaging (MRI) or computed tomography (CT) [79]. For instance, combining GCDs with magnetic nanoparticles allows for dual fluorescence-MRI imaging, which can provide comprehensive information about the tumor's location, size, and structural details [47]. This multimodal capability is advantageous in precision diagnostics, as it allows clinicians to gather more information using a single imaging agent, thereby reducing the number of procedures required for an accurate diagnosis and improving patient comfort. Alzheimer's disease (AD) in Figure 4 is a neurological condition that cannot be cured and becomes worse over time. It gradually erodes a person's memory and basic cognitive functioning. According to the Alzheimer's Association, there is presently no cure for Alzheimer's, making it the sixth most common cause of death in the United States. CDs that shown inhibitory effects on multiple pathogenic aspects of AD, including inflammation, tau, and amyloid, are intriguing nanodrugs in this context. B-CDs suppressed amyloid-beta (Aβ) 42 and 40 fibrillations. Molecular dynamic simulations (Figure 3) demonstrated that B-CDs' hydrophilic surface enhances their ability to suppress Aβ fibrillation. Additionally, it was discovered that B-CDs successfully prevented beta-secretase 1 activity and postponed the production of Aβ. Following their penetration of cell membranes and entry into cytosolic compartments, Y-CDs markedly reduced the expression and release of Aβ and amyloid precursor protein (APP), respectively (Figure 3). Additionally, CNDs demonstrated a dose-dependent inhibition of microtubule-associated protein tau (MAPT) aggregation. Molecular dynamic simulations revealed a hydrophobic interaction between MAPT and CNDs (Figure 4). CNDs have been shown in a photocatalysis experiment to deactivate reactive oxygen species (ROS) and consume those that are already in the environment, potentially lowering the oxidative stress on the brains of AD patients [80].

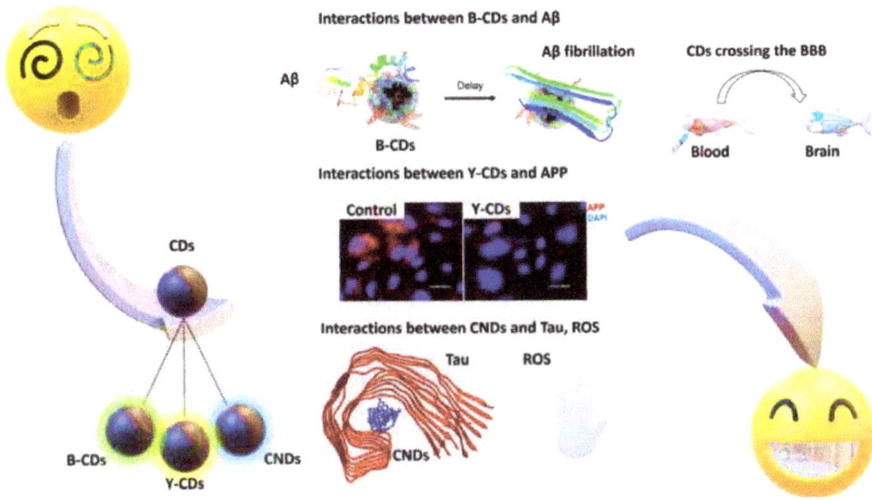

Figure 4. The interactions of various CDs involved in inhibiting different pathogeneses of Alzheimer's disease (AD) [80]. Copyright 2025, Genetic Engineering and Biotechnology News.

4.2. Therapy Selection and Monitoring

GCDs are emerging as pivotal tools in precision medicine, particularly in therapy selection and monitoring [47]. This ability to tailor and closely observe therapeutic responses aligns with the core objectives of precision medicine: to optimize treatments based on an individual's unique biological profile. Among the key applications of GCDs in precision therapy are their roles in drug delivery systems and theranostics. These two applications enable GCDs to function as both therapeutic and diagnostic tools, offering a seamless transition from diagnosis to treatment and facilitating real-time monitoring of therapeutic outcomes [13,66].

Drug Delivery Systems

In precision medicine, drug delivery systems are vital for ensuring that therapeutic agents reach their intended target while minimizing side effects [81]. GCDs offer significant advantages as drug delivery vehicles, including their small size, tunable surface chemistry, and biocompatibility [47]. As seen in Figure 5a, drugs are incorporated into nanoparticle carriers by encapsulation, adsorption, or binding in nanomaterial-based drug delivery, enabling stable and secure administration within the body [82]. Various types of antibiotic or anticancer medications have been administered successfully thus far [83]. CDs are useful vehicles for the targeted administration of anti-cancer medications (such as doxorubicin (DOX), cisplatin, paclitaxel, docetaxel, camptothecin, daunorubicin, etc.) for cancer treatment because they are simple to modify [84,85]. Because the majority of CDs@drug complexes have cleavable chemical bonds, medicines can be explicitly released in response to various stimuli or the acidic environment of tumor locations [86,87]. CDs can help cancer cells or other diseased cells overcome their drug resistance and improve delivery and controlled release at the target spot.

As seen in Figure 5b, one frequent first-line treatment for many malignancies is DOX, which can be coupled with DNA-related enzymes of cancer cells to intercalate DNA base pairs to prevent their synthesis and replication [88]. The targeted release of DOX in the tumor microenvironment via biocompatible CDs has garnered much interest in recent years. Most of these groundbreaking studies have concentrated on improving the EPR (Enhanced

Penetration and Retention) impact to boost medication loading capacity and increase tumor accumulation. Composites of CDs and DOX that exhibit notable tumor targeting, improved anti-tumor activity, and fewer side responses have been created by building CDs with a hydrophilic shell and a hydrophobic carbon core. Because normal CDs have trouble combining with DOX, CDs must be modified on the surface. The changeability of CDs@DOX systems must be carefully considered due to the acidic nature of the tumor site microenvironment [89]. Human glioblastoma cell line (U87MG) tumors were selectively targeted by CDs@DOX by surface charge modification using an octylamine-modified polyacrylic acid (cRGD-PAA-OA) and cRGD coating. Conversely, CDs-based composite materials demonstrated an excellent release profile of DOX over time in in vitro drug release studies (Figure 5b) [90]. Cisplatin (Pt(IV)), in addition to DOX (Figure 5c), is frequently used to treat a variety of cancers [91]. However, the effectiveness of cisplatin-based cancer treatments is typically hampered by acquired and/or intrinsic drug resistance, as well as the severe adverse effects. As a result, creating drug targeting systems with high site specificity and managing drug release rate is crucial. CDs make an excellent delivery system for these medications because of their compact size and simplicity of use. It is important to carefully evaluate using CDs in combination with Pt(IV)-based medications [92]. CDs can fix Pt(IV)-based medications when their surfaces are functionalized appropriately. PEGylation is a popular surface functionalization technique. Via an electrostatic repulsion mechanism, the cisplatin-based CDs platform (CDs-Pt (IV)@PEG-(PAH/DMMA)) can encourage the release of positively charged CDs-Pt (IV) seen in Figure 5c [93].

Controlled and gradual release of pharmaceuticals are still problems with CD-based systems that need to be resolved (Figure 5d), in addition to the quantity of drugs-loaded. Drugs that are rapidly removed or broken down when taken alone should be able to be delivered by an optimal sustained-release mechanism [94]. Additionally, the therapeutic agent must be released consistently during treatment to prevent the drug from being damaged during the release time. According to the previously described reports, medications can be placed onto CDs using various techniques, and they can subsequently be stimulated or released from the CDs in a regulated way [95]. Therapeutic agent typically used in CD-based controlled release systems, polymer matrices, hydrogel networks, or mesoporous silica work to enhance local therapeutic benefits and reduce adverse effects. In Figure 5e, reservoir-controlled release systems limit the release of trapped therapeutic agents through a membrane that controls the rate at which the therapeutic agent diffuses out of the reservoir; (3) The mesh size of the swelling polymer network controls the rate of drug release from hydrogel-based release systems; and (4) Therapeutic molecules diffuse through a tortuous network of interconnected pores created during the phase separation of the drug/excipient and the polymer. As an illustration of the latter, a pH-responsive biodegradable hydrogel-nanocomposite was created using CDs as the cross-linker and an agarose-poly(vinyl alcohol) copolymer. Norfloxacin (NFX), an antibiotic medication, was released from the hydrogel-nanocomposite using zero-order kinetics [96].

Furthermore, GCDs have demonstrated the ability to bypass efflux pumps on cancer cell membranes, often responsible for drug resistance in tumors. By penetrating cancer cells more effectively, GCD-based drug delivery systems can overcome resistance mechanisms and improve the intracellular accumulation of therapeutic agents, thus enhancing treatment success rates. These unique properties make GCDs invaluable in precision medicine, where the ability to precisely target and deliver drugs to specific locations within the body can significantly influence patient outcomes.

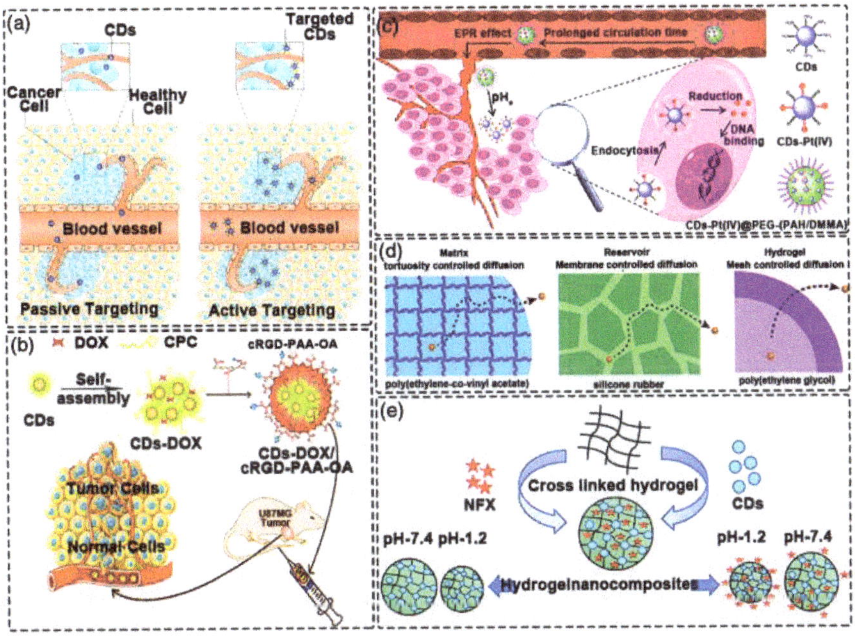

Figure 5. Drug delivery with CDs. (**a**) CDs utilizing the EPR effect to enter a lesion. (**b**) CDs as a delivery platform for DOX. (**c**) CDs being used to deliver cisplatin. (**d**) Mechanisms of controlled drug release. (**e**) CDs-containing hydrogel for the controlled release NFX [97]. Copyright 2022, Wiley Online Library.

5. Challenges in the Application of GCDs

5.1. Standardization and Scalability of Synthesis

A critical challenge in applying GCDs in clinical settings is the lack of standardized synthesis protocols. While numerous methods for synthesizing GCDs exist, each may yield GCDs with slightly different physicochemical properties, such as size, surface charge, and fluorescence [98]. These variations affect the performance of GCDs, including their bio-distribution, toxicity, and efficacy in targeting specific molecules [99]. Furthermore, the scalability of GCD synthesis remains a concern; ensuring consistent quality, stability, and bioactivity at large production scales is essential for regulatory approval and safe clinical application.

5.2. Understanding and Minimizing Toxicity

Although GCDs are generally considered biocompatible and have lower toxicity than traditional carbon-based nanomaterials, their long-term safety remains unclear [100]. The impacts of chronic exposure, accumulation in different organs, and potential for cellular or genetic toxicity are areas that need thorough investigation. Additionally, GCD surface modification, often necessary for targeted delivery or improved stability, may introduce new toxicity risks or unforeseen biological interactions [78]. A comprehensive understanding of GCD toxicity across various biological systems, alongside targeted studies on clearance mechanisms, is essential for addressing regulatory and safety concerns.

5.3. Limited Targeting Precision and Delivery Efficiency

While GCDs have shown promise as carriers for drug delivery and gene therapy, achieving high targeting specificity and efficient delivery remains challenging. Often,

GCDs struggle with avoiding off-target effects, as their small size and functionalization properties may lead them to accumulate in non-target tissues [101]. GCDs need advanced surface modifications or loading techniques to improve their precision to ensure selective binding to disease-specific receptors. Increasing targeting accuracy enhances therapeutic efficacy and minimizes side effects, which is critical for applications in sensitive areas such as oncology or neurodegenerative disorders [102].

6. Future Directions for GCDs in Precision Medicine

6.1. Developing Targeted and Functionalized GCDs

Future research on GCDs will likely focus on developing more targeted and functionalized GCDs for improved disease detection and treatment precision [45]. Researchers can enhance GCD selectivity by modifying GCDs with ligands, antibodies, or aptamers specific to disease biomarkers, enabling more accurate localization to disease sites and more effective therapeutic delivery. Additionally, developing stimuli-responsive GCDs that release therapeutic payloads in response to environmental changes, like pH or temperature, could significantly improve the control of drug release, optimizing treatment efficacy and minimizing side effects [103].

6.1.1. Enhancing Biocompatibility and Safety Profiles

As the long-term effects of GCDs in the body are still under research, future studies should prioritize biocompatibility and safety [78]. One approach is to design biodegradable GCDs that break down into non-toxic byproducts after completing their function. Further research into GCD surface chemistry will also be essential, aiming to minimize immune response, improve clearance from the body, and ensure that GCDs remain safe for repeated or prolonged use. This research will be instrumental in moving GCD-based technologies closer to clinical trials and eventual regulatory approval [47].

6.1.2. Advancing Bioinformatics Integration and Real-Time Monitoring Systems

With further advancements in bioinformatics, GCD-based platforms could enable real-time monitoring and adaptive treatment responses [104]. For instance, integrating GCD-generated data with machine learning models could provide predictive insights on treatment efficacy or disease progression, allowing for personalized adjustments [105]. Developing secure, interoperable data systems will ensure that GCD data can be effectively utilized alongside genomic, proteomic, and clinical data to create a comprehensive patient profile, enhancing the precision of diagnoses and therapies. Finally, GCDs represent a promising frontier in precision medicine, with applications ranging from drug delivery and diagnostics to real-time patient monitoring. However, challenges in standardization, biocompatibility, targeting precision, and bioinformatics integration must be addressed before these materials can be widely adopted in clinical practice. By focusing future research on functionalization, biocompatibility, multi-modal applications, and data integration, the medical community can move closer to realizing the potential of GCDs in delivering highly personalized, efficient, and safe patient care [106]. As seen in Figure 6, the cross-linked polymeric characteristics and easy manufacturing of CDs have stimulated the development of prospective biomedical research applications (e.g., drug delivery system, photodynamic treatment, bioimaging, etc.). Bioinformatics leverages computational tools and algorithms to analyze biological data, transforming raw data into actionable insights [107]. In precision medicine, bioinformatics helps make sense of data from a range of omics fields, including genomics, proteomics, and metabolomics, creating comprehensive patient profiles [108]. By integrating these datasets, bioinformatics enables clinicians to predict disease risk, personalize treatments, and monitor patient responses, all based on a thorough understanding of a patient's unique biological makeup. In this context, bioinformatics is essential for maximizing the utility of GCDs by interpreting the data they generate in real time [109].

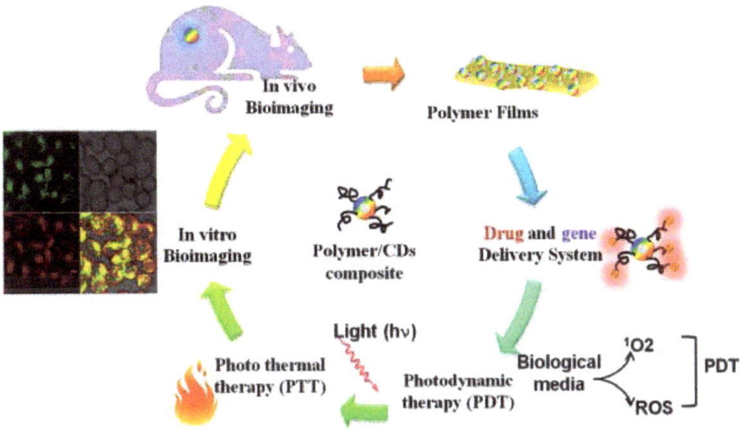

Figure 6. Proposed biomedical application of CD/polymer composite materials [110]. Copyright 2022, mdpi.com.

7. Conclusions

In conclusion, Green GCDs are a significant advancement in nanotechnology, particularly for personalized diagnostics and precision medicine. Their unique properties, such as small size, biocompatibility, and tunable fluorescence, make them ideal for various biomedical applications. GCDs are synthesized through eco-friendly methods and can be functionalized to target specific biomolecules, enhancing their versatility. In personalized diagnostics, GCDs excel in imaging and biosensing, providing high-resolution insights into disease states and enabling early and precise detection through molecular probing. For therapy selection and monitoring, GCDs offer innovative solutions like drug delivery systems and theranostics, which combine diagnostic and therapeutic functionalities for real-time treatment monitoring. Integrating GCDs with genomics, proteomics, and bioinformatics is advancing personalized treatment strategies by enabling comprehensive molecular profiling and optimizing patient-specific interventions through data analytics and machine learning. Despite their potential, GCDs face challenges such as the need for standardized synthesis methods, long-term biocompatibility assessments, and efficient targeting capabilities. Addressing these challenges through future research is crucial to fully realize the potential of GCDs in clinical settings. Overall, GCDs hold great promise for revolutionizing personalized diagnostics and precision medicine, offering targeted, effective, and responsive healthcare solutions tailored to individual patient profiles. By overcoming current limitations, GCDs could play a pivotal role in advancing precision medicine and improving patient outcomes.

Author Contributions: H.F.E.: Writing and original draft, Visualization, Methodology, Review and editing, Conceptualization, Investigation, and Formal analysis; F.B.D.: Supervising, Approving, Financial support, and Administrative support. All authors have read and agreed to the published version of the manuscript.

Funding: This research received no external funding.

Institutional Review Board Statement: Not applicable.

Informed Consent Statement: Not applicable.

Data Availability Statement: Not applicable.

Acknowledgments: The authors are grateful to Walter Sisulu University (WSU) for the support.

Conflicts of Interest: The authors declare no conflict of interest.

References

1. Jangjou, A.; Zareshahrabadi, Z.; Abbasi, M.; Talaiekhozani, A.; Kamyab, H.; Chelliapan, S.; Vaez, A.; Golchin, A.; Tayebi, L.; Vafa, E.; et al. Time to Conquer Fungal Infectious Diseases: Employing Nanoparticles as Powerful and Versatile Antifungal Nanosystems Against a Wide Variety of Fungal Species. *Sustainability* **2022**, *14*, 12942. [CrossRef]
2. Krishnaiah, P.; Atchudan, R.; Gangadaran, P.; Perumal, S.; Rajendran, R.L.; Sundramoorthy, A.K.; Kumar, R.S.; Ramalingam, S.; Ahn, B.-C.; Lee, S.W.; et al. A sustainable synthesis and applications of biomass waste-derived tunable fluorescent carbon dots: In vitro and in vivo fluorescent imaging. *J. Photochem. Photobiol. A Chem.* **2025**, *458*, 115944. [CrossRef]
3. Moniruzzaman, M.; Kim, J. Synthesis and post-synthesis strategies for polychromatic carbon dots toward unique and tunable multicolor photoluminescence and associated emission mechanism. *Nanoscale* **2023**, *15*, 13858–13885. [CrossRef] [PubMed]
4. Li, X.; Lin, Q.; Song, J.; Shen, H.; Zhang, H.; Li, L.S.; Li, X.; Du, Z. Quantum-dot light-emitting diodes for outdoor displays with high stability at high brightness. *Adv. Opt. Mater.* **2020**, *8*, 1901145. [CrossRef]
5. Shen, H.; Gao, Q.; Zhang, Y.; Lin, Y.; Lin, Q.; Li, Z.; Chen, L.; Zeng, Z.; Li, X.; Jia, Y.; et al. Visible quantum dot light-emitting diodes with simultaneous high brightness and efficiency. *Nat. Photonics* **2019**, *13*, 192–197. [CrossRef]
6. Dasari, B.L.; Nouri, J.M.; Brabazon, D.; Naher, S. Graphene and derivatives—Synthesis techniques, properties and their energy applications. *Energy* **2017**, *140*, 766–778. [CrossRef]
7. Peng, W.; Li, P.; Zia, M.; Bhuiyan, S.A.; Liu, Y.; Chen, D.; Tudi, M.; Gao, Y.; Yan, X.; Jia, Y.; et al. A Comparative Review on Carbon Nanomaterials and Their Impact on Plant Growth with the Lens of Green Chemistry Principles. *Glob. Chall.* **2024**, *8*, 2400008. [CrossRef]
8. Biranje, A.; Azmi, N.; Tiwari, A.; Chaskar, A. Quantum dots based fluorescent probe for the selective detection of heavy metal ions. *J. Fluoresc.* **2021**, *31*, 1241–1250. [CrossRef]
9. Sun, F. *Synthesis and Device Applications of Graphene Derivatives and Quantum Dots*; Springer Nature: London, UK, 2024.
10. Khan, A.U.; Liu, Y.; Wang, S.; Ullah, M.W.; Chen, Q.; Zhang, D.; Kang, Z.; Mao, B. Advancements in the green synthesis of carbon dots for sustainable development. *Sustain. Mater. Technol.* **2024**, *41*, e01004. [CrossRef]
11. Pu, Y.; Cai, F.; Wang, D.; Wang, J.-X.; Chen, J.-F. Colloidal synthesis of semiconductor quantum dots toward large-scale production: A review. *Ind. Eng. Chem. Res.* **2018**, *57*, 1790–1802. [CrossRef]
12. Yang, W.; Li, X.; Fei, L.; Liu, W.; Liu, X.; Xu, H.; Liu, Y. A review on sustainable synthetic approaches toward photoluminescent quantum dots. *Green Chem.* **2022**, *24*, 675–700. [CrossRef]
13. Lamba, R.; Yukta, Y.; Mondal, J.; Kumar, R.; Pani, B.; Singh, B. Carbon dots: Synthesis, characterizations, and recent advancements in biomedical, optoelectronics, sensing, and catalysis applications. *ACS Appl. Bio Mater.* **2024**, *7*, 2086–2127. [CrossRef]
14. Hatimuria, M.; Phukan, P.; Bag, S.; Ghosh, J.; Gavvala, K.; Pabbathi, A.; Das, J. Green carbon dots: Applications in development of electrochemical sensors, assessment of toxicity as well as anticancer properties. *Catalysts* **2023**, *13*, 537. [CrossRef]
15. Abdellatif, A.A.; Younis, M.A.; Alsharidah, M.; Al Rugaie, O.; Tawfeek, H.M. Biomedical applications of quantum dots: Overview, challenges, and clinical potential. *Int. J. Nanomed.* **2022**, *17*, 1951–1970. [CrossRef] [PubMed]
16. Kumar, Y.R.; Deshmukh, K.; Sadasivuni, K.K.; Pasha, S.K.K. Graphene quantum dot based materials for sensing, bio-imaging and energy storage applications: A review. *RSC Adv.* **2020**, *10*, 23861–23898. [CrossRef] [PubMed]
17. Ghosh, D.; Sarkar, K.; Devi, P.; Kim, K.-H.; Kumar, P. Current and future perspectives of carbon and graphene quantum dots: From synthesis to strategy for building optoelectronic and energy devices. *Renew. Sustain. Energy Rev.* **2021**, *135*, 110391. [CrossRef]
18. Gidwani, B.; Sahu, V.; Shukla, S.S.; Pandey, R.; Joshi, V.; Jain, V.K.; Vyas, A. Quantum dots: Prospectives, toxicity, advances and applications. *J. Drug Deliv. Sci. Technol.* **2021**, *61*, 102308. [CrossRef]
19. Wambu, E.W.; Huang, J. *Chemistry of Graphene: Synthesis, Reactivity, Applications and Toxicities*; BoD—Books on Demand: Norderstedt, Germany, 2024.
20. Zeng, H.; Peng, H.; He, H.; Feng, J.; Sun, Y.; He, H.; Li, L. Green and low-temperature synthesis of carbon dots for simple detection of kaempferol. *J. Fluoresc.* **2023**, *33*, 1971–1979. [CrossRef]
21. Tian, X.T.; Yin, X.B. Carbon dots, unconventional preparation strategies, and applications beyond photoluminescence. *Small* **2019**, *15*, 1901803. [CrossRef]
22. Sharma, V.; Tiwari, P.; Mobin, S.M. Sustainable carbon-dots: Recent advances in green carbon dots for sensing and bioimaging. *J. Mater. Chem. B* **2017**, *5*, 8904–8924. [CrossRef]
23. Pho, Q.H.; Escriba-Gelonch, M.; Losic, D.; Rebrov, E.V.; Tran, N.N.; Hessel, V. Survey of synthesis processes for N-doped carbon dots assessed by green chemistry and circular and EcoScale metrics. *ACS Sustain. Chem. Eng.* **2021**, *9*, 4755–4770. [CrossRef]

24. Wu, Z.L.; Liu, Z.X.; Yuan, Y.H. Carbon dots: Materials, synthesis, properties and approaches to long-wavelength and multicolor emission. *J. Mater. Chem. B* **2017**, *5*, 3794–3809. [CrossRef] [PubMed]
25. Waskaas, M. Synthesis and Comparative Analysis of Carbon Dots: A Study on Continuous Flow, Microwave, and Furnace Methods. Master's Thesis, The University of Bergen, Bergen, Norway, 2024.
26. Singh, I.; Arora, R.; Dhiman, H.; Pahwa, R. Carbon quantum dots: Synthesis, characterization and biomedical applications. *Turk. J. Pharm. Sci.* **2018**, *15*, 219–230. [CrossRef] [PubMed]
27. Adewuyi, A.O.; Awodumi, O.B. Renewable and non-renewable energy-growth-emissions linkages: Review of emerging trends with policy implications. *Renew. Sustain. Energy Rev.* **2017**, *69*, 275–291. [CrossRef]
28. Güney, T. Renewable energy, non-renewable energy and sustainable development. *Int. J. Sustain. Dev. World Ecol.* **2019**, *26*, 389–397. [CrossRef]
29. Majhi, K.C.; Karfa, P.; Kumar, S.; Madhuri, R. Water as the green solvent in organic synthesis. *Ind. Appl. Green Solvents* **2019**, *2*, 182–201.
30. Hessel, V.; Tran, N.N.; Asrami, M.R.; Tran, Q.D.; Long, N.V.D.; Escribà-Gelonch, M.; Tejada, J.O.; Linke, S.; Sundmacher, K. Sustainability of green solvents—Review and perspective. *Green Chem.* **2022**, *24*, 410–437. [CrossRef]
31. Maraveas, C. Production of sustainable and biodegradable polymers from agricultural waste. *Polymers* **2020**, *12*, 1127. [CrossRef]
32. Velvizhi, G.; Shanthakumar, S.; Das, B.; Pugazhendhi, A.; Priya, T.S.; Ashok, B.; Nanthagopal, K.; Vignesh, R.; Karthick, C. Biodegradable and non-biodegradable fraction of municipal solid waste for multifaceted applications through a closed loop integrated refinery platform: Paving a path towards circular economy. *Sci. Total Environ.* **2020**, *731*, 138049. [CrossRef]
33. Georgakilas, V.; Perman, J.A.; Tucek, J.; Zboril, R. Broad family of carbon nanoallotropes: Classification, chemistry, and applications of fullerenes, carbon dots, nanotubes, graphene, nanodiamonds, and combined superstructures. *Chem. Rev.* **2015**, *115*, 4744–4822. [CrossRef]
34. Yao, J.; Yang, M.; Duan, Y. Chemistry, biology, and medicine of fluorescent nanomaterials and related systems: New insights into biosensing, bioimaging, genomics, diagnostics, and therapy. *Chem. Rev.* **2014**, *114*, 6130–6178. [CrossRef] [PubMed]
35. Phan, L.M.T.; Cho, S. Retracted: Fluorescent Carbon Dot-Supported Imaging-Based Biomedicine: A Comprehensive Review. *Bioinorg. Chem. Appl.* **2022**, *2022*, 9303703. [CrossRef]
36. Paramasivam, G.; Palem, V.V.; Meenakshy, S.; Suresh, L.K.; Gangopadhyay, M.; Antherjanam, S.; Sundramoorthy, A.K. Advances on Carbon Nanomaterials and Their Applications in Medical Diagnosis and Drug Delivery. *Colloids Surf. B Biointerfaces* **2024**, *241*, 114032. [CrossRef] [PubMed]
37. Sahana, S.; Gautam, A.; Singh, R.; Chandel, S. A recent update on development, synthesis methods, properties and application of natural products derived carbon dots. *Nat. Prod. Bioprospect.* **2023**, *13*, 51. [CrossRef] [PubMed]
38. Yadav, S.; Daniel, S. Green synthesis of zero-dimensional carbon nanostructures in energy storage applications—A review. *Energy Storage* **2024**, *6*, e500. [CrossRef]
39. Juengst, E.; McGowan, M.L.; Fishman, J.R.; Settersten, R.A., Jr. From "personalized" to "precision" medicine: The ethical and social implications of rhetorical reform in genomic medicine. *Hastings Cent. Rep.* **2016**, *46*, 21–33. [CrossRef]
40. YahyaAlmakrami, I.; Al Omorat, T.; GhannamShreaf, M.M.; Al-Yami, S.A.S.; Alyami, K.A. Tailoring Treatment to the Individual: A Critical Examination of Precision Medicine and Personalized Healthcare Through the Lens of Genetics, Lifestyle, and Environmental Factors. *Chelonian Res. Found.* **2023**, *18*, 550–564.
41. Chakraborty, R.; Vilya, K.; Pradhan, M.; Nayak, A.K. Recent advancement of biomass-derived porous carbon based materials for energy and environmental remediation applications. *J. Mater. Chem. A* **2022**, *10*, 6965–7005. [CrossRef]
42. Mu, B.; Wang, A. Fabrication and Applications of Carbon/Clay Mineral Nanocomposites. In *Nanomaterials from Clay Minerals*; Elsevier: Amsterdam, The Netherlands, 2019; pp. 537–587.
43. Unnikrishnan, B.; Wu, R.-S.; Wei, S.-C.; Huang, C.-C.; Chang, H.-T. Fluorescent carbon dots for selective labeling of subcellular organelles. *ACS Omega* **2020**, *5*, 11248–11261. [CrossRef]
44. Ramesh, G.B.; Singh, P.; Biswas, K. Potentialities of Bio-Functionalized Carbon Nanotubes for Different Anti-Cancerous Activities. *J. Inorg. Organomet. Polym. Mater.* **2024**, *34*, 2325–2350. [CrossRef]
45. Li, L.; Zhang, Q.; Li, J.; Tian, Y.; Kang, Y.; Ren, G.; Liu, W.; Wang, H.; Wang, B.; Yan, L.; et al. Targeted delivery of doxorubicin using transferrin-conjugated carbon dots for cancer therapy. *ACS Appl. Bio Mater.* **2021**, *4*, 7280–7289. [CrossRef] [PubMed]
46. Minervini, G.; Panniello, A.; Dibenedetto, C.N.; Madonia, A.; Fanizza, E.; Curri, M.L.; Striccoli, M. Exploring Carbon Dots: Green Nanomaterials for Unconventional Lasing. *Small* **2024**, *20*, 2403653. [CrossRef] [PubMed]
47. Jing, H.H.; Shati, A.A.; Alfaifi, M.Y.; Elbehairi, S.E.I.; Sasidharan, S. The future of plant based green carbon dots as cancer Nanomedicine: From current progress to future perspectives and beyond. *J. Adv. Res.* **2024**, *67*, 133–159. [CrossRef] [PubMed]
48. Ma, J.; Chen, Q.; He, H.; Jiang, H.; Hu, J.; Liu, Y.; Yao, L.; Mao, H.; Li, J.; Li, B.; et al. Carbon dots-based materials and their applications in regenerative medicine. *MedComm–Biomater. Appl.* **2024**, *3*, e98. [CrossRef]
49. Genc, M.T.; Yanalak, G.; Aksoy, I.; Aslan, E.; Patır, I.H. Green carbon dots (GCDs) for photocatalytic hydrogen evolution and antibacterial applications. *ChemistrySelect* **2021**, *6*, 7317–7322. [CrossRef]

50. Lakkakula, J.; Kalra, P.; Mallick, G.; Mittal, H.; Uddin, I. Revolutionizing cancer treatment: Enhancing photodynamic therapy with cyclodextrin nanoparticles and synergistic combination therapies. *Mater. Today Sustain.* **2024**, *28*, 100958. [CrossRef]
51. Lin, X.; Xiong, M.; Zhang, J.; He, C.; Ma, X.; Zhang, H.; Kuang, Y.; Yang, M.; Huang, Q. Carbon dots based on natural resources: Synthesis and applications in sensors. *Microchem. J.* **2021**, *160*, 105604. [CrossRef]
52. Thota, S.P.; Kurdekar, A.; Vadlani, P.V.; Kumar, B.S. Green Synthesis of Highly Fluorescent Carbon Dots from Groundnut Shell Biomass for Bioimaging Applications. *J. Fluoresc.* **2024**, 1–11. [CrossRef]
53. Juntree, N.; Sangjan, A.; Pongchaikul, P.; Arjfuk, P.; Wanmolee, W.; Khemthong, P.; Srifa, A.; Posoknistakul, P.; Laosiripojana, N.; Wu, K.C.; et al. Unveiling role of carbon dots for non-invasive and ultra-sensitive glucose detection in biofluids for personal preventive care testing. *Green Anal. Chem.* **2024**, *11*, 100165. [CrossRef]
54. Li, X.; Yu, L.; He, M.; Chen, C.; Yu, Z.; Jiang, S.; Wang, Y.; Li, L.; Li, B.; Wang, G.; et al. Review on carbon dots: Synthesis and application in biology field. *BMEMat* **2023**, *1*, e12045. [CrossRef]
55. Etefa, H.F.; Tessema, A.A.; Dejene, F.B. Carbon dots for future prospects: Synthesis, characterizations and recent applications: A review (2019–2023). *C* **2024**, *10*, 60. [CrossRef]
56. Chen, L.; Wang, C.; Liu, C.; Chen, S. Facile access to fabricate carbon dots and perspective of large-scale applications. *Small* **2023**, *19*, 2206671. [CrossRef] [PubMed]
57. Usman, M.; Cheng, S. Recent Trends and Advancements in Green Synthesis of Biomass-Derived Carbon Dots. *Eng* **2024**, *5*, 2223–2263. [CrossRef]
58. Singh, P.; Rani, N.; Kumar, S.; Kumar, P.; Mohan, B.; Pallavi; Bhankar, V.; Kataria, N.; Kumar, R.; Kumar, K. Assessing the biomass-based carbon dots and their composites for photocatalytic treatment of wastewater. *J. Clean. Prod.* **2023**, *413*, 137474. [CrossRef]
59. Varadharajan, S.; Vasanthan, K.S.; Mathur, V.; Hariperumal, N.; Mazumder, N. Green synthesis and multifaceted applications: Challenges and innovations in carbon dot nanocomposites. *Discov. Nano* **2024**, *19*, 205. [CrossRef]
60. Li, C.; Sun, X.; Li, Y.; Liu, H.; Long, B.; Xie, D.; Chen, J.; Wang, K. Rapid and green fabrication of carbon dots for cellular imaging and anti-counterfeiting applications. *ACS Omega* **2021**, *6*, 3232–3237. [CrossRef]
61. Kanwal, A.; Bibi, N.; Hyder, S.; Muhammad, A.; Ren, H.; Liu, J.; Lei, Z. Recent advances in green carbon dots (2015–2022): Synthesis, metal ion sensing, and biological applications. *Beilstein J. Nanotechnol.* **2022**, *13*, 1068–1107. [CrossRef]
62. Chen, Y.; Bai, Y.; Han, Z.; He, W.; Guo, Z. Photoluminescence imaging of Zn^{2+} in living systems. *Chem. Soc. Rev.* **2015**, *44*, 4517–4546. [CrossRef]
63. Lu, R.; Wen, Z.; Zhang, P.; Chen, Y.; Wang, H.; Jin, H.; Zhang, L.; Chen, Y.; Wang, S.; Pan, S. Color-Tunable Perovskite Nanomaterials with Intense Circularly Polarized Luminescence and Tailorable Compositions. *Small* **2024**, *20*, 2311013. [CrossRef]
64. Shim, H.S.; Choi, J.; Jeong, S.; Nam, S.; Kim, J.; Song, J.K. Fluorophore-Dependent Optical Properties of Multicolor Carbon Dots for Bioimaging and Optoelectronic Devices. *ACS Appl. Nano Mater.* **2023**, *6*, 17120–17129. [CrossRef]
65. Huang, Z.; Lin, F.; Hu, M.; Li, C.; Xu, T.; Chen, C.; Guo, X. Carbon dots with tunable emission, controllable size and their application for sensing hypochlorous acid. *J. Lumin.* **2014**, *151*, 100–105. [CrossRef]
66. Feng, Z.; Adolfsson, K.H.; Xu, Y.; Fang, H.; Hakkarainen, M.; Wu, M. Carbon dot/polymer nanocomposites: From green synthesis to energy, environmental and biomedical applications. *Sustain. Mater. Technol.* **2021**, *29*, e00304. [CrossRef]
67. Jing, H.H.; Bardakci, F.; Akgöl, S.; Kusat, K.; Adnan, M.; Alam, M.J.; Gupta, R.; Sahreen, S.; Chen, Y.; Gopinath, S.C.; et al. Green carbon dots: Synthesis, characterization, properties and biomedical applications. *J. Funct. Biomater.* **2023**, *14*, 27. [CrossRef] [PubMed]
68. Nocito, G.; Calabrese, G.; Forte, S.; Petralia, S.; Puglisi, C.; Campolo, M.; Esposito, E.; Conoci, S. Carbon dots as promising tools for cancer diagnosis and therapy. *Cancers* **2021**, *13*, 1991. [CrossRef]
69. Zheng, M.; Ruan, S.; Liu, S.; Sun, T.; Qu, D.; Zhao, H.; Xie, Z.; Gao, H.; Jing, X.; Sun, Z. Self-targeting fluorescent carbon dots for diagnosis of brain cancer cells. *ACS Nano* **2015**, *9*, 11455–11461. [CrossRef]
70. Chen, Q.; Xia, X.; Liang, Z.; Zuo, T.; Xu, G.; Wei, F.; Yang, J.; Hu, Q.; Zhao, Z.; Tang, B.Z.; et al. Self-Assembled DNA Nanospheres Driven by Carbon Dots for MicroRNAs Imaging in Tumor via Logic Circuit. *Small* **2024**, *20*, 2310728. [CrossRef]
71. Gnanasegaran, G.; Shabo, G.; Buscombe, J.R. Ischaemic and Inflammatory Biomarkers in Cardiovascular Disease. In *Integrating Cardiology for Nuclear Medicine Physicians: A Guide to Nuclear Medicine Physicians*; Springer: Berlin/Heidelberg, Germany, 2009; pp. 77–85.
72. Théroux, P. *Acute Coronary Syndromes: A Companion to Braunwald's Heart Disease E-Book*; Elsevier Health Sciences: Amsterdam, The Netherlands, 2010.
73. Radhakrishnan, K.; Suriyaprakash, R.; Balamurugan, S.; Kumar, J.V.; Albeshr, M.F.; Mythili, R.; Srinivasan, P.; Nunna, G.P.; Ko, T.J. Fluorometric detection of copper and imidacloprid using nitrogen-doped graphitic carbon dots: A promising method for environmental monitoring. *Luminescence* **2024**, *39*, e4849. [CrossRef]
74. Li, G.; Wu, S.; Chen, W.; Duan, X.; Sun, X.; Li, S.; Mai, Z.; Wu, W.; Zeng, G.; Liu, H.; et al. Designing intelligent nanomaterials to achieve highly sensitive diagnoses and multimodality therapy of bladder cancer. *Small Methods* **2023**, *7*, 2201313. [CrossRef]

75. Ratre, P.; Nazeer, N.; Kumari, R.; Thareja, S.; Jain, B.; Tiwari, R.; Kamthan, A.; Srivastava, R.K.; Mishra, P.K. Carbon-based fluorescent nano-biosensors for the detection of cell-free circulating MicroRNAs. *Biosensors* **2023**, *13*, 226. [CrossRef]
76. Ratre, P.; Jain, B.; Kumari, R.; Thareja, S.; Tiwari, R.; Srivastava, R.K.; Goryacheva, I.Y.; Mishra, P.K. Bioanalytical applications of graphene quantum dots for circulating cell-free nucleic acids: A review. *ACS Omega* **2022**, *7*, 39586–39602. [CrossRef]
77. Sengupta, J.; Hussain, C.M. Early detection of cancer utilizing biosensors based on "Green Graphene": An innovative and sustainable methodology for advancing cancer diagnosis. *Trends Anal. Chem.* **2023**, *167*, 117254. [CrossRef]
78. Asghar, Z.; Ahmad, F.; Khan, M.I.; Shanableh, A.; Lashari, M.H. Advancements in Gadolinium-Doped Carbon Quantum Dots for Dual-Modal Bioimaging: Synthesis and Applications. *MatSci Express* **2025**, *2*, 77–92.
79. Guo, S.; Zhang, R.; Liu, Y.; Zhang, Q.; Liu, X.; Wu, X.; Li, B. Synthesis, applications in therapeutics, and bioimaging of traditional Chinese medicine-derived carbon dots. *Carbon Lett.* **2024**, *34*, 545–564. [CrossRef]
80. Zhou, Y. Carbon Dots for Nanomedicine: C-Dots and C-Dots Nanotec describe how carbon dots can aid high-performance drug delivery in the future treatment of human diseases. *Genet. Eng. Biotechnol. News* **2022**, *42*, 50–51. [CrossRef]
81. Herlem, G.; Picaud, F.; Girardet, C.; Micheau, O. Carbon nanotubes: Synthesis, characterization, and applications in drug-delivery systems. In *Nanocarriers for Drug Delivery*; Elsevier: Amsterdam, The Netherlands, 2019; pp. 469–529.
82. Riley, R.; June, C.H.; Langer, R.; Mitchell, M.J. Delivery technologies for cancer immunotherapy. *Nat. Rev. Drug Discov.* **2019**, *18*, 175–196. [CrossRef]
83. Wang, N.; Jiang, X.; Zhang, S.; Zhu, A.; Yuan, Y.; Xu, H.; Lei, J.; Yan, C. Structural basis of human monocarboxylate transporter 1 inhibition by anti-cancer drug candidates. *Cell* **2021**, *184*, 370–383.e13. [CrossRef]
84. Wang, J.; Teng, X.; Wang, Y.; Si, S.; Ju, J.; Pan, W.; Wang, J.; Sun, X.; Wang, W. Carbon dots based fluorescence methods for the detections of pesticides and veterinary drugs: Response mechanism, selectivity improvement and application. *Trends Anal. Chem.* **2021**, *144*, 116430. [CrossRef]
85. Luo, W.-K.; Zhang, L.-L.; Yang, Z.-Y.; Guo, X.-H.; Wu, Y.; Zhang, W.; Luo, J.-K.; Tang, T.; Wang, Y. Herbal medicine derived carbon dots: Synthesis and applications in therapeutics, bioimaging and sensing. *J. Nanobiotechnol.* **2021**, *19*, 320. [CrossRef]
86. Zhang, W.; Sigdel, G.; Mintz, K.J.; Seven, E.S.; Zhou, Y.; Wang, C.; Leblanc, R.M. Carbon Dots: A future Blood-Brain Barrier Penetrating Nanomedicine and Drug Nanocarrier. *Int. J. Nanomed.* **2021**, *16*, 5003–5016. [CrossRef]
87. Calabrese, G.; De Luca, G.; Nocito, G.; Rizzo, M.G.; Lombardo, S.P.; Chisari, G.; Forte, S.; Sciuto, E.L.; Conoci, S. Carbon dots: An innovative tool for drug delivery in brain tumors. *Int. J. Mol. Sci.* **2021**, *22*, 11783. [CrossRef]
88. Huang, K.M.; Thomas, M.Z.; Magdy, T.; Eisenmann, E.D.; Uddin, M.E.; DiGiacomo, D.F.; Pan, A.; Keiser, M.; Otter, M.; Xia, S.H.; et al. Targeting OCT3 Attenuates Doxorubicin-Induced Cardiac Injury. *Proc. Natl. Acad. Sci. USA* **2021**, *118*, e2020168118. [CrossRef] [PubMed]
89. Zhou, F.; Gao, J.; Xu, Z.; Li, T.; Gao, A.; Sun, F.; Wang, F.; Wang, W.; Geng, Y.; Zhang, F.; et al. Overcoming immune resistance by sequential prodrug nanovesicles for promoting chemoimmunotherapy of cancer. *Nano Today* **2021**, *36*, 101025. [CrossRef]
90. Liu, J.; Zhao, Z.; Qiu, N.; Zhou, Q.; Wang, G.; Jiang, H.; Piao, Y.; Zhou, Z.; Tang, J.; Shen, Y. Co-delivery of IOX1 and doxorubicin for antibody-independent cancer chemo-immunotherapy. *Nat. Commun.* **2021**, *12*, 2425. [CrossRef] [PubMed]
91. Johnstone, T.C.; Suntharalingam, K.; Lippard, S.J. The next generation of platinum drugs: Targeted Pt (II) agents, nanoparticle delivery, and Pt (IV) prodrugs. *Chem. Rev.* **2016**, *116*, 3436–3486. [CrossRef]
92. Chaft, J.E.; Rimner, A.; Weder, W.; Azzoli, C.G.; Kris, M.G.; Cascone, T. Evolution of systemic therapy for stages I–III non-metastatic non-small-cell lung cancer. *Nat. Rev. Clin. Oncol.* **2021**, *18*, 547–557. [CrossRef]
93. Jiang, D.M.; Gupta, S.; Kitchlu, A.; Meraz-Munoz, A.; North, S.A.; Alimohamed, N.S.; Blais, N.; Sridhar, S.S. Defining cisplatin eligibility in patients with muscle-invasive bladder cancer. *Nat. Rev. Urol.* **2021**, *18*, 104–114. [CrossRef]
94. Rosenkrans, Z.T.; Sun, T.; Jiang, D.; Chen, W.; Barnhart, T.E.; Zhang, Z.; Ferreira, C.A.; Wang, X.; Engle, J.W.; Huang, P.; et al. Selenium-Doped Carbon Quantum Dots Act as Broad-Spectrum Antioxidants for Acute Kidney Injury Management. *Adv. Sci.* **2020**, *7*, 2000420. [CrossRef]
95. Le, T.N.Q.; Tran, N.N.; Escribà-Gelonch, M.; Serra, C.A.; Fisk, I.; McClements, D.J.; Hessel, V. Microfluidic encapsulation for controlled release and its potential for nanofertilisers. *Chem. Soc. Rev.* **2021**, *50*, 11979–12012. [CrossRef]
96. Shen, Y.; Zhang, N.; Tian, J.; Xin, G.; Liu, L.; Sun, X.; Li, B. Advanced approaches for improving bioavailability and controlled release of anthocyanins. *J. Control. Release* **2022**, *341*, 285–299. [CrossRef]
97. Wang, B.; Cai, H.; Waterhouse, G.I.N.; Qu, X.; Yang, B.; Lu, S. Carbon dots in bioimaging, biosensing and therapeutics: A comprehensive review. *Small Sci.* **2022**, *2*, 2200012. [CrossRef]
98. He, F.; Liu, X.; Yang, S.; Tan, H.; Yang, L.-P.; Wang, L.-L. Guanidinium-Functionalized Carbon Dots: An Efficient Antibacterial Agent Against Multidrug-Resistant ESKAPE Pathogens. *ACS Appl. Mater. Interfaces* **2024**, *16*, 65955–65969. [CrossRef] [PubMed]
99. Germain, N.; Samb, R.; Côté, É.; Toulouse-Fournier, A.; Robitaille, J.; Turcotte, S.; Morin, M.; Audet, M.; Bert, L.; Rivard, J.; et al. Impact of a geriatric emergency management nurse on thirty-day emergency department revisits: A retrospective propensity score matched case-control study. *medRxiv* **2024**. [CrossRef]

100. Liu, S.; Xu, Y.; Wang, X.; Zhou, H.; Zhang, T. Insight Into the Synthetic Strategies of Carbon Dots and Its Structure-Property Interplay for Next-Generation Technologies. *Chem. Eng. J.* **2024**, *496*, 153914. [CrossRef]
101. Mitchell, M.J.; Billingsley, M.M.; Haley, R.M.; Wechsler, M.E.; Peppas, N.A.; Langer, R. Engineering precision nanoparticles for drug delivery. *Nat. Rev. Drug Discov.* **2021**, *20*, 101–124. [CrossRef]
102. Zhao, Z.; Ukidve, A.; Kim, J.; Mitragotri, S. Targeting strategies for tissue-specific drug delivery. *Cell* **2020**, *181*, 151–167. [CrossRef]
103. Zhao, D.; Deng, Y.; Jiang, X.; Bai, Y.; Qian, C.; Shi, H.; Wang, J. Advances in Carbon Dot Based Enhancement of Photodynamic Therapy of Tumors. *ACS Appl. Bio Mater.* **2024**, *7*, 8149–8162. [CrossRef]
104. Chaudhari, K.G.; Patil, P.O.; Nangare, S.N.; Khan, M.G.; Bari, S.B.; Khan, Z.G. Exploring the Significance of Carbon Quantum Dots and Graphene Quantum Dots in Sarcosine Sensing: A Key Biomarker for Prognosticating Prostate Cancer. *ChemistrySelect* **2024**, *9*, e202401643. [CrossRef]
105. Desai, D.; Shende, P. Integration of Internet of Things with Quantum Dots: A State-of-the-Art of Medicine. *Curr. Pharm. Des.* **2021**, *27*, 2068–2075. [CrossRef]
106. Horstmann, C.; Davenport, V.; Zhang, M.; Peters, A.; Kim, K. Transcriptome profile alterations with carbon nanotubes, quantum dots, and silver nanoparticles: A review. *Genes* **2021**, *12*, 794. [CrossRef]
107. Liu, J. Develop the Disease Specific Bioinformatics Platforms with Integrated Bioinformatics Data. Ph.D. Thesis, Indiana University-Purdue University Indianapolis, Indianapolis, IN, USA, 2022.
108. Liu, X.; Zhang, W. Bioinformatics in the Age of Big Data: Leveraging Computational Tools for Biological Discoveries. *Comput. Mol. Biol.* **2024**, *14*, 173–181. [CrossRef]
109. Pereira, R.; Oliveira, J.; Sousa, M. Bioinformatics and computational tools for next-generation sequencing analysis in clinical genetics. *J. Clin. Med.* **2020**, *9*, 132. [CrossRef]
110. Adam, G.O.; Sharker, S.M.; Ryu, J.H. Emerging biomedical applications of carbon dot and polymer composite materials. *Appl. Sci.* **2022**, *12*, 10565. [CrossRef]

MDPI AG
Grosspeteranlage 5
4052 Basel
Switzerland
Tel.: +41 61 683 77 34

International Journal of Molecular Sciences Editorial Office
E-mail: ijms@mdpi.com
www.mdpi.com/journal/ijms

www.ingramcontent.com/pod-product-compliance
Lightning Source LLC
Chambersburg PA
CBHW061839210926
51850CB00006B/929